Trees of Life

Theodore W. Pietsch

Trees of Life

A VISUAL HISTORY OF EVOLUTION

JOHNS
HOPKINS
UNIVERSITY
PRESS
BALTIMORE

This book has been brought to publication with the generous assistance of the University of Washington.

Printed in the United States of America on acid-free paper
Johns Hopkins Paperback edition, 2013
9 8 7 6 5 4 3 2

Johns Hopkins University Press
2715 North Charles Street
Baltimore, Maryland 21218-4363
www.press.jhu.edu

The Library of Congress has cataloged the hardcover edition of this book as follows:
Pietsch, Theodore W.
Trees of life : a visual history of evolution / Theodore W. Pietsch.
p. cm.
"Published with the assistance of the University of Washington"—T.p. verso.
Includes bibliographical references and index.
ISBN-13: 978-1-4214-0479-0 (hardcover : acid-free paper)
ISBN-10: 1-4214-0479-6 (hardcover : acid-free paper)
1. Evolution (Biology)—History. 2. Evolution (Biology)—Charts, diagrams, etc.—History. 3. Tree of life—History. 4. Naturalists—Biography. 5. Biologists—Biography. I. University of Washington. II. Title.
QH361.P53 2012
576.8—dc23 2011023115

A catalog record for this book is available from the British Library.

ISBN-13: 978-1-4214-1185-9
ISBN-10: 1-4214-1185-7

Special discounts are available for bulk purchases of this book. For more information, please contact Special Sales at 410-516-6936 or specialsales@press.jhu.edu.

Johns Hopkins University Press uses environmentally friendly book materials, including recycled text paper that is composed of at least 30 percent post-consumer waste, whenever possible.

To my former major professor,
BASIL G. NAFPAKTITIS,
with fond memories of an exciting
and fulfilling time as a graduate student
under his watchful eye

CONTENTS

PREFACE

This is a book about trees—not the transpiring, photosynthesizing kind, but tree-like branching diagrams that attempt to show the interrelationships of organisms, from viruses and bacteria to birds and mammals, both living and fossil. It is not intended as a treatise about the philosophy or science behind tree construction, nor is it a defense or refutation of the various relationships depicted among organisms. It is rather a celebration of the manifest beauty, intrinsic interest, and human ingenuity revealed in trees of life through time. The emphasis is on the images, arranged chronologically, two hundred and thirty chosen from among thousands of possibilities, dating from the mid-sixteenth century to the present day. The descriptive text is kept to a minimum—just enough to provide context.

The focus of this book is on diagrams that resemble trees in the botanical sense, images with parts analogous to trunks, limbs, and terminal twigs, but other configurations are also explored as precursors and variations on the theme of biosystematic iconography. These various related images include bracketed tables—trees laid on their side—similar to modern-day analytical keys;[1] maps, or so-called archipelagos, that hypothesize relationships analogous to the juxtaposition of geographical territories;[2] webs or networks, in which individual taxa or chains of taxa are interconnected by lines of affinity or resemblance; and various numerical, symmetrical and geometric systems. While their choice of imagery varied considerably, most all eighteenth- and early nineteenth-century naturalists were working toward the same goal: to construct classifications of plants and animals that were "natural." Their thought was that organisms brought together in "natural classifications" ought to share "natural affinities."[3] But just exactly what was meant by "natural affinity," remained an unresolved question.[4] It was Darwin's theory of evolutionary change by means of natural selection that provided the missing context and unified the work of biosystematists in their pursuit of a natural system of classification. The phylogenetic tree as we know it today was one conspicuous result.

Grateful thanks are extended to Tim M. Berra, Professor Emeritus of Evolution, Ecology and Organismal Biology at the Ohio State University, for his enthusiasm and early encouragement to pursue this project. Others who helped along the way, by providing references, images, or comments, include J. David Archibald of San Diego State University; Ralf Britz, James Maclaine, and Rosemary Lowe-McConnell, The Natural History Museum, London; Rebecca L. Cann, University of Hawaii at Manoa; Mary E. Endress, Managing Editor, *Taxon;* Alan Feduccia, University of North Carolina, Chapel Hill; Matthew Friedman, University of Oxford, Oxford; Michael T. Ghiselin, California Academy of Sciences, San Francisco; Karsten Hartel and Scarlett R. Huffman, Harvard University, Cambridge; David M. Hillis, Derrick J. Zwickl, Robin R. Gutell, and Timothy B. Rowe, University of Texas at Austin; T. Derrick Iles, St. Andrews Biological Station, Fisheries and Oceans Canada; Lucas Leclère, Université Paris; Agnès Dettai and Guillaume Lecointre, Museum National d'Histoire Naturelle, Paris; Ruth M. Long and Adam J. Perkins, University of Cambridge Library, Cambridge; Christiane Michaelis, Universität Rostock; Richard G. Olmstead, Beatrice Marx, and Louise Richards, University of Washington, Seattle; Norman R. Pace, University of Colorado at Boulder; Jerome C. Regier, University of Maryland, College Park; Vincent M. Sarich, University of California at Berkeley; Nalanie Schnell, Virginia Institute of Marine Science, Gloucester Point; Paul C. Sereno, University of Chicago; Peter F. Stevens, Missouri Botanical Garden and University of Missouri, St. Louis; and Heidi Nance and her staff, Interlibrary Loan & Document Delivery Services, University of Washington. Special thanks to Ray Troll for permission to use "The Family Tree" on the cover of the book.

The entire draft manuscript was critically read by Christopher P. Kenaley, University of Washington, Seattle, and James W. Orr and Duane E. Stevenson, both of NOAA Fisheries, Alaska Fisheries Science Center, Seattle. At the Johns Hopkins University Press, sincere thanks are extended to Vincent J. Burke, executive editor of biology and life sciences, for skillfully directing the publication of this volume, and to the publishing team: Jennifer E. Malat, acquisitions assistant; Michele T. Callaghan, copy editor; Omega Clay, designer; and Kathy Alexander, publicist.

Finally, I thank David A. Armstrong, Director of the School of Aquatic and Fishery Sciences, and Lisa J. Graumlich, Dean of the College of the Environment, University of Washington, for providing a subsidy to support publication of this volume.

Trees of Life

Introduction

The tree as an iconographic metaphor is perhaps the most universally widespread of all great cultural symbols.[1] Trees appear and reappear throughout human history to illustrate nearly every aspect of life. The structural complexity of a tree—its roots, trunk, bifurcating branches, and leaves—has served as an ideal symbol throughout the ages to visualize and map hierarchies of knowledge and ideas[2]: images on parchment or paper or carved in stone that date back to the beginnings of recorded human history. Thus, we have trees of logic; of principles, disciplines, morals, and intellect; of cosmology, languages, genealogy, and social groupings; and, perhaps the most enduring, trees of life.

The idea of a tree as a metaphor for understanding and graphically displaying relationships among organisms is most often attributed to Charles Darwin (1809–1882), who fully articulated the notion in his 1859 *Origin of Species by Means of Natural Selection.*[3] But the concept is considerably older, dating back to at least the mid-eighteenth century. Nearly a century before Darwin, Swiss naturalist and philosopher Charles Bonnet (1720–1793) is thought to have been the first to explicitly apply the tree metaphor to organisms.[4] In his *Contemplation de la Nature* (Contemplation of nature) of 1764, when attempting to understand the relationships among insects, mollusks, and crustaceans, and while never actually drawing a tree, he posed a number of questions that invoke the idea of a tree-like branching diagram:

> Does the scale of nature become branched as it arises?
>
> Are the insects and mollusks two parallel and lateral branches of this great trunk?

> Do the crayfish and the crab likewise branch off from the mollusks? We still cannot answer these questions.[5]

A short time later, Peter Simon Pallas (1741–1811), a German naturalist who spent most of his career working in Russia, was much more explicit. Like Bonnet, he apparently never constructed a tree of his own, but he proposed, in 1766, that gradation among organisms might best be described as a branching tree:

> But the system of organic bodies is best of all represented by an image of a tree that immediately from the root would lead forth out of the most simple plants and animals a double, variously contiguous animal and vegetable trunk; the first of which would proceed from mollusks to fishes, with a large side branch of insects sent out between these, hence to amphibians and at the farthest tip it would sustain the quadrupeds, but below the quadrupeds it would put forth birds as an equally large side branch.[6]

In that same year (1766), and still well before Darwin, Georges Louis Leclerc de Buffon (1707–1788) spoke in botanical terms when contemplating affinities among mammals, writing that some groups "appear to form families in which one ordinarily notices a principal and common trunk, from which seem to have issued different stems and the more numerous, as the individuals of each species are smaller and more prolific."[7]

In 1801, a little-known French naturalist, Augustin Augier (fl. 1800), published a diagram of a tree, complete with roots, trunk, branches, and leaves, to illustrate his thoughts about the relationships among members of the plant kingdom (see Figure 21). Working quietly in isolation, unaware of the earlier suggestions of Bonnet, Pallas, and Buffon, Augier was entirely original and well ahead of his time in not only proposing the tree metaphor, but in constructing a diagram to go along with it:

> A figure like a genealogical tree appears to be the most proper to grasp the order and gradation of the series or branches which form classes or families. This figure, which I call a botanical tree, shows the agreements which the different series of plants maintain amongst each other, although detaching themselves from the trunk; just as a genealogical tree shows the order in which different branches of the same family came from the stem to which they owe their origin.[8]

Jean-Baptiste Lamarck (1744–1829), who surely never saw Augier's botanical tree, wrote in 1809: "I do not mean that existing animals form a very simple series, regularly graded throughout; but I do mean that they form a branching series, irregularly graded and free from discontinuity, or at least once free from it."[9] And Alfred Russel Wallace (1823–

1913) as well, in a paper titled "On the Law Which has Regulated the Introduction of New Species," published in 1855:

> We are also made aware of the difficulty of arriving at a true classification, even in a small and perfect group;—in the actual state of nature it is almost impossible, the species being so numerous and the modifications of form and structure so varied, arising probably from the immense number of species which have served as antitypes for the existing species, and thus produced a complicated branching of the lines of affinity, as intricate as the twigs of a gnarled oak or the vascular system of the human body. Again, if we consider that we have only fragments of this vast system, the stem and main branches being represented by extinct species of which we have no knowledge, while a vast mass of limbs and boughs and minute twigs and scattered leaves is what we have to place in order, and determine the true position each originally occupied with regard to the others, the whole difficulty of the true Natural System of classification becomes apparent to us.[10]

Of these forward-thinking naturalists, only Lamarck and Wallace were contemplating life in terms of evolutionary change, while Bonnet, Pallas, and Augier, in all their writings, evoked the existence of a divine Creator. But it was Darwin who most explicitly and poetically placed the tree metaphor in an evolutionary context:

> The affinities of all the beings of the same class have sometimes been represented by a great tree. I believe this simile largely speaks the truth. The green and budding twigs may represent existing species; and those produced during former years may represent the long succession of extinct species. . . . As buds give rise by growth to fresh buds, and these, if vigorous, branch out and overtop on all sides many a feebler branch, so by generation I believe it has been with the great Tree of Life, which fills with its dead and broken branches the crust of the earth, and covers the surface with its ever-branching and beautiful ramifications.[11]

For most of us, the modern concept of a tree of life is a diagram that shows the evolutionary divergence or branching of groups of organisms through time, a product of the Darwinian revolution that took place during the second half of the nineteenth century. The intent and look of images that came before, however, are quite different. The ancient *scala naturae,* or "Great Chain of Being," that originated with Aristotle attempted to classify animals in relation to a hierarchical ladder or stairway of nature, in which matter and living things were arranged according to their structural complexity and function.[12] The ladder of nature was most often depicted graphically as a list of entities arranged vertically, with the most basic elements like air, water, earth, and fire at the bottom and man, angels, and God at the top (Figure 1). Other renderings on the same general theme are more elaborate; for example, Ramon Lull's "Ladder of Ascent and Descent of the Intellect" of 1512, which demonstrates ascendance from inferior to superior beings, and vice versa (Figure 2).[13]

Inadequate to make sense of the ever-increasing complexity of knowledge of biological diversity, the concept of the ladder of life, while continuing to enjoy a philosophical function well into the late eighteenth century, was gradually replaced by a fascinating and wide variety of iconography that are the subject of this book.[14]

FIGURE 1. The "Chain of Being," published by French mathematician Charles de Bovelles, also known by the Latin name Carolus Bovillus (c. 1475–after 1566) in his *Physicorum elementorum* (Elements of natural history) of 1512.

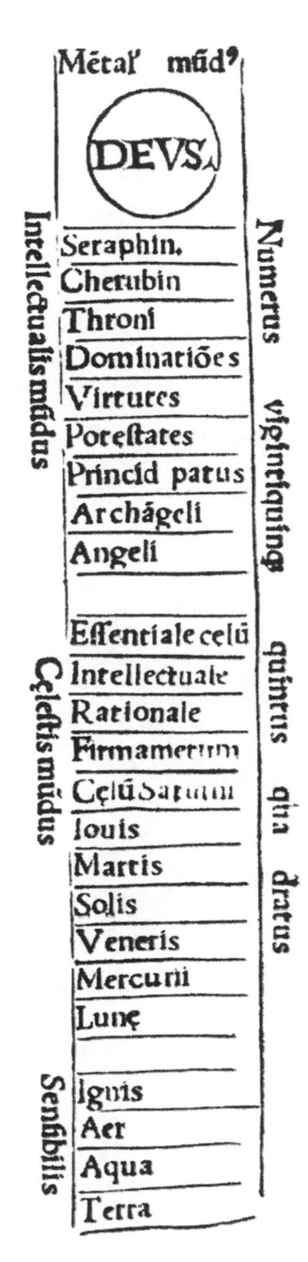

FIGURE 2. The Ladder of Ascent and Descent of the Intellect, not tree-like at first glance, but certainly branching dichotomously, the steps labeled from bottom to top, with representative figures on the right and upper left: *Lapis* (stone), *Flamma* (fire), *Planta* (plant), *Brutum* (beast), *Homo* (human), *Caelum* (sky), *Angelus* (angel), and *Deus* (God), a scheme that shows how one might ascend from inferior to superior beings and vice versa. After Ramon Lull (1232–1315), *Liber de ascensu et descensu intellectus,* written about 1305 but not published until 1512.

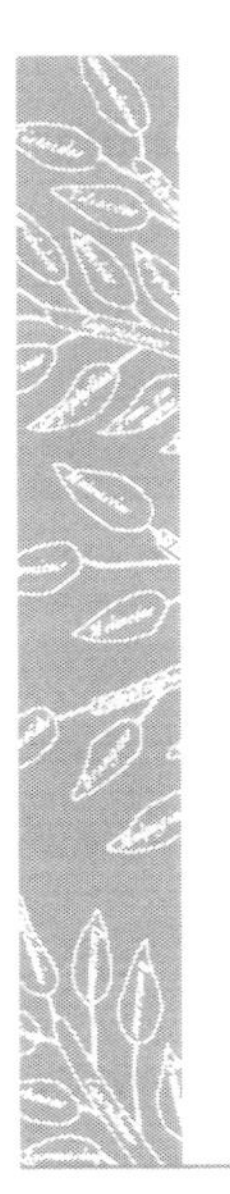

Brackets and Tables, Circles and Maps

1554–1872

Today's evolutionary, or phylogenetic, trees have their earliest beginnings in bracketed, tabular displays, which are essentially tree-like branching diagrams displayed horizontally. When first employed, they were meant to demonstrate perceived affinities or similarities between plants and animals, to reveal the imagined temporal order in which God created life on Earth, and otherwise to bring order into the chaos of names and kinds of living things that no one had paid much attention to since the time of Aristotle (384–322 BCE) and Theophrastus (c. 371–287 BCE). Serving also as the precursors of today's dichotomous analytical keys, they start with a general, all-encompassing taxon displayed on the left, followed by a splitting into two or more categories on the basis of a particular character. Each smaller category is then divided once again on the basis of another character, and so on, resulting in the most specific taxa of the group lined up along the right margin.

Although some authors have implied an origin of such bracketed tables among seventeenth-century authors[1]—such as Fredericus Caesius (fl. 1650), John Wilkins (1614–1672), Robert Morison (1620–1683), and Augustus Quirinus Rivinus (1652–1723)—they first appear at least a century earlier in the works of naturalist and bibliographer Conrad Gessner (1516–1565).[2] In his five-part *Historiae animalium* (History of animals) of 1551–1558, a monumental work considered to mark the beginning of modern zoology,[3] Gessner used bracketed tables to demonstrate his ideas about relationships among, for example, shorebirds, including snipes, plovers, curlews, and sandpipers (Figure 3). A series of similar diagrams—synoptic views of generic, specific, varietal, and sometimes family groups of plants—was published shortly thereafter by the celebrated French botanist Mathias de Lobel (1538–1616) and his fellow student Pierre Pena (fl. 1570) in their *Stirpium adversaria nova* (Plants, a new adversary) of 1570 (Figure 4). Other examples from the last quarter of the sixteenth century can be found, such as the elaborate classification of orchids constructed by Cornelius Gemma (1535–1578), no doubt inspired by the tables of Pena and Lobel and published by Lobel in his *Planarum seu stirpium historia* (History of plants) of 1576 (Figure 5).

Somewhat later, Fredericus Caesius made good use of bracketed tables in his *Phytosophicae tabulae* appended to Francisco Hernández's *Rerum medicarum Novae Hispaniae thesaurus* (Medical thesaurus of New Spain), prepared in 1628, but not published until 1651 (Figure 6).[4] John Wilkins followed in 1668 with a large folio volume titled *An Essay towards a Real Character, and a Philosophical Language,* in which he attempted to show the nature and relationships of all things. With subject matter outlined in bracketed form (Figures 7 and 8) and "conteining a regular enumeration and description of all those things and notions to which names are to be assigned," he explained his tables as follows:

> These Species are commonly joyned together by *pairs,* for the better helping of the Memory. . . . Those things which naturally have *Opposites,* are joyned with them, according to such Opposition, whether *Single* or *Double.* Those things that have no Opposites, are paired together with respect to some *Affinity* which they have one to another.[5]

In this way, Wilkins considered a whole host of subjects, including actions, discourse, elements, plants, animals, space, operation, private and public relations, judicial and military matters, and so on.[6]

Strongly influenced by Wilkins, the use of bracketed diagrams proliferated during the remaining quarter of the seventeenth century and throughout subsequent decades, becoming the most common way to display differences and similarities between plants and animals. Of numerous additional examples that could be cited, some of the more significant, for historical reasons, are those of Francis Willughby (1635–1672), who, with the help of his friend and colleague John Ray (1627–1705), published in 1686 the most important treatise on fishes of the seventeenth century (Figure 9); Carl Linnaeus (1707–1778), famous for, among many other things, his controversial 1735 sexual system of plant classification (Figure 10); Jean-Baptiste Lamarck (1744–1829), who articulated the first comprehensive theory of organic evolution (Figure 11); Henri Marie Ducrotay de Blainville (1777–1850), a worthy successor to Lamarck in the chair of natural history (1830) and later to Georges Cuvier (1769–1832) in comparative anatomy (1832) at the Museum d'Histoire Naturelle, Paris (Figure 12); and Theodore Nicholas Gill (1837–1914), said by David Starr Jordan (1851–1931) to be "the keenest interpreter of taxonomic facts yet known in the history of ichthyology" (Figure 13).[7] While not so prevalent today, such diagrams, in somewhat modified form, can still be found. The modern versions serve primarily as a means of identifying taxa and are usually referred to as keys.

One early proponent of bracketed diagrams, botanist Robert Morison, in dealing with the plant genus *Myrrhis* (Figure 14), or something more all inclusive that he called *Umbellae Semine rostrato,* took his figure a step further. His work, published in 1672, experimented with a very different looking image: a series of circles representing taxa, interconnected by simple, bifurcating, and trifurcating lines (Figure 15). Although not obvious at first glance, the upper part of Morison's circle diagram is an exact duplication of the relationships indicated in his bracket diagram, published in the same work (compare Figures 14 and 15).

Directly influenced by Linnaeus, who wrote that all "plants show affinities on all sides, like the territories on a geographical map,"[8] Paul Dietrich Giseke (1741–1796) published a circle diagram in 1792 reminiscent of that of Morison but without interconnecting lines (Figure 16). Like free-floating soap bubbles, the circles of Giseke's diagram represent the families of plants as conceived by Linnaeus,[9] their diameters roughly proportional to the number of contained genera and their relative position, contact, or degree of spacing between, a measure of relative affinity.

Following the geographic map approach, but with connecting lines now added between circles, a similar "genealogical tree" (perhaps better called a network) of the races of dogs (Figure 17) was published more than a half-century later by Georges Louis Leclerc de Buffon (1707–1788). In describing his tree, Buffon wrote:

> In order to give a clearer idea of the dog group, their modifications in different climates, and the mixture of their races, I include a figure, or, if one wishes to term it so, a sort of genealogical tree, wherein one can see at a glance all of the varieties. This figure is oriented as a geographical map, and in its construction the relative positions of the climates have been maintained to the extent possible.[10]

In a rare and early ecological connection between animal affinities and geography—perhaps the first example of what we now term "phylogeography"—one that certainly has evolutionary implications, Buffon surmised that climatic conditions in different parts of the world produced the great morphological diversity among dogs from a single ancestral race, which he believed to be the shepherd's dog (*chien de berger*):

> The shepherd's dog is the source of the tree. This dog, transported into the rigorous climates of the north became smaller and uglier in Lapland, but appears to have been maintained, and even perfected, in Iceland, Russia, and Siberia, where the climate is a little less rigorous and the inhabitants are a little more civilized. These changes have occurred only through the influence of climate.[11]

Buffon's diagram certainly implies evolutionary change, but he was not an evolutionist—in all his published work he displayed a strong belief in the fixity of species.[12] Yet, at the same time, he was interested in animal breeding and in the diversity of forms it could produce, but he considered breeding a special case because it could not occur without human intervention.[13]

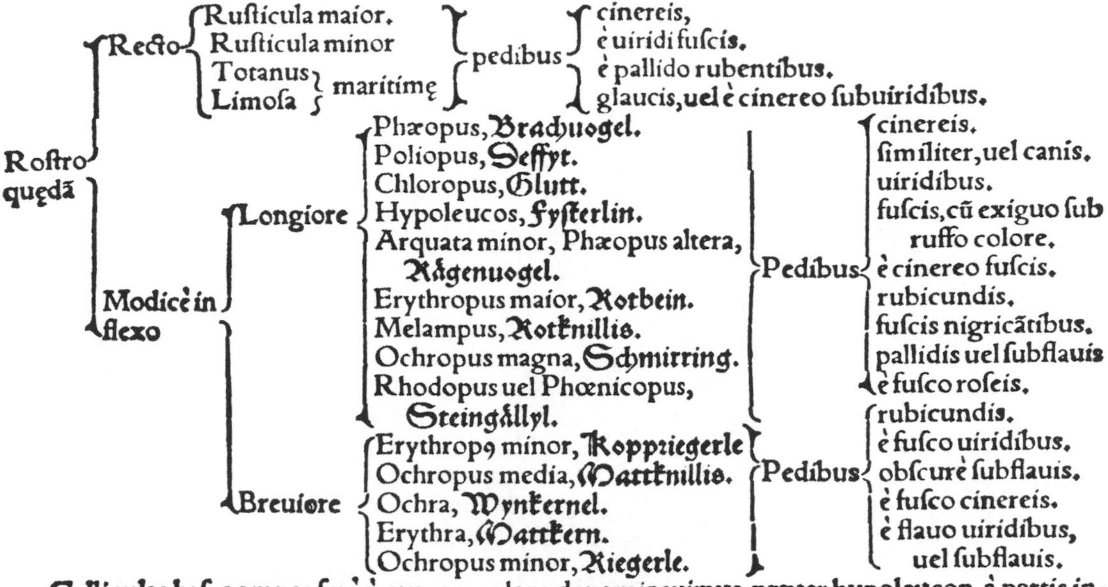

FIGURE 3. Conrad Gessner's classification of wading birds, from "De avium natura," book 3 of *Historia animalium* (History of animals), published in 1555, perhaps the earliest example of a bracketed table designed to show similarities and differences among organisms.

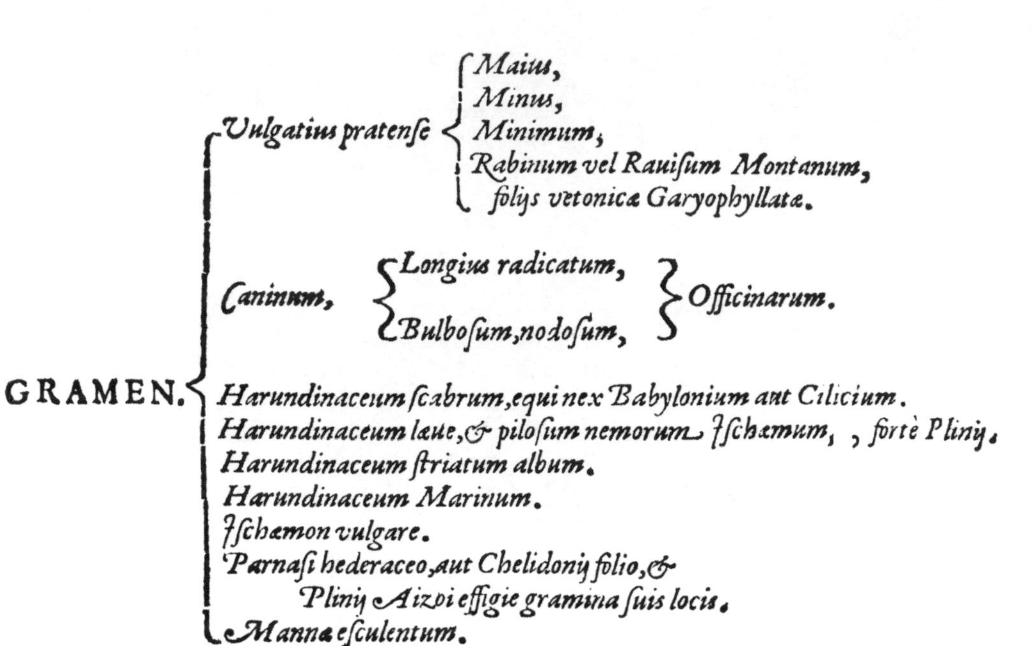

FIGURE 4. A classification of grasses, from *Stirpium adversaria nova* (Plants, a new adversary) of Pierre Pena and Mathias de Lobel, published in 1570.

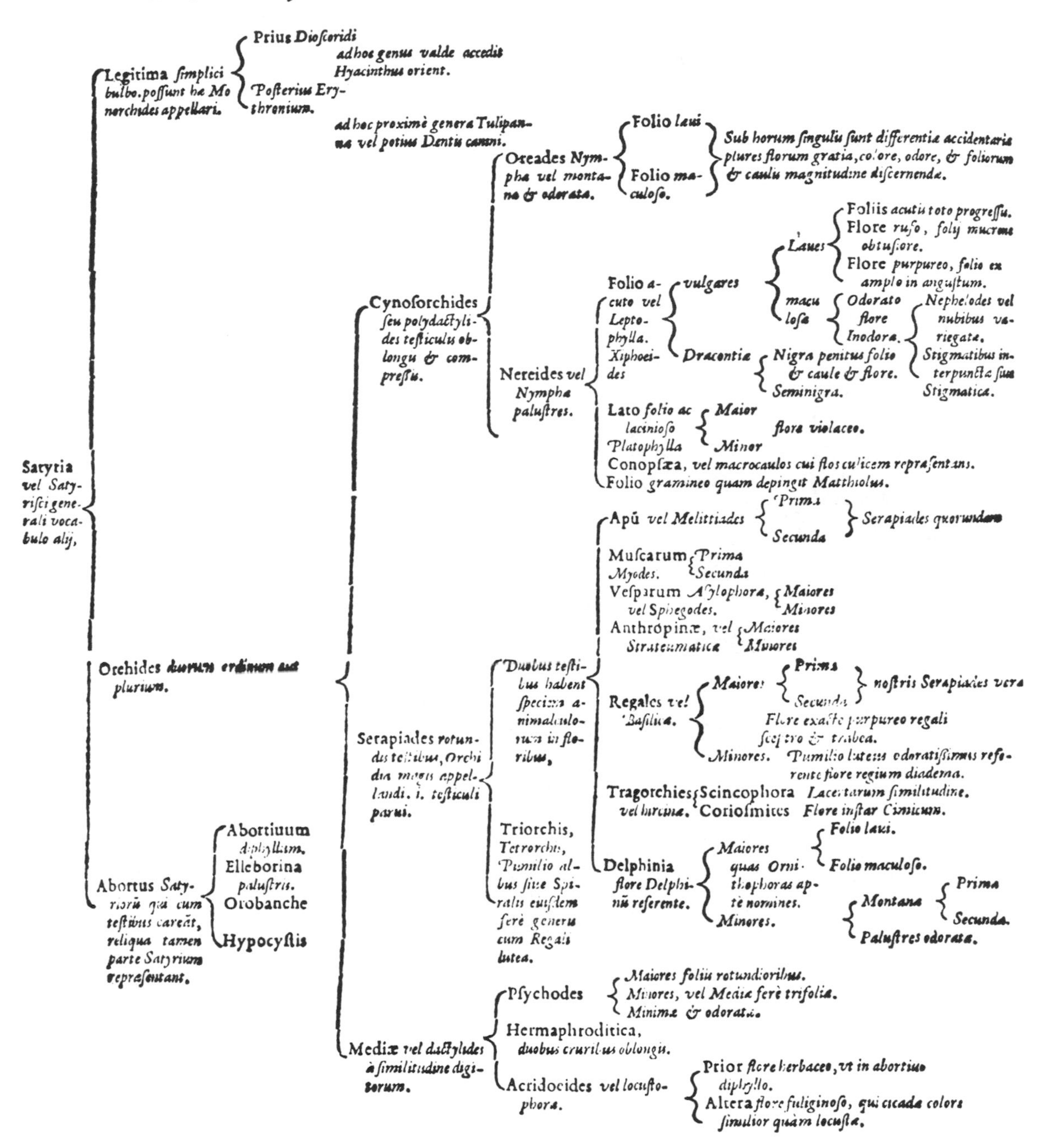

FIGURE 5. Cornelius Gemma's classification of orchids, published by Mathias de Lobel in *Planarum seu stirpium historia* (Plants or history of plants), published in 1576.

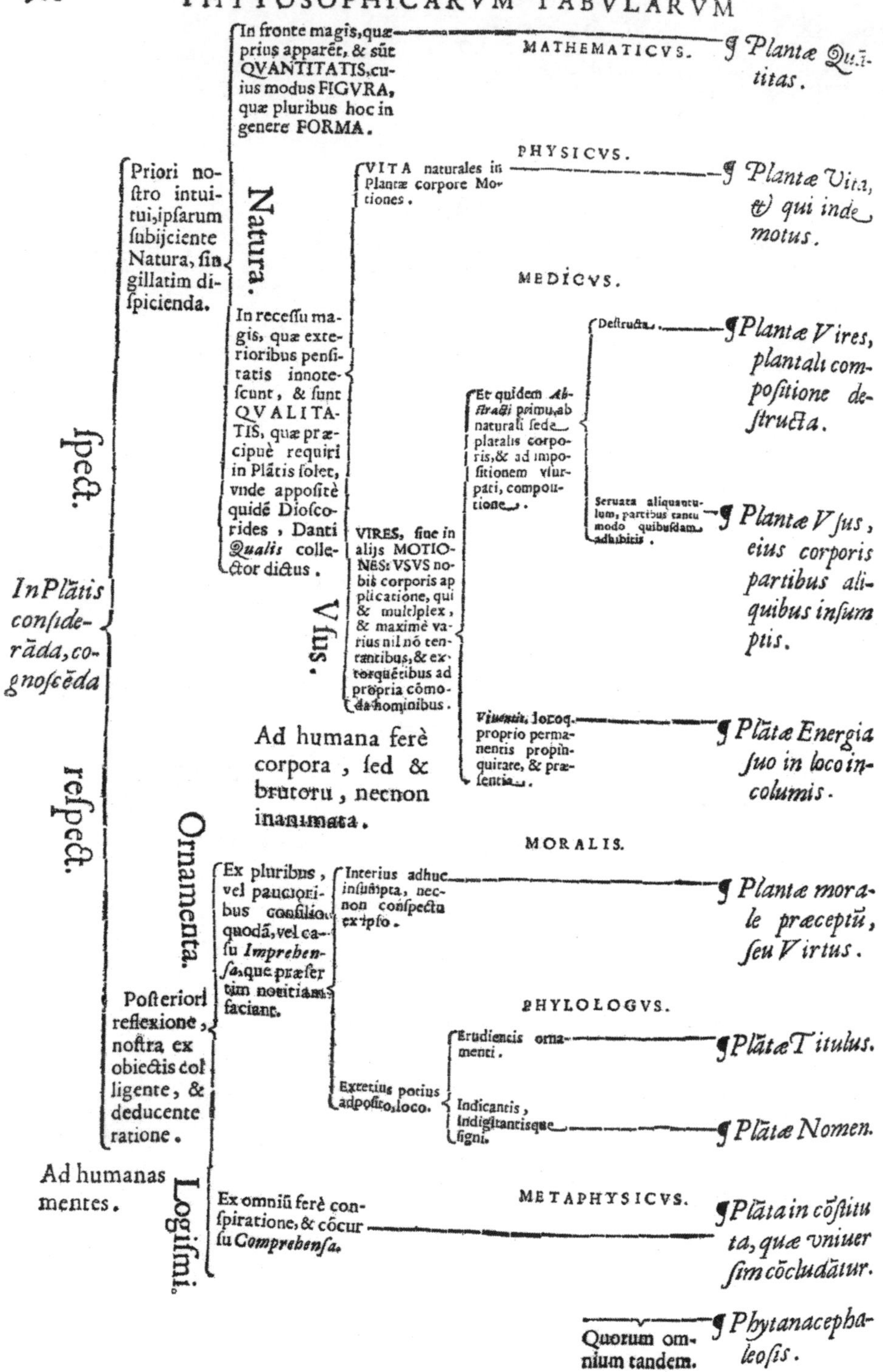

FIGURE 6. One of numerous bracketed diagrams taken from Italian botanist Fredericus Caesius's *Phytosophicarum tabularum* (Classification of plants), from the *Rerum medicarum Novae Hispaniae thesaurus* (Medical thesaurus of New Spain) of Francisco Hernández, prepared in 1628 but not published until 1651.

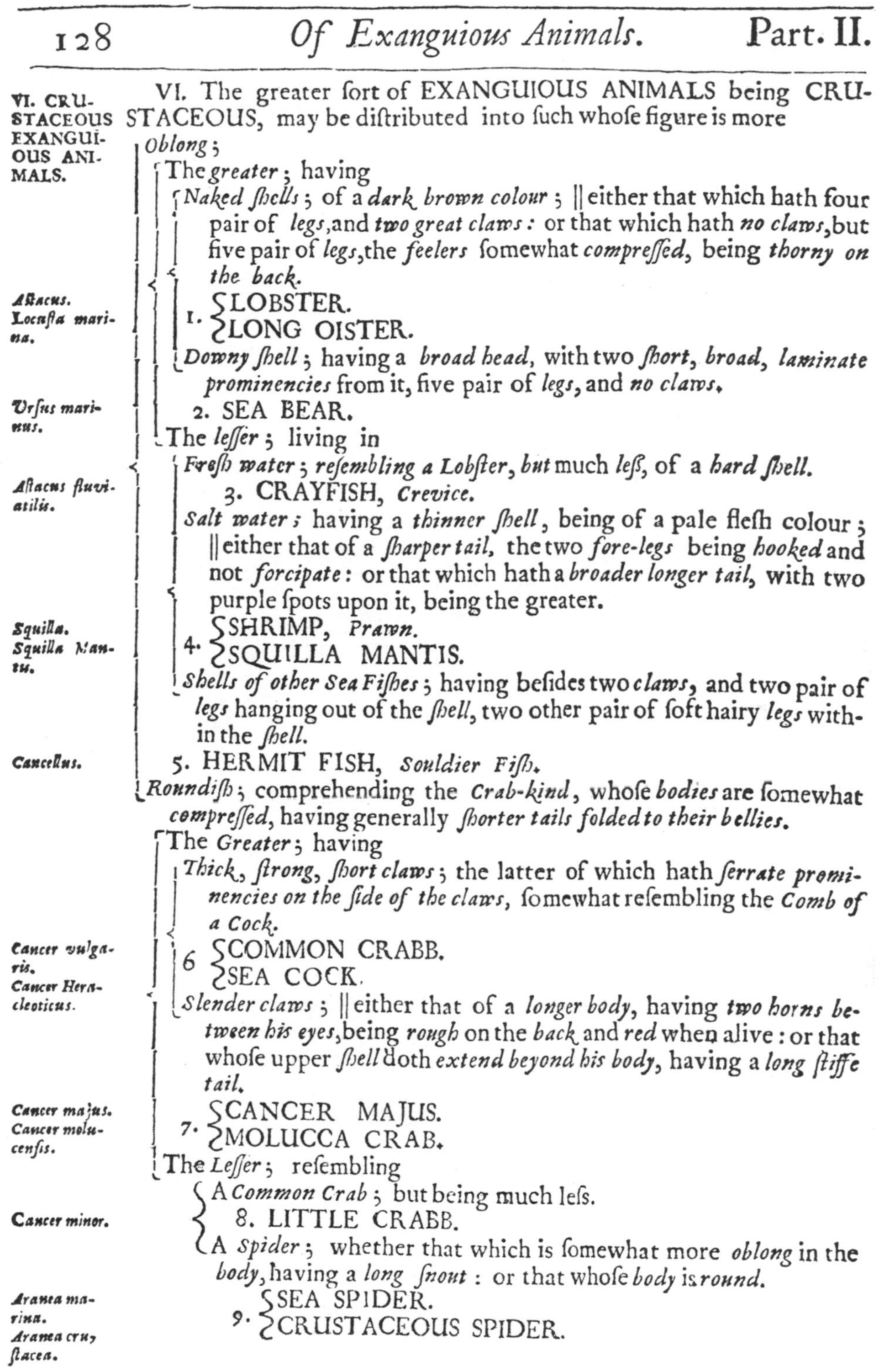

128 *Of Exanguious Animals.* Part. II.

VI. CRUSTACEOUS EXANGUIOUS ANIMALS.

VI. The greater ſort of EXANGUIOUS ANIMALS being CRUSTACEOUS, may be diſtributed into ſuch whoſe figure is more

Oblong;

- The *greater*; having
  - *Naked ſhells*; of a *dark brown colour*; ‖ either that which hath four pair of *legs*, and *two great claws*: or that which hath *no claws*, but five pair of *legs*, the *feelers* ſomewhat *compreſſed*, being *thorny on the back*.
    - 1. LOBSTER. / LONG OISTER. (*Aſtacus. Locuſta marina.*)
  - *Downy ſhell*; having a *broad head*, with two *ſhort*, *broad*, *laminate prominencies* from it, five pair of *legs*, and *no claws*.
    - 2. SEA BEAR. (*Urſus marinus.*)
- The *leſſer*; living in
  - *Freſh water*; *reſembling a Lobſter*, *but* much *leſs*, of a *hard ſhell*.
    - 3. CRAYFISH, *Crevice*. (*Aſtacus fluviatilis.*)
  - *Salt water*; having a *thinner ſhell*, being of a pale fleſh colour; ‖ either that of a *ſharper tail*, the two *fore-legs* being *hooked* and not *forcipate*: or that which hath a *broader longer tail*, with two purple ſpots upon it, being the greater.
    - 4. SHRIMP, *Prawn*. / SQUILLA MANTIS. (*Squilla. Squilla Mantu.*)
  - *Shells of other Sea Fiſhes*; having beſides two *claws*, and two pair of *legs* hanging out of the *ſhell*, two other pair of ſoft hairy *legs* within the *ſhell*.
    - 5. HERMIT FISH, *Souldier Fiſh*. (*Cancellus.*)

Roundiſh; comprehending the *Crab-kind*, whoſe *bodies* are ſomewhat *compreſſed*, having generally *ſhorter tails folded to their bellies*.

- The *Greater*; having
  - *Thick*, *ſtrong*, *ſhort claws*; the latter of which hath *ſerrate prominencies on the ſide of the claws*, ſomewhat reſembling the *Comb of a Cock*.
    - 6 COMMON CRABB. / SEA COCK. (*Cancer vulgaris. Cancer Heracleoticus.*)
  - *Slender claws*; ‖ either that of a *longer body*, having *two horns between his eyes*, being *rough* on the *back* and *red* when alive: or that whoſe upper *ſhell* doth *extend beyond his body*, having a *long ſtiffe tail*.
    - 7. CANCER MAJUS. / MOLUCCA CRAB. (*Cancer majus. Cancer moluccenſis.*)
- The *Leſſer*; reſembling
  - A *Common Crab*; but being much leſs.
    - 8. LITTLE CRABB. (*Cancer minor.*)
  - A *Spider*; whether that which is ſomewhat more *oblong* in the *body*, having a *long ſnout*: or that whoſe *body* is *round*.
    - 9. SEA SPIDER. / CRUSTACEOUS SPIDER. (*Aranea marina. Aranea cruſtacea.*)

FIGURE 7. Of exsanguinous, or bloodless, animals. An example of the hundreds of bracketed diagrams published by John Wilkins in part two of his 1668 *Essay Towards a Real Character, and a Philosophical Language.*

Of Birds.

§ IV. BIRDS may be diſtinguiſhed by their uſual place of living, their food, bigneſs, ſhape, uſe and other qualities, into

Terreſtrial; living chiefly on *dry land*; whether
CARNIVOROUS; feeding chiefly on *Fleſh*. I.
PHYTIVOROUS; feeding on *Vegetables*; whether
Of ſhort round wings; leſs fit for flight. II.
Of long wings; and ſwifter flight; having their *Bills*; either more
LONG AND SLENDER; comprehending the *Pidgeon* and *Thruſh-kind*. III.
SHORT AND THICK; comprehending the *Bunting* and *Sparrow-kind*. IV.
Inſectivorous; feeding chiefly on *Inſects*; (tho ſeveral of them do likewiſe ſometimes feed on *Seeds*) having *ſlender ſtreight bills* to thruſt into holes, for the pecking out of *Inſects*; whether the
GREATER KIND. V.
LEAST KIND. VI.
Aquatic; living either
About and NEAR WATERY PLACES. VII.
In waters; whether
FISSIPEDES; having the *toes of their feet* divided. VIII.
PALMIPEDES; having the *toes of their feet* united by a *membrane*. IX.

FIGURE 8. A classification of birds published by John Wilkins in 1668, distinguished by their usual place of living, food, size, shape, use, and other qualities.

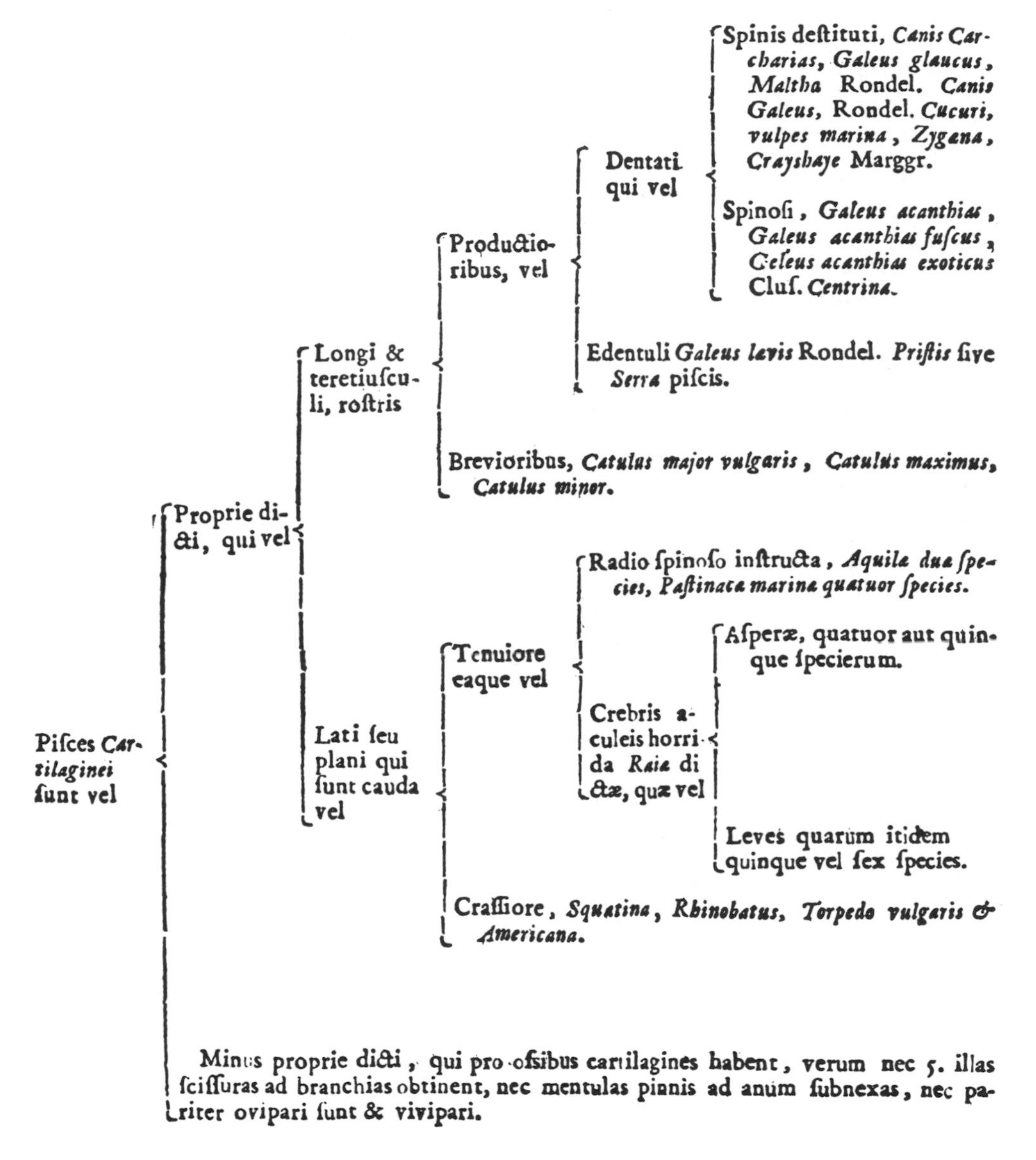

Piscium CARTILAGINEORUM Tabula.

Piſces *Cartilaginei* ſunt vel
- Proprie dicti, qui vel
  - Longi & teretiuſculi, roſtris
    - Productioribus, vel
      - Dentati qui vel
        - Spinis deſtituti, *Canis Carcharias*, *Galeus glaucus*, *Maltha* Rondel. *Canis Galeus*, Rondel. *Cucuri*, *vulpes marina*, *Zygana*, *Crayshaye* Marggr.
        - Spinoſi, *Galeus acanthias*, *Galeus acanthias fuſcus*, *Galeus acanthias exoticus* Cluſ. *Centrina*.
      - Edentuli *Galeus lævis* Rondel. *Priſtis* ſive *Serra* piſcis.
    - Brevioribus, *Catulus major vulgaris*, *Catulus maximus*, *Catulus minor*.
  - Lati ſeu plani qui ſunt cauda vel
    - Tenuiore eaque vel
      - Radio ſpinoſo inſtructa, *Aquila duæ ſpecies*, *Paſtinaca marina quatuor ſpecies*.
      - Crebris aculeis horrida *Raiæ* dictæ, quæ vel
        - Aſperæ, quatuor aut quinque ſpecierum.
        - Leves quarum itidem quinque vel ſex ſpecies.
    - Craſſiore, *Squatina*, *Rhinobatus*, *Torpedo vulgaris & Americana*.
- Minus proprie dicti, qui pro oſſibus cartilagines habent, verum nec 5. illas ſciſſuras ad branchias obtinent, nec mentulas pinnis ad anum ſubnexas, nec pariter ovipari ſunt & vivipari.

FIGURE 9. A classification of cartilaginous fishes published in 1686 by Francis Willughby in *De historia piscium libri quatuor* (History of fishes in four books).

CLAVIS SYSTEMATIS SEXUALIS.

Flos eft plantarum gaudium.

--- Sic planta propagat!

NUPTIÆ PLANTARUM
Actus generationis incolarum Regni Vegetabilis.
Florefcentia.

PUBLICÆ.
Nuptiæ coram totum mundum vifibilem apertè celebrantur.
Flores unicuique vifibiles funt.

MONOCLINIA.
Mariti & Uxores uno eodemque Thalamo gaudent.
Flores omnes hermaphroditi funt, & ftamina cum piftillis in eodem flore.

DIFFINITAS.
Mariti inter fe non cognati funt.
Stamina nullâ fuâ parte connata inter fe funt.

INDIFFERENTISMUS.
Mariti nullam fubordinationem inter fe invicem obfervant.
Stamina nullam accuratam proportionem longitudinis inter fe invicem habent.

MONANDRIA. à μόνος unicus, & ἀνὴρ maritus. — I.
Maritus unicus in matrimonio.
Stamen unicum in flore hermaphrodito.

DIANDRIA. — II.
Mariti duo in eodem conjugio.
Stamina duo in flore hermaphrodito.

TRIANDRIA. — III.
Mariti tres in eodem conjugio.
Stamina tria in flore hermaphrodito.

TETRANDRIA. — IV.
Mariti quatuor in eodem conjugio.
Stamina quatuor in eodem flore cum fructu.
Obf. *Si Stamina 2 proxima breviora funt, referatur ad Cl. 14.*

PENTANDRIA. — V.
Mariti quinque in eodem conjugio.
Stamina quinque in flore hermaphrodito.

HEXANDRIA. — VI.
Mariti fex in eodem conjugio.
Stamina fex in flore hermaphrodito.
Obf. *Si ex his Stamina 2 oppofita breviora, pertinet ad Cl. 15.*

HEPTANDRIA. — VII.
Mariti feptem in eodem conjugio.
Stamina feptem in flore eodem cum piftillo.

OCTANDRIA. — VIII.
Mariti octo in eodem thalamo cum femina.
Stamina octo in eodem flore cum piftillo.

ENNEANDRIA. — IX.
Mariti novem in eodem thalamo cum femina.
Stamina novem in flore hermaphrodito.

DECANDRIA. — X.
Mariti decem in eodem conjugio.
Stamina decem in eodem flore cum piftillo.

DODECANDRIA. — XI.
Mariti duodecim in eodem conjugio.
Stamina duodecim in flore hermaphrodito.

ICOSANDRIA. ab εἴκοσι viginti & ἀνὴρ. — XII.
Mariti viginti communiter, fæpe plures, raro pauciores.
Stamina (non receptaculo) *calicis lateri interno adnata.*

POLYANDRIA. à πολὺς & ἀνὴρ. — XIII.
Mariti viginti & ultra in eodem cum femina thalamo.
Stamina à 15 ad 1000 *in eodem, cum piftillo, flore.*

SUBORDINATIO.
Mariti certi reliquis præferuntur.
Stamina duo femper reliquis breviora funt.

DIDYNAMIA. à δὶς, bis, & δύναμις potentia. — XIV.
Mariti quatuor, quorum 2 longiores, & 2 breviores.
Stamina quatuor, quorum 2 proxima longiora funt.

TETRADYNAMIA. — XV.
Mariti fex, quorum 4 longiores in flore hermaphrodito.
Stamina fex, quorum 4 longiora, 2 autem oppofita breviora.

AFFINITAS.
Mariti propinqui & cognati funt.
Stamina cohærent vel inter fe invicem aliqua fua parte, vel cum piftillo.

MONADELPHIA. à μόνος unicus, & ἀδελφὸς frater. — XVI.
Mariti, ut fratres, ex una bafi proveniunt.
Stamina filamentis in unum corpus coalita funt.

DIADELPHIA. — XVII.
Mariti è duplici bafi, tamquam è duplici matre, oriuntur.
Stamina filamentis in duo corpora connata funt.

POLYADELPHIA. — XVIII.
Mariti ex pluribus, quam duabus, matribus orti funt.
Stamina filamentis in tria, vel plura, corpora coalita.

SYNGENESIA. à σὺν fimul, & γένεσις generatio. — XIX.
Mariti cum genitalibus fœdus conftituerunt.
Stamina antheris (raro filamentis) in cylindrum coalita.

GYNANDRIA. à γυνὴ femina, & ἀνὴρ maritus. — XX.
Mariti cum feminis monftrofe connati.
Stamina piftillis (non receptaculo) *infident.*

DICLINIA. à δὶς bis & κλίνη Lectus, Thalamus.
Mariti feu feminæ diftinctis thalamis gaudent.
Flores mafculini vel feminini in eadem fpecie.

MONOECIA. à μόνος unicus, & οἰκία domus. — XXI.
Mares habitant cum fem. in eadem domo, fed diverfo thalamo.
Flores mafculini & feminini in eadem planta funt.

DIOECIA. — XXII.
Mares & feminæ habitant in diverfis thalamis & domiciliis.
Flores mafculini in diverfa planta, à femininis nafcuntur.

POLYGAMIA. à πολὺς, & γάμος Nuptiæ. — XXIII.
Mariti cum uxoribus & innuptis cohabitant in diftinctis thal.
Flores Hermaphrodit. & mafculini l. femin. in eadem fpecie.

CLANDESTINÆ.
Nuptiæ clam inftituuntur.
Flores, oculis noftris nudis vix confpiciuntur.

CRYPTOGAMIA. à κρυπτὸς occultus, & γάμος Nuptiæ. — XXIV.
Nuptiæ clam celebrantur.
Florent intra fructum, vel parvitate oculos noftros fubterfugiunt.

FIGURE 10. The famous "sexual system" of plant relationships published by Carl Linnaeus in the first edition of *Systema naturae* (System of nature), 1735.

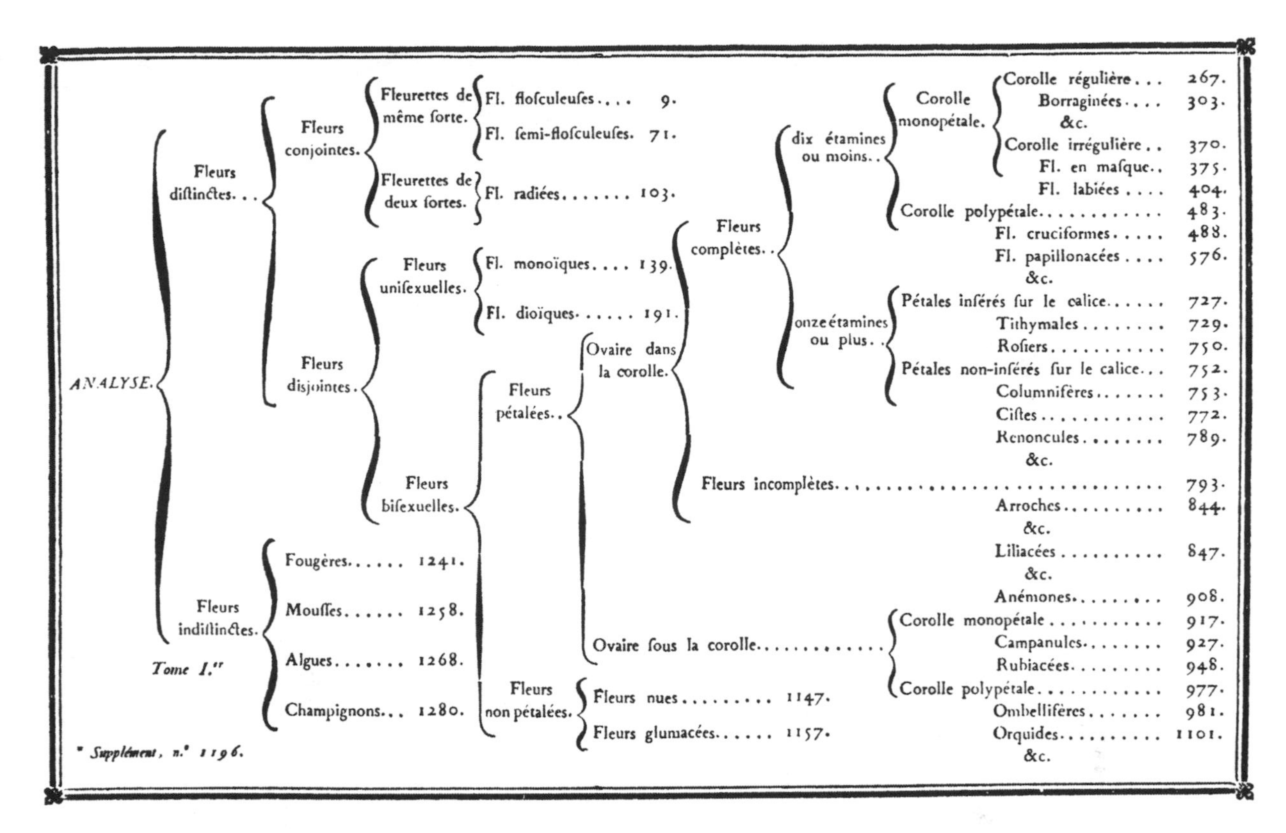

FIGURE 11. Jean-Baptiste Lamarck's 1778 bracketed diagram of the major groups of plants of France taken from the second volume of his *Flore françoise* (French flora) of 1778.

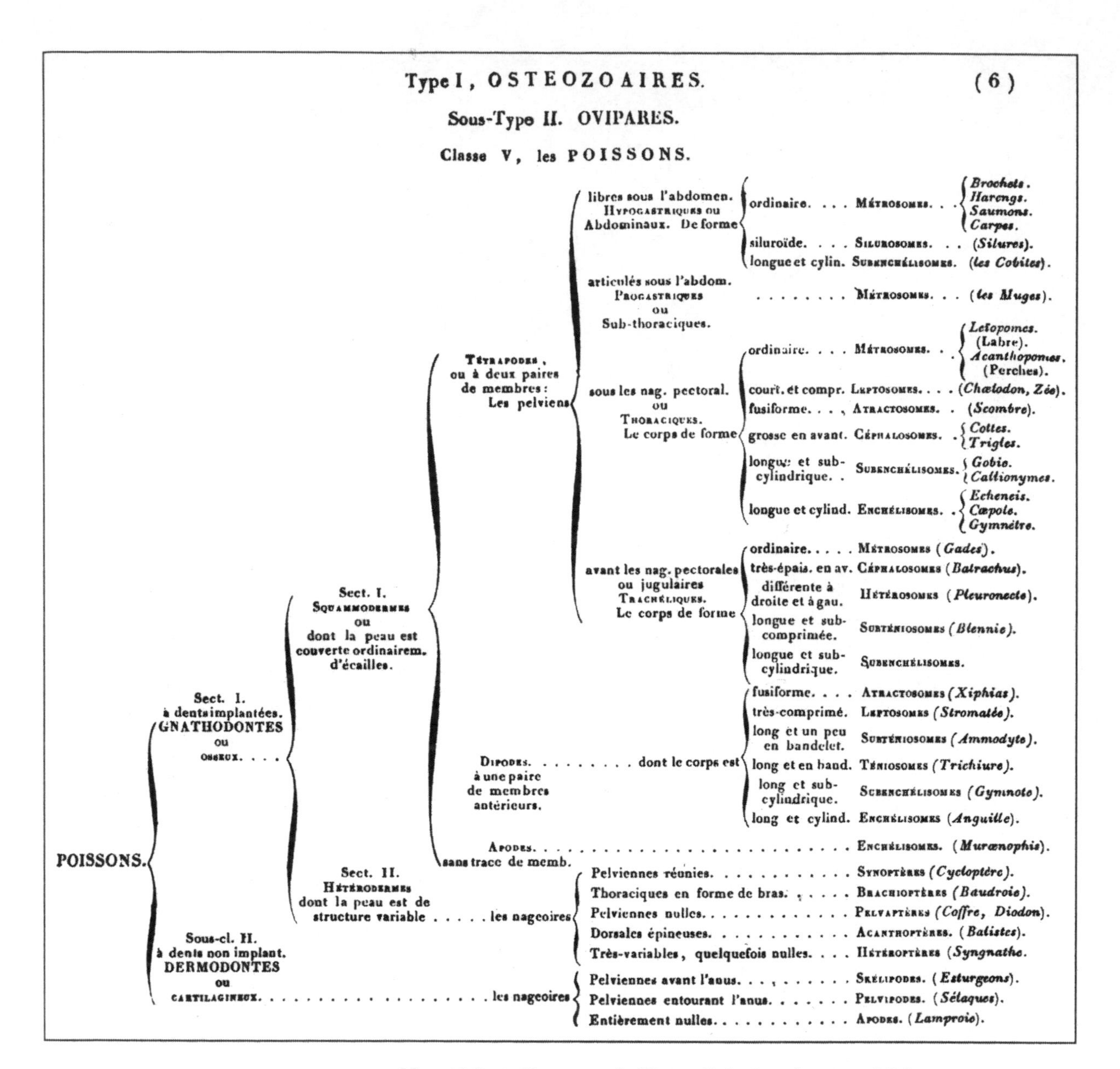

FIGURE 12. Henri Marie Ducrotay de Blainville's classification of fishes as presented in his *De l'organisation des animaux, ou Principes d'anatomie comparée* (Animal organization, or principles of comparative anatomy) of 1822.

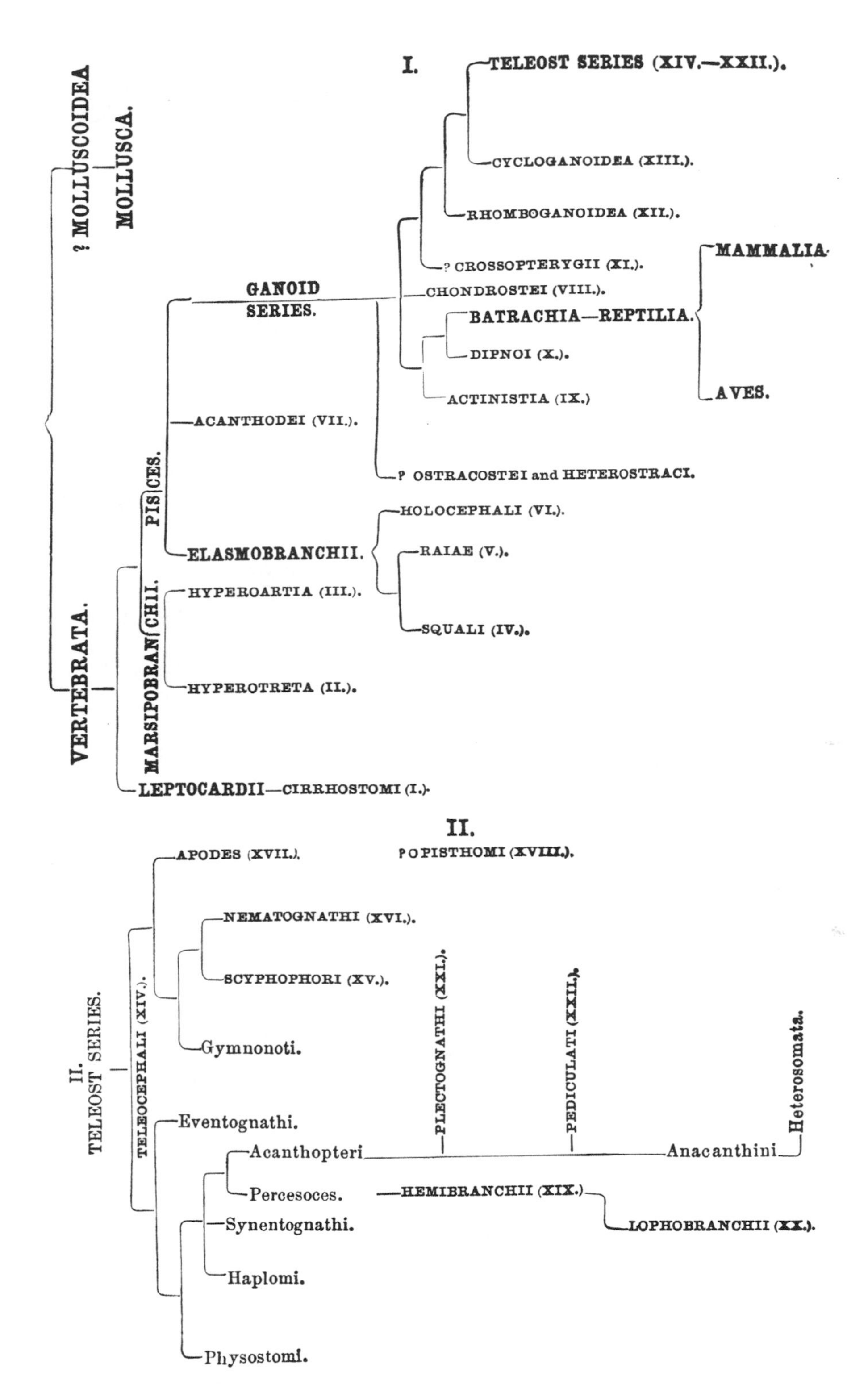

FIGURE 13. A classification of fishes published in 1872 by Theodore Nicolas Gill, showing vertebrates descending from mollusks, reminiscent of the classification of Lamarck published in 1809 (see figure 24).

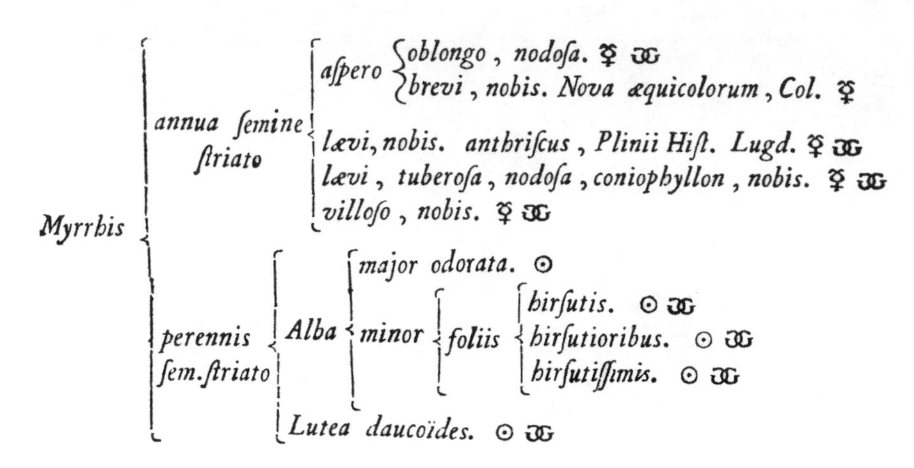

FIGURE 14. One of a number of bracketed diagrams constructed by Robert Morison and published in his *Plantarum umbelliferarum distribution nova* (New classification of umbelliferous plants) of 1672, this one showing the relationships of some umbelliferous plants of the genus *Myrrhis,* then thought to contain a number of species, but now recognized as monotypic, including only the sweet cicely or garden myrrh of modern common vernacular.

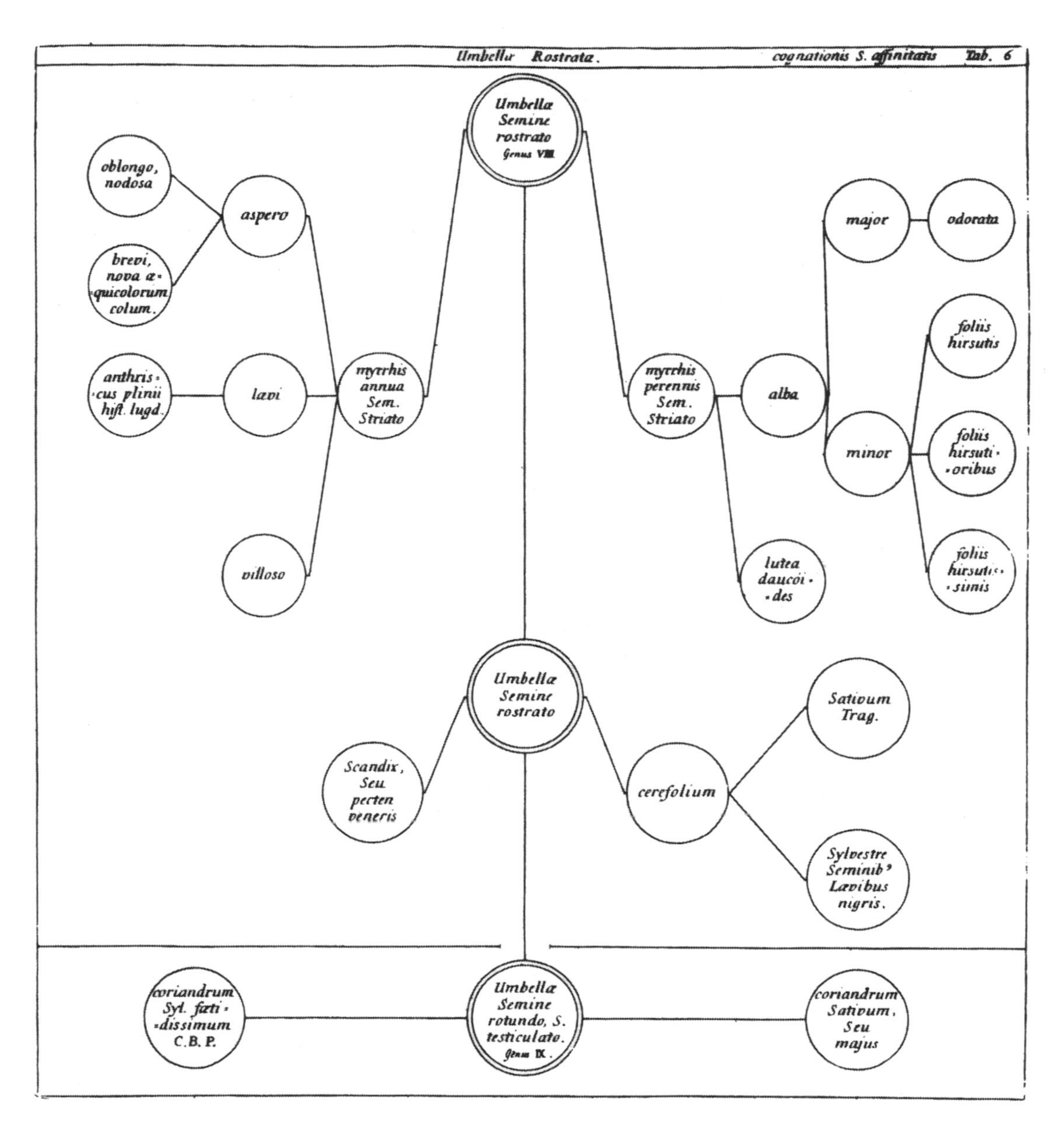

FIGURE 15. Robert Morison's (1672) recasting of his classification of some umbelliferous plants in the form of circles interconnected by simple, bifurcating, and trifurcating lines (compare with figure 14).

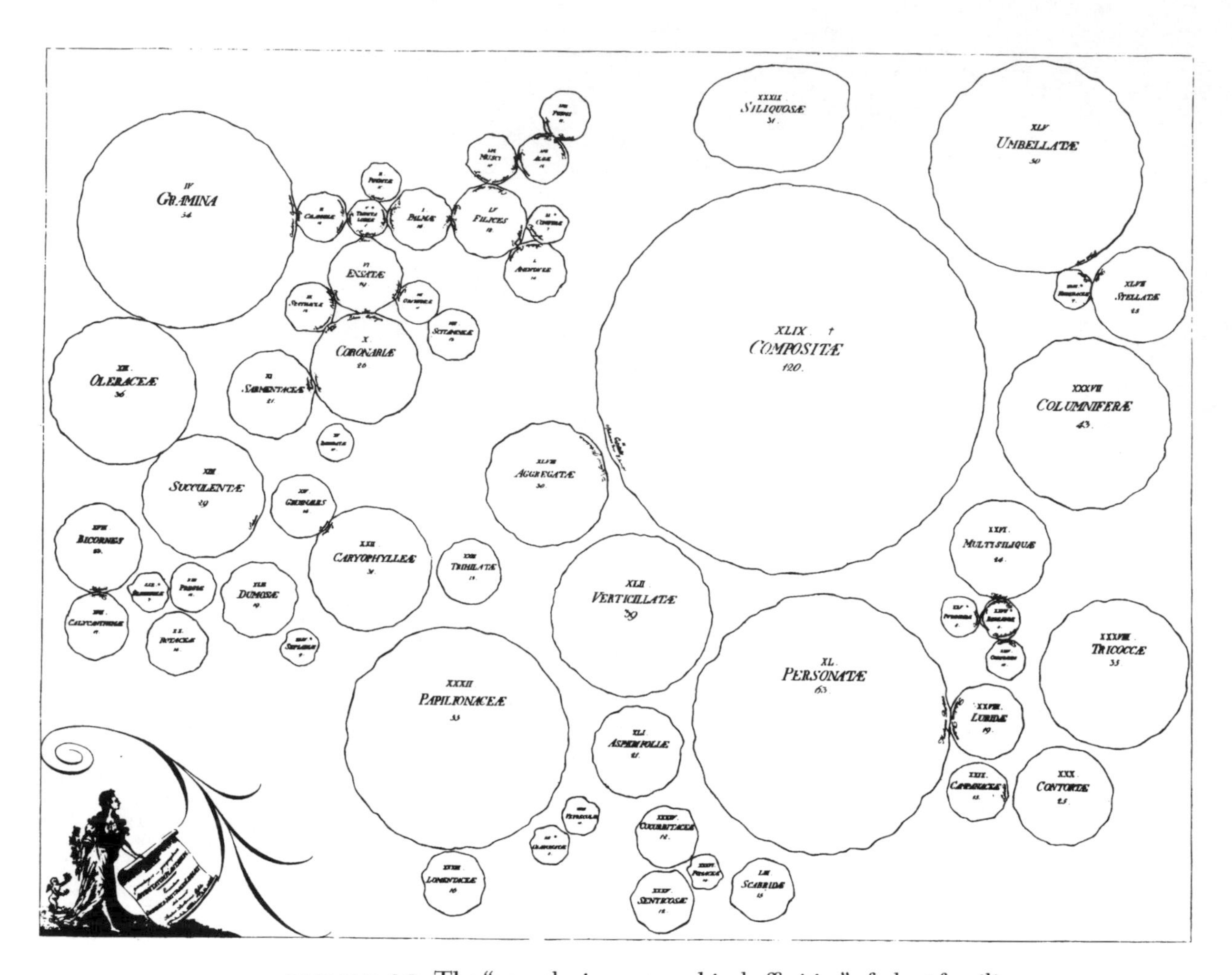

FIGURE 16. The "genealogico-geographical affinities" of plant families based on the natural orders of Carl Linnaeus (1751), published by Paul Dietrich Giseke in 1792. Each family is represented by a numbered circle (*roman numerals*), the diameter of which gives a rough measure of the relative number of included genera (*arabic numerals*).

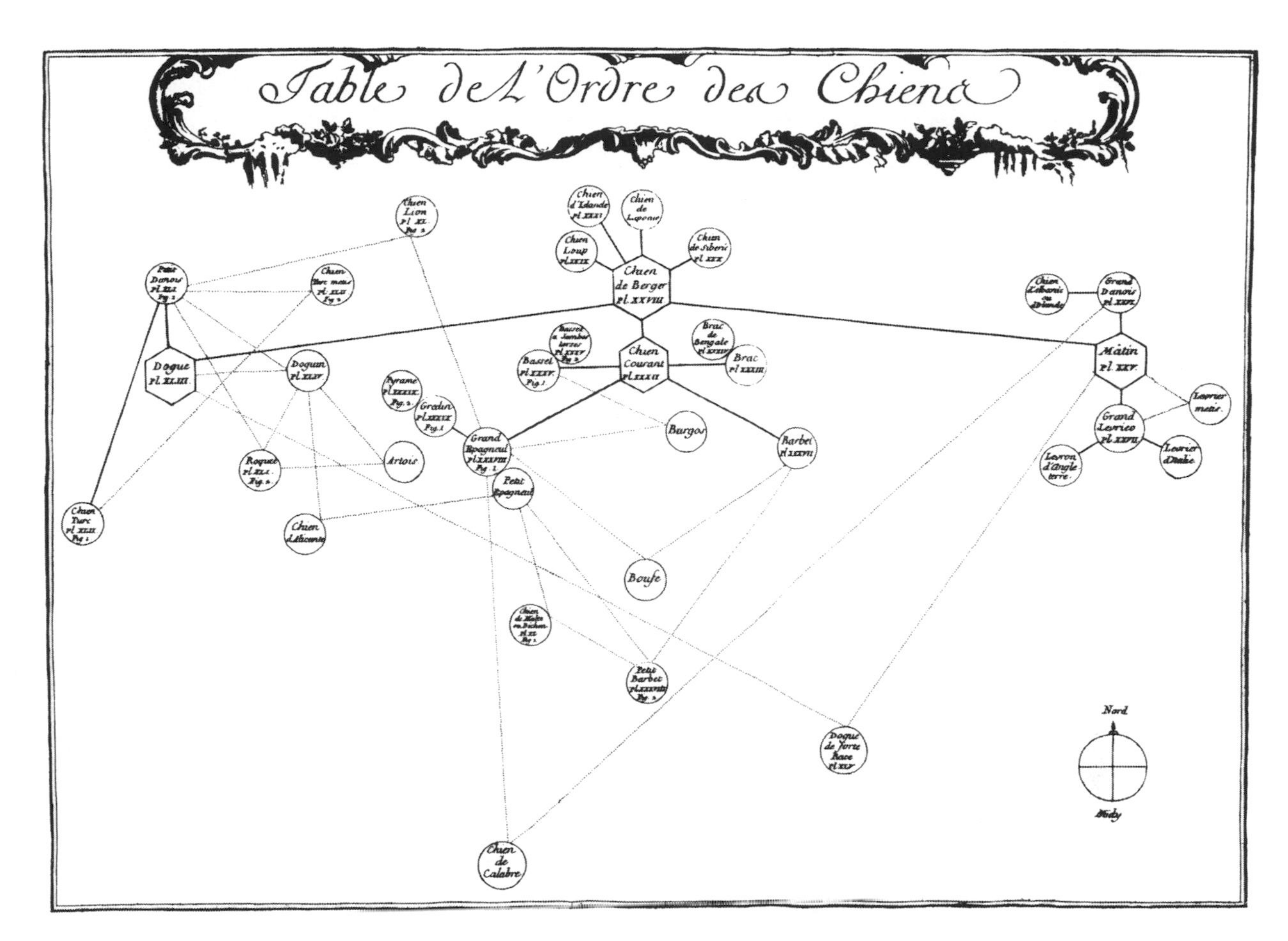

FIGURE 17. Georges Louis Leclerc de Buffon's "genealogical tree" of the races of dogs taken from *Histoire naturelle, générale et particulière, avec la description du cabinet du Roy* (Natural history, general and specific, with a description of the royal cabinet) published by Buffon and Louis Jean Marie Daubenton in 1755. The pure races are indicated by solid lines, the hybrids, by dotted lines. The tree is rooted with "*Chien de Berger*" (shepherd dog) placed within the large hexagon in the upper center of the diagram.

Early Botanical Networks and Trees

1766–1815

While naturalists of the mid-eighteenth century were busy constructing their bracket diagrams and a few were experimenting with other more or less related configurations, others, especially among botanists, were constructing trees more reminiscent of those we see today. Appearing somewhat like a hybrid between a bracketed diagram and a branching tree is German botanist Johann Philipp Rüling's (1741–?) complex network of affinities among the natural orders of plants, first published in 1766 (Figure 18). As shown in his diagram, Rüling believed, as did many of his contemporaries, that the products of nature existed in chains of being, not just one continuous chain including all life, but a series of numerous interconnecting progressions.[1] Chains of taxa of various lengths are shown in vertical series, linked horizontally by bracketed lines. A network of much greater complexity—described by one historian as "much like a demented spider's web"[2]—based on the same premise and again showing affinities within the vegetable kingdom,[3] was produced by August Johann George Carl Batsch (1761–1802) in 1802 (Figure 19).

An exceptional example of a tree-like diagram from this early period, is Antoine-Nicolas Duchesne's (1747–1827) "genealogy of strawberries" of 1766 (Figure 20). Horticulturalist to the king at the garden of Versailles and undoubtedly the world's authority on strawberries at the time, Duchesne was astonished to discover in 1761, in a bed of wood strawberries, a new variety that differed structurally from all those he knew in having leaves with one blade instead of two.[4] Not only was it different, but it maintained its uniqueness when multiplied by sowing. These surprising facts, gathered through detailed observation and repeated experimentation—rare activities for the time—led him to wonder about and question the fixity of species, and to speculate on the problems of defining the different categories of classification, genus, species, race, and variety.[5] Greatly puzzled, Duchesne pondered:

> How should it be considered . . .? Is it a species? . . . [or] is it but a variety? . . . In other genera, how many varieties are considered as species? I was long in this alternative . . . it appeared to me that there was something to correct in current ideas; but that the confusion was mainly caused by different authors' applying the same words to totally opposite ideas . . . . This reasoning has led me to consider strawberries, altogether, as forming a species distinct from all others, and each particular strawberry as a race or variety; it has also prompted me to investigate their genealogy.[6]

One result of this investigation was the arrangement of the races of strawberries in a branching diagram, in which the perpetual strawberry (*le Fraisier des mois*, which bears fruit in all months of the year) gives rise to all other varieties then known. Well ahead of his time, Duchesne concluded that "the genealogical order is the only one nature indicates, the only order that fully satisfies the mind."[7]

Even more remarkable than Duchese's strawberry tree is the *Arbre botanique* (Botanical tree) published in 1801 by Augustin Augier (Figure 21), a French naturalist from Lyon. As described by Peter Stevens, Augier's tree-like diagram depicts the relationships he believed to exist among members of the plant kingdom:

> He attempted to produce a natural classification based on several parts of the flower, rather than on a single one, that would also function like a key. The botanical tree was the result, and Augier described in considerable detail the methodology that he used to produce this tree, a way of representing natural relationships distinctly uncommon at the time. Augier used the botanical tree to distinguish between different kinds of relationships, two of which are similar to those currently called homology and analogy.[8]

Augier enthusiastically adopted the notion that a "figure like a genealogical tree appears to be the most proper to grasp the order and gradation of the series or branches that form classes or families."[9] Yet he was not an evolutionist, but convinced instead of the existence of a divine Creator: "It appears, and one can hardly doubt it, that the Creator, when making flowers, followed certain proportions and progressions in the number of their parts."[10] And while his leafy botanical image demonstrates a clear departure from the concept of a linear series of taxa—on which many of his contemporaries based their diagrams of relationship—it still represents an overall progression, from mosses and fungi at the bottom to the most perfect plants at the top. Nevertheless,

Augier's tree is an early example of some of the major changes that were shortly to transform pre-Darwinian systematics.[11]

Another tree extraordinary for its time, comparable in many ways to Augier's figure, was constructed by French physician and botanist Nicolas Charles Seringe (1776–1858), published in 1815 to illustrate a monograph on the willows (genus *Salix*) of Switzerland (Figure 22). It was intended not so much to show affinities between the species but rather to lay out differences and similarities among them and thus to function as a means of identification—a pictorial key of sorts.[12] If it were not for the obvious use of both primitive and derived (positive and negative) character states, written directly on all the bifurcating branches, it might be confused with a present-day cladogram (see Figures 190–192). Taxa are not dispersed along the lengths of the branches but indicated only at the terminus of each branch—a thoroughly modern approach.

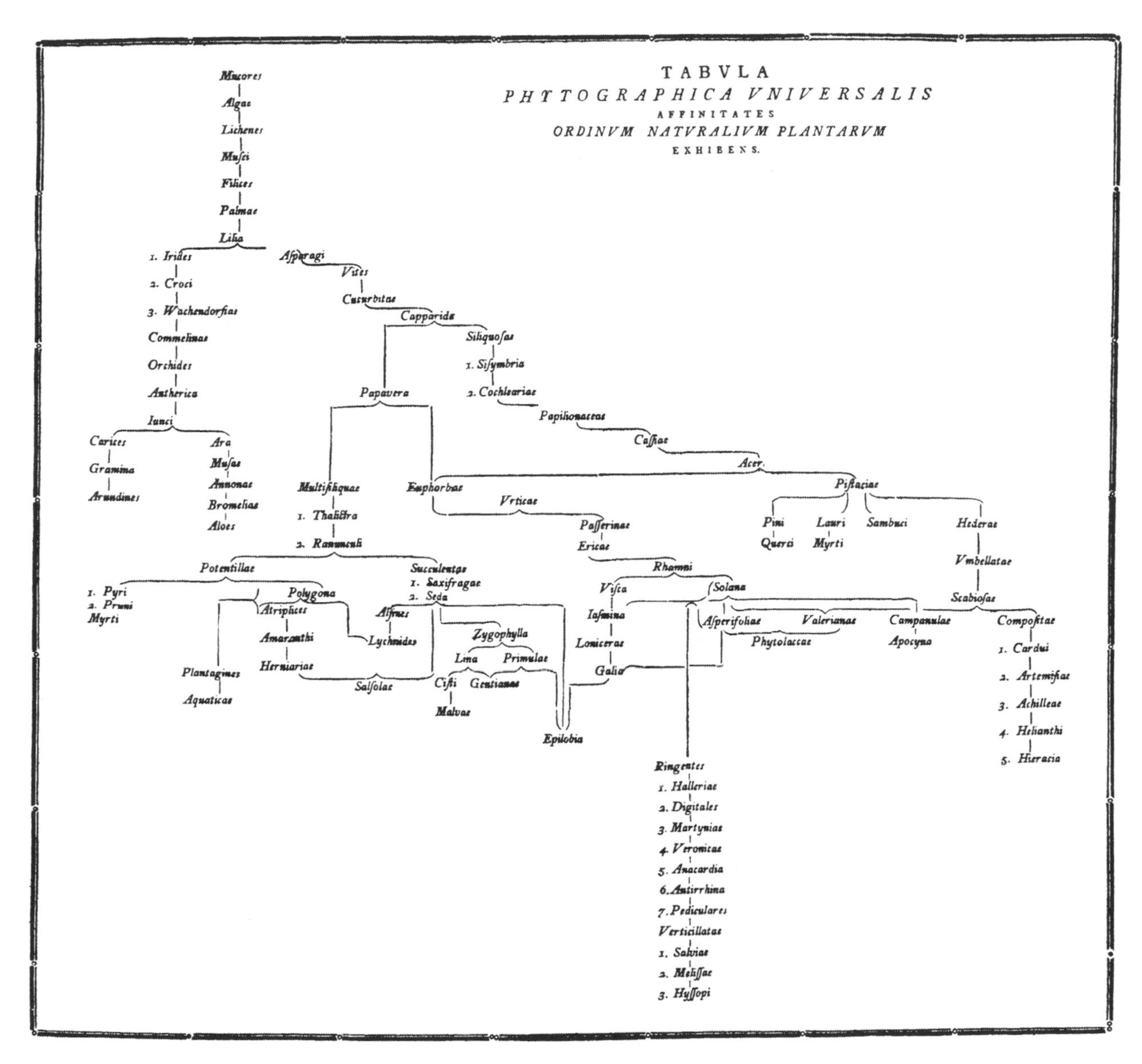

FIGURE 18. Johann Philipp Rüling's network of affinities among the natural orders of plants published in 1766.

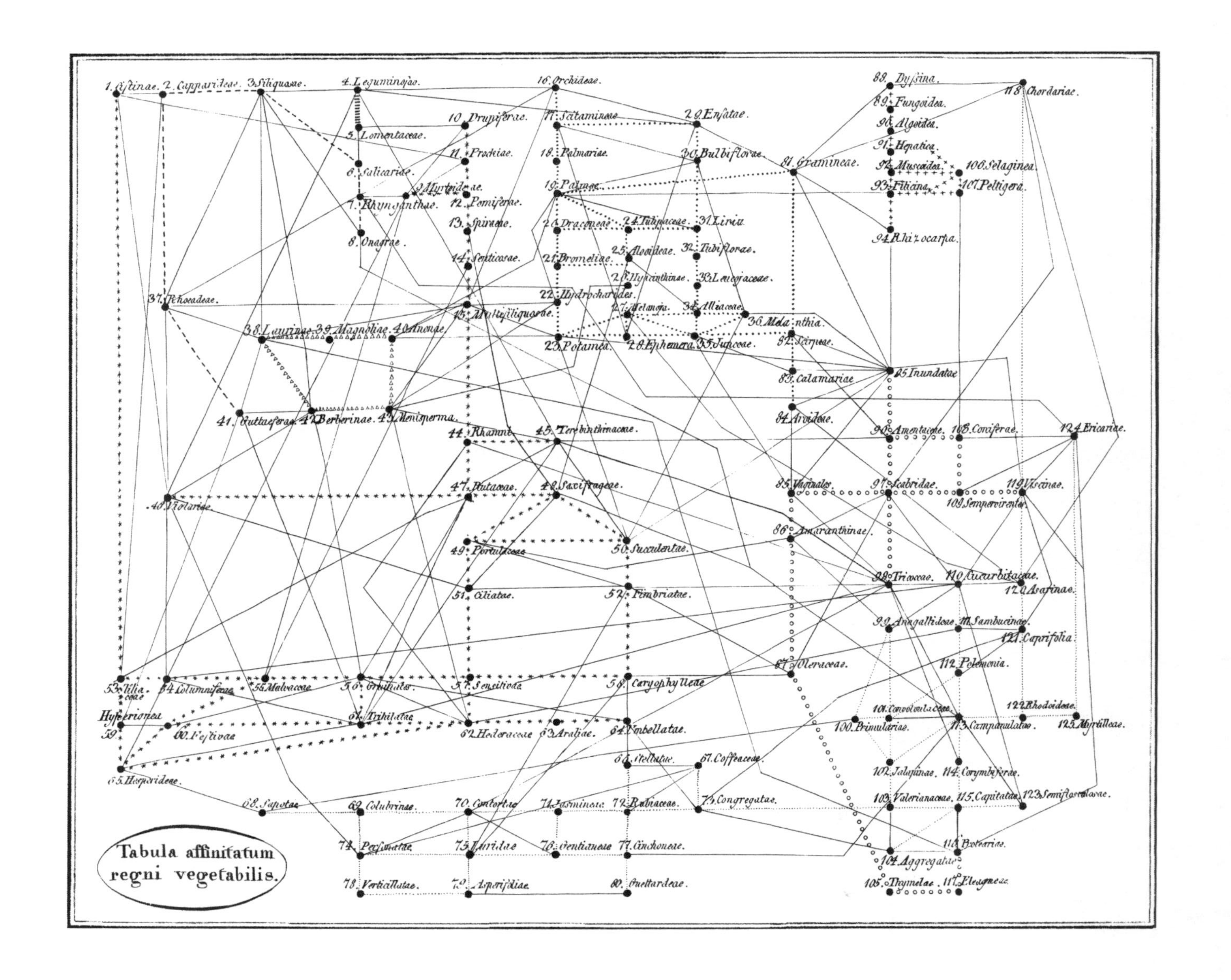

Tabula affinitatum regni vegetabilis.
1. Cistinae.
2. Capparideae.
3. Siliquosae.
4. Leguminosae.
5. Lomentaceae.
6. Salicariae.
7. Rhynganthae.
8. Onagrae.
10. Drupiferae.
11. Prakiae.
12. Pomiferae.
13. Spiraeae.
14. Senticosae.
15. Multisiliquae.
16. Orchideae.
17. Scitamineae.
18. Palmariae.
19. Palmae.
20. Draconeae.
21. Bromeliae.
22. Hydrocharides.
23. Potamea.
24. Tulipaceae.
25. Aloeideae.
28. Ephemera.
29. Ensatae.
30. Bulbiflorae.
31. Liriu.
32. Tubiflorae.
34. Alliaceae.
35. Junceae.
37. Rhoeadeae.
38. Laurinae.
39. Magnoliae.
40. Anonae.
41. Guttaeferae.
42. Berberinae.
43. Menisperma.
44. Rhamni.
45. Terebinthinaceae.
46. Violariae.
47. Rutaceae.
48. Saxifrageae.
49. Portulaceae.
50. Succulentae.
51. Ciliatae.
52. Fimbriatae.
55. Malvaceae.
57. Sensitivae.
58. Caryophylleae
60. Festivae
62. Hederaceae
63. Araliae.
64. Umbellatae.
65. Hesperideae.
67. Coffeaceae.
68. Sapotae
69. Colubrinae.
70. Contortae
71. Jasmineae
72. Rubiaceae.
73. Luridae
74. Congregatae.
75. Personatae.
76. Gentianeae
77. Cinchoneae.
78. Verticillatae.
79. Asperifoliae
80. Stellatae.
81. Gramineae.
83. Calamariae
84. Aroideae.
85. Inundatae
86. Amaranthinae.
87. Oleraceae.
88. Byssina.
89. Fungoidea.
90. Algoidea.
91. Hepatica.
92. Muscoidea.
93. Filicina.
94. Rhizocarpa.
96. Amentaceae.
97. Scabridae.
98. Tricoccae.
100. Primulariae.
101. Convolvulaceae.
102. Jalapinae.
103. Valerianaceae.
104. Aggregatae
105. Thymelae.
106. Selaginea.
107. Peltigera.
108. Coniferae.
109. Sempervirentes.
110. Cucurbitaceae.
111. Sambucinae.
112. Polemonia.
113. Campanulatae.
114. Corymbiferae.
115. Capitatae.
117. Eleagneae.
118. Chordariae.
119. Viscinae.
120. Asarinae.
121. Caprifolia.
122. Rhodoideae.
123. Semiflosculosae.
124. Ericariae.
125. Myrtilleae.

facing page

FIGURE 19. *Table of affinity of the vegetable kingdom, published by August Johann George Carl Batsch in 1802.*

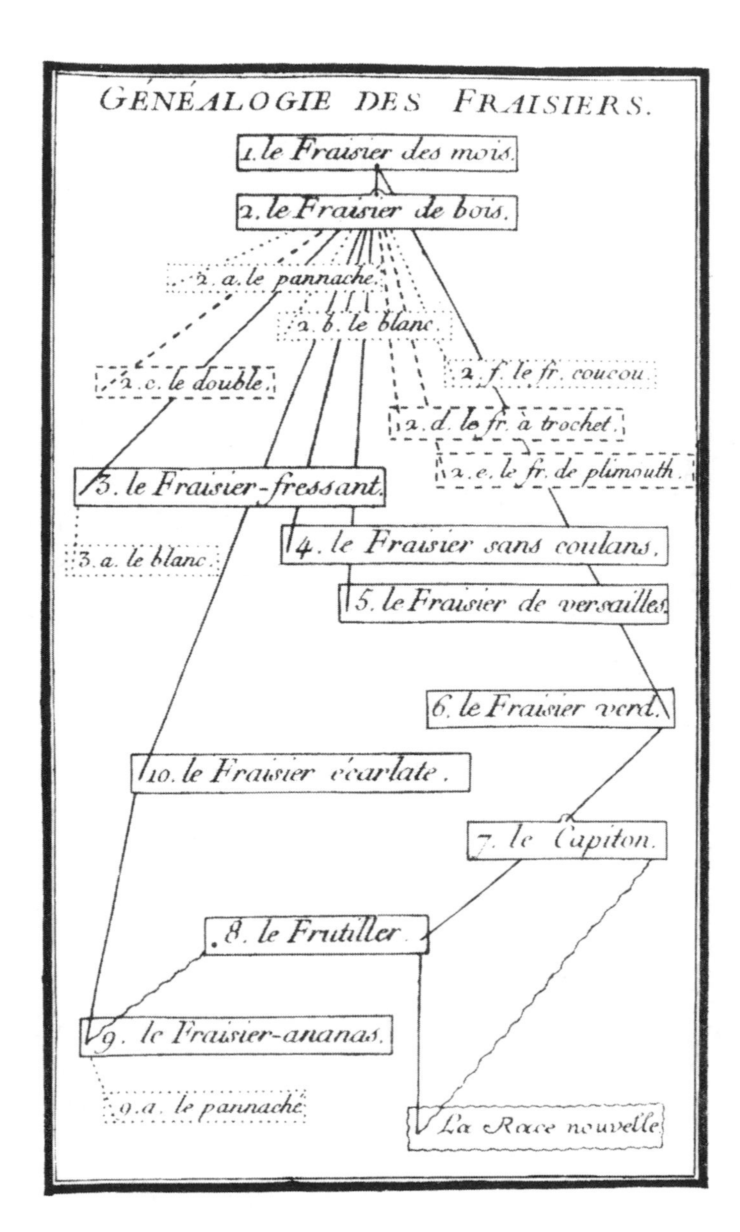

FIGURE 20. Genealogical relationships of strawberries as conceived by Antoine-Nicolas Duchesne in his *Histoire naturelle des fraisiers* (Natural history of strawberries) of 1766, an upside-down tree descending from the perpetual strawberry (*le Fraisier des mois*) and leading to nine numbered races or varieties (note Duchesne's "new race" on the bottom right): "The lines which link the strawberries together show the descent which I have observed or I presume. The dotted lines indicate those which have fecundated the seed from which these strawberries originated" (Duchesne, 1766:228).

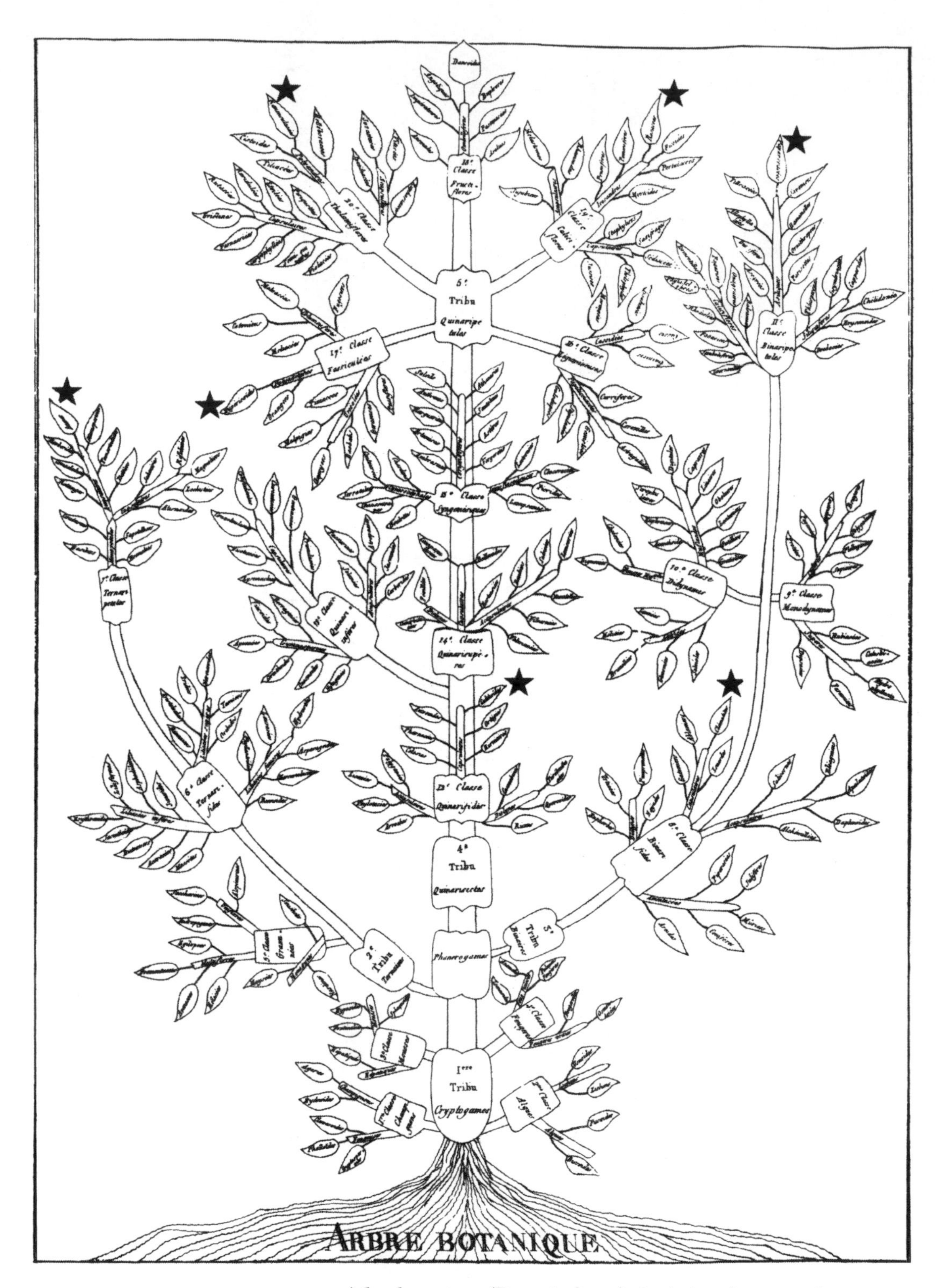

FIGURE 21. *Arbre botanique* (Botanical tree), depicting the natural relationships among members of the plant kingdom as conceived by Augustin Augier in 1801. The stars indicate families that show "relationship of analogy." Stevens, 1983, *Taxon,* 32(2):203–211, fig. 1; courtesy of Peter F. Stevens, Mary E. Endress, and the International Association for Plant Taxonomy. Used with permission.

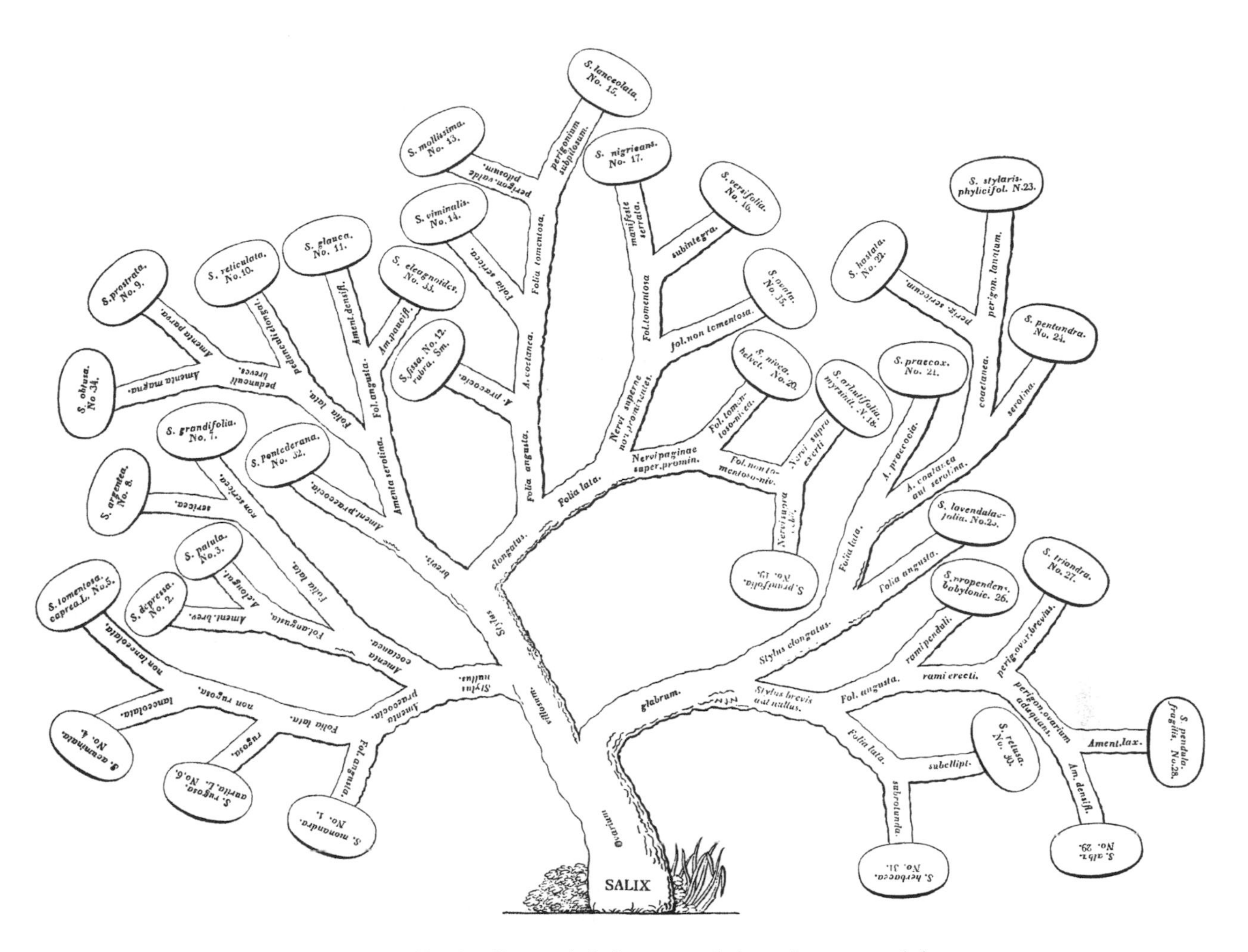

FIGURE 22. Nicolas Charles Seringe's dichotomously branching tree of the willows (genus *Salix*) of Switzerland published in 1815, with key characters indicated on the branches. The numbers within the ovals refer to the order in which the species are described in his text.

The First Evolutionary Tree

1786–1820

Jean-Baptiste Lamarck's famous branching diagram of 1809 stands in stark contrast to those of Augier and Seringe. From all evidence, these two men never had a thought of evolutionary change or transformation of species when constructing their botanical trees.[1] Lamarck made evolutionary change, "including the ongoing spontaneous creation of primitive forms and the upward transformation of existing life," central to his view of biology as a whole.[2] As Ernst Mayr argued, none of Charles Darwin's (1809–1882) predecessors can be more closely associated with the idea of evolution:

> All others before him had discussed evolution *en passant* and incidentally to other subjects or else in poetical or metaphorical terms. He was the first author to devote an entire book primarily to the presentation of a theory of organic evolution. He was the first to present the entire system of animals as a product of evolution.[3]

Lamarck had devised a table in 1786 (Figure 23) showing the "true order of gradation" among plants and showing further how this linear arrangement forms "a perfect counterpart with the large sections that divide the animal kingdom."[4] But in 1809, the year of Darwin's birth and a full half-century before publication of *On the Origin of Species* (1859), Lamarck published the first tree that can truly be called evolutionary (Figure 24). At the same time, he clearly made use of the tree metaphor, referring repeatedly to botanical terms:

> The table [Figure 24] may facilitate the understanding of what I have said. It is there shown that in my opinion the animal scale begins by at least two

> separate branches, and that as it proceeds it appears to terminate in several twigs in certain places. This series of animals begins with two branches, where the most imperfect animals are found; the first animals therefore of each of these branches derive existence only through direct or spontaneous generation.[5]

In his figure, an upside-down tree, something like Duchesne's (1766) view of strawberries (see Figure 20), Lamarck proposed two primary but unconnected lineages. The first includes single-celled organisms (*infusoires*) and coelenterates (*polypes* and *radiaires*); the second begins with unsegmented worms (*vers*), and leads to a group containing the insects, spiders, and crustaceans on the right; and to another containing the annelid worms, barnacles (*cirrhipèdes*), and mollusks on the left. The latter assemblage descends to the vertebrates, the fishes and reptiles (including amphibians), giving rise to the birds and egg-laying monotremes on the left and mammals on the right. Somewhat later, Lamarck (1815) advanced his ideas of invertebrate relationship with a revised proposal (a vertically oriented bracket diagram; Figure 25), again dividing the group into two unconnected lineages or series: the inarticulated and articulated animals, respectively, with the various groups somewhat rearranged. This time, however, the whole assemblage is divided horizontally into three parts, labeled inert insensitive animals (*Animaux apathiques*), sensate animals (*Animaux sensibles*), and intelligent animals (*Animaux intelligens*). While the first two categories are connected within both series by vertical lines (thus demonstrating a continuum of increasing complexity),[6] the "intelligent" vertebrates at the bottom of the figure remain isolated in space, Lamarck confessing that he could no longer find a way to connect the vertebrates at any point with the invertebrates. Still later, in 1820, he replaced the idea of multiple lineages with a single sequence in which all life derived from a single form.[7]

* *Etres organiques vivans, aſſujettis à la mort, & qui ont la faculté de ſe reproduire eux-mêmes.*

Animaux.	Végétaux.
1. *LES QUADRUPEDES.*	*LES POLYPÉTALÉES.* 1.
1. Terreſtres onguiculés.	Thalamiflores. 1.
2. Terreſtres ongulés.	Caliciflores. 2.
3. Marins.	Fructiflores. 3.
2. . . . . *LES OISEAUX.*	*LES MONOPÉTALÉES*, ... 2.
1. Terreſtres.	Fructiflores. 1.
2. Aquatiques à cuiſſes nues.	Caliciflores. 2.
5. Aquatiques nageants.	Thalamiflcres. 3.
3. . . . . . *LES AMPHIBIES.*	*LES COMPOSÉES.* 3
1. Tétrapodes.	Diſtinctes. 1
	Tubuleuſes. 2.
2. Apodes.	Ligulaires. 3.
4. . . . . *LES POISSONS.*	*LES INCOMPLETTES.* . . . . . . 4.
1. Cartilagineux.	Thalamiflores. 1.
	Caliciflores. 2.
2. Epineux.	Diclynes. 3.
	Gynandres. 4.
5 . . . . *LES INSECTES.*	*LES UNILOBÉES.* . . . . 5.
1. Tétraptères.	Fructiflores. 1.
2. Diptères.	Thalamiflores. 2.
3. Aptères.	
6. . . . . *LES VERS.*	*LES CRYPTOGAMES.* . . . . 6.
1. Nuds.	Epiphylloſpermes. 1.
2. Teſtacés.	Urnigères. 2.
3. Lithophytes.	Membraneuſes. 3.
4. Zoophytes.	Fongueuſes. 4.

Botanique. Tome II.

FIGURE 23. Jean-Baptiste Lamarck's 1786 table demonstrating the "true order of gradation" in the plant kingdom and the way this arrangement forms a "perfect counterpart" to gradations that segregate the animal kingdom. While admitting that this dual linear arrangement (plants on the right, animals on the left) might not be a perfect representation of nature's true plan, it seemed to him "well suited to human intellectual capabilities" (see Burkhardt, 1995:56).

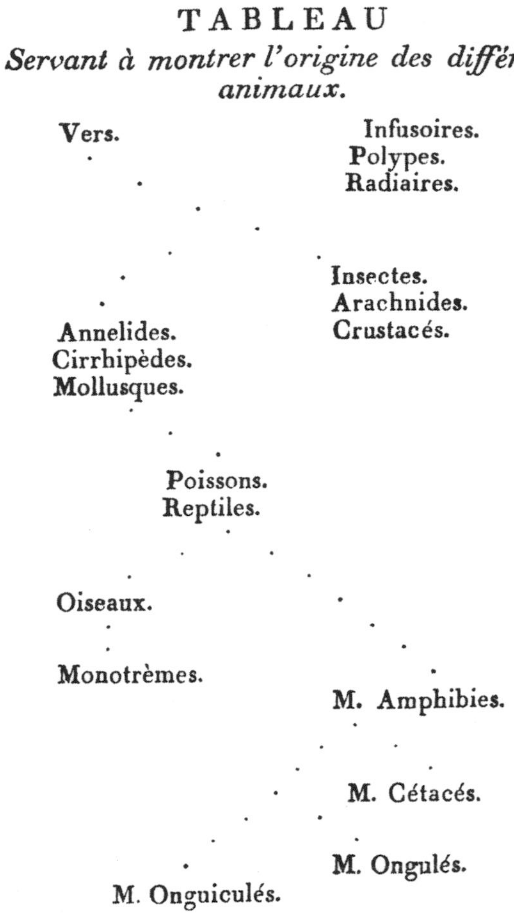

TABLEAU

Servant à montrer l'origine des différens animaux.

Vers.

Infusoires.
Polypes.
Radiaires.

Insectes.
Arachnides.
Crustacés.

Annelides.
Cirrhipèdes.
Mollusques.

Poissons.
Reptiles.

Oiseaux.

Monotrèmes.

M. Amphibies.

M. Cétacés.

M. Ongulés.

M. Onguiculés.

FIGURE 24. The earliest evolutionary tree: Lamarck's "table, serving to demonstrate the origin of the different animals," published in his *Philosophie zoologique* (Zoological philosophy) of 1809. By connecting the dots, note that the mammals, descending from a group containing the fishes and reptiles, form a group containing the seals, walruses, dugongs, and manatees (*M. Amphibies*), which bifurcates to form the cetaceans (*M. Cétacés*) on the right, and to hoofed mammals (*M. Ongulés*) and remaining mammals (*M. Onguiculés*) on the left.

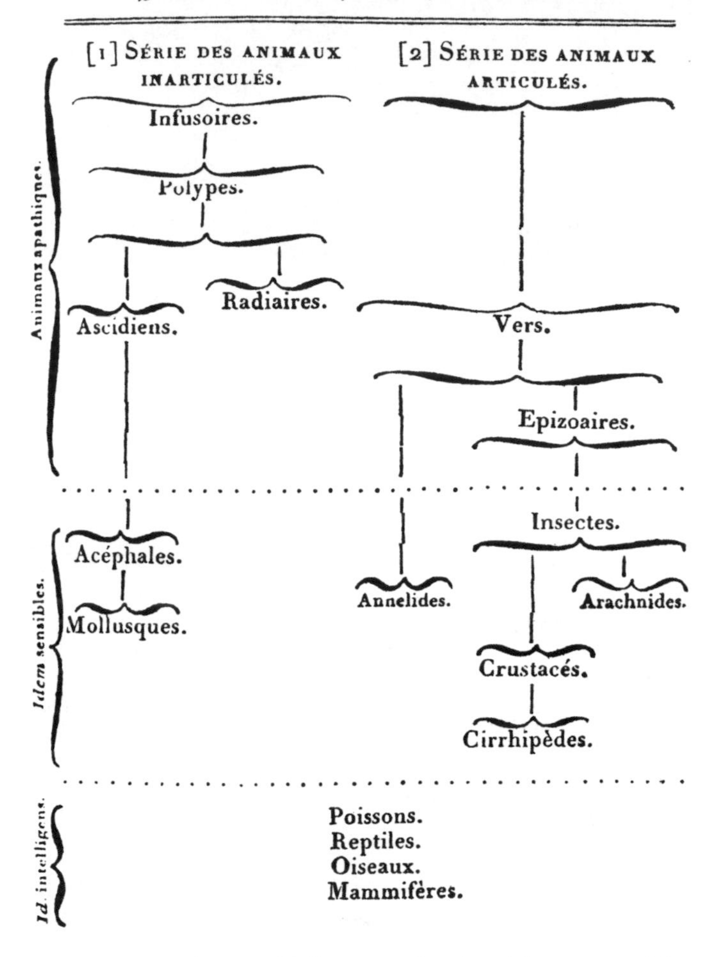

FIGURE 25. Lamarck's "presumed order of formation of the animals, presenting two separate, branching series," published in his *Histoire naturelle des animaux sans vertèbres* (Natural history of invertebrates) of 1815.

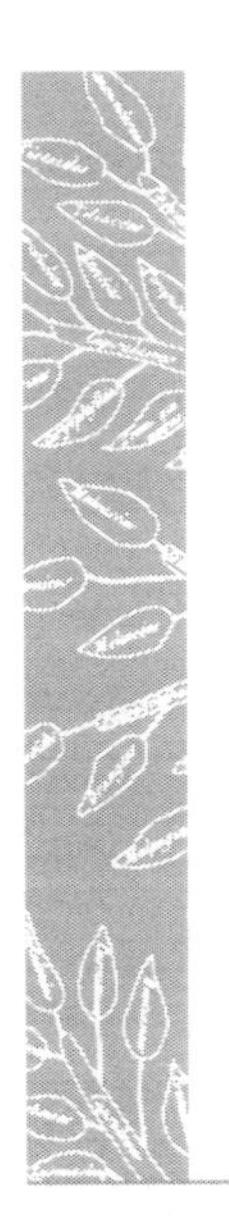

Diverse and Unusual Trees of the Early Nineteenth Century

1817–1834

Quite different from anything published before or after is French botanist Michel-Félix Dunal's (1789–1856) 1817 "table of affinities" within the Annonaceae, the custard apple family, a large group of flowering trees and shrubs or rarely lianas (Figure 26). The family is now thought to contain some 2,500 species in 130 genera, but Dunal seems to have recognized only nine genera. Although not easy to grasp at first glance, each genus in the figure, distinguished by structural differences in its fruiting body, is indicated by a labeled knotted bow, within a matrix of flowing ribbons, a concept likened by Gareth Nelson and Norman Platnick to "towns connected by railways."[1] As in the earlier circle maps and networks of Giseke and Buffon (see Figures 16 and 17), the relative position and distance between the knotted bows (genera) were apparently intended by Dunal to represent the degree of relative affinity between each. The assemblage of nine genera is enclosed or cut off by large bows on each side to emphasize the integrity of the group as a whole, but flanked on the outer margins by indications of closely related families, the Magnoliaceae (the magnolias) on the left and Menispermaceae (the moonseed family containing mostly woody climbing species of tropical forests) on the right.

Again very different and perhaps more bizarre is the unique "system of animals" published in 1817 by German zoologist Georg August Goldfuss (1782–1848): a nested series of ovals and circles placed within an outer egg-shaped form, the overall design perhaps meant to invoke an analogy between the egg and the birth and progression of life (Figure 27). While seemingly unrelated to contemporary diagrams of plant and animal relationships, a close look will show that Goldfuss has more or less repackaged, albeit with a bit more detail, the same linear progression seen many times before: single-celled animals (*Protozoa*) at the bottom, ascending through echinoderms and mollusks to vertebrates, with "sensible" humans (*Homo*) placed at the top.

Another unique design was published by Carl Edward von Eichwald (1795–1876) in his *Tabula transituum Animalium* (Table of animal transitions) of 1821—a highly geometric figure, reminiscent of a high-voltage electrical tower, with its numerous criss-crossing, interconnecting, supportive struts or lines between taxa (Figure 28). But once again, the same essential linear progression is repeated, the whole assemblage arising from a narrow base labeled "algae" and ascending through single-celled animals (*polypi* and *infusoria*) to spiders, insects, crustaceans, fishes, amphibians, reptiles, birds, and mammals.

A totally different image, published by Eichwald in 1829, described by one historian as a "bunch of asparagus-shoots,"[2] shows a leafless, dead-looking tree, its compound trunk thickly rooted in a swampy and otherwise almost featureless landscape, illuminated from above by sunbeams (Figure 29). It is said to be based on rules laid down much earlier by German naturalist Peter Simon Pallas (1741–1811), who, nearly a century before Darwin, proposed in his *Elenchus zoophytorum* (List of animals) of 1766 that gradation among organisms might best be described as a branching tree.[3] Like a modern-day phylogenetic tree, his fanciful drawing shows individual taxa represented by Roman numerals placed out on the tips of the twigs, rather than a linear arrangement of one group leading to another.

Nearly impossible to decipher, unless you know the plants well, is French botanist Adrien-Henri de Jussieu's (1797–1853) 1825 network of affinities among various groups of the Rutaceae (Figure 30), a family that contains, among other things, the species of *Citrus* that produce both the citrus fruits (such as lemons, oranges, mandarins, tangerines, limes, and kumquats) and the essential oils used in perfumery. The nine-sided polyhedron is divided into triangles and tetragons, each one labeled either within or along its outer margin with the appropriate Latinized name and each is devoted to a cluster of genera indicated by black dots. In a convention seen earlier, for example, with the circles of varying diameter in Buffon's (1755) genealogical tree of the races of dogs, the black dots vary in size depending on the number of included species, the smallest containing only a single species, the largest containing twelve or more. Note that the subgroups of the Aurantiaceae are divided into geographic realms, European, South African (*Capenses*), Australian, and American—another early example of animal affinity superimposed on geography (see Figure 17).[4]

A diagram published in 1828 (Figure 31) by Augustin-Pyramus de Candolle (1778–1841), showing the relationships of the Crassulaceae, or orpine family, a group of plants that stores large amounts of water in their succulent leaves, is in some ways similar to Jussieu's polyhedron (Figure 30) published three years before. But, instead of a subdivided, nine-sided figure, Candolle presents a much simpler arrangement: a circle divided into quadrants, each quadrant containing a cluster of more or less related genera:

> The central genera follow one another in the order of their affinities in a manner that appears to me very exact. This circular disposition, very adaptable to natural families, does, it seems to me, render the real analogies clearly apparent, and show the complete impossibility of establishing a regularly linear series.[5]

Like Dunal's (1817) knotted ribbons (see Figure 26), the integrity of the Crassulaceae, with all its included genera, is indicated by its enclosure within a double circle, while two closely related families, the latter shown within circles of their own, move off the page at the top. The three families are connected by dotted lines, stretching from apparently transitional genera: the genus *Tillaea* having affinity with the family Paronychieae on the left, and *Penthorum,* with the Saxifragaceae on the right.

Another Candolle diagram (Figure 32), also published in 1828, shows affinities among plants of the family Melastomataceae. This more complex image, reflecting the much greater size of the taxon, containing some two hundred genera and 4,500 species, is based on similar principles.[6] A close relationship with the much smaller Chariantheae is indicated by the overlapped circle on the left margin and still more distant relationships with the Lythrarieae and Myrtaceae listed above and below.

One last Candolle diagram published just a year earlier (1827) is very different from his other designs and still more complex, this one showing relationships among the legumes (Figure 33). Reminiscent of the egg-shaped diagram published by Goldfuss in 1817 (see Figure 27), groups are enclosed in ovals within ovals, most touching one another, with one or two cross-linkages, the latter apparently intended to indicate the sharing of analogous characters (see quinarian circular systems in the next chapter). Affinities with the Terebinthaceae—a plant family no longer recognized that was thought to include the cashew, pistachio, and mango trees—and the Rosaceae or rose family are indicated by open ovals at the top and bottom of the figure, respectively.

Perhaps the most ingenious and complex image from this period is Paul Fedorowitsch Horaninow's (1796–1865) circular diagram published in his *Primae lineae systematis naturae* (Primary system of nature) of 1834 (Figure 34). Difficult to interpret, but beautiful in its design, Horaninow's figure was meant to demonstrate the existence of a universal system that incorporates the relationships between and within all three kingdoms of nature: animals, plants, and minerals. Although appearing as circles within circles, closer inspection reveals a more or less continuous spiral, with humans (*Homo*) given the supreme position in the center of the natural universe, surrounded by everything else. From there, the remaining mammals, birds (*Aves*), amphibians, and fishes (*Pisces*), in that order, peel away, spiraling off clockwise, with the fishes eventually transitioning smoothly into crustaceans. The crustaceans are followed in turn by the insects, spiders (*Arachnida*), worms (*Annulata*), and so forth, eventually reaching the lowliest of animals, the single-celled creatures (*Infusoria*), which transition smoothly into the plants, the plants eventually giving way to nonmetallic minerals, and minerals to metals.

facing page

FIGURE 26. Michel-Félix Dunal's (1817) "ribbons and bows," showing the relationships among genera of the plant family Annonaceae. Although difficult to interpret at first glance, his approach was not unlike those of Morison (1672), Buffon (1755), and Giseke (1792) in figures 15–17.

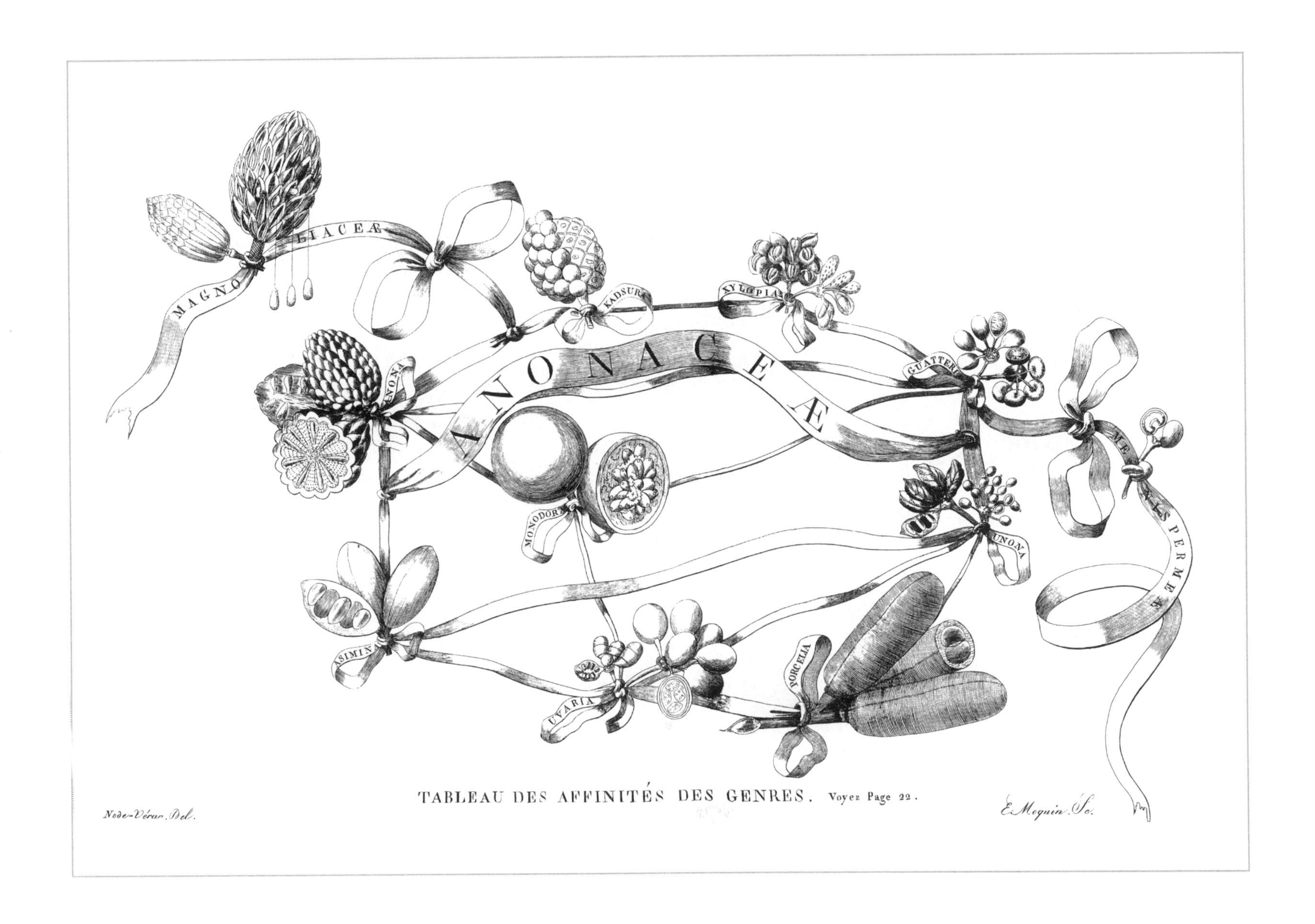

TABLEAU DES AFFINITÉS DES GENRES. Voyez Page 22.

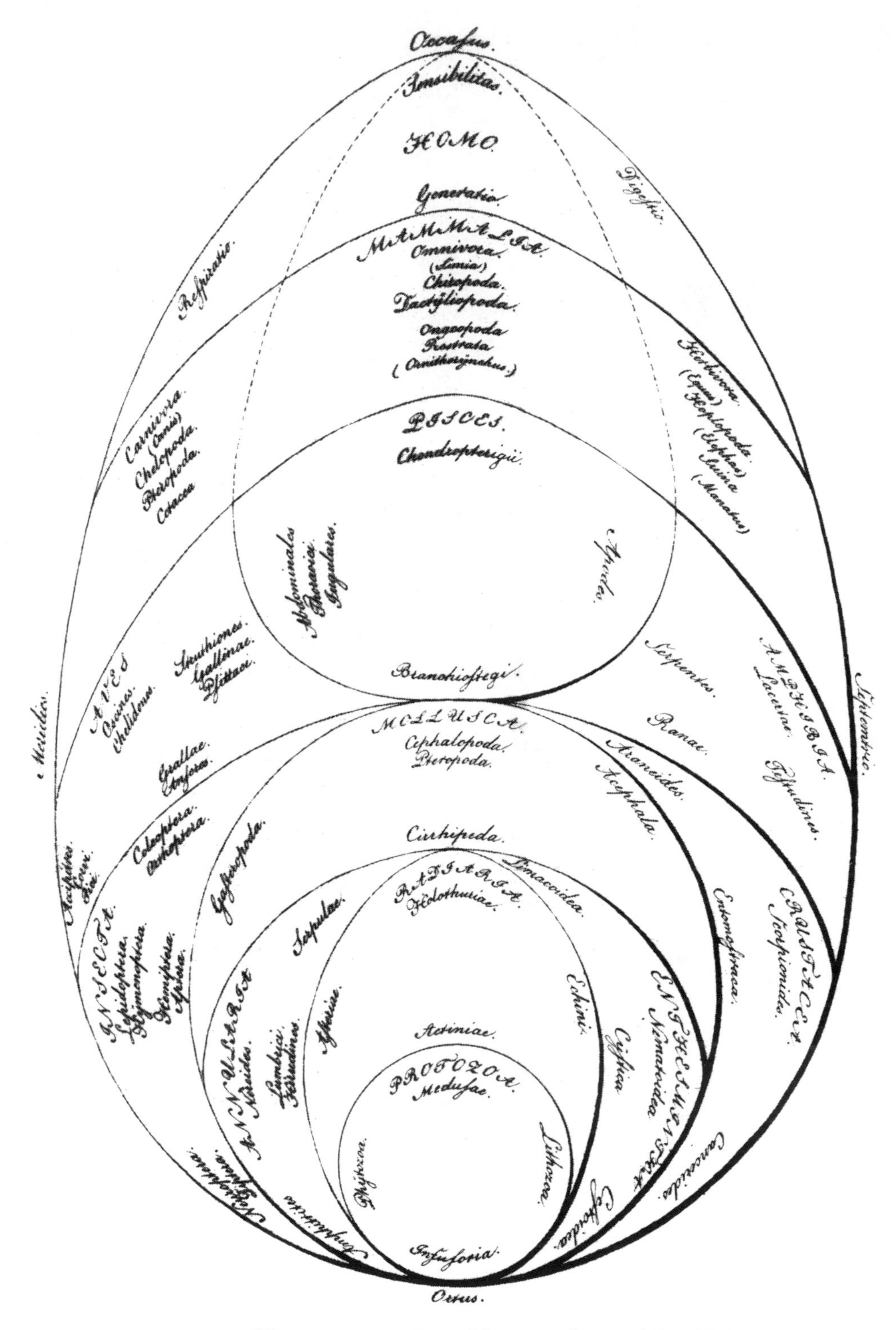

FIGURE 27. The unique egg-shaped "system of animals" published by German zoologist Georg August Goldfuss in his *Über de Entwicklungsstufen des Thieres* (On animal development) of 1817.

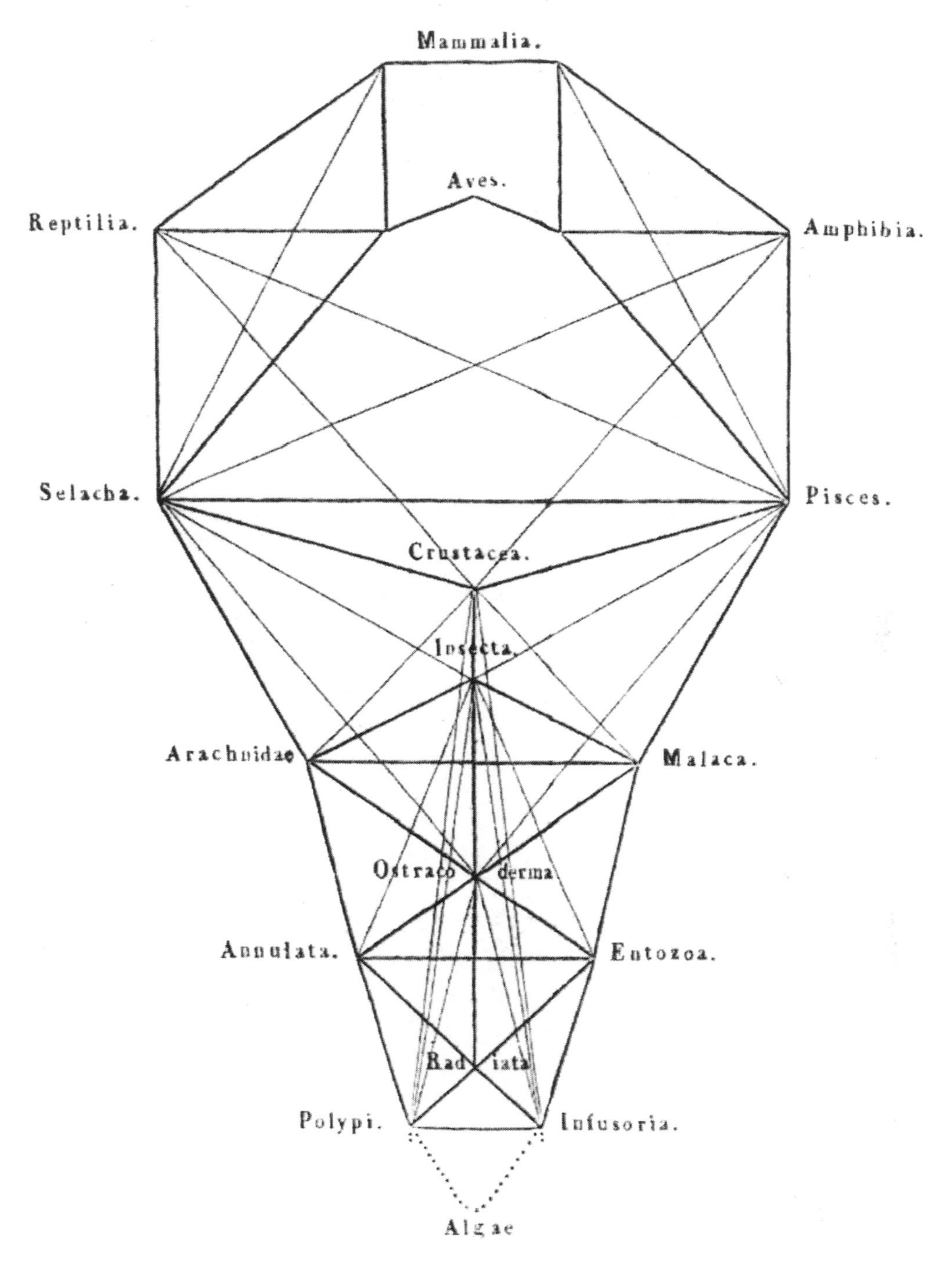

FIGURE 28. *Tabula transituum animalium* (Table of animal transitions), from Carl Edward von Eichwald's *De regni animalis limitibus atque evolutionis gradibus* (Evolution of the animal kingdom) published in 1821.

FIGURE 29. Tree of animal life, from the *Zoologia specialis* (Special zoology) of Carl Edward von Eichwald of 1829, thought to be based on a much earlier description published by Peter Simon Pallas in 1766. The tips of the branches are labeled with Roman numerals that indicate the major groups of animals.

18. *BAROSMA serratifolium.*

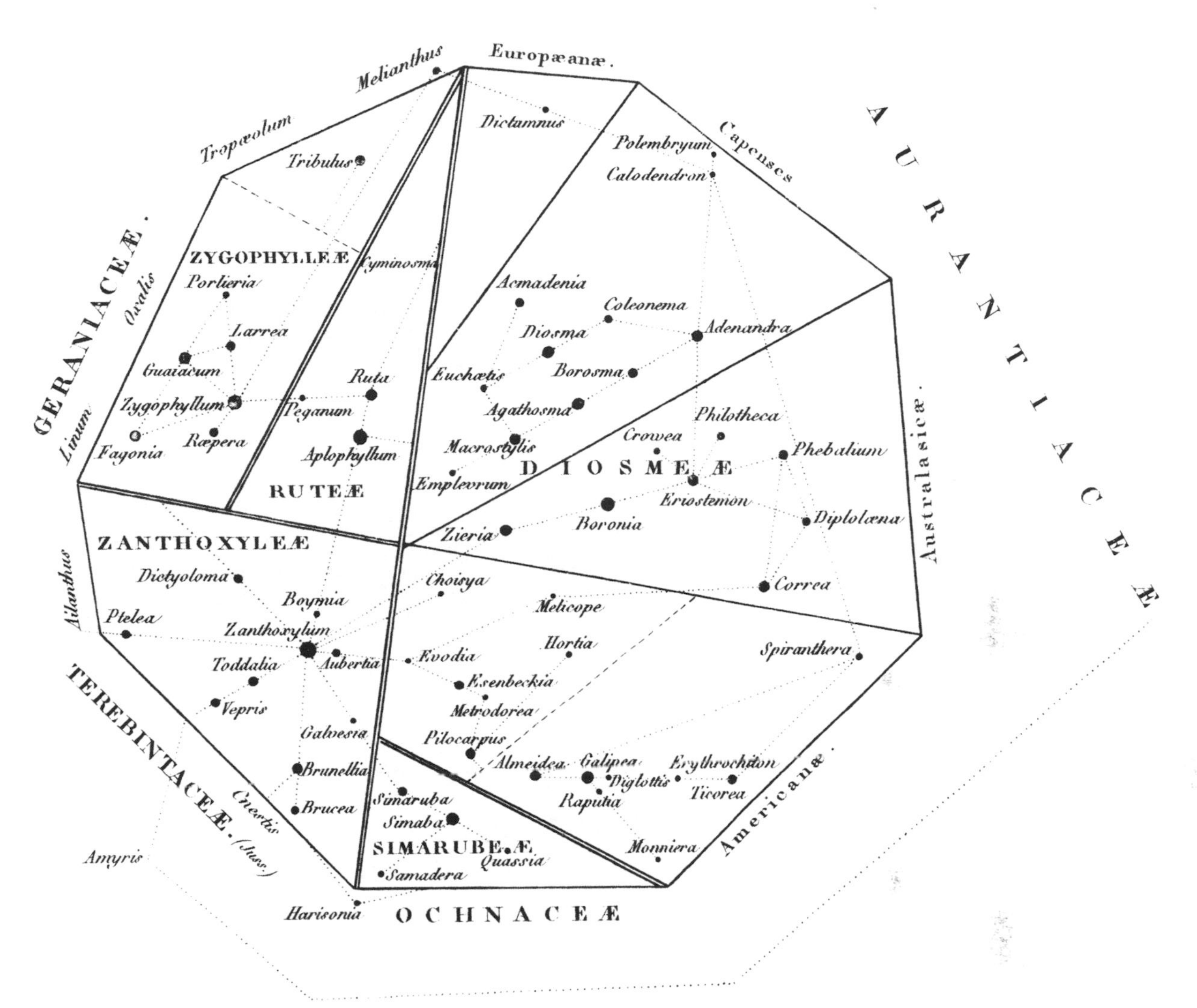

Tentamen tabulæ genera Rutacearum secundum mutuas affinitates disposita exhibentis.

•	*Genus complectens Species*	1.
•	————	2.-3.
•	————	4.-11.
•	————	12. aut plures.

FIGURE 30. Adrien-Henri de Jussieu's 1825 network of affinities among various subgroups of the plant family Rutaceae.

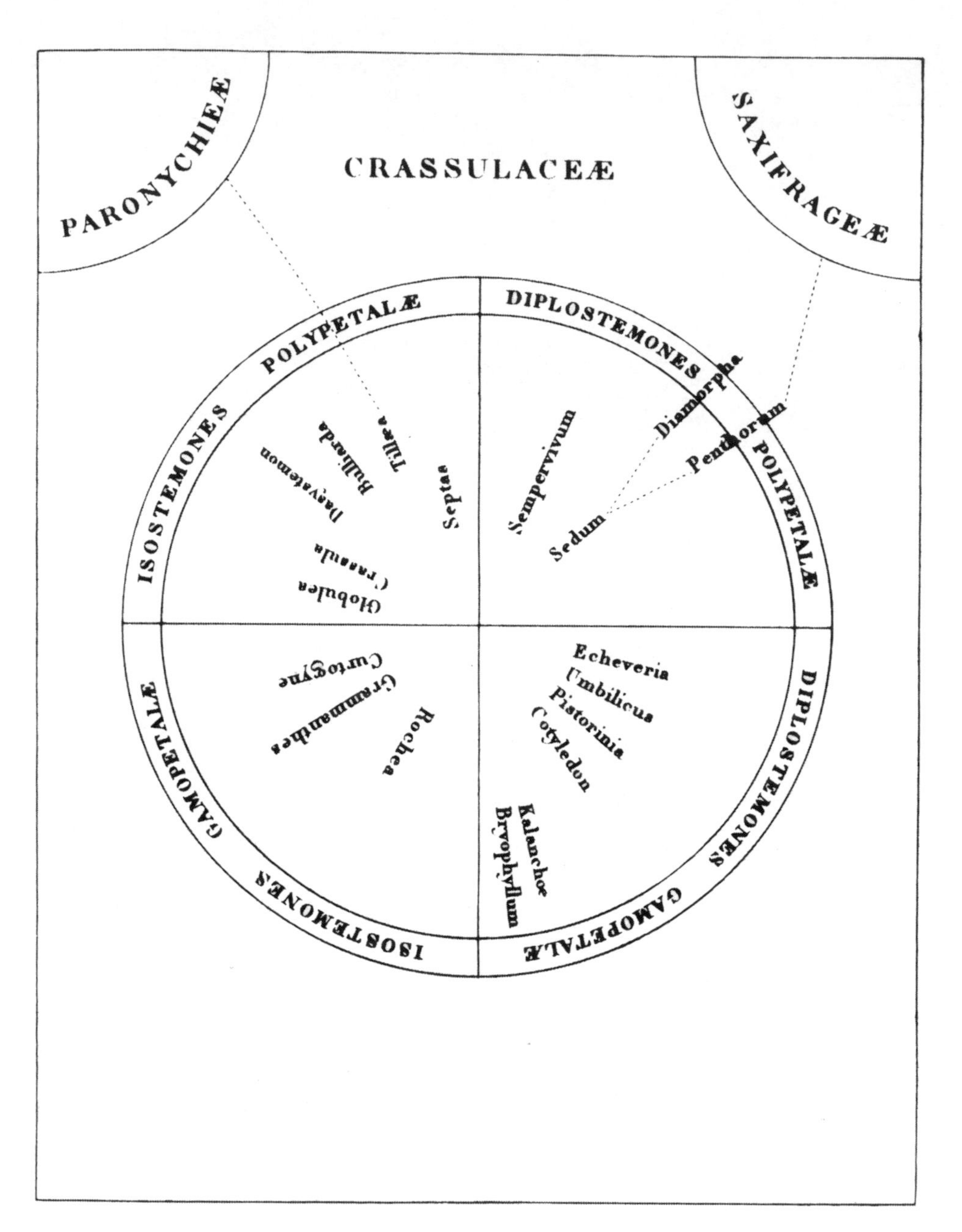

FIGURE 31. Relationships of genera of the plant family Crassulaceae published by Augustin-Pyramus de Candolle in 1828. Note how a close affinity of the Crassulaceae with two other plant families (the Paronychieae and Saxifragaceae, similarly enclosed within circles) is indicated by dotted lines extending from transitional genera, the genus *Tillaea* having affinity with the family Paronychieae on the left, and *Penthorum,* with the Saxifragaceae on the right.

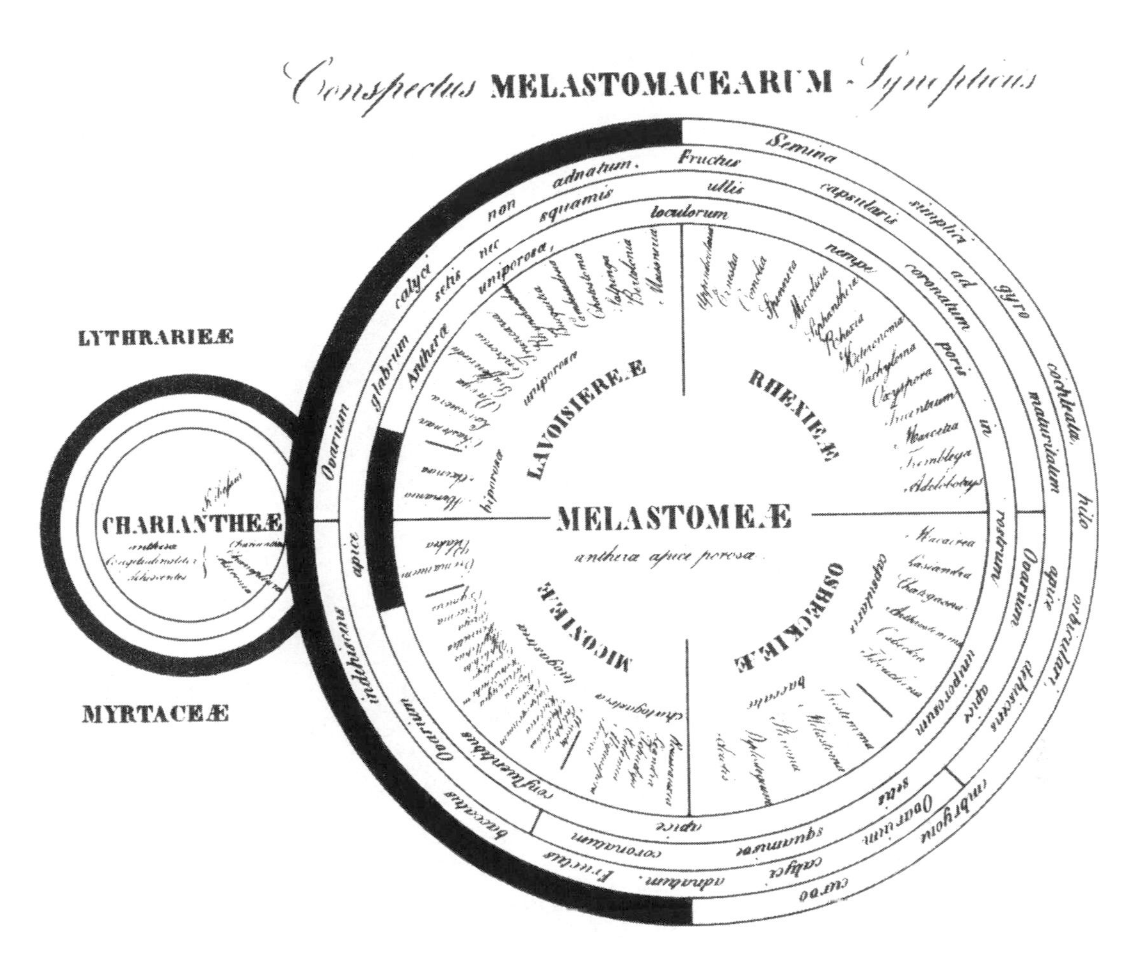

FIGURE 32. Augustin-Pyramus de Candolle's 1828 classification of plants of the family Melastomataceae. Affinity with the family Chariantheae is indicated by the smaller overlapped circle on the left margin, and, above and below the latter, more distant affinities with the families Lythrarieae and Myrtaceae.

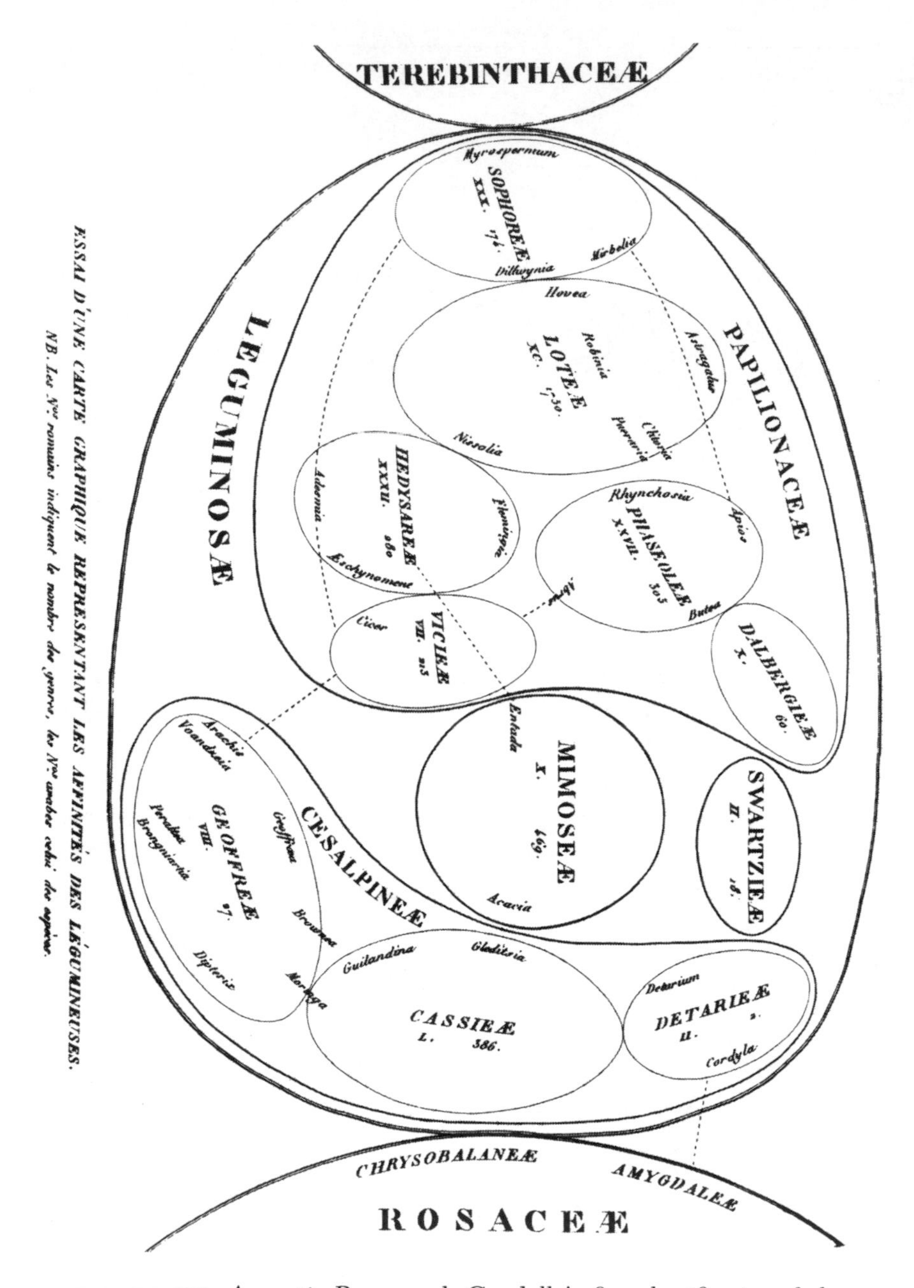

FIGURE 33. Augustin-Pyramus de Candolle's 1827 classification of plants of the legume family. Affinity with the families Terebinthaceae and Rosaceae are shown above and below.

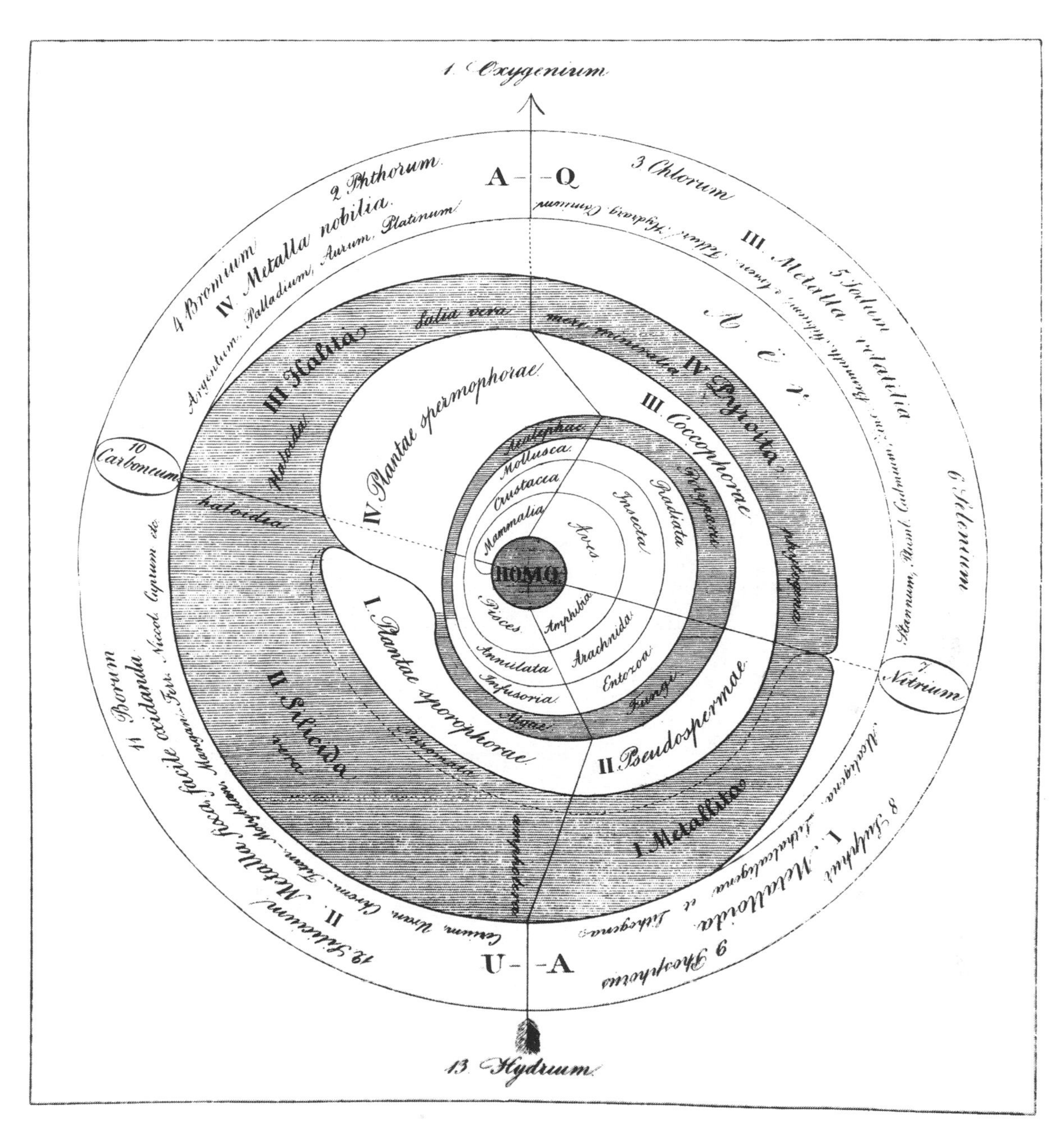

FIGURE 34. "Universal system of nature," from Paul Horaninow's *Primae lineae systematis naturae* (Primary system of nature) of 1834, an ingenious and seemingly indecipherable clockwise spiral that places animals in the center of the vortex, arranged in a series of concentric circles, surrounded in turn by additional nested circles that contain the plants, nonmetallic minerals, and finally metals within the outermost circle. Not surprisingly, everything is subjugated to humans (*Homo*) located in the center.

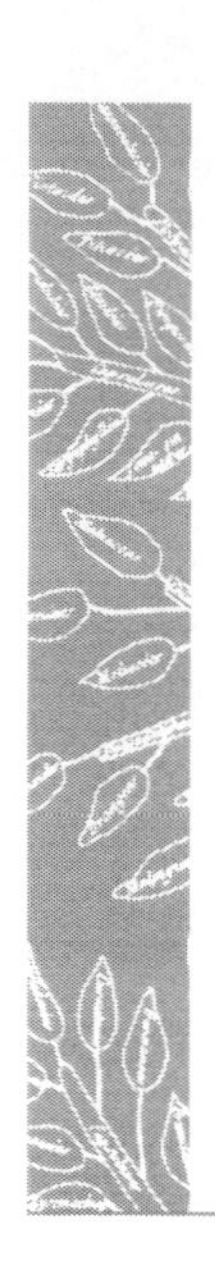

The Rule of Five

1819–1854

The circular arrangement of genera exemplified by the diagrams of Candolle is reminiscent of the so-called quinarian approach to classification that was embodied in the work of British entomologist William Sharpe Macleay (1792–1840) but taken up almost immediately by others, principally Nicholas Aylward Vigors (1787–1840) and William Swainson (1789–1855), during the 1820s and 30s. They somehow came to believe that living things existed in natural groups of five, that such groups of five are naturally divisible into five subgroups, each subgroup into five sub-subgroups and so on.[1] Affinities among taxa formed circular chains: if A shared affinity with B, and B with C, and C with D, and D with E, than E would always share affinity back to A.[2] Quinarians were also convinced that similarities between taxa based on affinity as well as analogy could be indicated on the same diagram—that circular chains of affinity connected taxa within each group of five and that relationships of analogy existed among taxa occupying corresponding positions in different circles of affinity.[3]

The earliest diagram based on these assumptions is Macleay's 1819 arrangement of families of scarab beetles (Figure 35): two circles, touching one another, enclosing natural groups, the Saprophaga and Thalerophaga, and each containing five families arranged circularly. The taxa enclosed within each circle are related by affinity, while those associated by horizontal lines are related by analogy. A bracket diagram from the same publication helps to understand these supposed relationships (Figure 36): the two lists of five families correspond to those within the two circles, respectively, and are united, three plus two, by characters of affinity; families across from one another on the horizontal shared analogous characters.

A much more complex and all inclusive diagram, based on the same circular system of five, was produced by Macleay in 1821 (Figure 37): five adjacent circles, each devoted to a major group of animals, and each, in turn, enclosing five subtaxa; radiating out from between the circles are five transitional groups. An adjacent half circle on the left indicates affinity with the "Least Organized beings of the Vegetable Kingdom." Asterisks in the circle of the Mollusca were placeholders for taxa not yet discovered but that surely must exist.

One more Macleay circle diagram is reproduced here (Figure 38) to show how this ingenious naturalist constructed his polygon of analogies among the so-called annulose animals (Figure 39), a group of invertebrates that then included, among other smaller taxa, insects, spiders, and crustaceans. At first glance the two diagrams look unrelated, but, as explained by Macleay, drawing lines between all points of analogy on the circle diagram and dropping out the circles themselves, the symmetrical set of nested polygons appears:

> Nothing in Natural History is, perhaps, more curious than that these analogies should be represented by a figure so strictly geometrical. One is almost tempted to believe that the science of the variations of animal structures may, in the end, come within the province of the mathematician.[4]

Nicholas Vigors, an enthusiastic follower of Macleay, published in 1824 a quinarian classification of birds: five orders within circles, with five families arranged circularly within each (Figure 40). In keeping with the assumptions implicit in the "rule of five,"[5] Vigors did his best to emphasize the continuity among taxa, believing, along with Macleay and other quinarians, that taxa formed unbroken circular chains of affinity. He was unhappy, however, with the final printed version of his article: "By an oversight of the printer's, the circles . . . were not made to touch each other . . . they thus seem to convey an erroneous idea of the series of affinity being incontinuous."[6]

These two kinds of similarity, affinity and analogy, so fundamental to quinarian systematics, were well differentiated by English ornithologist and all-around naturalist William Swainson. He, more than anyone else, promoted and popularized Macleay's novel approach to systematics during the 1830s:

> It is evident that all natural objects possess two different sorts of relationship: one which is *immediate*, and another which is remote. The goatsucker and the swallow exemplify the first of these relations. These genera are inti-

> mately connected by structure, habits, and economy . . . but the goatsucker, besides this relation, has evidently another to the bats,—by flying at the same hour of the day, and by feeding in the same manner. The first relation is *intimate*—the latter *remote*. Hence arises the necessity, imposed upon all who wish to develope the natural system, of possessing clear perceptions of these two sorts of relations; and of becoming well acquainted with the difference between *affinity* and *analogy*. The first is exemplified by the swallow and the goatsucker; the latter by the goatsucker and the bat.[7]

Swainson devoted several years to the development of a new general classification of animals based on quinarianism (for which he was later criticized),[8] applying it, for example, throughout his two-volume *Natural History of Birds* that appeared in 1836 and 1837. His diagram of relationships within and between the starling and crow families is typical (Figure 41), but, in the same publication, he took the approach a step further by enclosing three of the five circles within a circle of their own to show greater affinity between the three to the exclusion of the remaining two (Figure 42).

As the approach developed and became more widespread, some proponents of circle arrangements were not content to restrict the number to five. One early defector from the "rule" was English entomologist Edward Newman (1801–1876) who advocated a septenary system, producing in 1837 a cluster of seven contiguous circles to show relationships among the insects (Figure 43).[9] Others were not content with circles: John Lindley (1799–1865) preferred five radiating hexagons to demonstrate his ideas about relationships among the exogenous plants, gymnosperms plus angiosperms (Figure 44). Still others preferred stars as in Johann Jakob Kaup's (1803–1873) proposed affinities among genera and subfamilies of crows, published in 1854 (Figure 45).

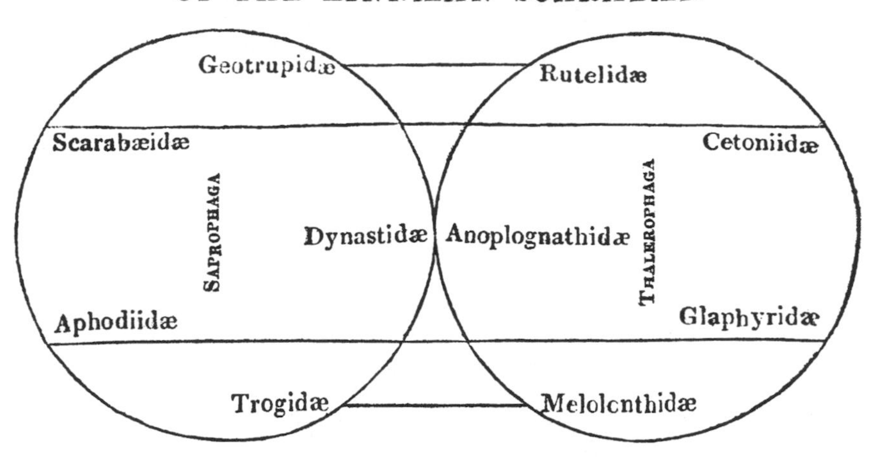

FIGURE 35. The earliest diagram of relationships based on the quinarian system published by William Sharpe Macleay in the first volume of his *Horae Entomologicae* (Entomological hours) of 1819, showing relationships among families of scarab beetles. The families within each circle form a continuous circular chain of affinity, while the circles themselves are connected by affinity at their points of contact; the horizontal lines indicate relationships of analogy.

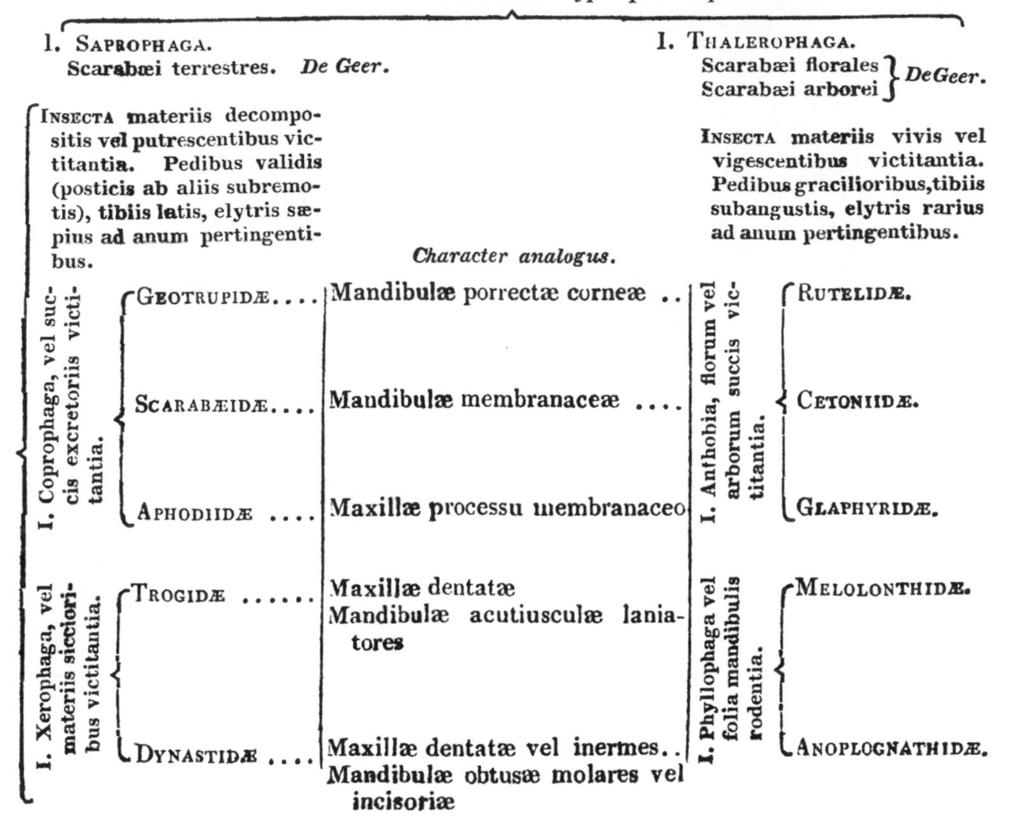

Insecta PETALOCERA. Dumeril.
Scarabæus. Linnæus.
Antennæ rectæ, capitulo flabellato;
Mandibulæ clypeo plerumque obtectæ vel raro exsertæ.
I. Saprophaga.
Scarabæi terrestres. De Geer.
Insecta materiis decompositis vel putrescentibus victitantia. Pedibus validis (posticis ab aliis subremotis), tibiis latis, elytris sæpius ad anum pertingentibus.
I. Thalerophaga.
Scarabæi florales
Scarabæi arborei
DeGeer.
Insecta materiis vivis vel vigescentibus victitantia. Pedibus gracilioribus, tibiis subangustis, elytris rarius ad anum pertingentibus.
Character analogus.
I. Coprophaga, vel succis excretoriis victitantia.
Geotrupidæ.... Mandibulæ porrectæ corneæ ..
Scarabæidæ.... Mandibulæ membranaceæ ....
Aphodiidæ .... Maxillæ processu membranaceo
I. Anthobia, florum vel arborum succis victitantia.
Rutelidæ.
Cetoniidæ.
Glaphyridæ.
I. Xerophaga, vel materiis sicciori-bus victitantia.
Trogidæ ...... Maxillæ dentatæ
Mandibulæ acutiusculæ laniatores
Dynastidæ .... Maxillæ dentatæ vel inermes..
Mandibulæ obtusæ molares vel incisoriæ
I. Phyllophaga vel folia mandibulis rodentia.
Melolonthidæ.
Anoplognathidæ.

FIGURE 36. William Sharpe Macleay's 1819 bracketed table of relationships among families of scarab beetles, giving characters of affinity on the vertical and analogous characters on the horizontal, and providing clarity to his circle diagram (figure 35).

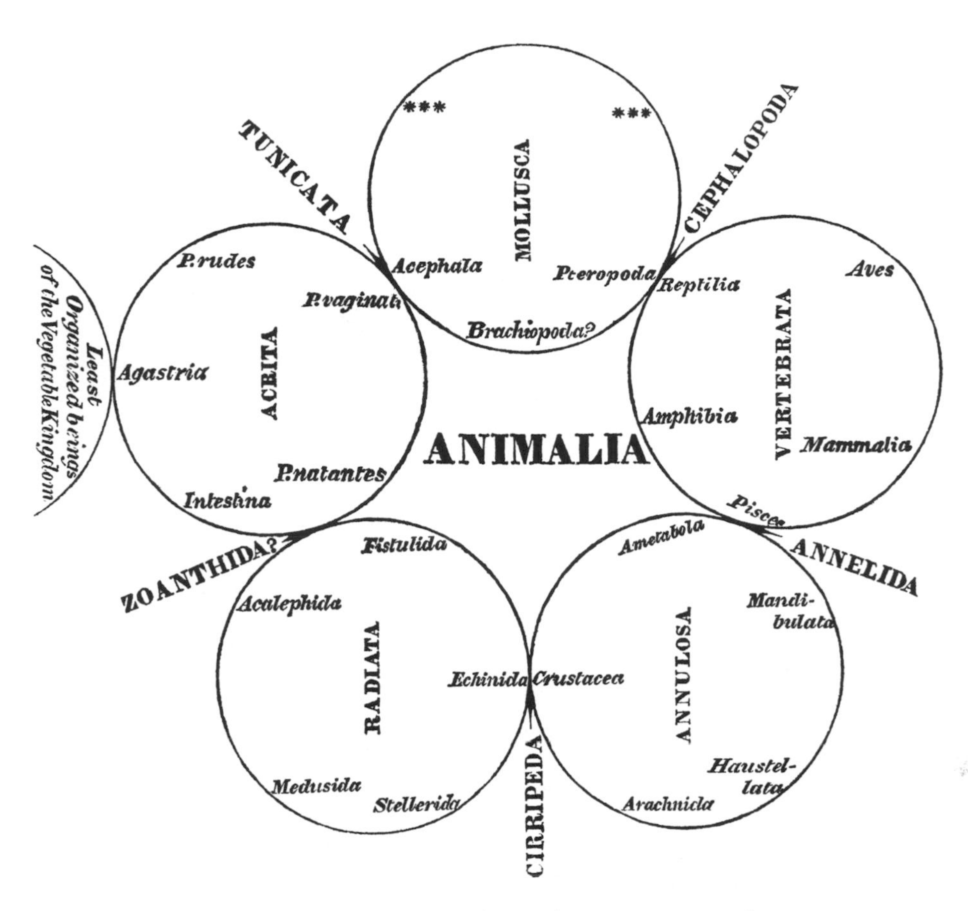

FIGURE 37. A quinarian approach to relationships among animals, published by William Sharpe Macleay in the second volume of his *Horae Entomologicae* (Entomological hours) of 1821. The asterisks in the uppermost circle represent taxa not yet discovered but that must exist to fulfill the assumptions of the "rule of five."

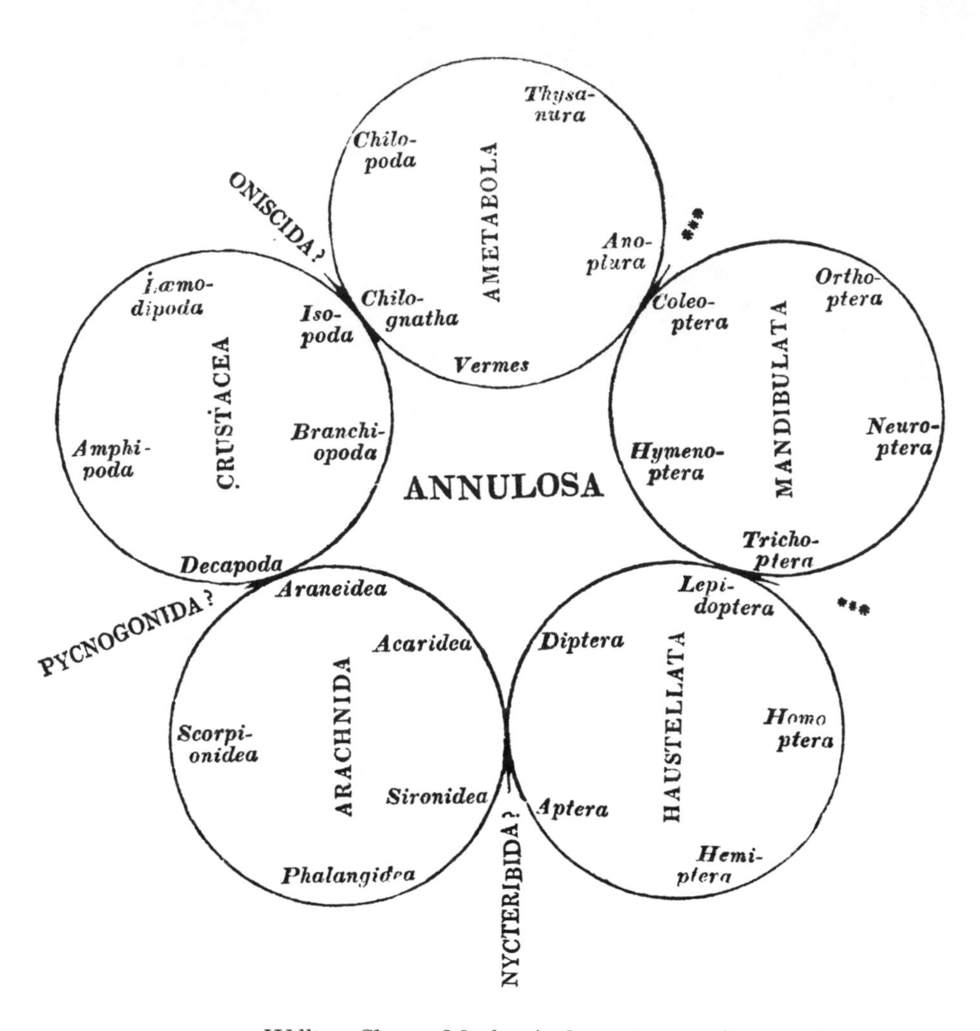

FIGURE 38. William Sharpe Macleay's 1821 quinarian diagram showing relationships among various groups of invertebrate animals.

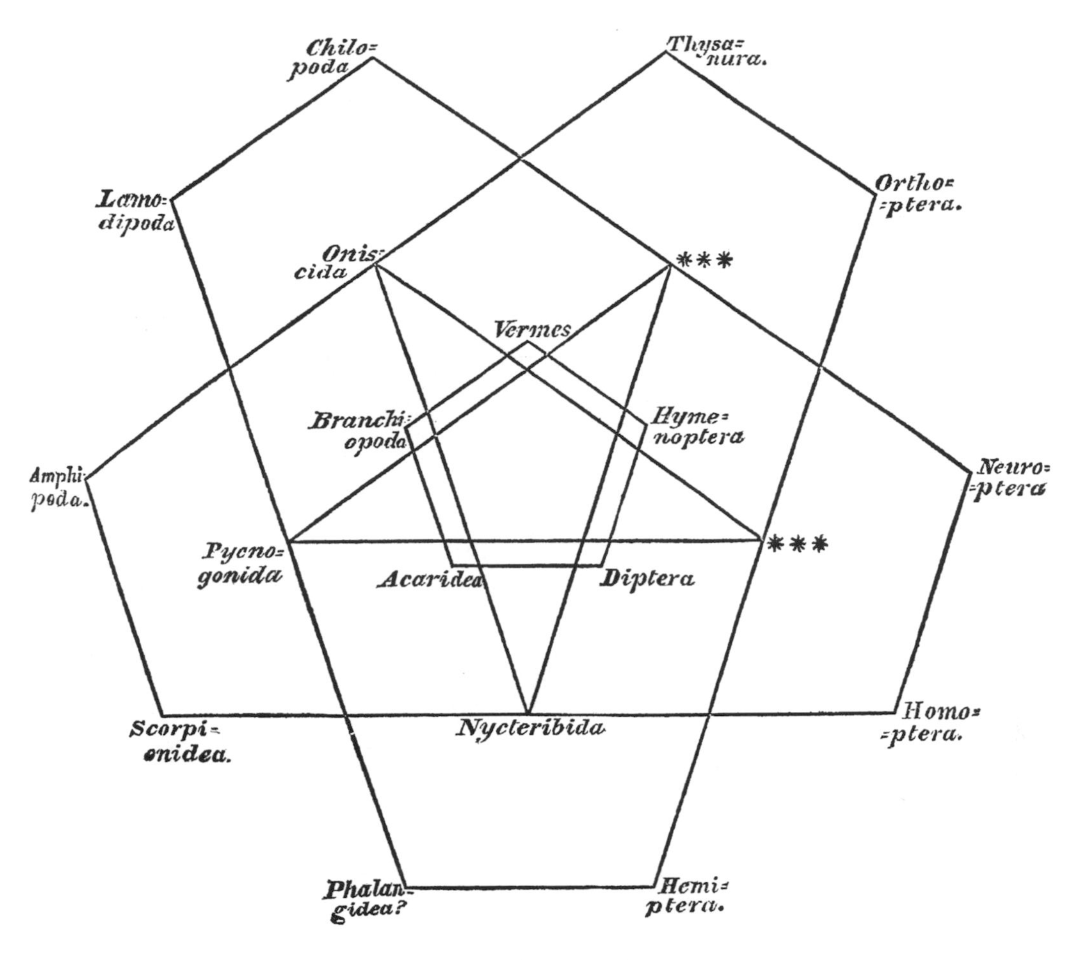

FIGURE 39. William Sharpe Macleay's 1821 "polygon of analogies" among annulose invertebrates, constructed by drawing lines between all points of analogy on the circle diagram (figure 38).

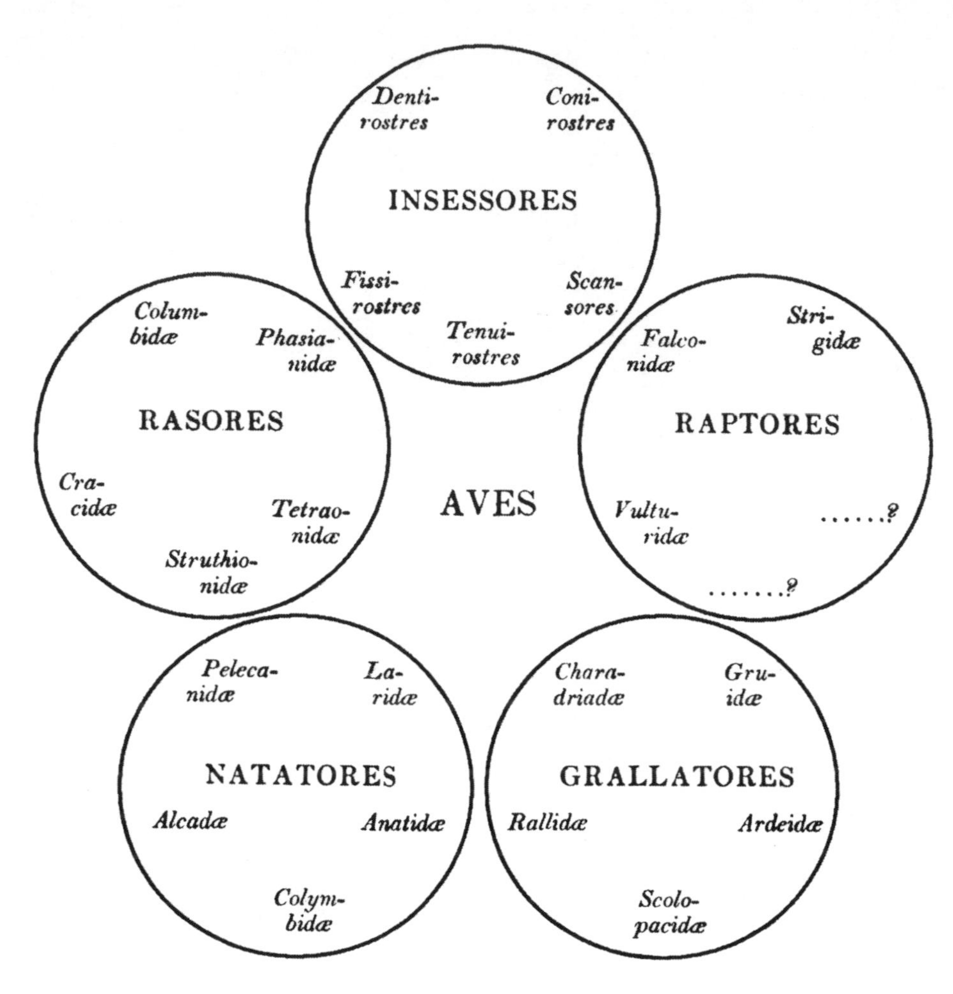

FIGURE 40. "Observations of the natural affinities that connect the orders and families of birds," Nicholas Aylward Vigors's 1824 quinarian diagram of bird relationships. Question marks within the circle of raptors are placeholders for taxa as yet undiscovered.

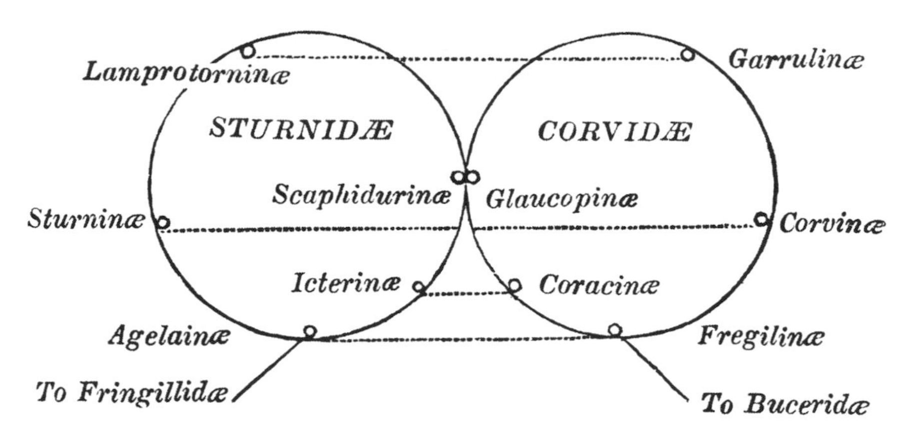

FIGURE 41. William Swainson's 1837 quinarian diagram showing circular affinities within and analogies between subfamilies of starlings and crows. More distant affinities with the bird families Fringillidae and Buceridae are indicated by lines radiating out from the subfamilies Agelainae and Fregilinae, respectively. The horizontal dotted lines indicate relationships of analogy.

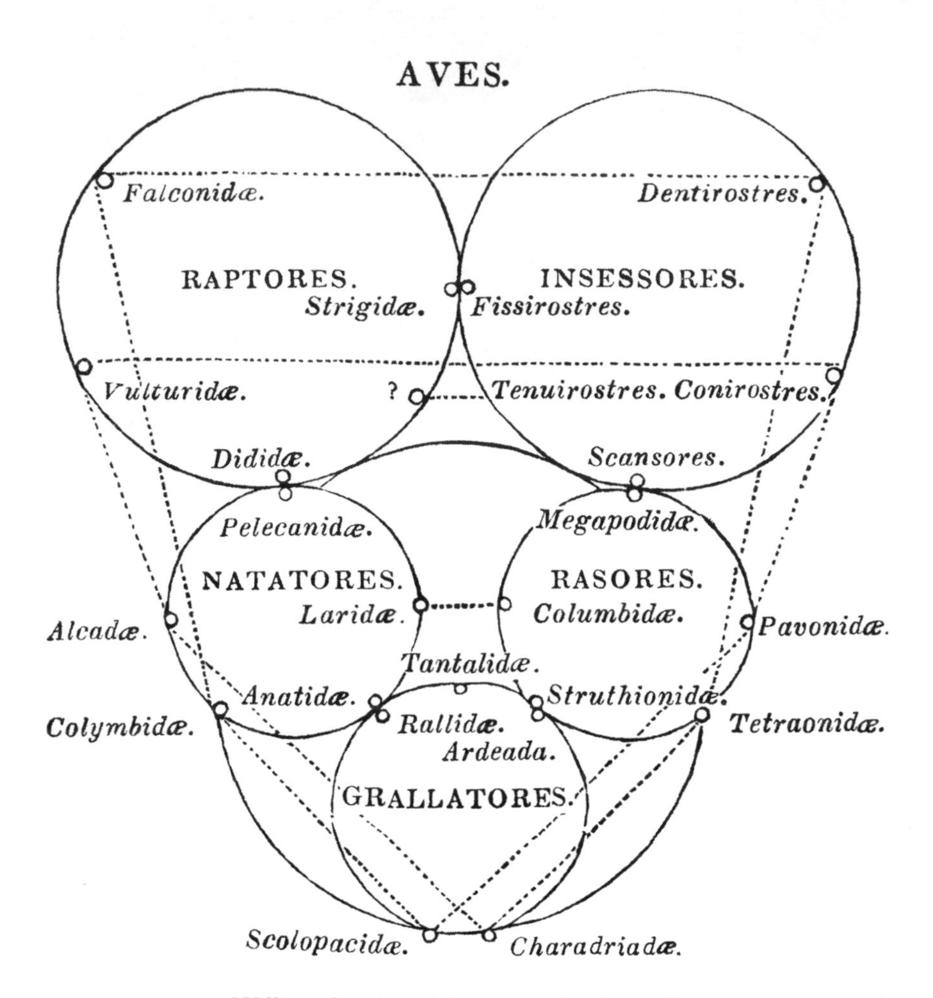

FIGURE 42. William Swainson's five natural orders of birds, each enclosed within a circle and each circle in turn containing five families, published in 1837. The lowermost three orders are enclosed in circles of smaller diameter, and together contained within a circle of their own to indicate greater affinity between the three to the exclusion of the remaining two. The dotted lines indicate relationships of analogy.

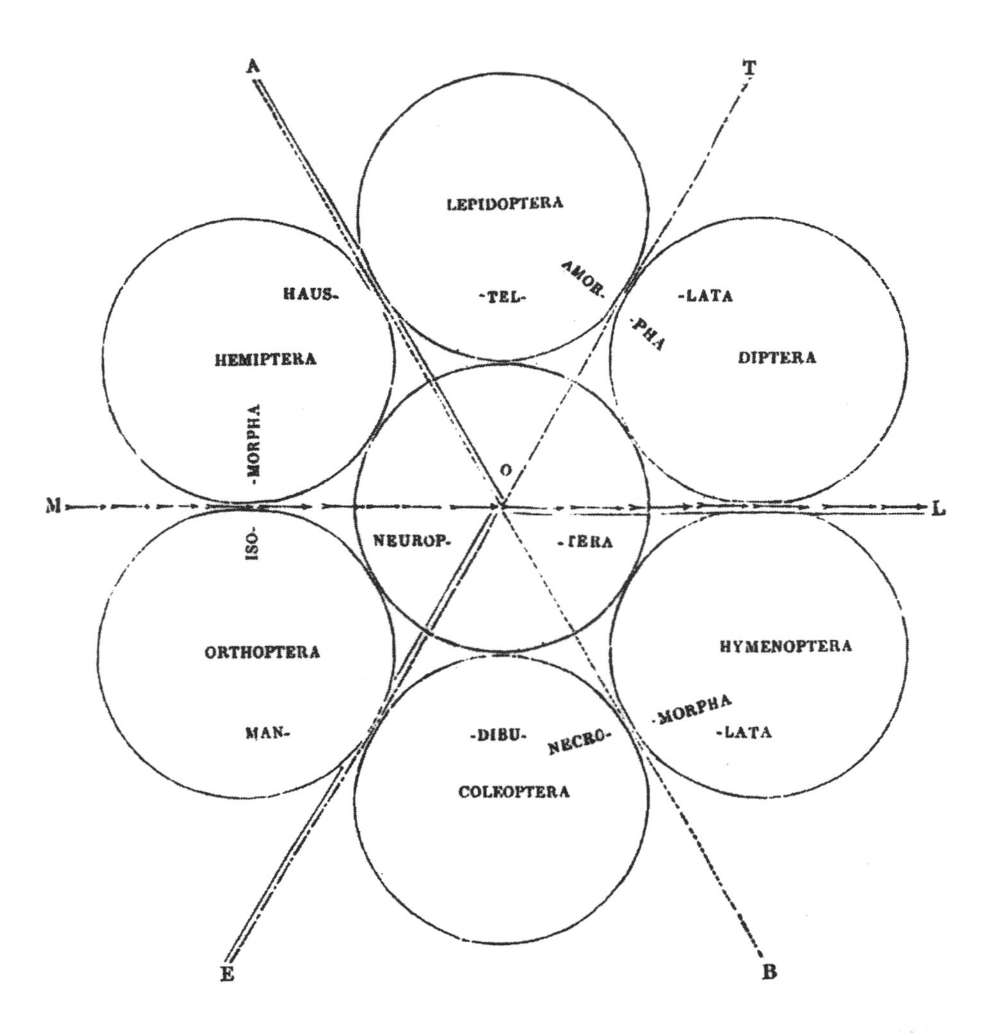

FIGURE 43. Edward Newman's seven-cycle system demonstrating his ideas about affinities within the insects, published in 1837.

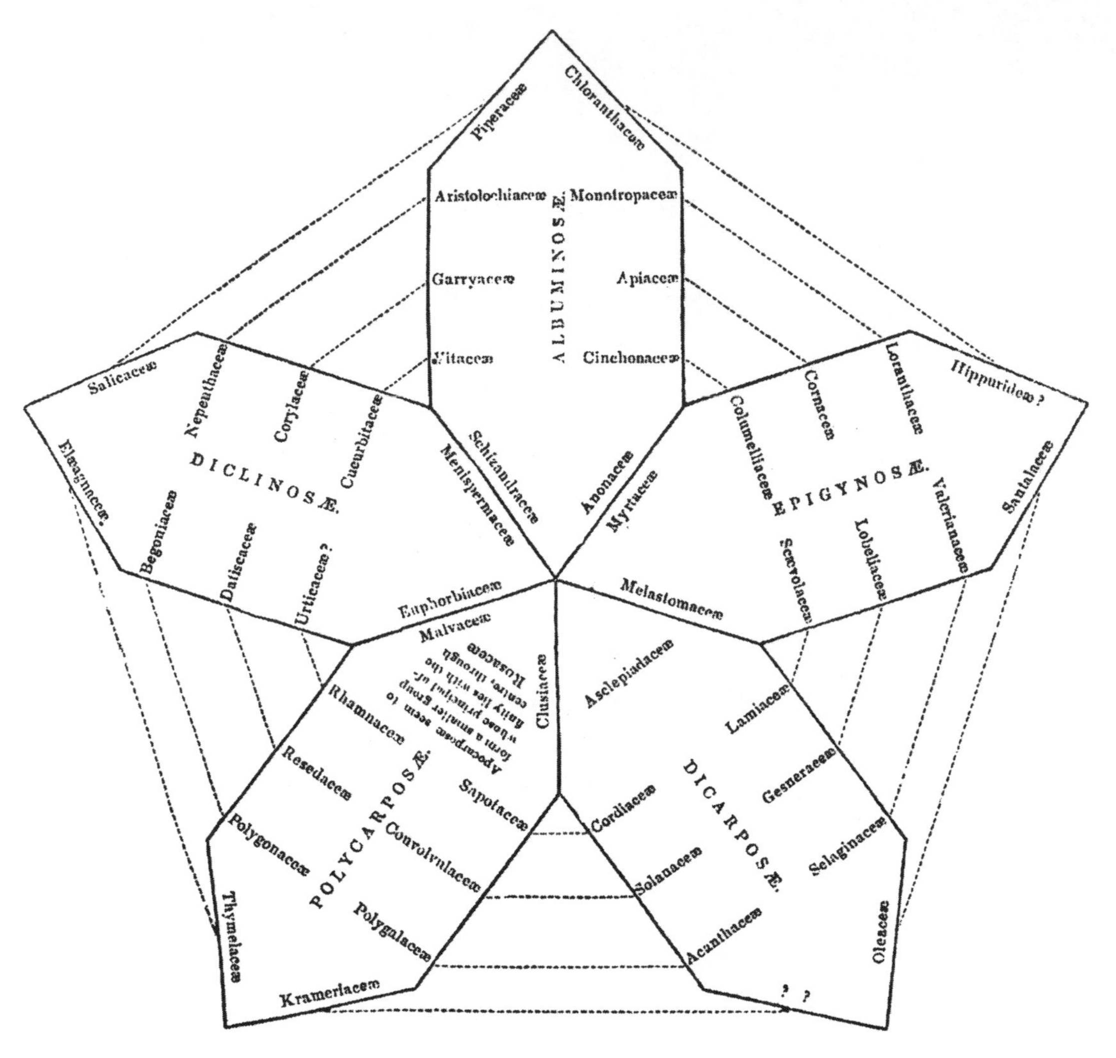

FIGURE 44. Circles have been replaced with elongate radiating hexagons to form a five-pointed star in this arrangement of plant affinities published by John Lindley in 1838. In contrast to the "rule of five," each radiating hexagon contains ten taxa. Dotted lines extending between hexagons indicate relationships of analogy.

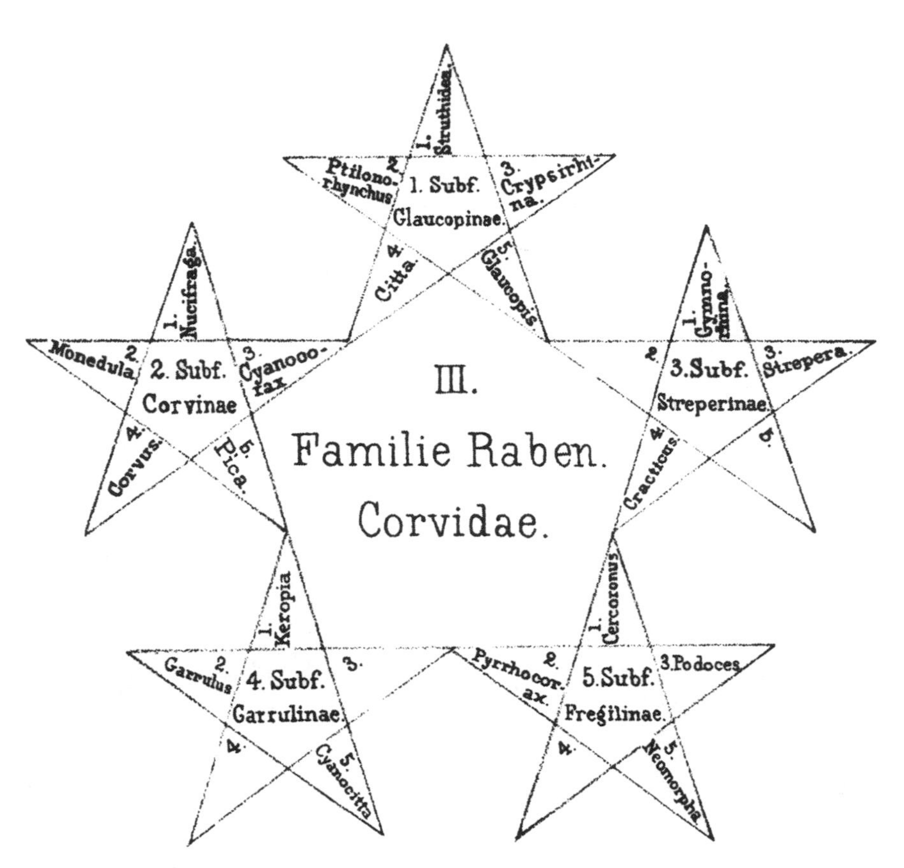

FIGURE 45. Stars have replaced circles in this quinarian view of affinities among crows, published in 1854 by Johann Jakob Kaup. Empty triangles stand for groups not yet discovered.

Pre-Darwinian Branching Diagrams

1828–1858

While the Quinarians were busy constructing their circular diagrams, others were experimenting with a host of divergent diagrammatic approaches to demonstrate their own ideas about plant and animal relationships. A figure jotted down by Richard Owen (1804–1892) during a lecture given by Joseph Henry Green (1791–1863) in 1828, showing "the 13 orders of mammals arranged in an ascending series" (Figure 46), is surprisingly modern in appearance and not unlike Darwin's famous "I think" diagram drawn in 1837 (see Figure 58, in the next chapter), yet neither Owen nor Green were evolutionists.[1]

A tangle of dichotomous branching, with a three-dimensional quality, might well describe Scottish physician Martin Barry's (1819–1872) attempt in 1837 to illustrate Karl Ernst von Baer's (1792–1876) theory of embryology (1828), suggesting that animals should be classified according to attributes observed in embryological development (Figure 47).[2] Barry was convinced that von Baer's embryological studies had provided proof "that in all classes of animals, from Infusoria to Man, germs at their origin are *essentially the same in character*; and that they have in common a homogenous or general structure."[3] In other words, every animal begins life structurally the same and acquires differentiating characteristics as it develops. Embryology therefore provides the essential key to the classification of organisms.[4]

The notions of von Baer, made popular by Martin Barry and others, became widely accepted among naturalists during the late 1830s and early 40s. But for some, the almost incomprehensible complexity of diagrams like that published by Barry was too much, and, in a successful attempt to simplify things, physiologist William Benjamin Carpenter (1813–1885) produced in 1841 a strikingly plain image of five lines, one vertical and four branching off at right angles (Figure 48):

> [It has been] stated as a general law, that all the higher animals in the progress of their development pass through a series of forms analogous to those encountered in ascending the animal scale. But this is not correct; . . . the correspondence is much closer between the embryonic Fish and the foetal Bird or Mammal, than between these and the adult Fish. . . . The view here stated may perhaps receive further elucidation from a simple diagram [Figure 48]. Let the vertical line represent the progressive change of type observed in the development of the foetus, commencing from below. The foetus of the Fish only advances to the stage F; but it then undergoes a certain change in its progress towards maturity, which is represented by the horizontal line FD. The foetus of the Reptile passes through the condition which is characteristic of the *foetal* Fish; and then, stopping short at the grade R, it changes to the perfect Reptile. The same principle applies to Birds and Mammalia; so that A, B, and C,—the *adult* conditions of the higher groups,—are seen to be very different from *the foetal,* and still more from the *adult,* forms of the lower; whilst between the embryonic forms of all the classes, there is, at certain periods, a very close correspondence, arising from the law of gradual progress from a general to a special condition, already so much dwelt upon.[5]

A similar diagram (Figure 49) was published three years later by Robert Chambers (1802–1871), in the first edition of his *Vestiges of the Natural History of Creation* of 1844, a publication that was so controversial that his authorship was not acknowledged until after his death.[6] Providing still greater clarity to the notions embodied in Carpenter's interpretation of Barry's tree, Chambers made two changes: "he tilted the first three horizontal lines upward and completely omitted the fourth. In this way, he gave the ladder of progressive development a dynamic twist. Organisms were striving toward a goal—it seemed as though, deep down, every fish, every reptile, and every bird wanted to be a mammal."[7]

Others attempted to apply von Baer's embryological ideas to the problems of classification but with quite different iconographic results. Eminent French zoologist Henri Milne-Edwards (1800–1885), also thoroughly convinced that animal classification should be based on embryonic characters, carried out von Baer's ideas in a more complete and thorough manner.[8] His branching diagram of 1844, intended to show the natural affinities of vertebrate animals, is a case in point (Figure 50). Vertebrates are divided initially into two primary groups based on egg type: reptiles, birds, and mammals, which possess an allantois (*vertébrés allantoidiens*)—which, along with other fetal membranes, the amnion and chorion, identifies them as amniotes—and fishes and amphibians, which lack this structure (*vertébrés anallantoidiens*). Primary divisions among mammals are based on characters of the placenta, its absence in marsupials and the egg-laying monotremes, and its shape and the degree of contact between fetal and maternal tissue in the remaining mammals. In an apparent fallback to quinarian notions, lines crossing the major divisions of Milne-Edwards's classification are intended to show analogous relationships, between such unlikely pairs as sharks (*Chondroptérygiens*) and whales (*Cétacés*), between frogs (*Anoures*) and turtles (*Chélonians*), between herbivorous marsupials (*M. herbivores*) and rodents (*Rongeurs*), and so on.[9] Note also that he placed the lungfish (*Lepidosiren*) among the amphibians, while indicating only an analogous relationship to bony fishes (*P. osseux*).[10]

In a strong reaction against Macleay's quinarian circular approach, which turned out to be a relatively short-lived fad during the 1820s and 30s,[11] English ornithologist Hugh Edwin Strickland (1811–1853) presented a paper titled "On the true method of discovering the natural system in zoology and botany" at the 1840 annual meeting of the British Association for the Advancement of Science in Glasgow. In it, he declared that "all systems, circular, quinary, dichotomous, etc., are not natural, but artificial and only of use in arranging museums." He further rejected the idea that relationships of analogy have a place in systematics, stating that only affinity determines "the place of a species in the natural system." He then proposed instead a kind of mapping approach described as follows:

> The plan proposed is to take any species, A, and ask the question, What are its nearest affinities? If after examination of its points of resemblance to all other known species, it should appear that there are two other species, B and C, which closely approach it in structure, and that A is intermediate between them, the question is answered, and the formula BAC would express a portion of the natural system. . . . Then take C, and ask the same question. One of the affinities that of C to A, is already determined; and we will suppose that D is found to form its nearest affinity on the other side. Then BACD will represent four species, the relative affinities of which are determined. By repetition of this process . . . the whole of organized creation might be ultimately arranged in order of its affinities, and our survey of the natural system would then be finally effected. Now, if each species never had more than two affinities, and those in opposite directions, as in the above example, the natural system would form a straight line, as some authors have assumed it to be. But we shall often find, in fact, that a species has only one direct affinity, and in other cases that it has three or more,

showing the existence of lateral ramifications instead of a simple line; as shown in this example, where C, besides its affinity to A and D, has an affinity to a third species, E, which therefore forms a lateral ramification.[12]


```
B—A—C—D
    |
    E
```

Strickland then applied this method to an analysis of relationships among genera of kingfishers (Figure 51). Using the same technique a short time later, he put together a large "Chart of the Natural Affinities of the Class of Birds" (Figure 52), which, although completed in 1843, was not published until 1858 by William Jardine (1800–1874) in his *Memoirs of Hugh Edwin Strickland.*

Alfred Russel Wallace (1823–1913) also strongly rejected quinarian approaches, attacking Macleay and Swainson's "belief in the universal existence in nature of a numerical and circular arrangement."[13] In contrast, he praised Strickland's method of mapping by affinities[14] and included two rather simple diagrams based on this approach in his 1856 "Attempts at a natural arrangement of birds" (Figure 53). But more than accepting Strickland on face value, Wallace provided details of his own thoughts about tree construction, accepting that "there exists a main axis along which many taxa can be arranged in a series of affinities, while affinities of other taxa can be represented as branches and subbranches to the left and right off this main axis."[15] He also stated: "It is intended that the distances between the several names should show to some extent the relative amount of affinity existing between them; and the connecting lines show in what direction the affinities are supposed to lie."[16] By this time, in 1856, Wallace was obviously thinking in terms of transmutation of species, seeking supporting evidence for evolutionary change in his studies of birds. In fact, by 1855 he had fully adopted the tree metaphor as a means to describe evolutionary change: "Again, if we consider that we have only fragments of this vast system, the stem and main branches being represented by extinct species of which we have no knowledge, while a vast mass of limbs and boughs and minute twigs and scattered leaves is what we have to place in order, and determine the true position each originally occupied with regard to the others, the whole difficulty of the true Natural System of classification becomes apparent to us."[17]

Unlike any diagram of animal or plant relationship published to date, branching or not, is that produced by the American geologist Edward Hitchcock (1793–1864): the earliest paleontological-based tree of life (Figure 54),[18] appearing initially in 1840 in the first edition of his popular geology textbook. No one before Hitchcock thought to superimpose geological time on a branching diagram of plant and animal relationships. Reproduced in numerous subsequent editions up until 1859, this "Paleontological Chart" shows two bush-like trees, their trunks with roots embedded within various kinds of rock—quartz, mica, slate, granite, gneiss, and limestone—and each branching almost immediately from the base to produce a cluster of ascending branches. The tree on the left is dedicated to plants, including both extinct and living groups, while that on the right shows the same for animals. Each lineage terminates with a crowned "king": a group containing the "Palms" is given premier place among the plants and "Man" likewise among the animals. Both trees are set within a context of geological time, with the names of the various periods in vogue at the time indicated along the vertical on each side of the figure. By varying the widths of the various stems and branches of each tree, Hitchcock attempted to show the relative number of taxa within each group at any given point in time. Based on this history of life portrayed as a branching tree, with explicit reference to multiple episodes of extinction, one might suspect that Hitchcock was an evolutionist. But, on the contrary, he believed strongly in a deity as the agent of change and repeatedly criticized the prevailing hypotheses of transmutation promoted by Lamarck, Chambers, and eventually Darwin.[19]

Shortly after Hitchcock's paleontological trees appeared, Louis Agassiz (1807–1873), Swiss paleontologist, geologist, and later Harvard professor, who is often credited as a founding father of the modern American scientific tradition,[20] published a branching diagram in his *Recherches sur poissons fossils* (Research on fossil fishes) of 1844 (Figure 55). He described his figure as a "family tree" that

> represents the history of the development of the class of fish through all the geological formations and which expresses at the same time the degrees of affinity between the various families. . . . Finally the convergence of all these vertical lines indicates the affinity of families with the principal stock of each kind. I however did not bind the side branches to the principal trunks because I have the conviction that they do not descend the ones from the others by way of direct procreation or successive transformation, but that they are materially independent one from the other, though forming integral part of a systematic unit, whose connection can be sought only in the creative intelligence of its author.[21]

Appearing a full fifteen years before the publication of Darwin's *Origin of Species,* it shows his ideas about relationships between various groups of fishes. His order Placoïdes reasonably unites sharks and their allies, with cyclostomes (hagfishes and lampreys), unrepresented in the fossil record, situated by themselves on the far right. His order Ganoïdes contains a mix of primitive and derived taxa that are now recognized to be only very distantly related: the relatively primitive "Acipenserides" (sturgeons), with the "Siluroïdes" (catfishes) of intermediate derivation, and the highly evolved "Lophobranches," "Gymnodontes," and "Selérodermes" (pipefishes and seahorses, triggerfishes, and pufferfishes). The remaining two assemblages, the "Cycloïdes" and "Ctenoïdes" on the left, have similar problems, but overall the image has a surprisingly modern look. It cannot, however, be called evolutionary.[22] Agassiz was a staunch creationist who saw the Divine Plan of God everywhere in nature. A fierce opponent of Darwinism, he rejected evolutionary theory even in old age when it was widely accepted by his scientific contemporaries.[23] Agassiz's tree instead attempts to document the presence of various taxa in the fossil record, as indicated by the geological time scale provided along the right margin. Unlike the wavering width of the branches of Hitchcock's trees, the smoothly tapering spindle-shaped elements of Agassiz's diagram do not so much reflect the relative abundance of fossil material found within various geological strata through time, but only provide an indication of initial increase and, in the case of extinct taxa, eventually decrease of taxa contained within each group.[24]

A striking figure illustrating the "history of life" was published in 1848 by Louis Agassiz in his *Principles of Zoology,* a popular textbook coauthored by Augustus Addison Gould (1805–1866).[25] As described by Agassiz and Gould,

> The four Ages of Nature, are represented by four zones, of different shades, each of which is subdivided by circles, indicating the number of formations of which they are composed [Figure 56]. The whole disc is divided by radiating lines into four segments, to include the four great departments of the Animal Kingdom; the Vertebrates, with Man at their head, are placed in the upper compartment, the Articulates at the left, the Mollusks at the right, and the Radiates below, as being the lowest in rank. Each of these compartments is again subdivided to include the different classes belonging to it, which are named at the outer circle. At the centre is placed a figure to represent the primitive egg, with its germinative vesicle and germinative dot, indicative of the universal origin of all animals, and the epoch of life when all are apparently alike. Surrounding this, at the point from which each department radiates, are placed the symbols of the several departments. The zones are traversed by rays which represent the principal types of animals, and their origin and termination indicates the age at which they first appeared or disappeared, all those which reach the circumference being still

in existence. The width of the ray indicates the greater or less prevalence of the type at different geological ages.[26]

German paleontologist Heinrich Georg Bronn (1800–1862) is said to have produced the first tree-like diagram demonstrating evolutionary history (Figure 57) since Lamarck published his "table serving to demonstrate the origin of the different animals" in 1809 (see Figure 24).[27] Bronn's spindly, dead-looking tree, published in 1858, is a theoretical diagram, labeled only with letters. The letters indicate a series of species of increasing "structural perfection," succeeding one another as time moves along, the oldest at the bottom, arranged along the trunk and bases of the primary limbs, the progressively younger ones out on the secondary branches and twigs. The progression continues on all the branches, but, at any point in time, new species arise that are more perfect than others situated on the lower branches.[28] Bronn's tree made no lasting impact on evolutionary biology—he is remembered instead for his 1860 translation of Darwin's *Origin of Species* into German, to which he added notes of his own, including a final chapter of criticisms.[29] Nevertheless, he quickly became an avid supporter of Darwinism, despite some initial misgivings and a former adherence to a creationist view of life.[30]

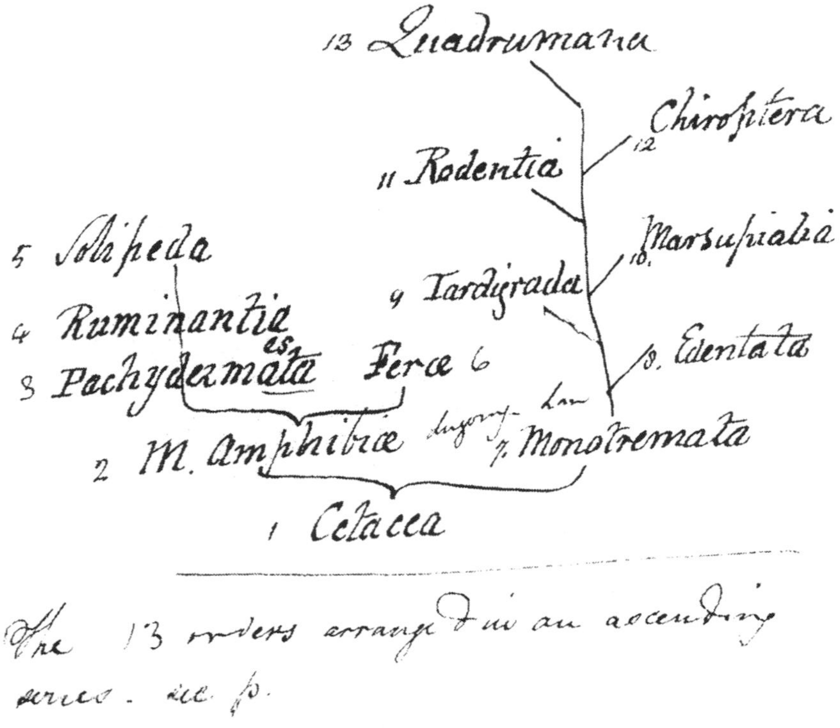

13 Quadrumana

12 Chiroptera

11 Rodentia

10. Marsupialia

5 Solipeda

9 Tardigrada

4 Ruminantia

8. Edentata

3 Pachydermata Ferae 6

2 M. Amphibiae 7. Monotremata

1 Cetacea

The 13 orders arranged in an ascending series. see p.

FIGURE 46. Joseph Henry Green's diagram of thirteen orders of mammals "arranged in an ascending series."

From the papers of Joseph Henry Green, lecture 3, notes of lectures by Professor Green titled "Mr Hunter's Notion of Life Anterior to Organisation," 1828, Archives Ref. MS0122/4, from the Archives of the Royal College of Surgeons of England; courtesy of Katherine Tyte and Louise King. Reproduced with permission.

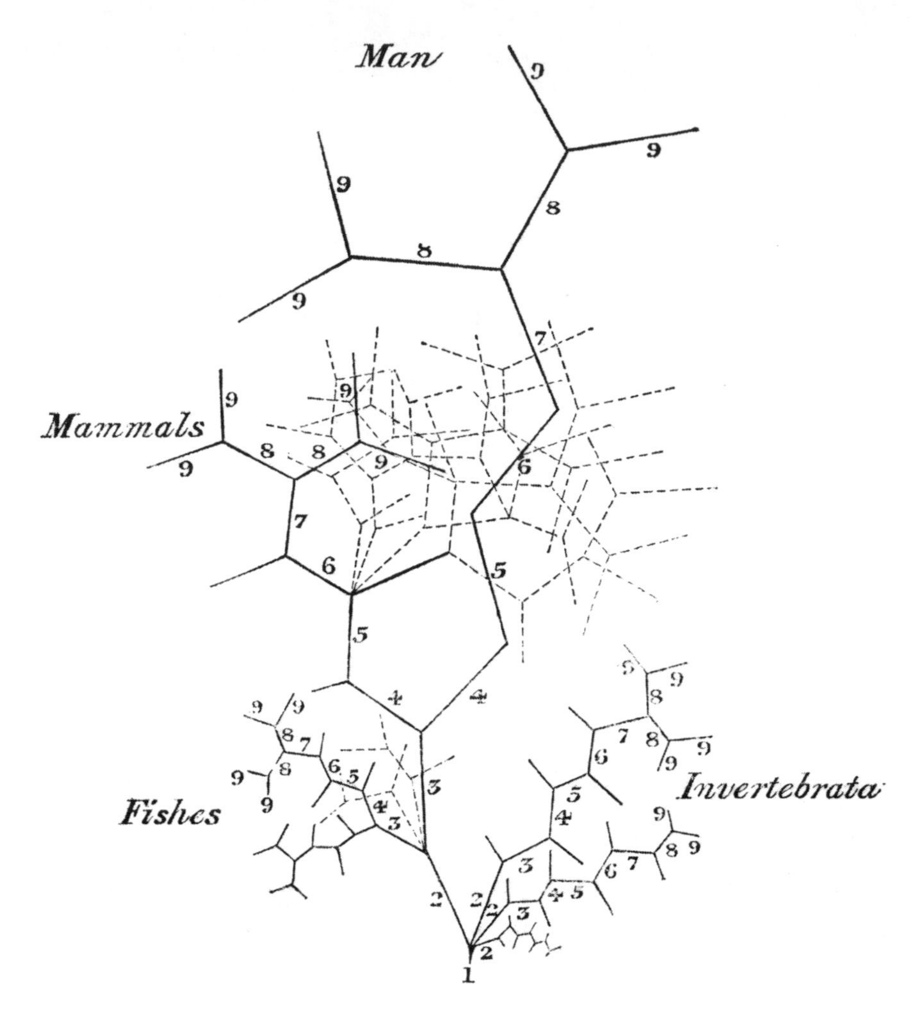

FIGURE 47. Scottish physician Martin Barry's 1837 "Tree of Animal Development," intended to illustrate Karl Ernst von Baer's theory of embryology.

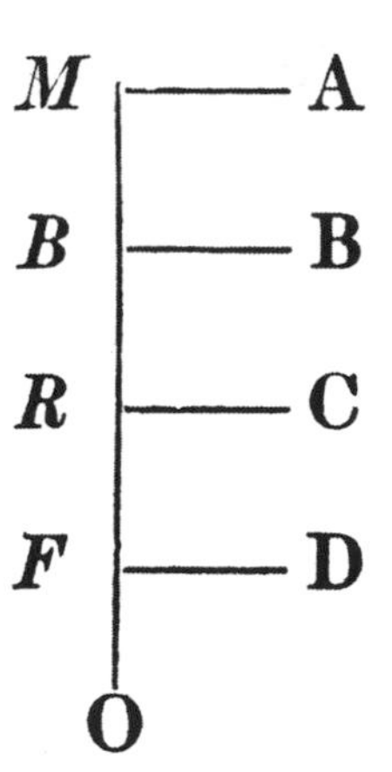

FIGURE 48. A branching diagram of the major groups of vertebrates published by William Benjamin Carpenter in his 1841 *Principles of General and Comparative Physiology* based on the 1828 embryological work of Karl Ernst von Baer.

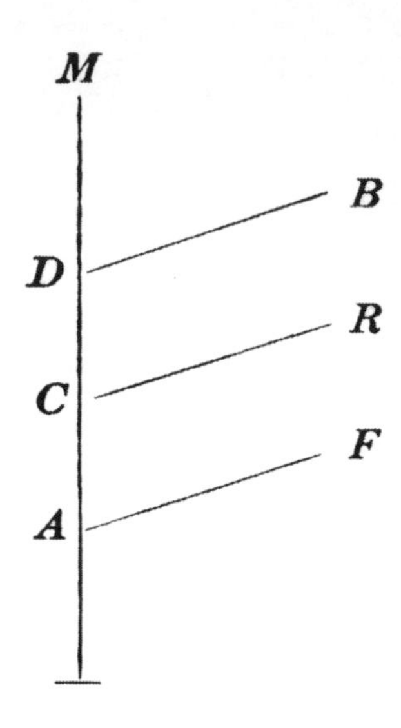

FIGURE 49. "Hypothesis of the development of the vegetable and animal kingdoms," a reinterpretation of William Carpenter's "tree" (figure 48), published by Robert Chambers in his famous 1844 *Vestiges of the Natural History of Creation*: "this diagram shews only the main ramifications; but the reader must suppose minor ones, representing the subordinate differences of orders, tribes, families, genera, etc., if he wishes to extend his views to the whole varieties of being in the animal kingdom" (Chambers, 1844:212–113). Letters A, C, and D, are intended simply to label the nodes, while F, R, B, and M (as in figure 48), refer to fish, reptile, bird, and mammal.

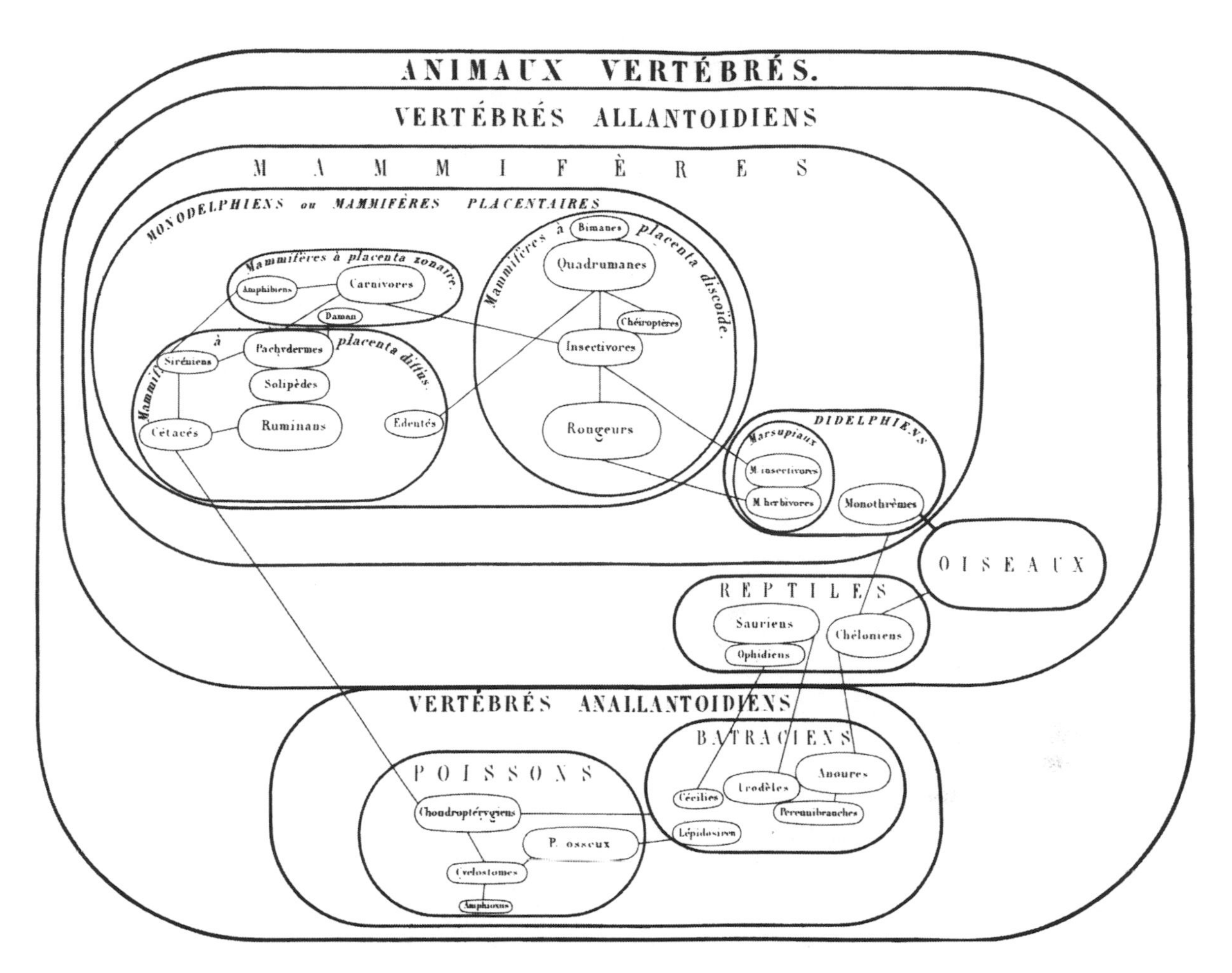

FIGURE 50. Henri Milne-Edwards's 1844 embryological classification of the vertebrates, a nested set of ovals that indicates relationships of both affinity and analogy in an apparent fallback to quinarian notions.

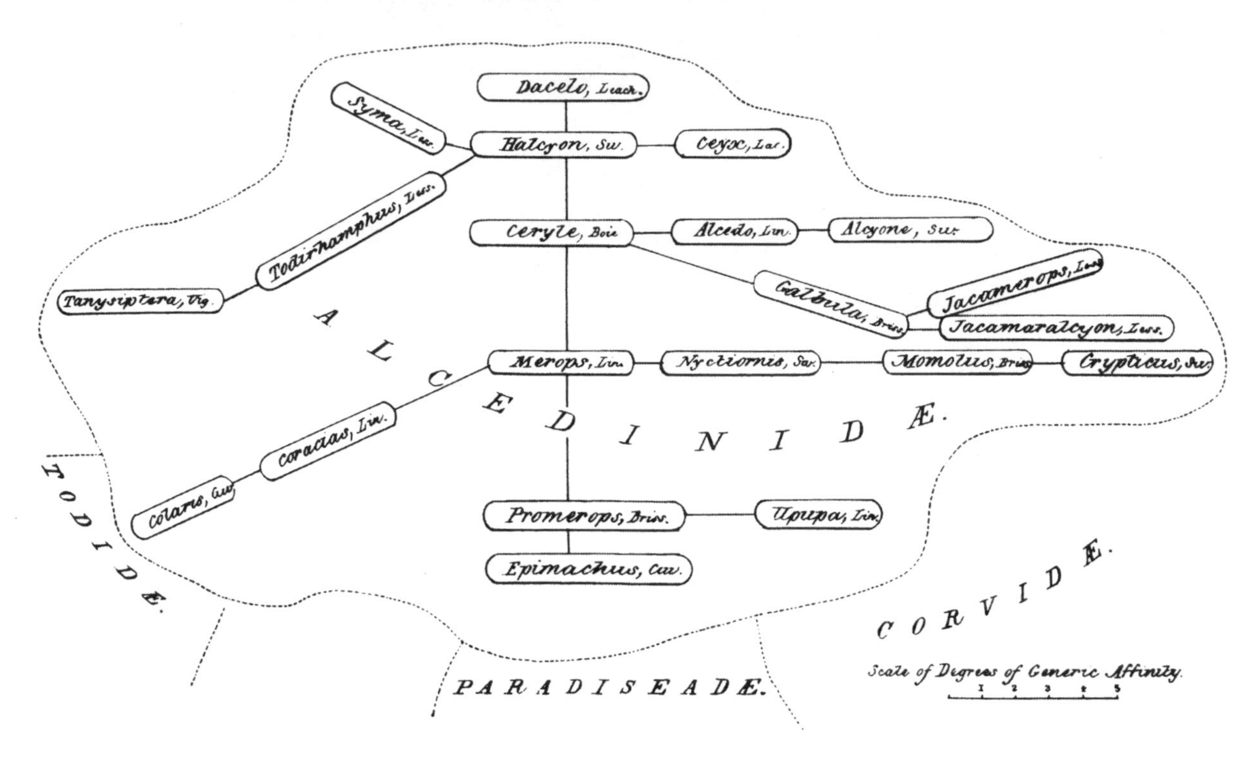

FIGURE 51. Hugh Edwin Strickland's 1841 "map" of relationships among the genera of the kingfishers and their allies, indicating the distances between the taxa by the "Scale of Degrees of Generic Affinity" shown on the lower right.

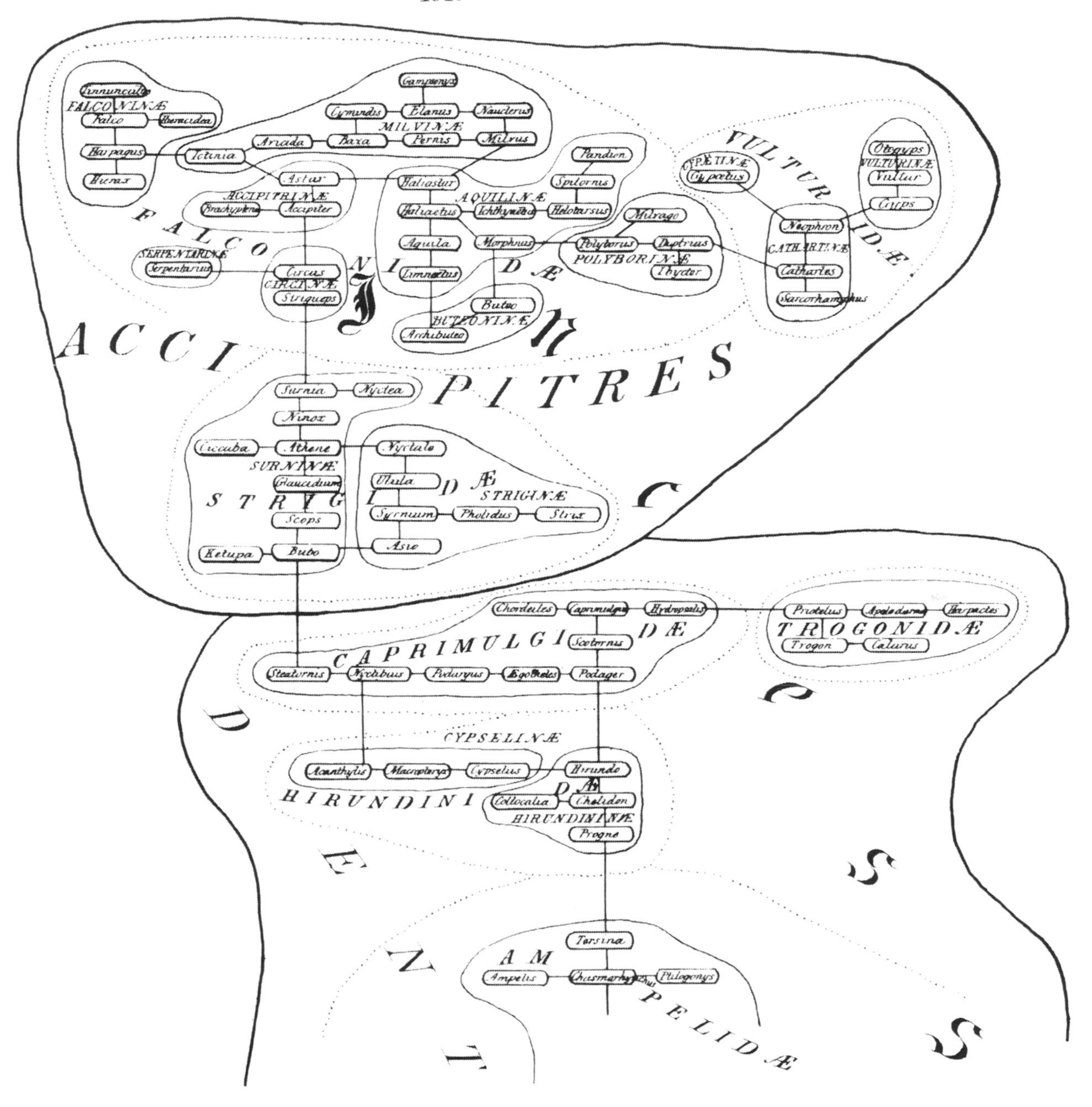

FIGURE 52. Part of Hugh Edwin Strickland's "chart of the natural affinities of the class of birds," displayed at the 1843 annual meeting of the British Association for the Advancement of Science, and published after his death by his father-in-law, William Jardine, in 1858.

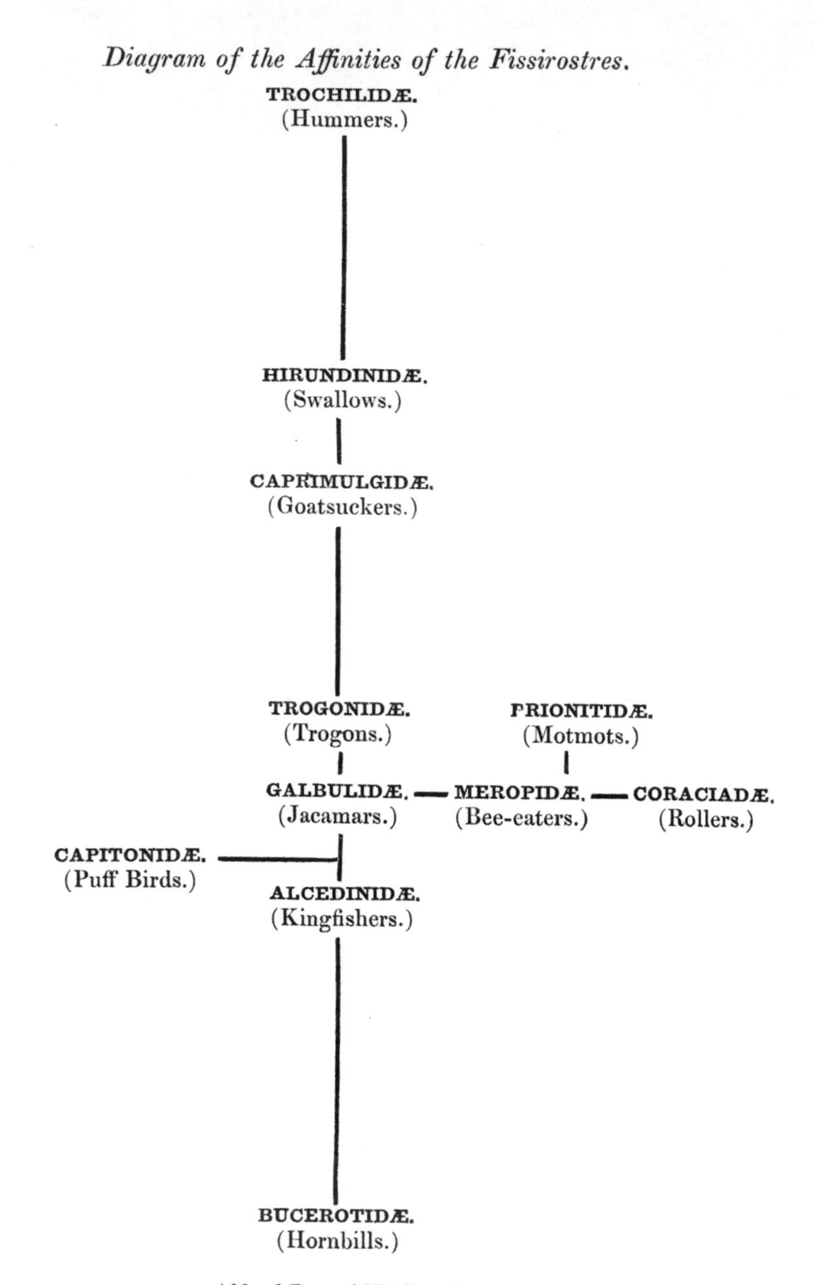

FIGURE 53. Alfred Russel Wallace's 1856 diagram of the affinities of the fissirostral birds, based on the mapping technique originated by Hugh Strickland.

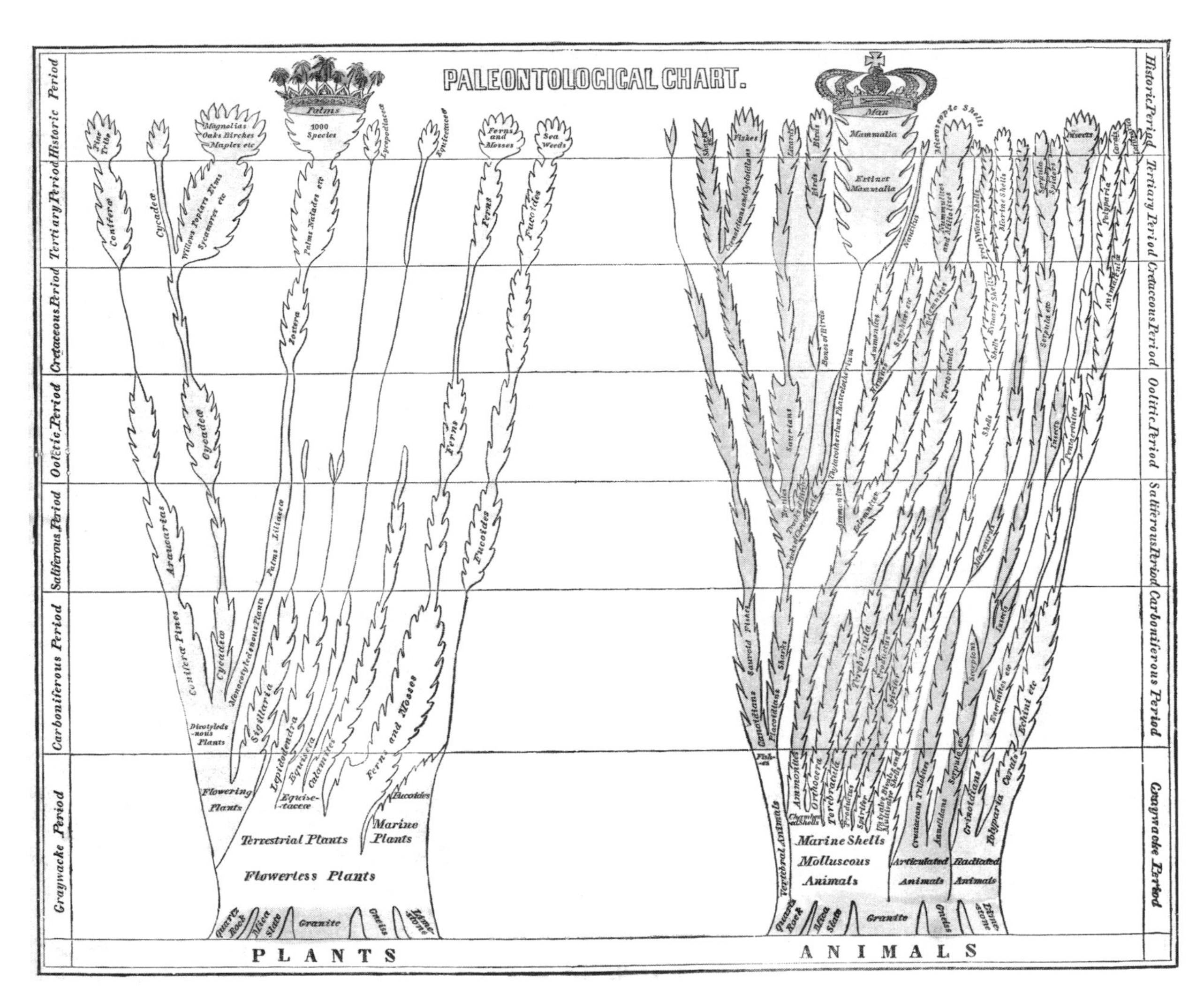

FIGURE 54. Edward Hitchcock's "Paleontological Chart" that appeared in the first edition of his *Elementary Geology* of 1840.

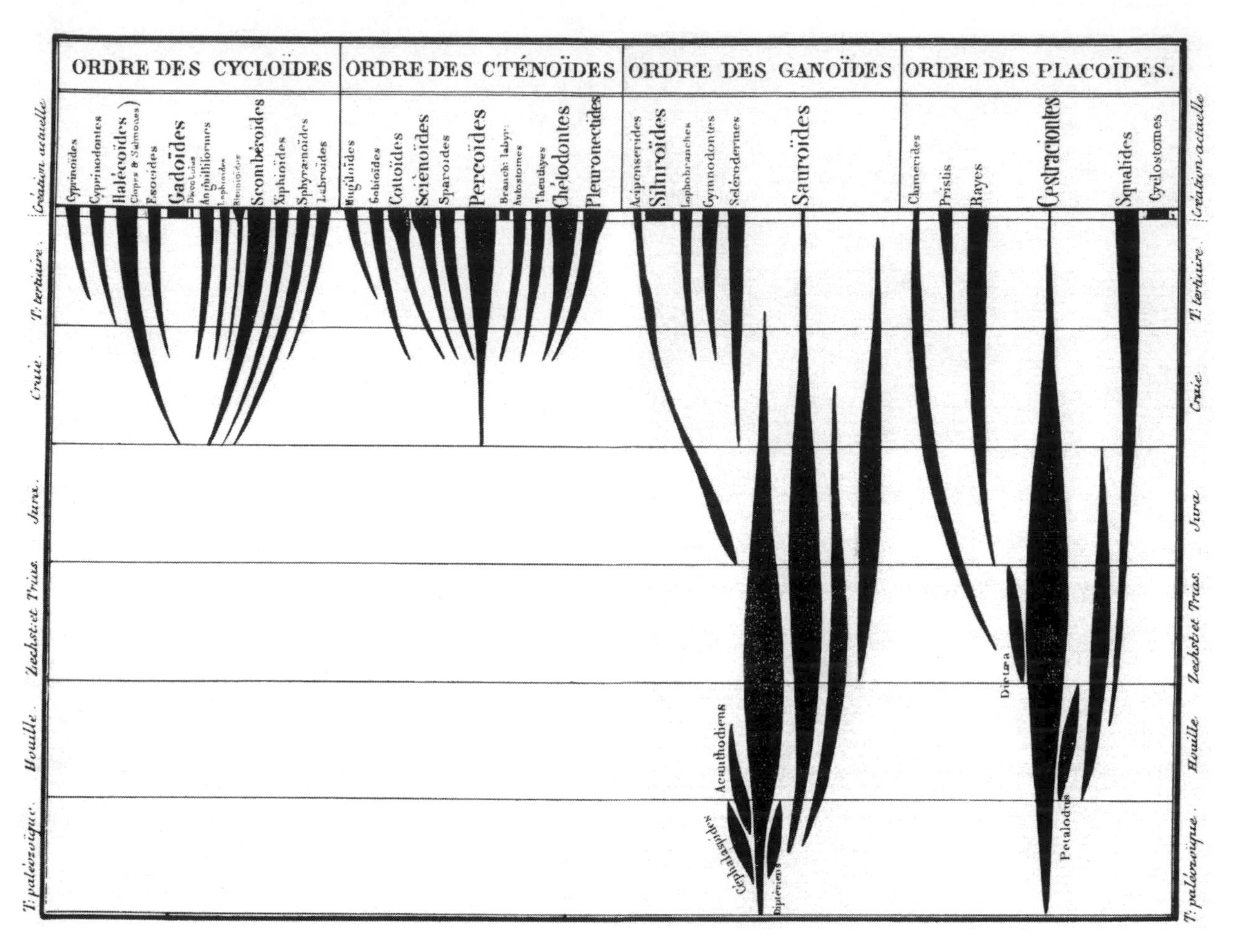

FIGURE 55. "Genealogy of the class of fishes" published by Louis Agassiz in his *Recherches sur les poissons fossiles* (Research on fossil fishes) of 1844.

---

facing page

FIGURE 56. The "Crust of the Earth as Related to Zoology," presenting, at one glance, the "distribution of the principle types of animals, and the order of their successive appearance in the layers of the earth's crust," published by Louis Agassiz and Augustus Addison Gould as the frontispiece of their 1848 *Principles of Zoölogy.* The diagram is like a wheel with numerous radiating spokes, each spoke representing a group of animals, superimposed over a series of concentric rings of time, from pre-Silurian to the "modern age." According to a divine plan, different groups of animals appear within the various "spokes" of the wheel and then, in some cases, go extinct. Humans enter only in the outermost layer, at the very top of the diagram, shown as the crowning achievement of all Creation.

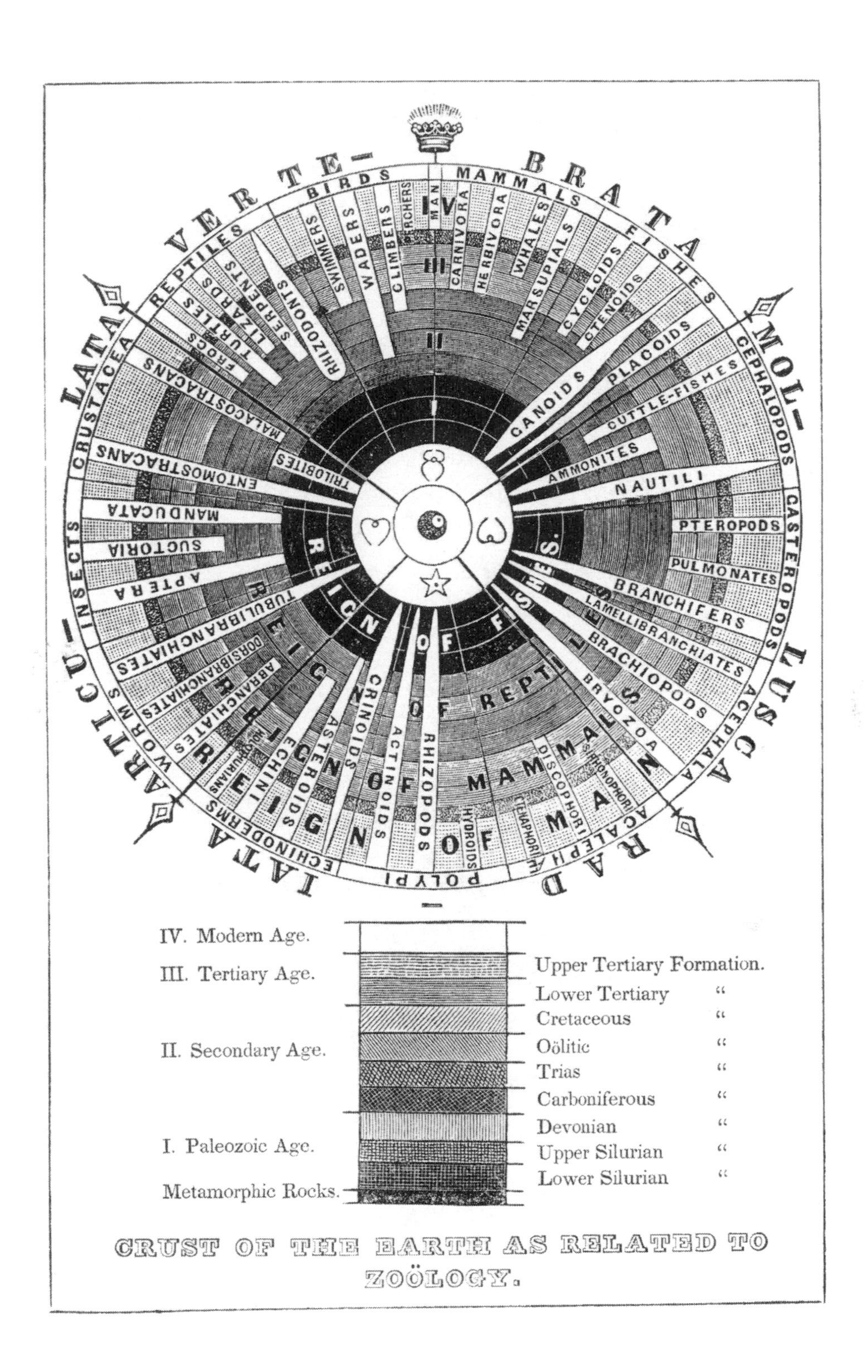

CRUST OF THE EARTH AS RELATED TO ZOÖLOGY.

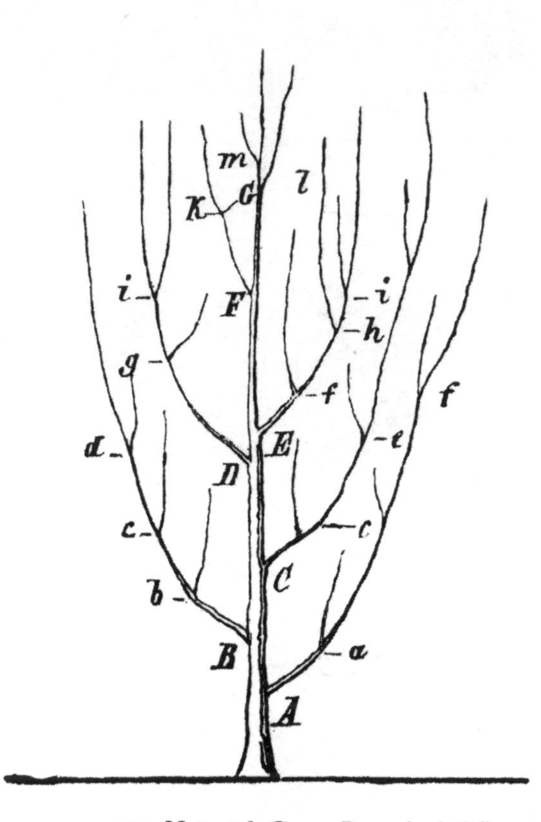

FIGURE 57. Heinrich Georg Bronn's 1858 "system of animals" based on the fossil record. The letters indicate a series of species of increasing structural perfection, succeeding one another as time moves along, the oldest at the bottom, arranged along the trunk and bases of the primary limbs, the progressively younger ones out on the secondary branches and twigs.

Evolution and the Trees of Charles Darwin

1837–1868

While others in the biosystematic world were pondering the best way to graphically display their ideas of plant and animal classification —experimenting with maps, networks, circles within circles, and all manner of other geometric configurations—Charles Darwin (1809–1882), just recently back from his five-year voyage aboard H.M.S. *Beagle* (December 1831–October 1836), and working more or less in isolation, was quietly thinking in another direction. In July 1837, he sketched his now famous "I think" diagram (Figure 58) in his leather-bound Notebook B, the first notebook on the "Transmutation of Species."[1] With just a few quick strokes and a couple of sentences written in telegraphic style, he demonstrated how a genus of related species might originate by divergence from a single starting point, thus revealing his early theoretical thoughts about phylogeny:

> I think [the] case must be that one generation then should be as many living as now. To do this & to have many species in [the] same genus (as is) *requires* extinction. Thus between A & B immense gap of relation. C & B the finest gradation, B & D rather greater distinction. Thus genera would be formed. —bearing relation to ancient types with several extinct forms.[2]

For the first time, Darwin clearly revealed the pattern of biological diversity to be hierarchically arranged in nested sets of taxa, all having descended from a single common ancestor.

As suggested by Julia Voss, in her recent book *Darwin's Pictures,*[3] it is intriguing to wonder if Darwin had the thirteen species of finches of the Galápagos Islands in mind when he constructed his first tree.

Although some of the branches are designated by letters, most are unnamed; yet, taken together, the tree indicates thirteen living species, on twigs ending in a cross-stroke and clustered at the terminus of two major branches. At the same time, twelve extinct species are indicated by twigs terminating abruptly and emerging at lower points along the branches.

Between 1837 and 1868, when he constructed his last known branching diagram (see below), Darwin drew a number of additional trees, but most were never published and remain today in manuscript form among his private papers in the Cambridge University Library. Of special interest, hidden in the Darwin Archive, is a tree dating from the early 1850s (Figure 59), based on Agassiz and Gould's circular "Crust of the Earth as Related to Zoology" diagram (see Figure 56). Applying the heading, "Dot means new form," Darwin replaced the stiff radiating "spokes" of Agassiz and Gould's figure with multiple branching lines emanating from a central point. The lines pass through concentric half circles that represent the boundaries between major time periods in Earth's history, which correspond to the "ages" indicated in Agassiz and Gould's diagram: "Paleozoic," "Secondary," and "Tertiary." Many of the lines terminate early, indicating extinction, while others extend to the outer margin, indicating their survival to the "modern age."[4]

Another unpublished Darwin tree, this one also constructed sometime during the 1850s, is prefaced by the phrase, "Let dots represent genera???" (Figure 60). Focusing this time on mammals, the root of the tree is labeled "Parent of Marsupials & Placentals." As in Darwin's modified Agassiz and Gould diagram (see Figure 59), lightly penciled, concentric half-circles represent the boundaries of geological time, many taxa becoming extinct along the way, others, such as living "marsupials" and "rodents," passing through the ages and reaching modern times.

Darwin greatly expanded his ideas about tree construction during the late 1850s, putting together four, much more elaborate figures (Figure 61, labeled Diagram I–IV), intended to accompany what he called his "Big Species Book," a manuscript that was already several hundred pages in length, but only two-thirds finished when he set it aside to begin work on an abbreviated version that was soon to become his *Origin of Species*.[5] The original manuscript version of this theoretical figure, dated 1857, still exists in the Darwin Archive, maintained by the Cambridge University Library,[6] but it has since been set to type and published in 1975.[7] In explaining his tree, Darwin wrote:

> The complex action of these several principles, namely, natural selection, divergence & extinction, may be best, yet very imperfectly, illustrated by the following Diagram [Figure 61], printed on a folded sheet for convenience of reference. This diagram will show the manner, in which I believe species descend from each other & therefore shall be explained in detail: it will, also, clearly show several points of doubt & difficulty.[8]

Darwin's 1857 diagrams were revised substantially—condensed into a single figure and now ascending rather than descending on the page (Figure 62)—for inclusion in the *Origin of Species,* which was published on 24 November 1859, but his explanatory text remained largely the same:

> Let A to L represent the species of a genus large in its own country; these species are supposed to resemble each other in unequal degrees, as is so generally the case in nature, and as is represented in the diagram by the letters standing at unequal distances. . . . Let (A) be a common, widely-diffused, and varying species. . . . The little fan of diverging dotted lines of unequal lengths proceeding from (A), may represent its varying offspring. The variations are supposed to be extremely slight, but of the most diversified nature; they are not supposed all to appear simultaneously, but often after long intervals of time; nor are they all supposed to endure for equal periods. Only those variations which are in some way profitable will be preserved or naturally selected. And here the importance of the principle of benefit being derived from divergence of character comes in; for this will generally lead to the most different or divergent variations (represented by the outer dotted lines) being preserved and accumulated by natural selection. When a dotted line reaches one of the horizontal lines, and is there marked by a small numbered letter, a sufficient amount of variation is supposed to have been accumulated to have formed a fairly well-marked variety, such as would be thought worthy of record in a systematic work.[9]

And so Darwin continued to describe in great detail—taking up eight pages of his chapter on "Natural selection"—the intricacies of his tree as an aid in understanding what he called the "rather perplexing subject" of plant and animal descent with modification.[10] The first edition of his book went on for 502 pages, but it was largely this diagram and his carefully worded, step-by-step explanation that sold his revolutionary ideas to an initially skeptical audience.

On Sunday, 23 September 1860, Darwin wrote to his close friend and colleague Charles Lyell (1797–1875), the foremost geologist of the day, to report on his latest thoughts about the evolution of mammals: "My dear Lyell . . . I have a very decided opinion that all mammals must have descended from a *single* parent . . . [the] large amount of similarity

I must look at as certainly due to inheritance from a common stock." With this letter, published posthumously in 1887 by his son Francis,[11] Darwin enclosed two branching diagrams (Figure 63), expressing an inability to decide which of the two more closely reflects evolutionary history: whether marsupial and placental mammals descended from some unknown common ancestor that was "intermediate between Mammals, Reptiles and Birds, as intermediate as [the lungfish] *Lepidosiren* now is between Fish and Batrachians"; or whether the whole assemblage arose from some true marsupial that had gone through a progression from "lowly" to "highly developed."[12]

Darwin's last known tree, dated 21 April 1868, is a roughly drawn look at primate evolution, a kind of working diagram, with multiple revisions, in pen and ink and later in pencil, that seem to indicate his uncertainty about the relationships depicted (Figure 64). From the trunk of the tree labeled "Primates," an initial divergence leads to the "lemurs," followed by a major bifurcation, leading to the "new world monkeys" on the right and the "old world monkeys" on the left. Ascending farther up along the main stem, a branch to the right gives rise to the cercopithecid genus *Cercopithecus,* then including all the baboons and macaques; and *Semnopithecus,* the Asian langurs. Diverging off to the left is an uncertain lineage—initially indicated by a dotted line, but then penciled over—leading to "Man," and in the center at the top of the tree, a triumvirate of gorillas and chimpanzees, orangutans, and gibbons (genus Hylobates). In his placement of "Man," Darwin differs drastically from all tree-like diagrams presented before, and many that were to come afterward, in giving humans a position that is no more elevated in stature than our close relatives the gorillas and chimps—he presents us as simply part of the primate family tree.[13]

Shortly after publication of the *Origin,* Darwin received a manuscript from William Charles Linnaeus Martin (1798–1864)—a former curator of the museum of the Zoological Society of London, until he was let go due to financial cutbacks—titled "Comments on Mr. Darwin's grand theory," which contained a family tree of the birds (Figure 65):

> As may easily be perceived, this is a very rude "*Ebauche*" a mere rough sketch (on rough paper), intended to convey an imperfect, & as yet, not maturely studied idea of the development of the leading forms of birds divaricating from a primeval, extinct, & of course unknown root. . . . I suppose each circle (or most of the circles) to represent a group which in the wear of time has proved its stability & contended successfully against contingences in the great struggle of Life—I suppose them to have given off successors, still better fitted for this struggle according to the alteration of influencing

> conditions. . . . Between the rude circles I suppose a long but variable period to intervene—a period mostly perhaps to be measured by ages—geological ages,—for the modifications must be slow— Between these circles I have placed nothing—but the circles represent forms of longer or shorter duration— The small dots, like flower stamens, convey an idea of the ornithological multitude of families & genera.[14]

Martin was generally supportive of Darwin's theory, but he argued that flightless birds were primitive to birds as a whole, contradicting Darwin's contention that the loss of wings was a secondary event in bird evolution, caused by disuse.[15] Unfortunately in poor physical condition when he wrote to Darwin—"I am not in health & every trifle is a labour"[16]—he died in 1864 before he could get his ideas and his tree published.

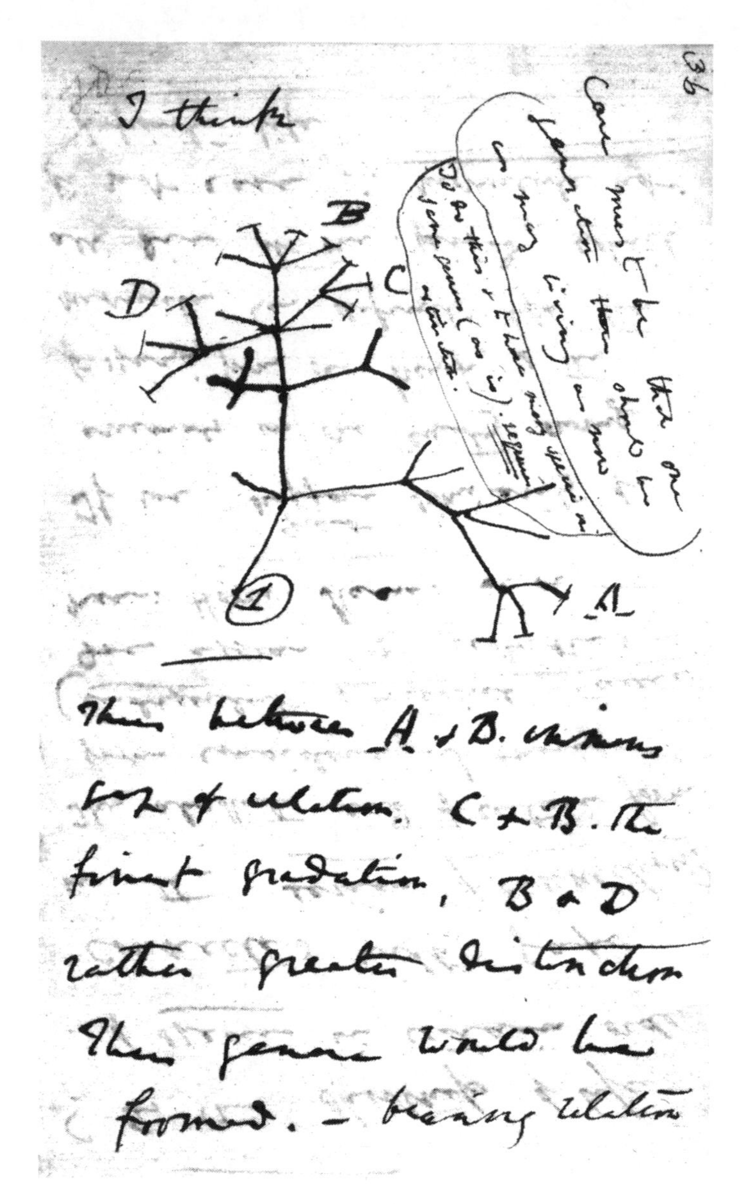

FIGURE 58. Charles Darwin's famous "I think" diagram drawn in July 1837, taken from his first notebook on the "Transmutation of Species." The main trunk of the tree is labeled number 1. Twigs ending in a cross-stroke and clustered at the terminus of major branches (labeled A, B, C, and D) indicate living species; extinct species are indicated by twigs terminating abruptly and emerging at lower points along the branches.

Darwin Archive, DAR 121.36; courtesy of Ruth M. Long and Don Manning. Reproduced by the kind permission of the Syndics of the Cambridge Library.

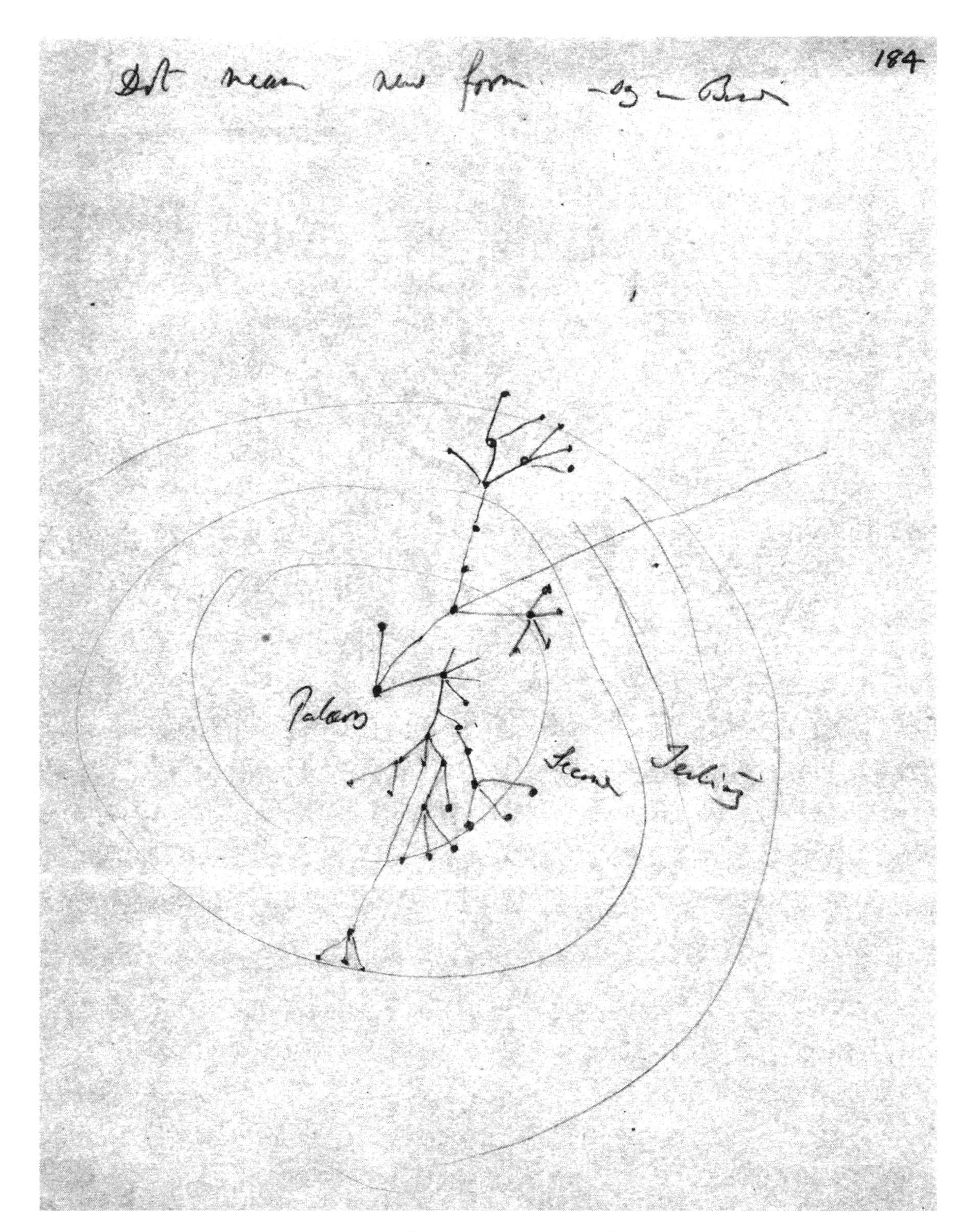

FIGURE 59. "Dot means new form": Charles Darwin's sketch from the early 1850s, showing the relationships of fishes through time, based on Agassiz and Gould's image of 1848 (see figure 56).

Darwin Archive, DAR 205.5.184r; courtesy of Ruth M. Long and Don Manning. Reproduced by the kind permission of the Syndics of the Cambridge Library.

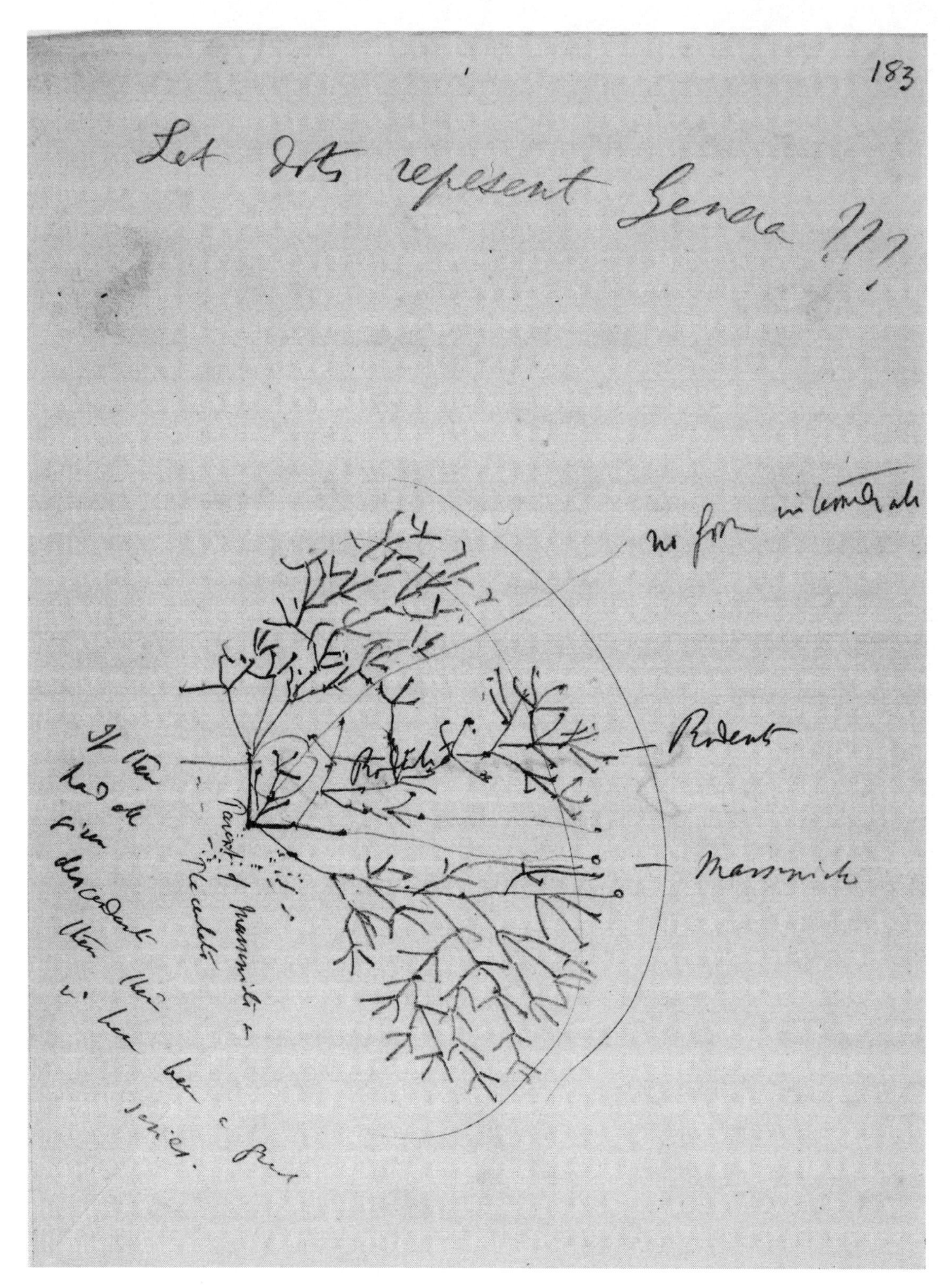

FIGURE 60. "Let dots represent genera???": evolutionary relationships of the rodents and marsupials, drawn by Charles Darwin in the 1850s.

Darwin Archive, DAR 205.5.183r; courtesy of Ruth M. Long and Don Manning. Reproduced by the kind permission of the Syndics of the Cambridge Library.

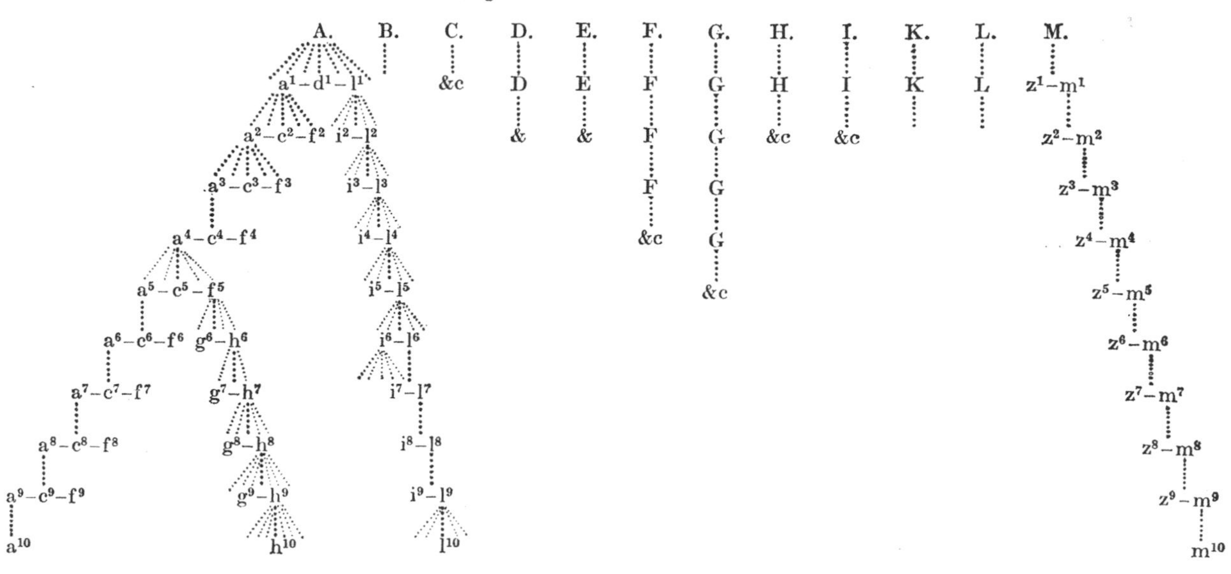

Diagram II

A. B. C. D. E. F. G. H. I. K. L. M.

$a^{1}-d^{1}-l^{1}$ &c D E F G H I K & $z^{1}-m^{1}$

$a^{2}-c^{2}-f^{2}$ $i^{2}-l^{2}$ &c &c F G &c & & $z^{2}-m^{2}$

$a^{3}-c^{3}-f^{3}$ $i^{3}-l^{3}$ F G $z^{3}-m^{3}$

$a^{4}-c^{4}-f^{4}$ $i^{4}-l^{4}$ &c G $z^{4}-m^{4}$

$a^{5}-c^{5}-f^{5}$ $i^{5}-l^{5}$ &c $z^{5}-m^{5}$

$a^{6}-c^{6}-f^{6}$ $i^{6}-l^{6}$ $z^{6}-m^{6}$

$a^{7}-c^{7}-f^{7}$ $i^{7}-l^{7}$ $z^{7}-m^{7}$

$a^{8}-c^{8}-f^{8}$ $i^{8}-l^{8}$ $z^{8}-m^{8}$

$a^{9}-c^{9}-f^{9}$ $g^{9}-h^{9}$ $i^{9}-l^{9}$ $z^{9}-m^{9}$

f^{10} h^{10} l^{10} m^{10}

Original Species in Diag. I	A.			B.	C.	D.	E.	F.	G.	H.	I.	K.	L.	M.
Diagram III	a^{10}.	h^{10}.	l^{10}.	B.	C.	D.	E.	F.	G.	H.	I.	K.	L.	M^{10}
Diag. IV	a^{20} k^{20} n^{20}		p^{20} t^{20} l^{20}				E^{20}	F^{20}						x^{20} z^{20} m^{20}

FIGURE 61. The printed version of Darwin's tree planned for inclusion in his "Big Species Book," a manuscript that was set aside in 1858 to begin work on an abbreviated version that was soon to become his *Origin of Species*.

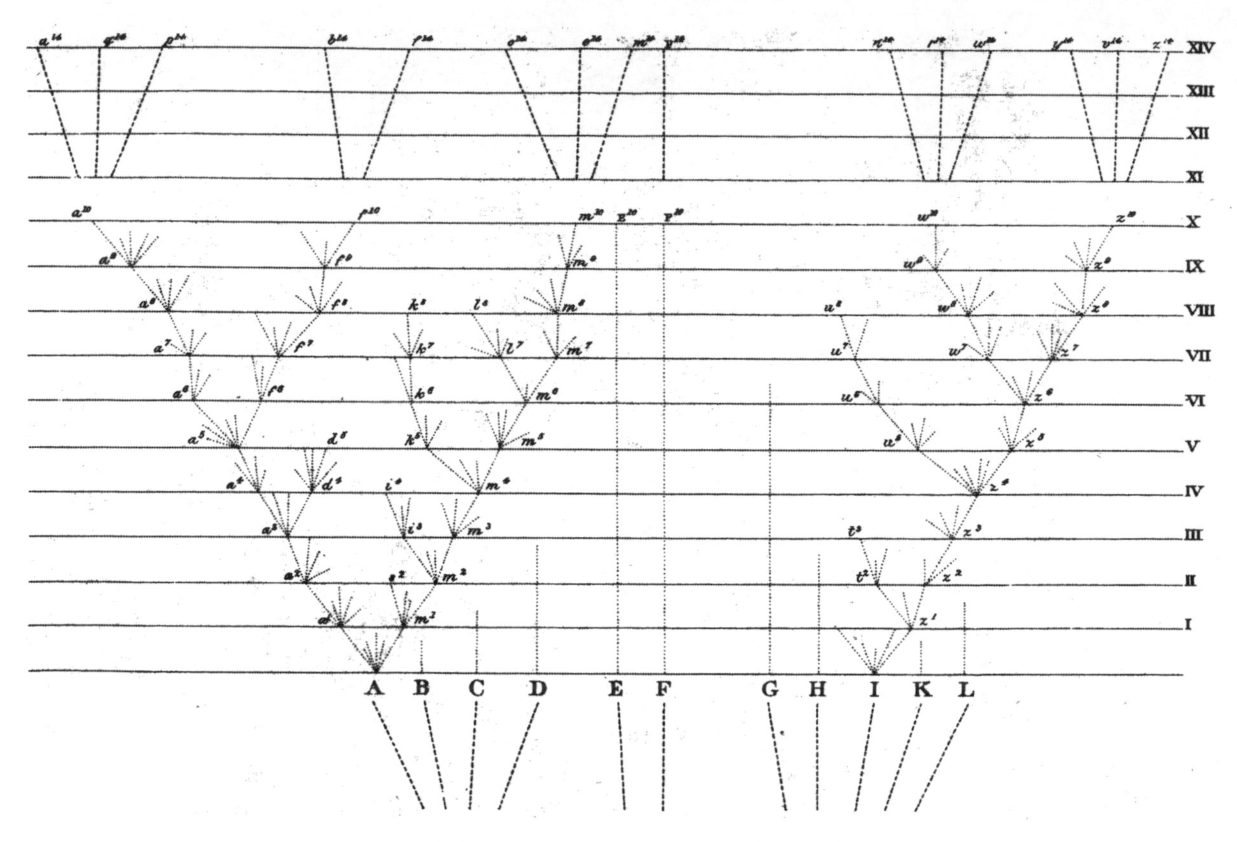

FIGURE 62. The final version of Darwin's famous branching diagram, constructed to explain "the probable effects of the action of natural selection through divergence of character and extinction, on the descents of a common ancestor," as published in his *Origin of Species* of 1859.

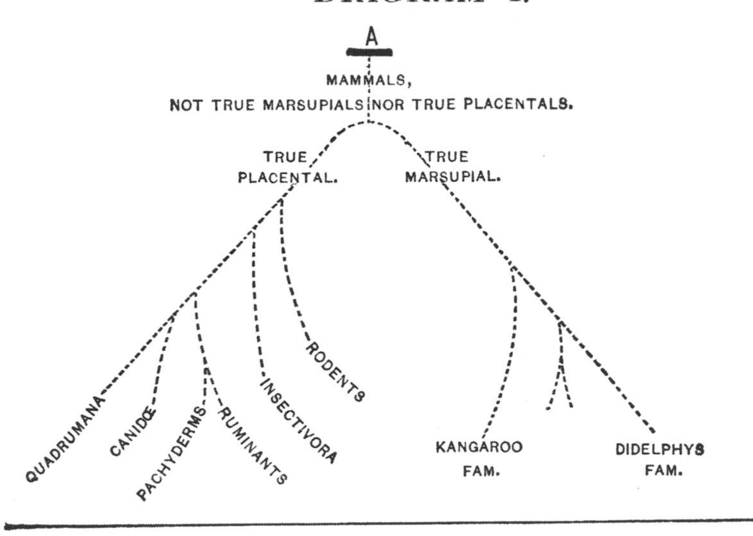

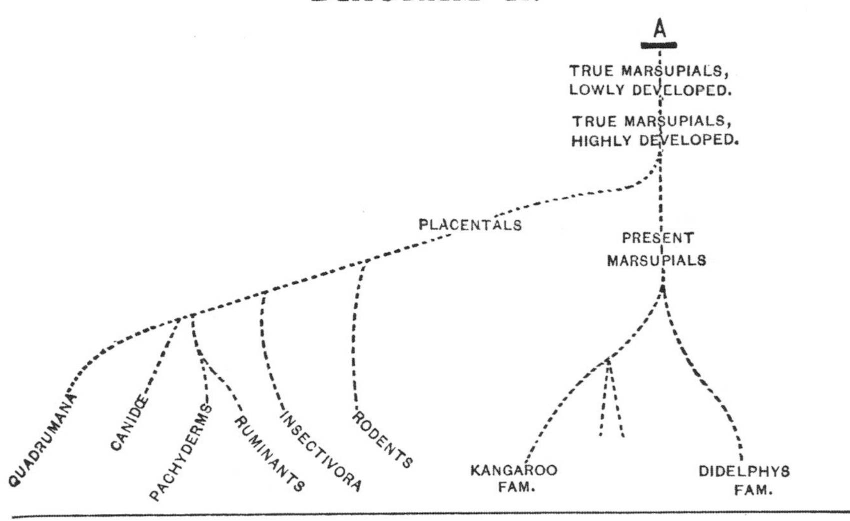

FIGURE 63. Darwin's trees of mammal evolution originally enclosed in a letter to British geologist Charles Lyell dated 23 September 1860 and later published by Francis Darwin in his 1887 *Life and Letters of Charles Darwin.*

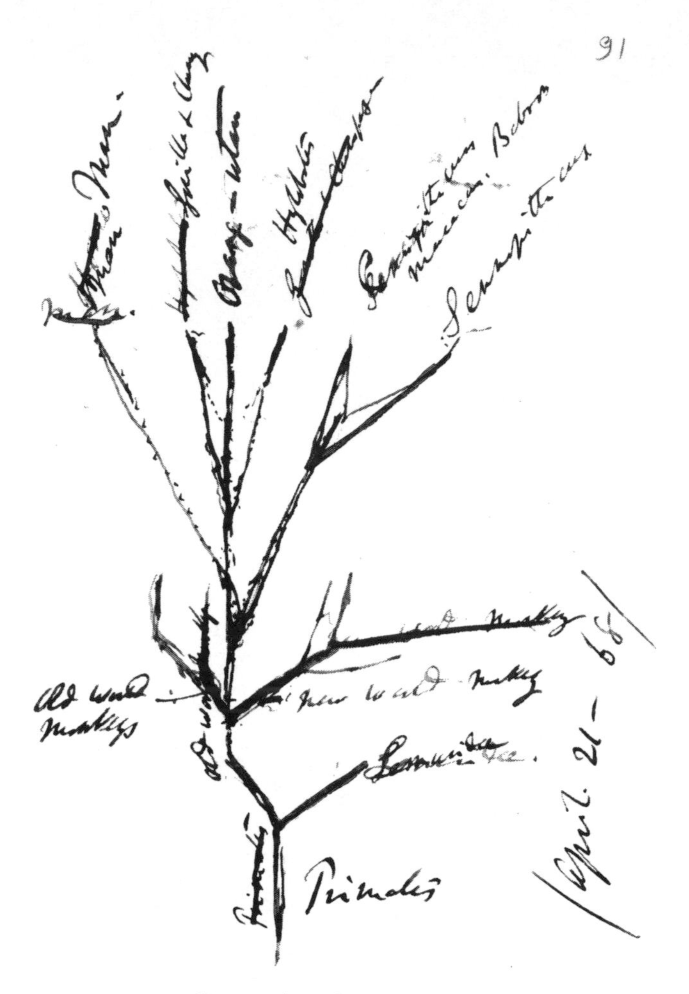

FIGURE 64. Darwin's evolutionary tree of primate relationships, sketched in 1868, showing "man" on the upper far left next to a lineage containing gorillas and chimps.

Darwin Archive, DAR 80.91r; courtesy of Ruth M. Long and Don Manning. Reproduced by the kind permission of the Syndics of the Cambridge Library.

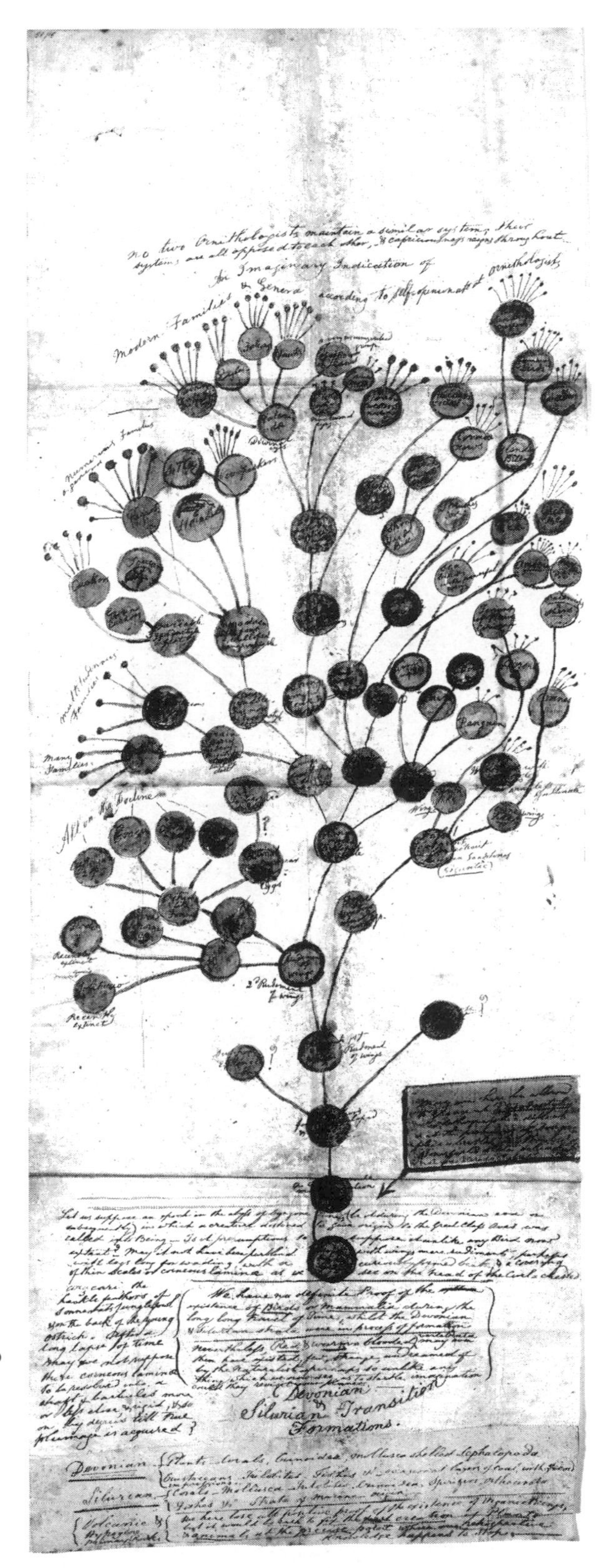

FIGURE 65. A family tree of the birds constructed by William Charles Linnaeus Martin and sent to Darwin for his approval, soon after publication of the *Origin of Species*. Darwin Archive, DAR 171:56.15r; courtesy of Ruth M. Long and Don Manning. Reproduced by the kind permission of the Syndics of the Cambridge Library.

The Trees of Ernst Haeckel

1866–1905

Ernst Heinrich Philipp August Haeckel (1834–1919) was an eminent German biologist, naturalist, philosopher, physician, professor, and scientific illustrator who discovered, described, and named thousands of new species, and coined many terms in biology, including ecology, phylum, phylogeny, ontogeny, monophyletic, polyphyletic, metamerism, Protista, and Metazoa.[1] Sometimes called the German Darwin, Haeckel got his first detailed overview of Darwinism in 1861 when he read Heinrich Bronn's German translation of the *Origin,* which had been published in 1860, only a few months after Darwin's original version appeared.[2] Although he had difficulty accepting the idea that natural selection was the only mechanism for generating biological diversity, Haeckel quickly became one of Darwin's most prolific and vociferous supporters, recognizing early on that the best way to demonstrate organic change through time was to construct trees.[3]

Haeckel was the first great tree-maker. During his long career, he drew hundreds of visually striking genealogical trees to which specific biological taxa were assigned, unlike Darwin's purely hypothetical trees. Included among the earliest of these was the first tree relating all major forms of life, postulating a common ancestor among what he called monerans, "simple, homogenous, structureless and formless little lumps of mucous or albuminous matter,"[4] now recognized as one-celled organisms that have no nucleus or organelles (Figure 66). He was also the first to publish that the ancestors of humans were among the great apes, placing "man" at the top of his branching diagrams as the pinnacle of evolution. In his first major publication as an evolutionist, a book in two volumes titled *Generelle morphologie der organismen* (General morphology of organisms) published in 1866, he included a series of *Monophyletischer Stammbäume* (monophyletic family trees) that are thoroughly botanical in appearance, all with sturdy trunks, major limbs, branches, and terminal twigs—all but the leaves are there. Each one is unique and rarely seen, most not having been reproduced for well over a century, thus all eight are included here (Figures 66–73).

In later publications and in revised editions of old ones, Haeckel continued to include trees, refining and extending his treatment to numerous additional taxa and inventing new designs as he went along.[5] In his wildly popular *Natürliche Schöpfungsgeschichte* (Natural history of creation), which, first published in 1868, went through thirteen editions during his own lifetime, Haeckel employed a variety of new stylistic approaches, turning to much simpler, much less botanical designs: some consisting of thin curved or wavy lines, each terminating in a dense cluster of wiggly twigs (Figures 74 and 75); others dominated by strange, tapering, triangular shapes, each containing numerous diverging or converging lines (the latter only to fit the lineages in a pleasing way on the page) meant to indicate bundles of closely related animal lineages (Figure 76). In contrast, however, are two intricate paleontological trees of plants and vertebrate animals, showing feathery groups of organisms diverging from one another through time, with the geological epochs and periods indicated along the left margin (Figures 77 and 78). Notable in his tree of the animal kingdom is the position of the "Drachen" (dinosaurs), shown to be ancestral to the birds (Vögel, Aves), a relationship that is often thought to be a recent discovery.

In the second edition of *Natürliche Schöpfungsgeschichte,* which appeared in 1870, Haeckel went simpler still, resurrecting the old bracketed diagram motif (albeit arranged vertically rather than horizontally) that was popular among Renaissance naturalists (Figure 79; compare with Figures 3–14). But at the same time, he constructed the first diagrams that show multiple independent plant and animal origins (examples of what he called "polyphyly") as well as numerous extinctions, each of the latter indicated by a dagger, the earliest appearance of a convention quickly adopted by paleontologists and still universally applied today (Figure 80). Finally, in this 1870 edition, unlike a tree but rather more resembling a complex, sprawling, branching bush superimposed over a map of the world, is Haeckel's 1870 "hypothetical sketch of the monophyletic origin and spread [throughout the world] of the twelve species of man," the earliest example of true evolutionary-based phylogeography (Figure 81). Convinced that all human races arose from a common ancestor,[6] this map, with its meandering pathways of human migration, was his attempt to address a major question that had puzzled evolutionists, including Darwin, up until that time: where on Earth did humans originate? Darwin at one point correctly hypothesized Africa (on biogeographical evidence),[7] while Haeckel thought perhaps it was somewhere in the western Indian Ocean (the site of the hypothetical continent of Lemuria) or in the Dutch East Indies (now Indonesia, see below).[8]

In 1874, Haeckel published *Anthropogenie oder Entwickelungsgeschichte des menschen* (The evolution of man) in which he traced in great detail the origin of all life on Earth back to lowly single-celled organisms and at the same time argued that the ancestry of humans was among the anthropoid apes, the gorilla and chimpanzee on one hand and the orangutan and gibbons on the other. To illustrate these relationships, he constructed his celebrated *Stammbaum des Menschen* (Family tree of man), his big gnarly oak, with its thick, bark-encrusted trunk, knotted limbs, and tiny twigs—the best known and by far the most often reproduced example of a tree of life (Figure 82). As in all his trees that include humans ("man"), they are placed at the apex as fits their assumed superior place among animals, but, unfortunately, a hierarchy of relative superiority among races of humans was assumed as well:

> We as yet know of no fossil remains of the hypothetic primaeval man (*Homo primigenius*) who developed out of anthropoid apes during the tertiary period, either in Lemuria or in southern Asia, or possibly in Africa. But considering the extraordinary resemblance between the lowest woolly-haired men, and the highest man-like apes, which still exist at the present day, it requires but a slight stretch of the imagination to conceive an intermediate for connecting the two, and to see in it an approximate likeness to the supposed primaeval men, or ape-like men [Figure 83].[9]

Haeckel was an enthusiastic, flamboyant figure, whose stellar performances at scientific meetings had a magnetic effect on the audience,[10] but his reputation often suffered for making grand assumptions based on little or no evidence.[11] When Darwin published his *Origin of Species* in 1859, no fossil remains of human ancestors had yet been found,[12] but Haeckel, ever confident in himself, boldly hypothesized in 1868 that evidence of human evolution would eventually be found in the Dutch East Indies (now Indonesia). He described these hypothetical remains in great detail, imagining a form that walked erect and had a greater intellectual development than the anthropoid apes, but lacked the faculty of speech.[13] Without physical evidence of any kind, he named the species *Pithecanthropus alalus* (speechless ape-man), giving it a place in his genealogical trees (Figures 84 and 85) and encouraging his students to go out and find it. Many critics complained that Haeckel had gone too far, but this time he was vindicated: in 1891, twenty-four years after Haeckel's prophecy, a young Dutch paleontologist, Marie Eugène François Thomas Dubois (1858–1940), discovered on the Indonesian island of Java the fossil remains of what would soon became known around the world as Java Man, a species described as intermediate between apes and humans,[14] recognized today as *Homo erectus*. Haeckel was understandably elated with the news: "The phylogenetic hypothesis of the organisation of this 'Ape-man' which I advanced [in 1868] was brilliantly confirmed."[15]

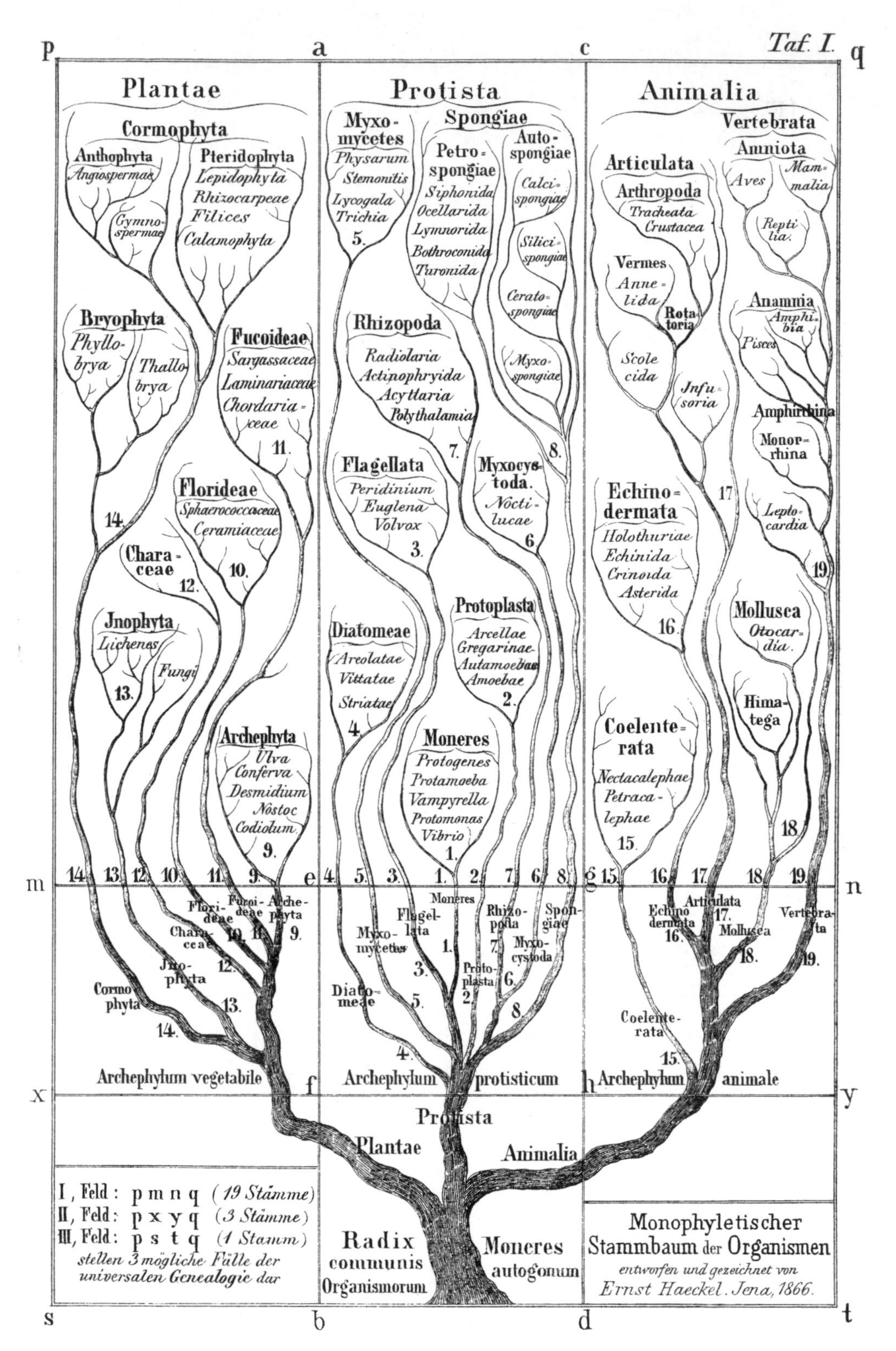

FIGURE 66. Ernst Haeckel's earliest published branching diagram, a "monophyletic family tree of organisms," recognizing three kingdoms, plants, protists, and animals, from the first edition of his *Generelle morphologie der organismen* (General morphology of organisms) published in 1866.

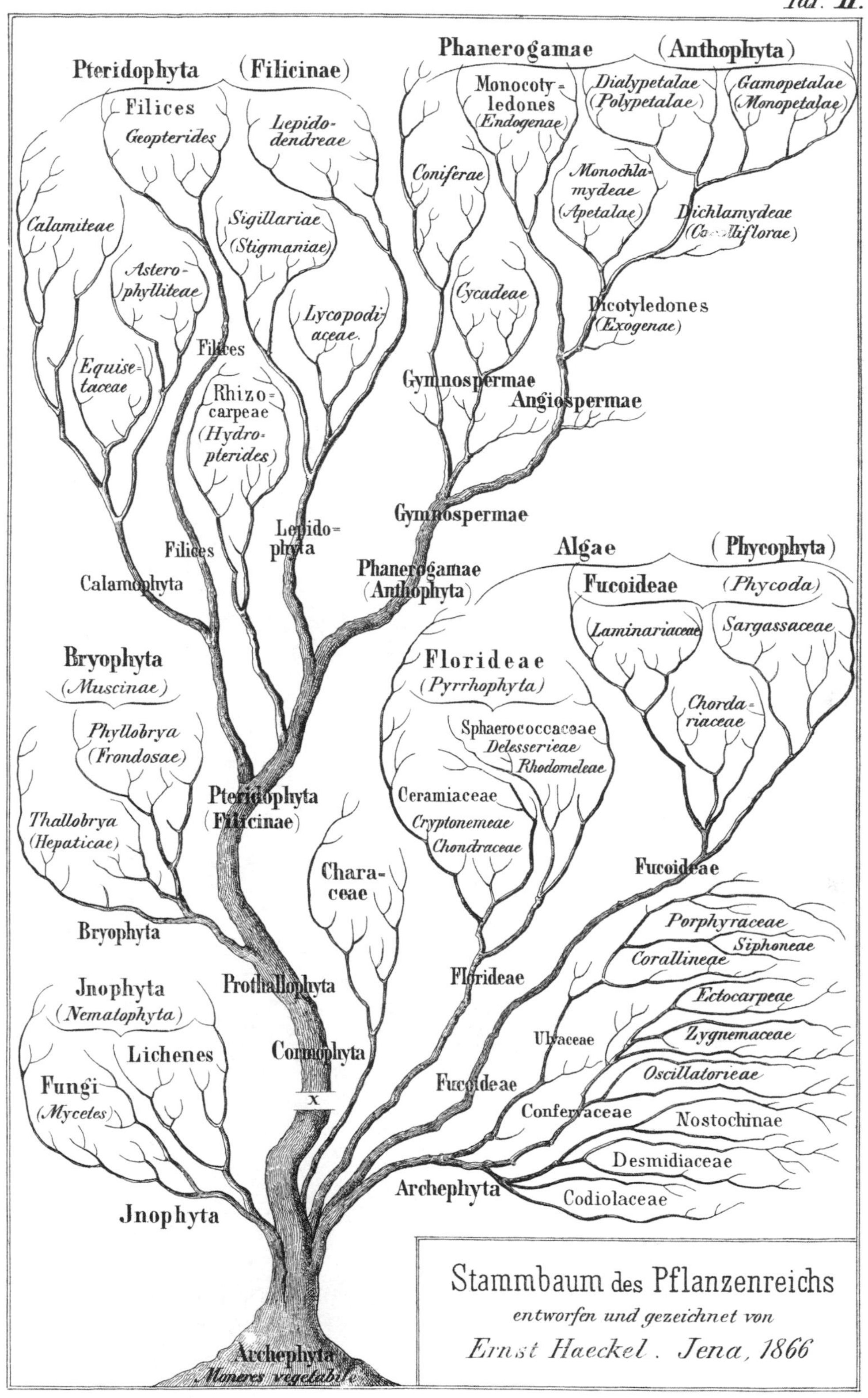

FIGURE 67. Ernst Haeckel's family tree of the plant kingdom, from 1866.

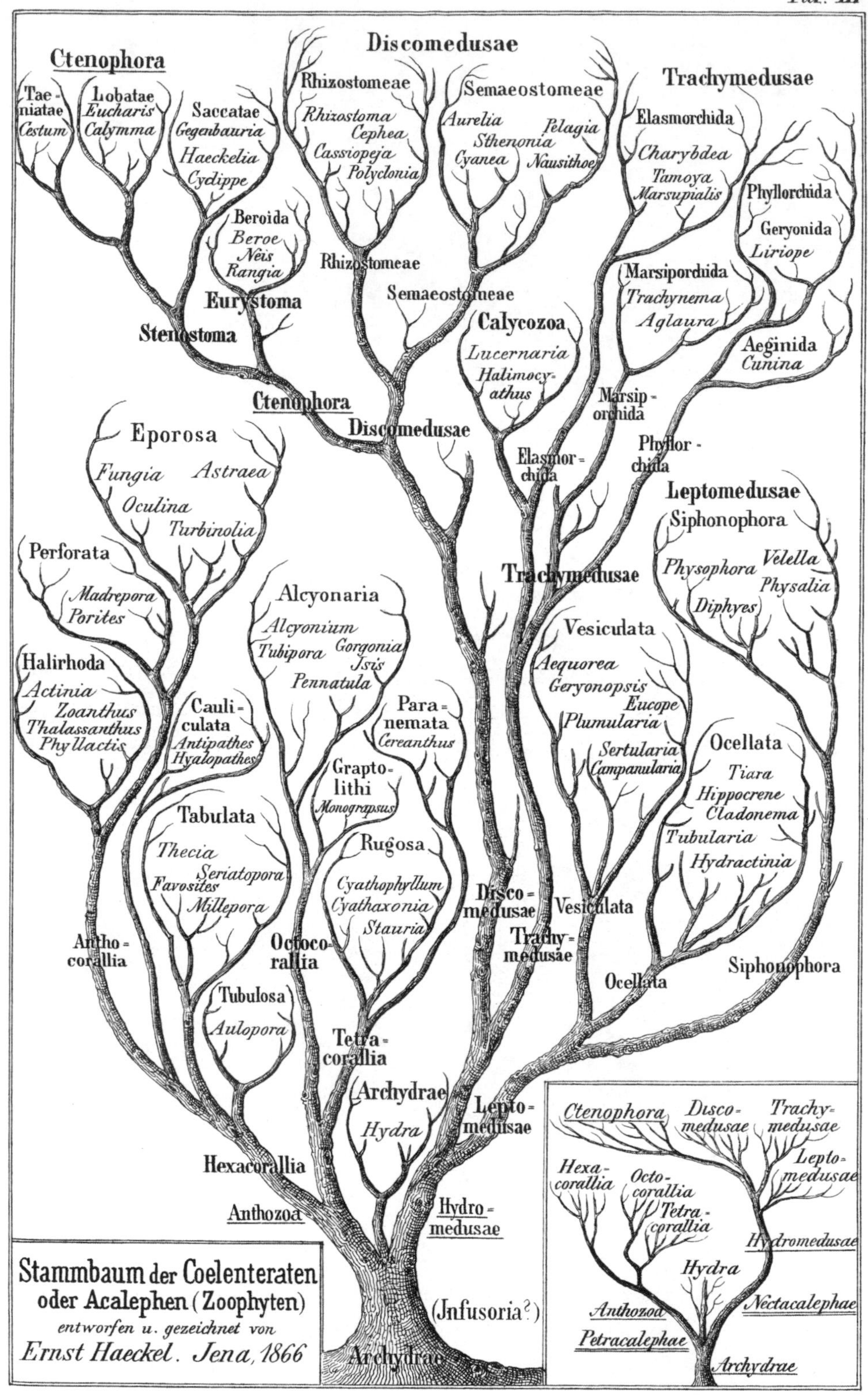

FIGURE 68. Ernst Haeckel's family tree of the coelenterates or medusa (jellyfishes), from 1866.

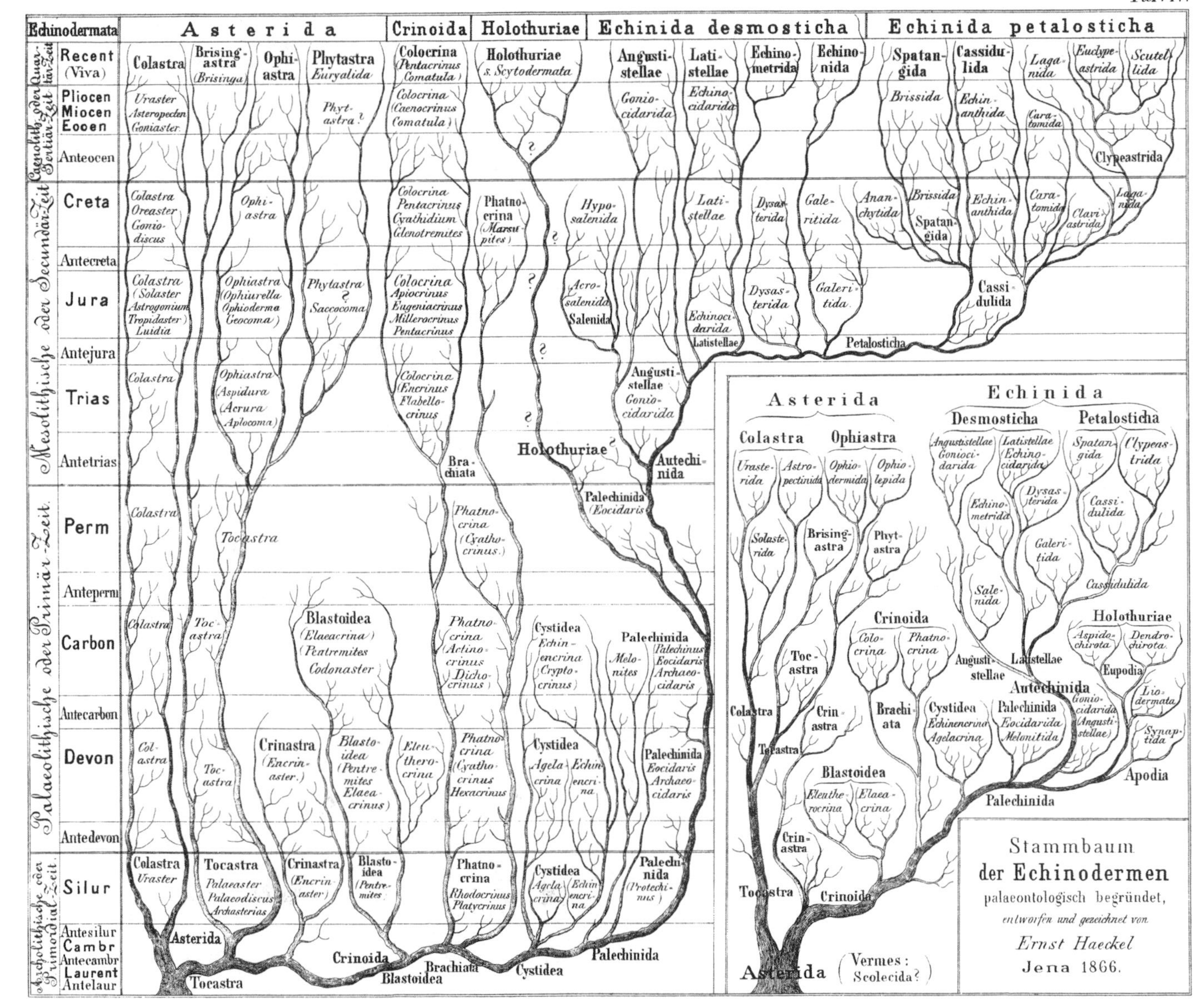

FIGURE 69. Ernst Haeckel's paleontological family tree of the echinoderms, from 1866.

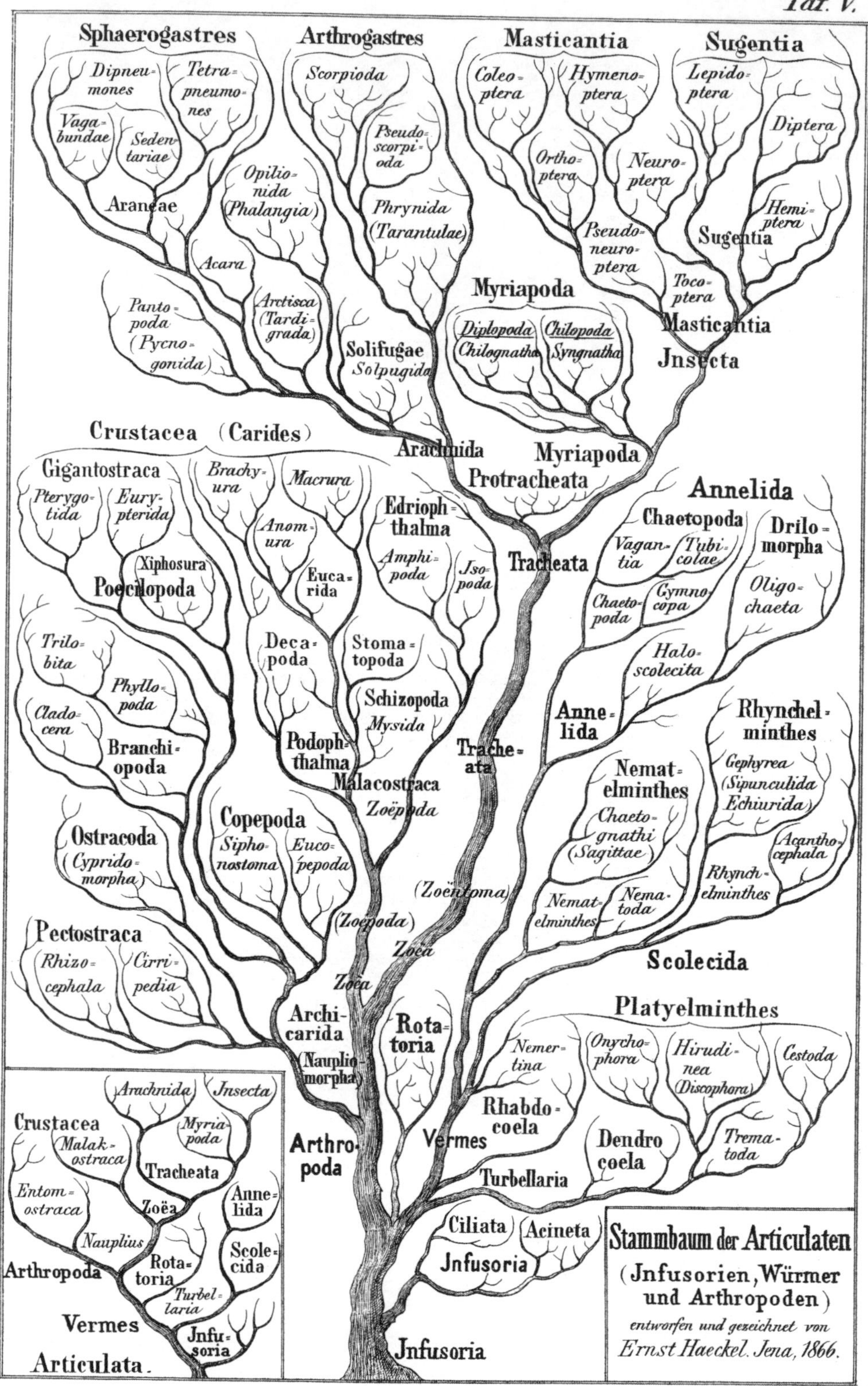

FIGURE 70. Ernst Haeckel's family tree of the articulate animals (single-celled organisms, worms, and arthropods), from 1866.

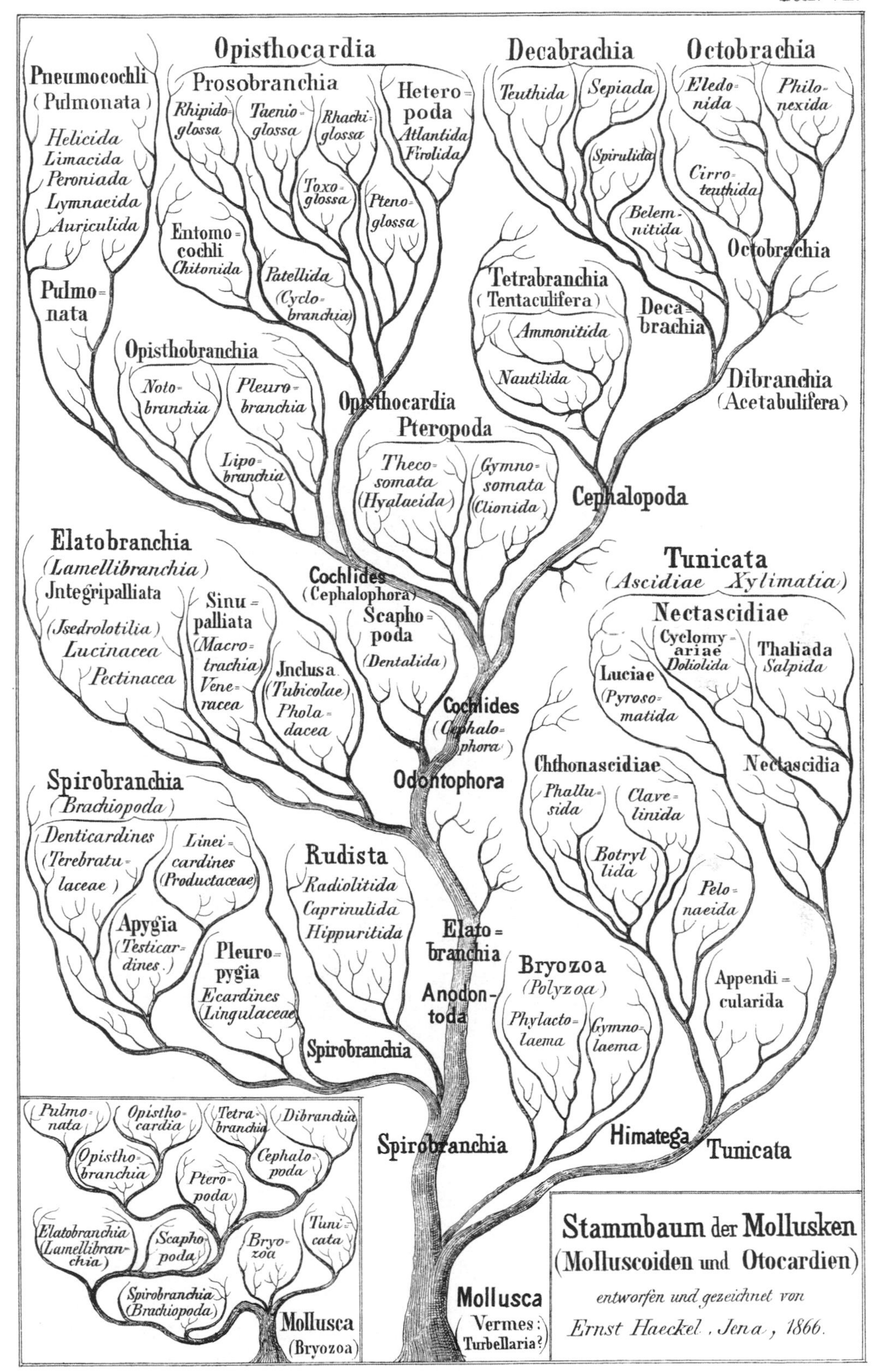

FIGURE 71. Ernst Haeckel's family tree of the mollusks, from 1866.

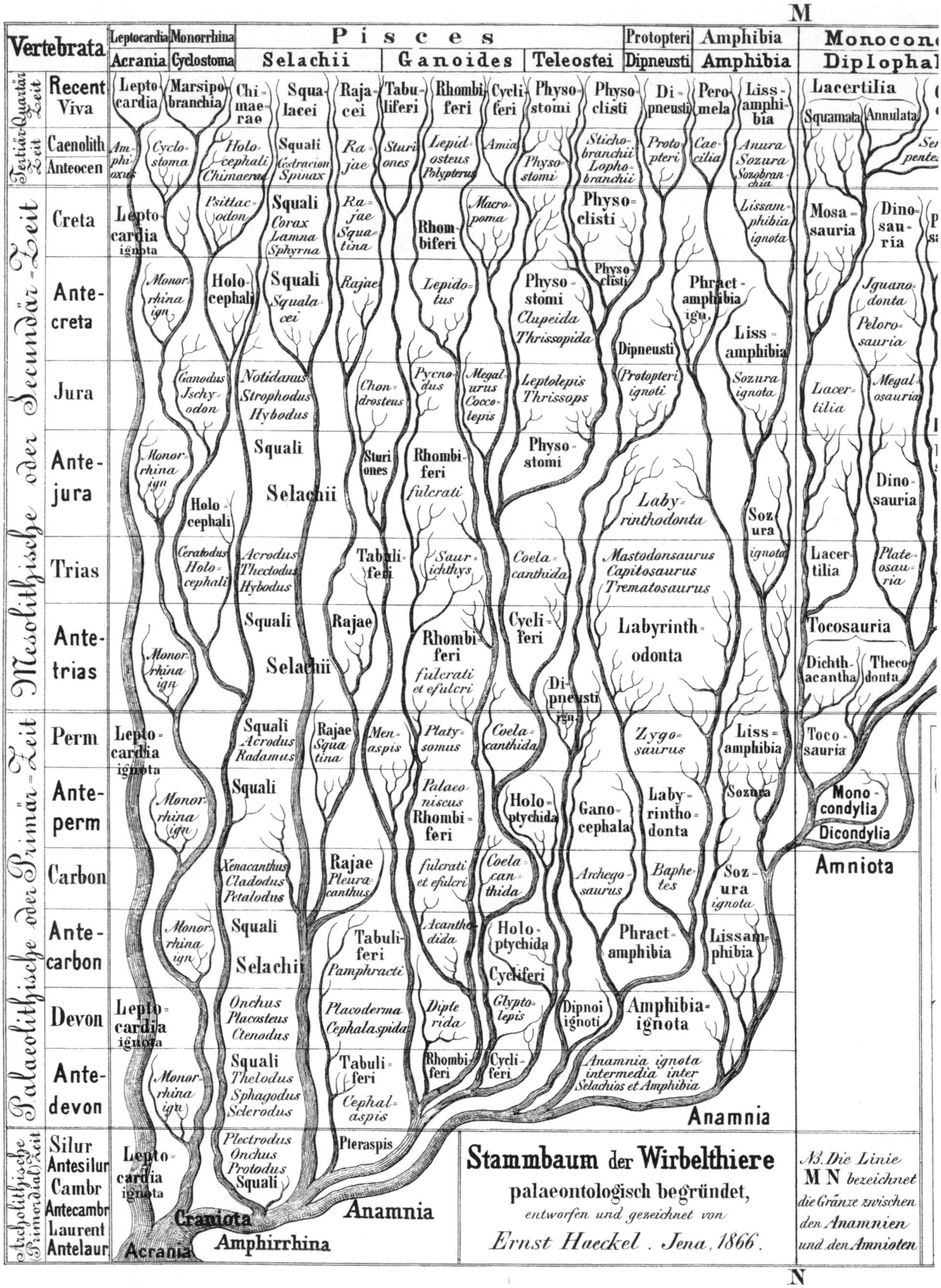

M

Vertebrata	Leptocardia	Monorrhina	Pisces				Protopteri	Amphibia	Monocon[dylia]
	Acrania	Cyclostoma	Selachii	Ganoides		Teleostei	Dipneusti	Amphibia	Diplophal[...]

Quartär Zeit — Recent, Viva: Leptocardia, Marsipobranchia, Chimaerae, Squalacei, Rajacei, Tabuliferi, Rhombiferi, Cycliferi, Physostomi, Physoclisti, Dipneusti, Peromela, Lissamphibia, Lacertilia (Squamata, Annulata)

Tertiär Zeit — Caenolith, Anteocen: Amphioxus, Cyclostoma, Holocephali Chimaerae, Squali Cestracion Spinax, Rajae, Sturiones, Lepidosteus Polypterus, Amia, Physostomi, Stichobranchii Lophobranchii, Protopteri, Caecilia, Anura Sozura Sozobranchia, Serpentes

Mesolithische oder Secundär-Zeit

Creta: Leptocardia ignota, Psittacodon, Squali Corax Lamna Sphyrna, Rajae Squatina, Rhombiferi, Macropoma, Physoclisti, Lissamphibia ignota, Mosasauria, Dinosauria

Antecreta: Monorrhina ign., Holocephali, Squali Squalacei, Rajae, Lepidotus, Physostomi Clupeida Thrissopida, Physoclisti, Phractamphibia ign., Dipneusti, Lissamphibia, Iguanodonta, Pelorosauria

Jura: Ganodus Ischyodon, Notidanus Strophodus Hybodus, Chondrosteus, Pycnodus, Megalurus Coccolepis, Leptolepis Thrissops, (Protopteri ignoti), Sozura ignota, Lacertilia, Megalosauria

Antejura: Monorrhina ign., Holocephali, Squali, Selachii, Sturiones, Rhombiferi fulcrati, Physostomi, Labyrinthodonta, Sozura ignota, Dinosauria

Trias: Ceratodus Holocephali, Acrodus Thectodus Hybodus, Tabuliferi, Saurichthys, Coelacanthida, Mastodonsaurus Capitosaurus Trematosaurus, Lacertilia, Plateosauria

Antetrias: Monorrhina ign., Squali, Selachii, Rajae, Rhombiferi fulcrati et efulcri, Cycliferi, Dipneusti ign., Labyrinthodonta, Tocosauria, Dichthacantha, Thecodonta

Palaeolithische oder Primär-Zeit

Perm: Leptocardia ignota, Squali Acrodus Radamus, Rajae Squatina, Menaspis, Platysomus, Coelacanthida, Zygosaurus, Lissamphibia, Tocosauria

Anteperm: Monorrhina ign., Squali, Palaeoniscus Rhombiferi, Holoptychida, Ganocephala, Labyrinthodonta, Sozura, Monocondylia, Dicondylia

Carbon: Xenacanthus Cladodus Petalodus, Rajae Pleuracanthus, fulcrati et efulcri, Coelacanthida, Archegosaurus, Baphetes, Sozura ignota, Amniota

Antecarbon: Monorrhina ign., Squali Selachii, Tabuliferi Pamphracti, Acanthodida, Holoptychida, Cycliferi, Phractamphibia, Lissamphibia

Devon: Leptocardia ignota, Onchus Placosteus Ctenodus, Placoderma Cephalaspida, Dipterida, Glyptolepis, Dipnoi ignoti, Amphibia ignota

Antedevon: Monorrhina ign., Squali Thelodus Sphagodus Sclerodus, Tabuliferi, Cephalaspis, Rhombiferi, Cycliferi, Anamnia ignota intermedia inter Selachios et Amphibia, Anamnia

Archolithische (Primordial) Zeit — Silur, Antesilur, Cambr, Antecambr, Laurent, Antelaur: Leptocardia ignota, Plectrodus Onchus Protodus Squali, Pteraspis, Anamnia, Craniota, Acrania, Amphirrhina

Stammbaum der Wirbelthiere
palaeontologisch begründet,
entworfen und gezeichnet von
Ernst Haeckel. Jena, 1866.

NB. Die Linie **M N** *bezeichnet die Gränze zwischen den Anamnien und den Amnioten*

N

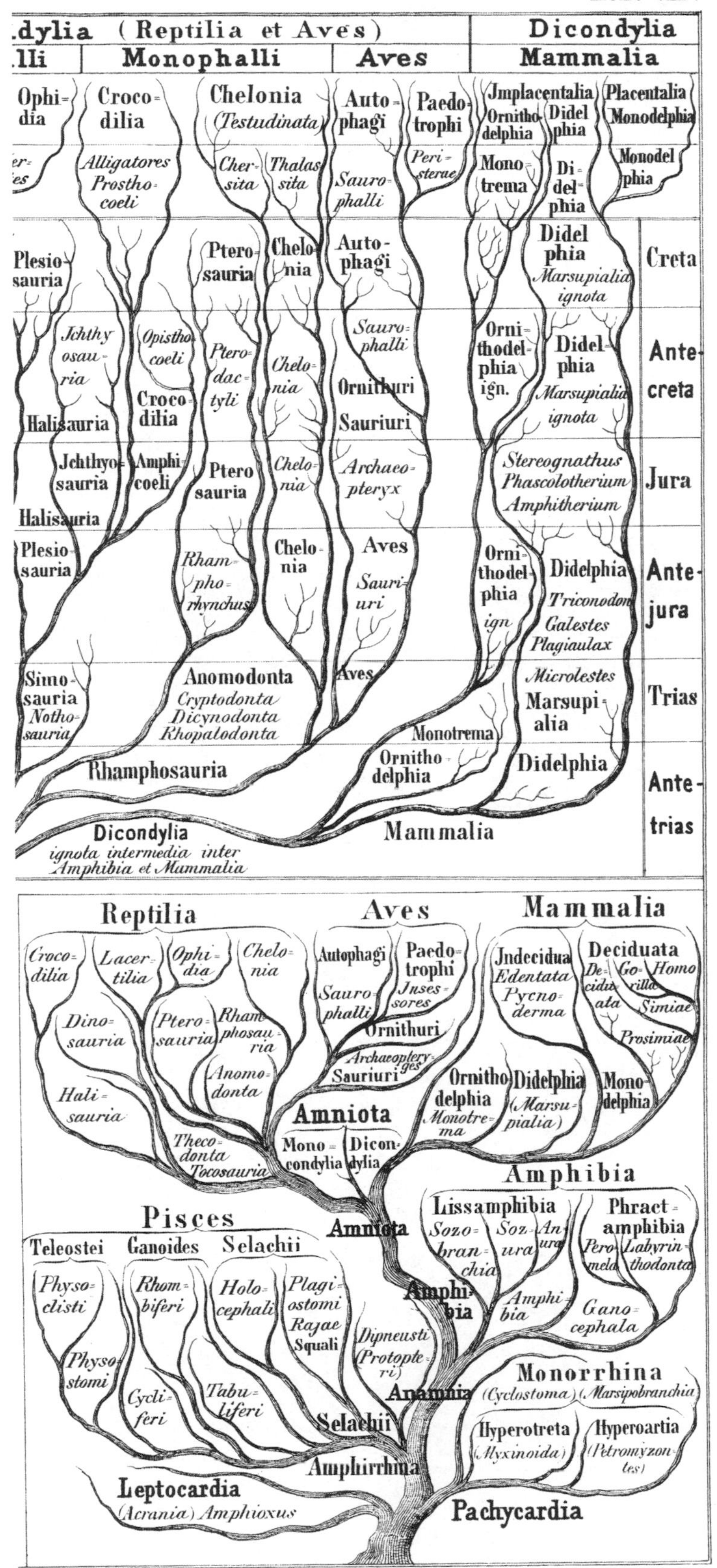

FIGURE 72. Ernst Haeckel's paleontological family tree of the vertebrates, from 1866. The thin vertical line labeled M-N that bisects the diagram separates anamniote vertebrates (fishes and amphibians, which lack the amnion during fetal development) on the left from amniotes (reptiles, birds, and mammals) on the right.

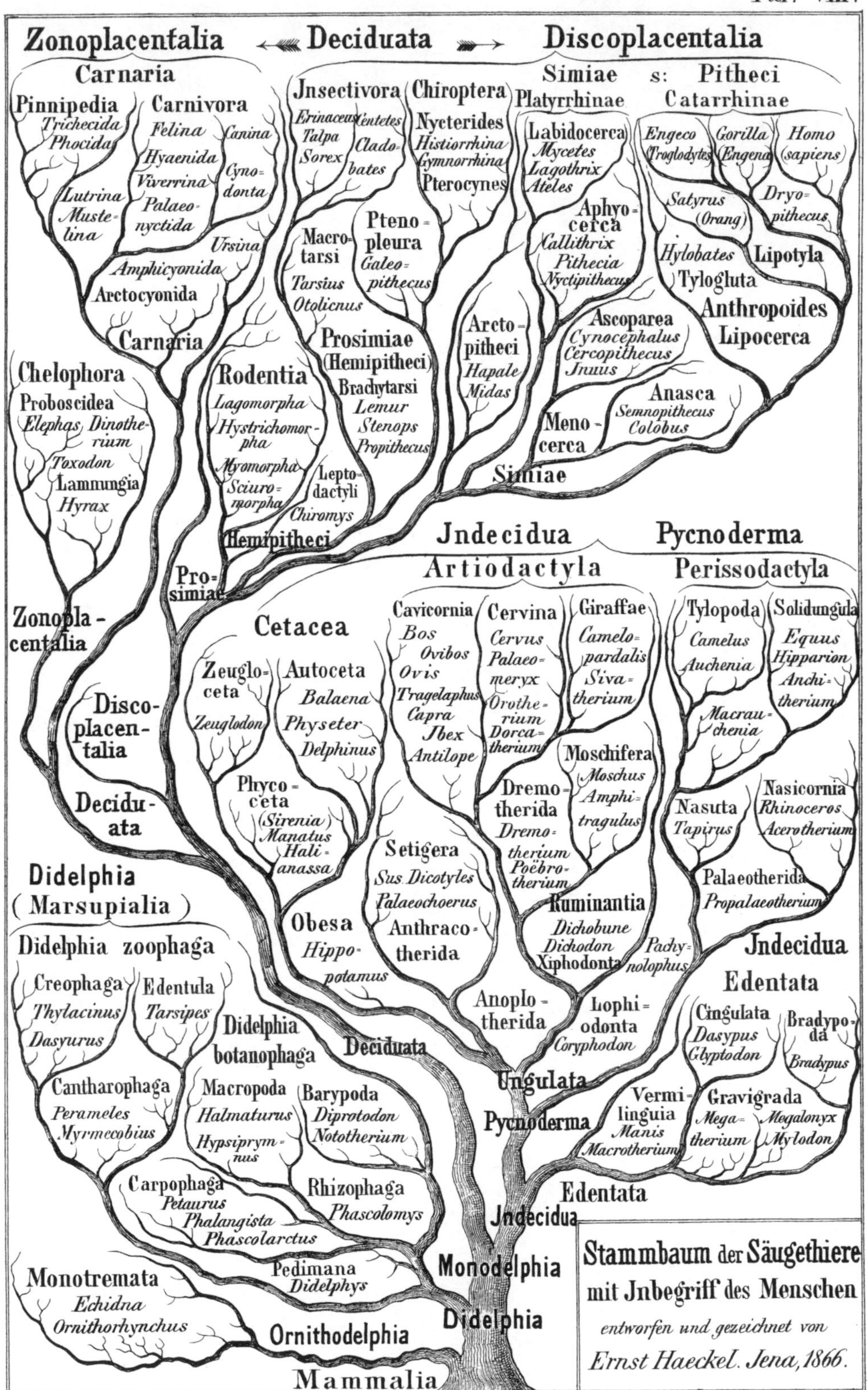

FIGURE 73. Ernst Haeckel's family tree of the mammals, from 1866.

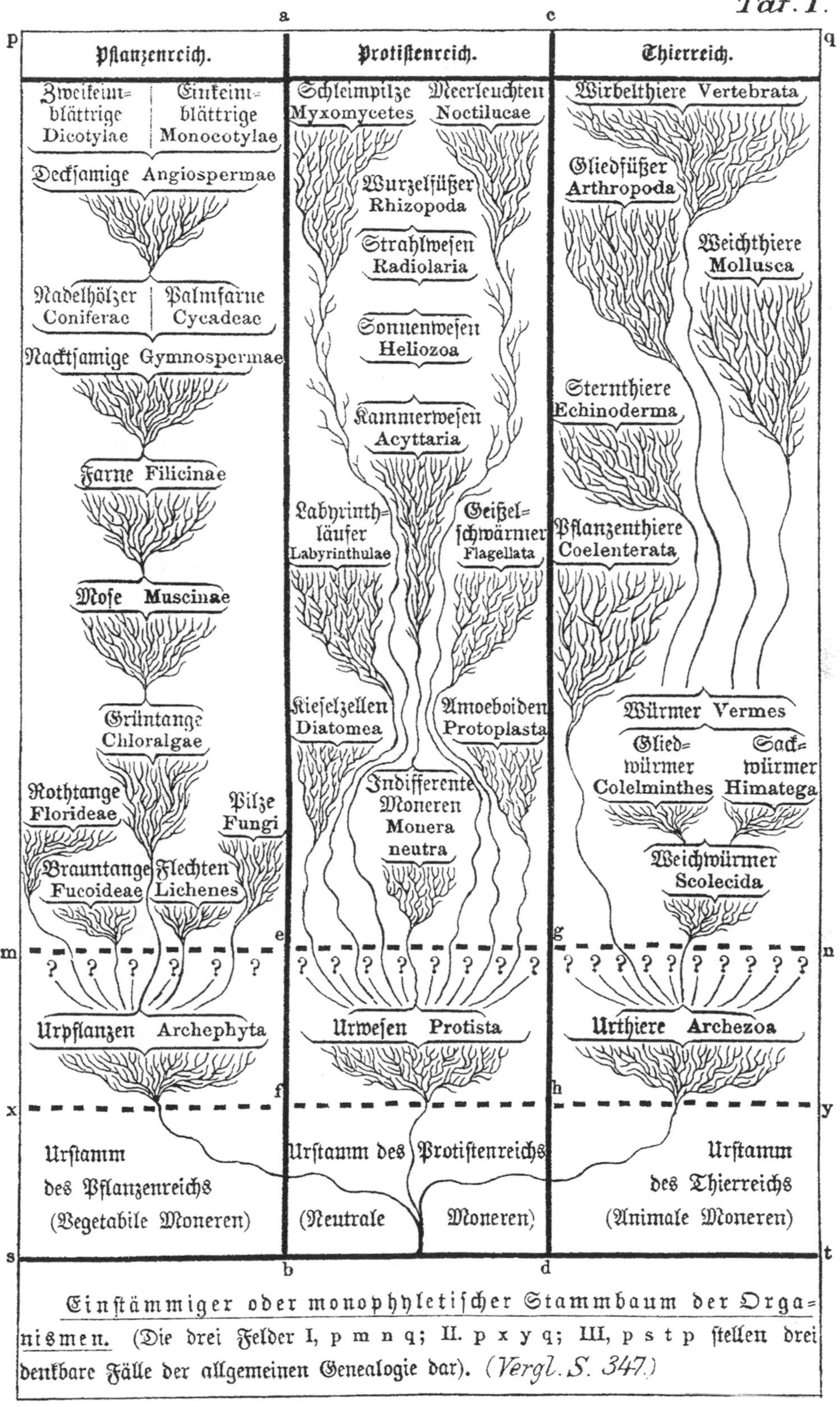

FIGURE 74. A monophyletic family tree of organisms, from Ernst Haeckel's 1868 *Natürliche Schöpfungsgeschichte* (Natural history of creation).

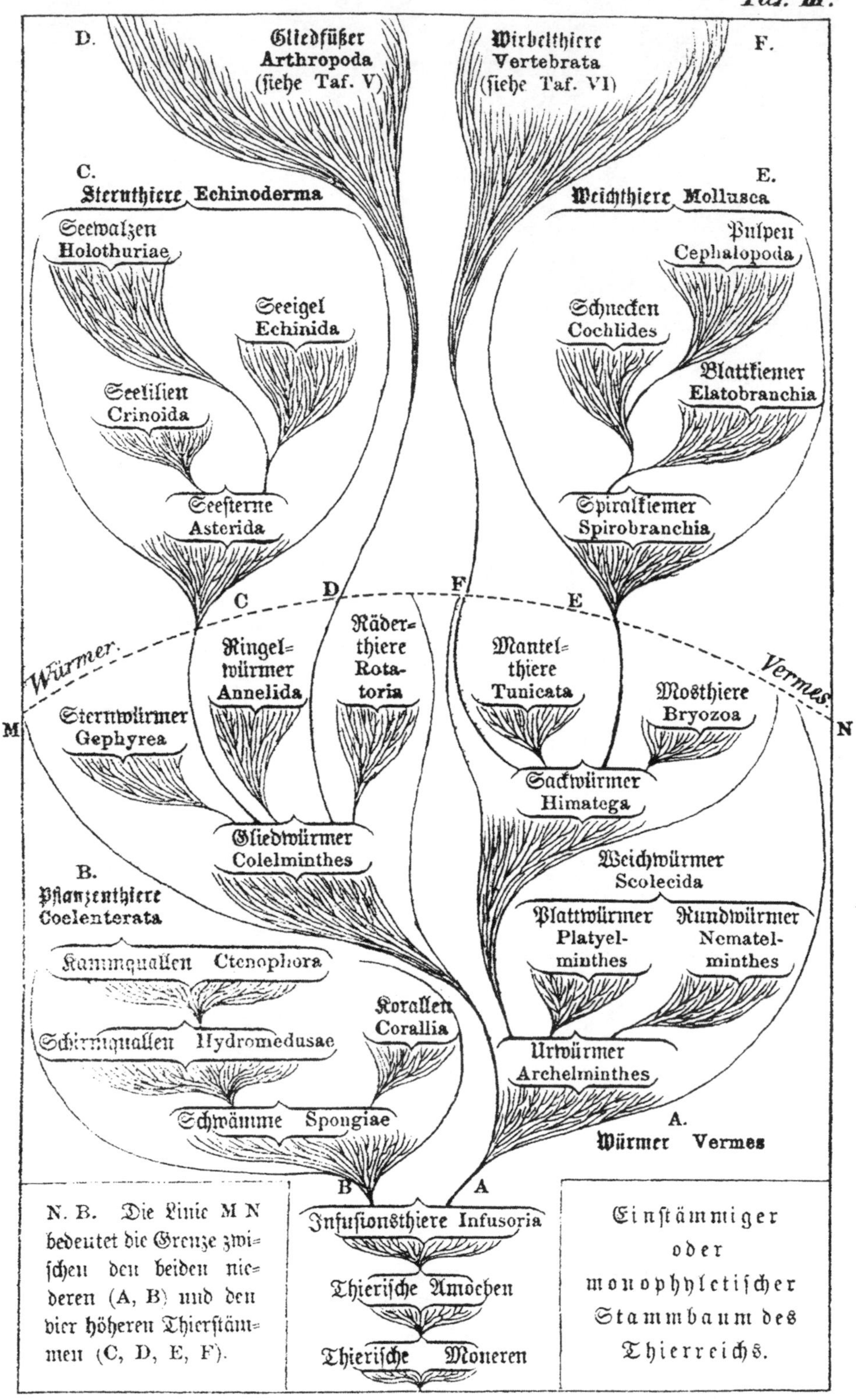

FIGURE 75. A monophyletic family tree of animals, published by Ernst Haeckel in 1868.

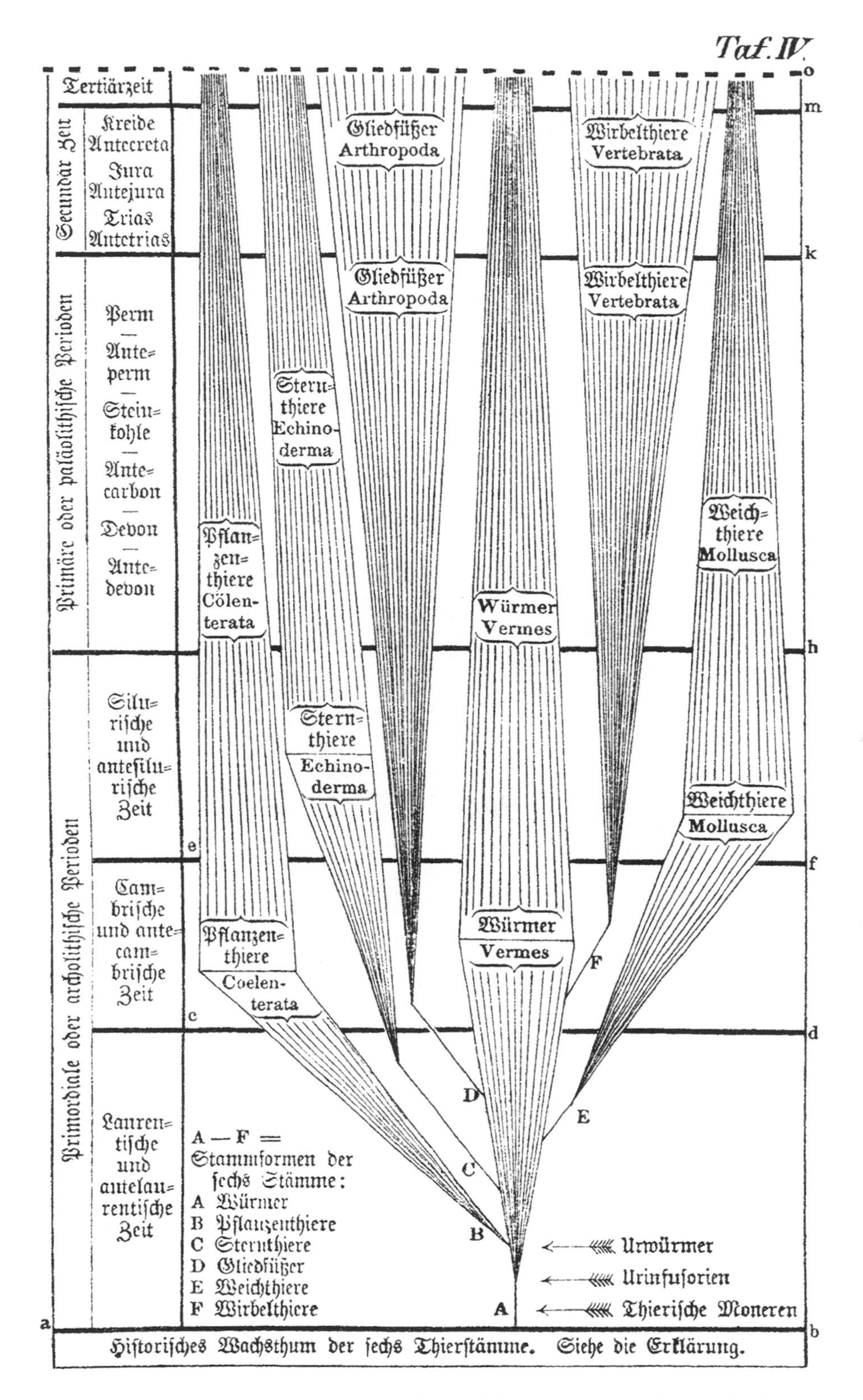

FIGURE 76. A family tree of relationships of six major groups of animals, the zoophytes (or plant-like animals, equivalent to the coelenterates), echinoderms, arthropods, worms, vertebrates, and mollusks, from Ernst Haeckel's 1870 second edition of *Natürliche Schöpfungsgeschichte* (Natural history of creation).

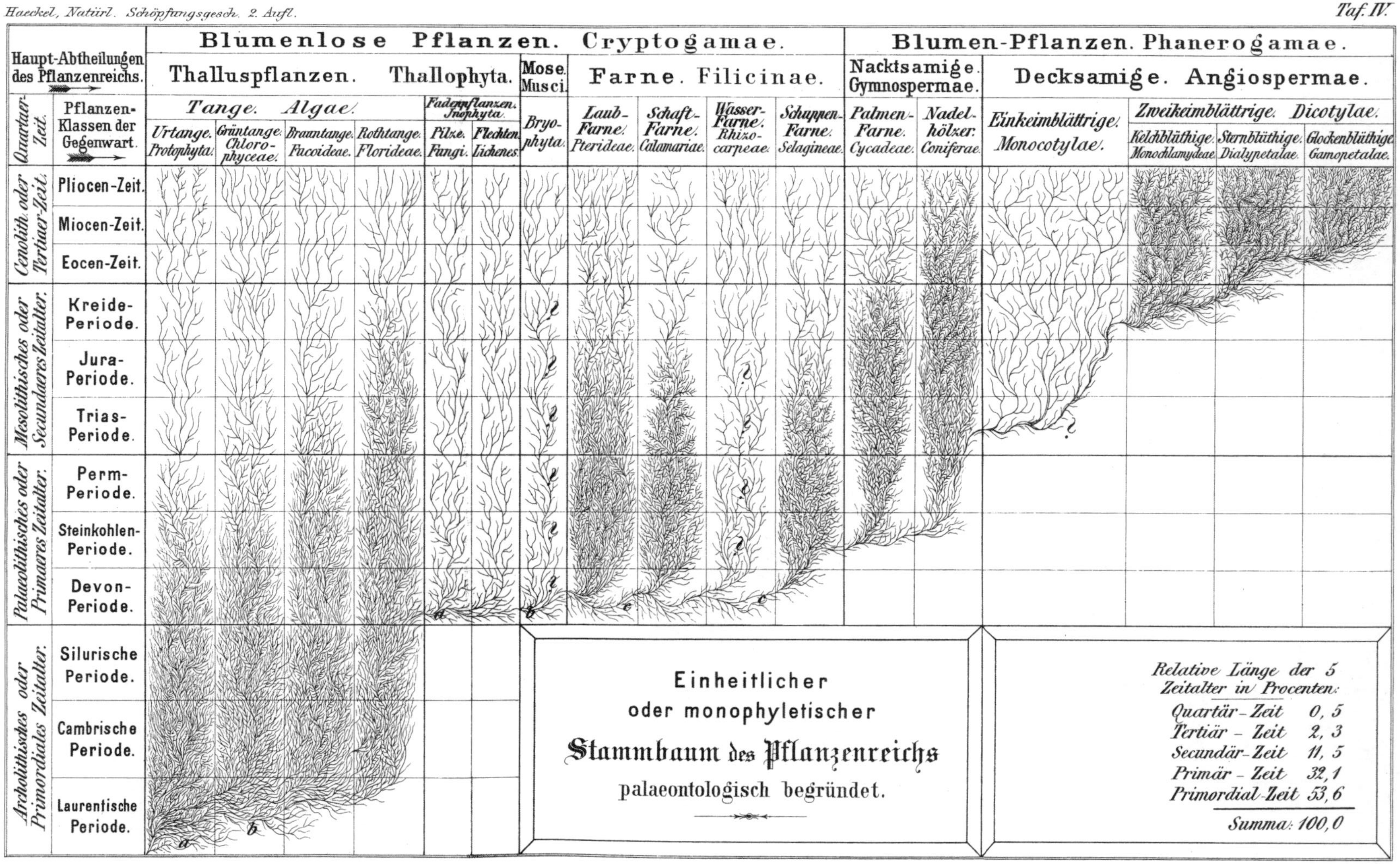

Haeckel, Natürl. Schöpfungsgesch. 2. Aufl.

Taf. IV.

Blumenlose Pflanzen. Cryptogamae.

Blumen-Pflanzen. Phanerogamae.

Haupt-Abtheilungen des Pflanzenreichs.

Thalluspflanzen. Thallophyta.

Mose. Musci.

Farne. Filicinae.

Nacktsamige. Gymnospermae.

Decksamige. Angiospermae.

Pflanzen-Klassen der Gegenwart.

Tange. Algae.

Fadenpflanzen. Inophyta.

Urtange. Protophyta.

Grüntange. Chlorophyceae.

Brauntange. Fucoideae.

Rothtange. Florideae.

Pilze. Fungi.

Flechten. Lichenes.

Bryophyta.

Laub-Farne. Pterideae.

Schaft-Farne. Calamariae.

Wasser-Farne. Rhizocarpeae.

Schuppen-Farne. Selagineae.

Palmen-Farne. Cycadeae.

Nadelhölzer. Coniferae.

Einkeimblättrige. Monocotylae.

Zweikeimblättrige. Dicotylae.

Kelchblüthige. Monochlamydeae.

Sternblüthige. Dialypetalae.

Glockenblüthige. Gamopetalae.

Quartär-Zeit.

Cenolith. oder Tertiaer-Zeit.

Pliocen-Zeit.

Miocen-Zeit.

Eocen-Zeit.

Mesolithisches oder Secundaeres Zeitalter.

Kreide-Periode.

Jura-Periode.

Trias-Periode.

Palaeolithisches oder Primaeres Zeitalter.

Perm-Periode.

Steinkohlen-Periode.

Devon-Periode.

Archolithisches oder Primordiales Zeitalter.

Silurische Periode.

Cambrische Periode.

Laurentische Periode.

a

b

c

Einheitlicher
oder monophyletischer
Stammbaum des Pflanzenreichs
palaeontologisch begründet.

Relative Länge der 5 Zeitalter in Procenten:

Quartär-Zeit	0, 5
Tertiär - Zeit	2, 3
Secundär-Zeit	11, 5
Primär - Zeit	32, 1
Primordial-Zeit	53, 6
Summa:	100, 0

FIGURE 77. A paleontological family tree of the plant kingdom, published by Ernst Haeckel in 1870.

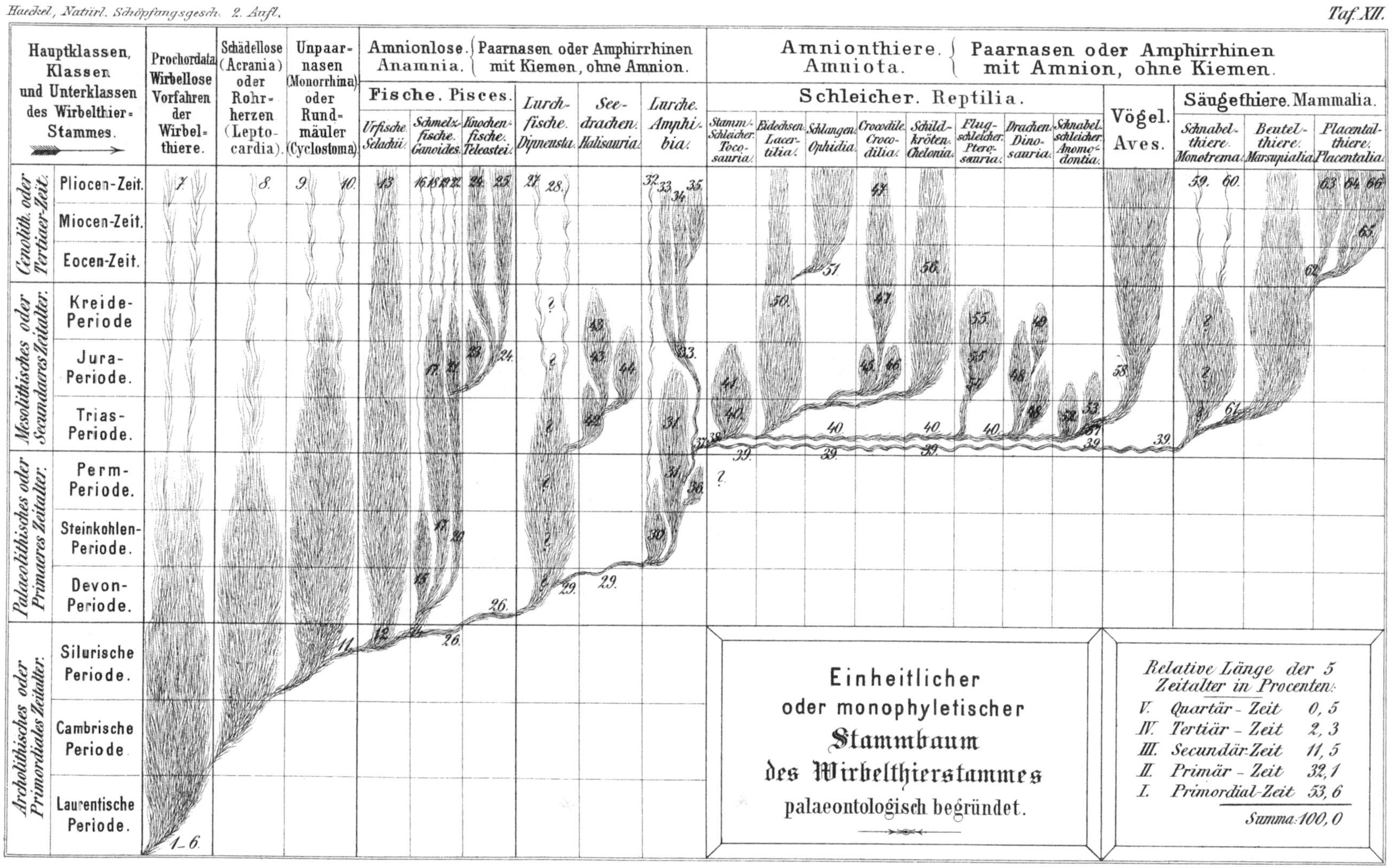

Haeckel, Natürl. Schöpfungsgesch. 2. Aufl.

Taf. XII.

Hauptklassen, Klassen und Unterklassen des Wirbelthier-Stammes.

Prochordata. Wirbellose Vorfahren der Wirbelthiere.

Schädellose (Acrania) oder Rohrherzen (Leptocardia).

Unpaarnasen (Monorrhina) oder Rundmäuler (Cyclostoma).

Amnionlose. Anamnia. { Paarnasen oder Amphirrhinen mit Kiemen, ohne Amnion.

Fische. Pisces.

Urfische. Selachii.

Schmelzfische. Ganoides.

Knochenfische. Teleostei.

Lurchfische. Dipneusta.

Seedrachen. Halisauria.

Lurche. Amphibia.

Amnionthiere. Amniota. { Paarnasen oder Amphirrhinen mit Amnion, ohne Kiemen.

Schleicher. Reptilia.

Stamm-Schleicher. Tocosauria.

Eidechsen. Lacertilia.

Schlangen. Ophidia.

Crocodile. Crocodilia.

Schildkröten. Chelonia.

Flugschleicher. Pterosauria.

Drachen. Dinosauria.

Schnabelschleicher. Anomodontia.

Vögel. Aves.

Säugethiere. Mammalia.

Schnabelthiere. Monotrema.

Beutelthiere. Marsupialia.

Placentalthiere. Placentalia.

Cenolith. oder Tertiaer-Zeit.: Pliocen-Zeit. Miocen-Zeit. Eocen-Zeit.

Mesolithisches oder Secundaeres Zeitalter.: Kreide-Periode. Jura-Periode. Trias-Periode.

Palaeolithisches oder Primaeres Zeitalter.: Perm-Periode. Steinkohlen-Periode. Devon-Periode.

Archolithisches oder Primordiales Zeitalter.: Silurische Periode. Cambrische Periode. Laurentische Periode.

Einheitlicher
oder monophyletischer
Stammbaum
des Wirbelthierstammes
palaeontologisch begründet.

Relative Länge der 5 Zeitalter in Procenten:

V. Quartär-Zeit 0,5
IV. Tertiär-Zeit 2,3
III. Secundär-Zeit 11,5
II. Primär-Zeit 32,1
I. Primordial-Zeit 53,6
Summa: 100,0

A. Schütze Lith. Inst. in Berlin.

FIGURE 78. A paleontological family tree of the vertebrates, published by Ernst Haeckel in 1870.

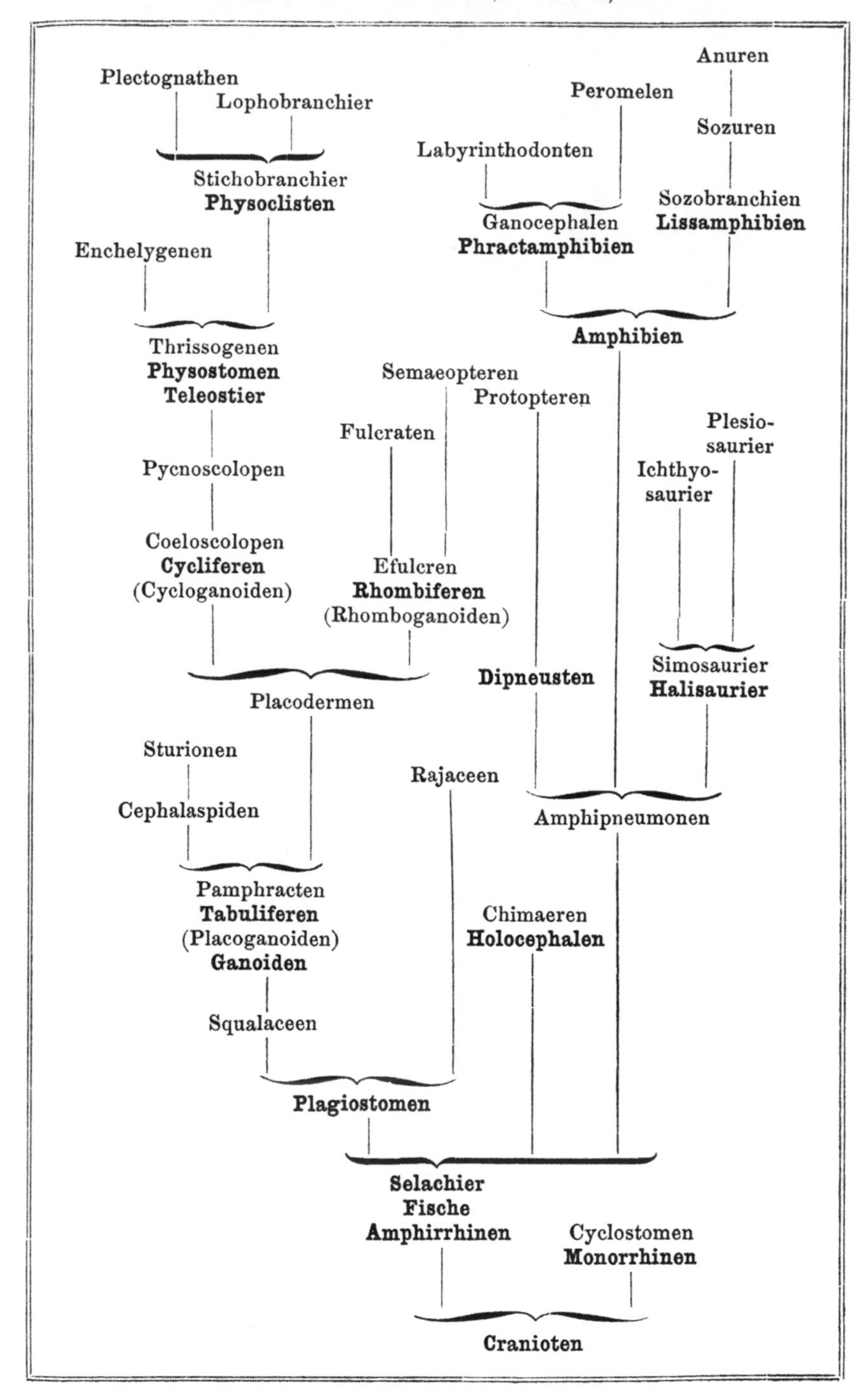

FIGURE 79. A family tree of amniote vertebrates employing the bracketed table style of the Renaissance naturalists, published by Ernst Haeckel in 1870.

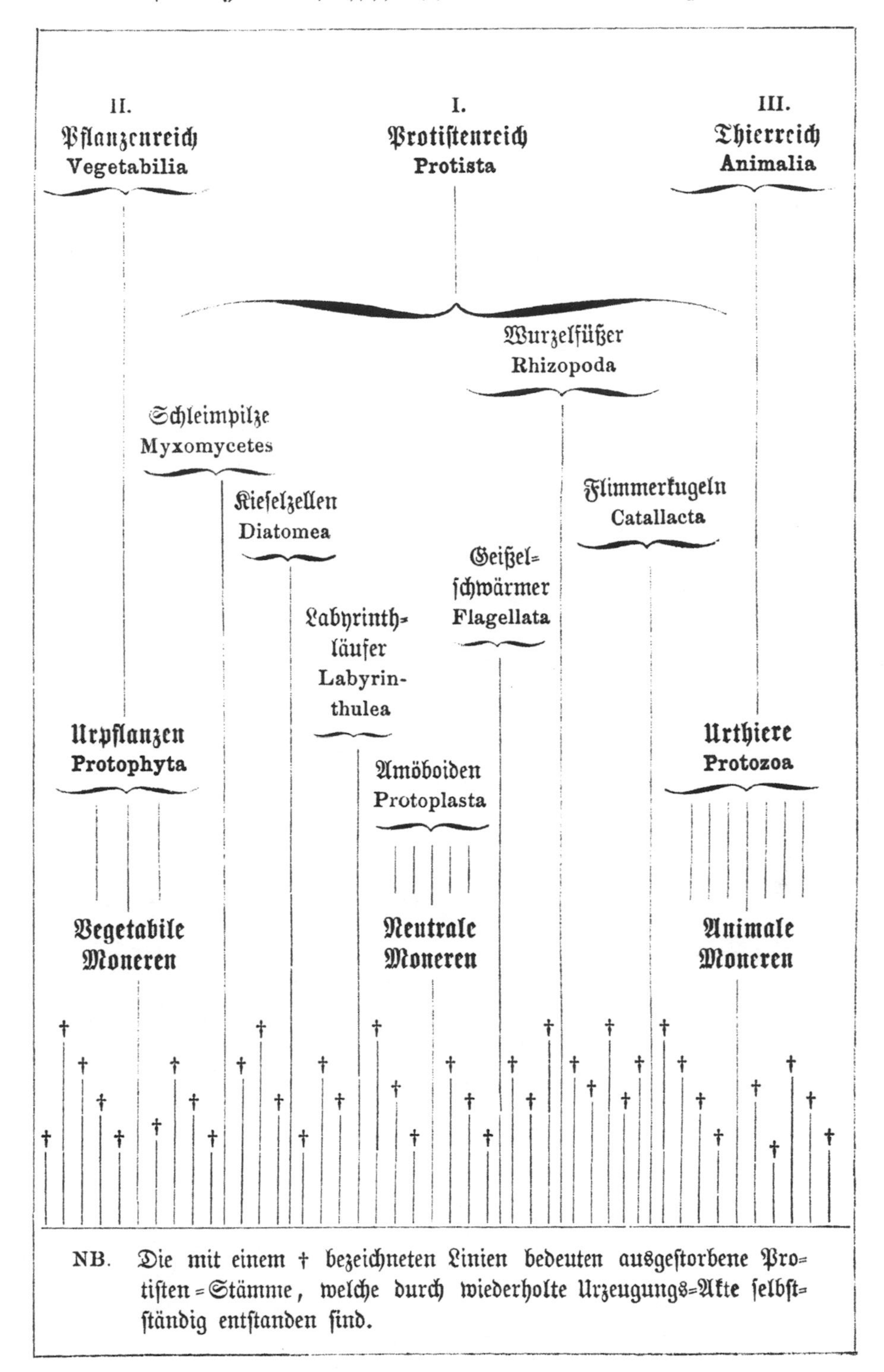

FIGURE 80. A polyphyletic family tree of organisms indicating multiple origins by way of spontaneous generation, and a far greater number of extinction events indicated by lineages terminating with the sign of a dagger, published by Ernst Haeckel in 1870.

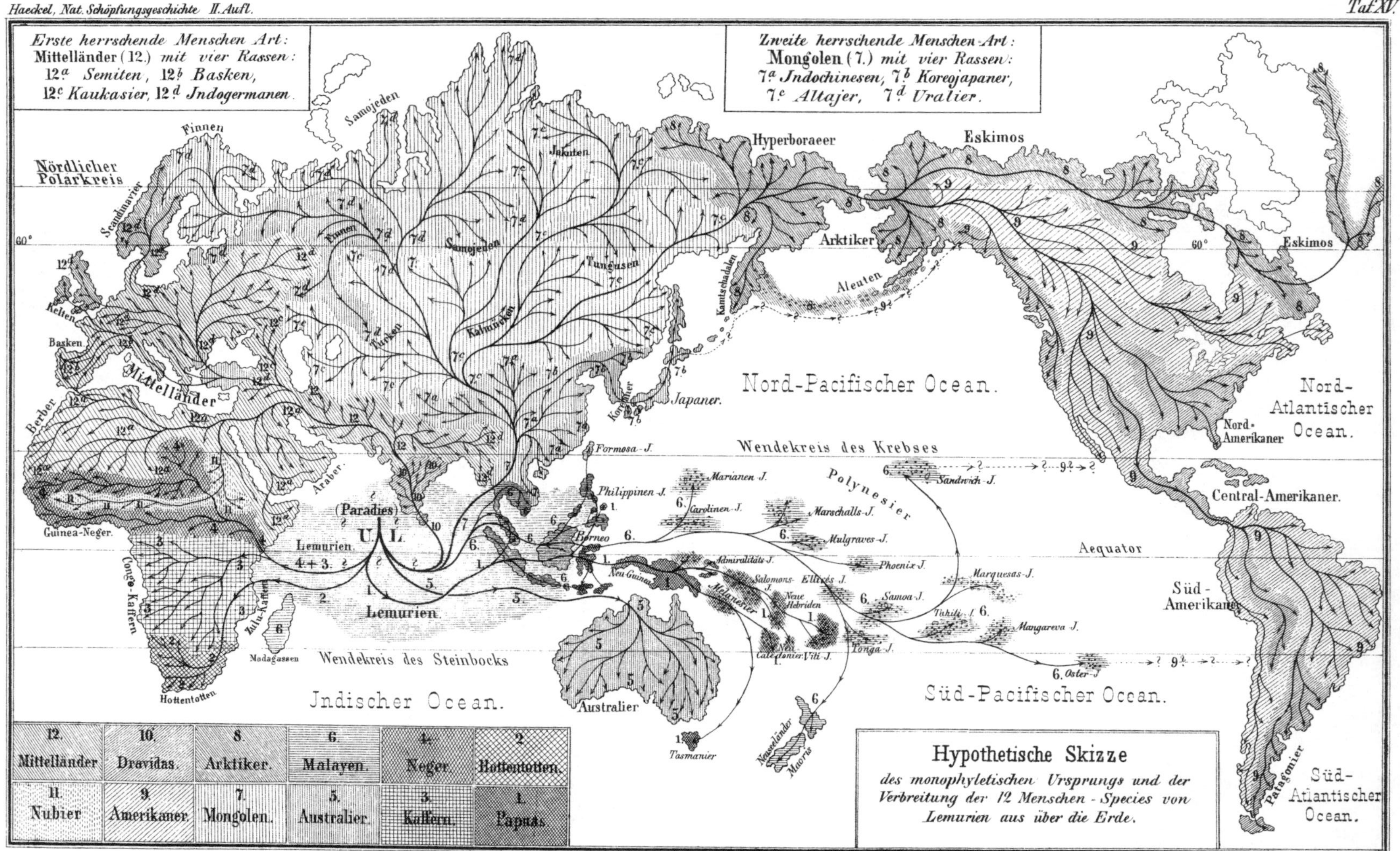

FIGURE 81. Ernst Haeckel's 1870 hypothetical sketch of the monophyletic origin and pattern of migration of what he argued were the twelve species and thirty-six races of humans. The root of the "tree" is situated at the site of the mythological, lost continent of Lemuria, described as "paradise," in the western Indian Ocean.

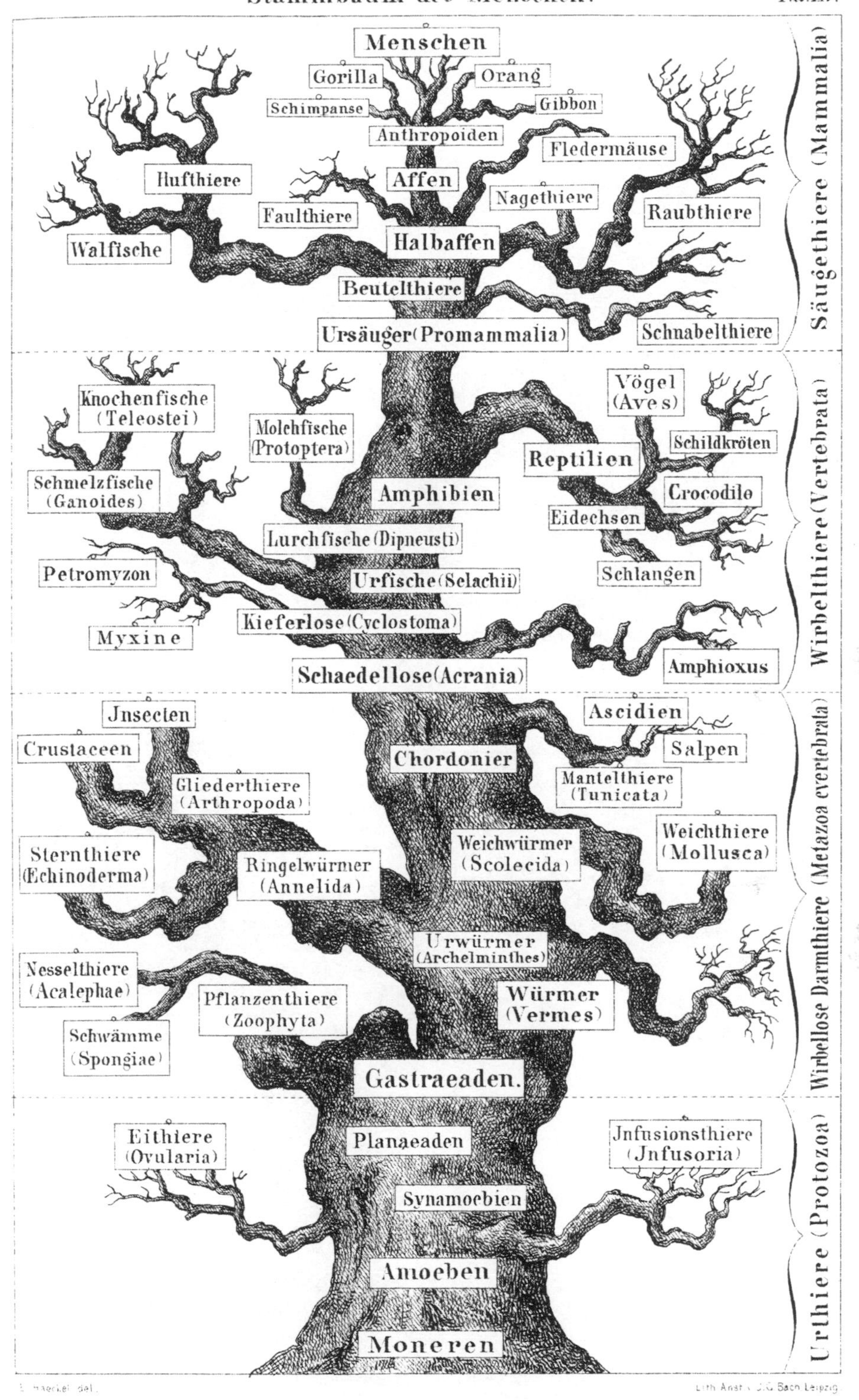

FIGURE 82. Ernst Haeckel's famous "great oak," a family tree of animals, from the first edition of his 1874 *Anthropogenie oder Entwickelungsgeschichte des menschen* (The evolution of man).

FIGURE 83. A blatantly racist tree of primate relationships, published by Ernst Haeckel in 1874.

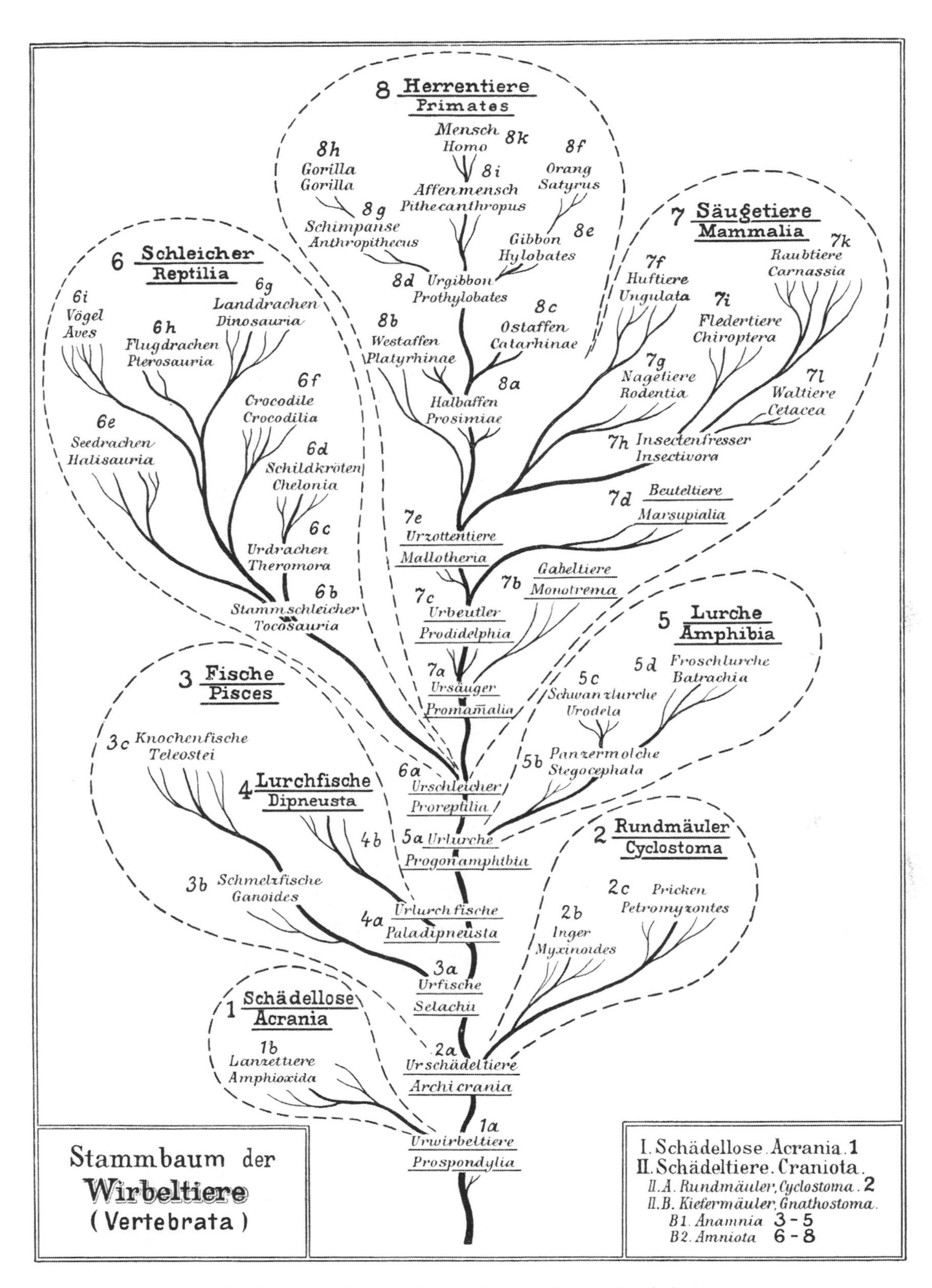

FIGURE 84. A family tree of the vertebrates, showing Ernst Haeckel's hypothetical ape-man, *Pithecanthropus,* ancestral to humans, from *Der Kampf um den Entwickelungs-Gedanken* (The struggle over evolutionary thought), 1905.

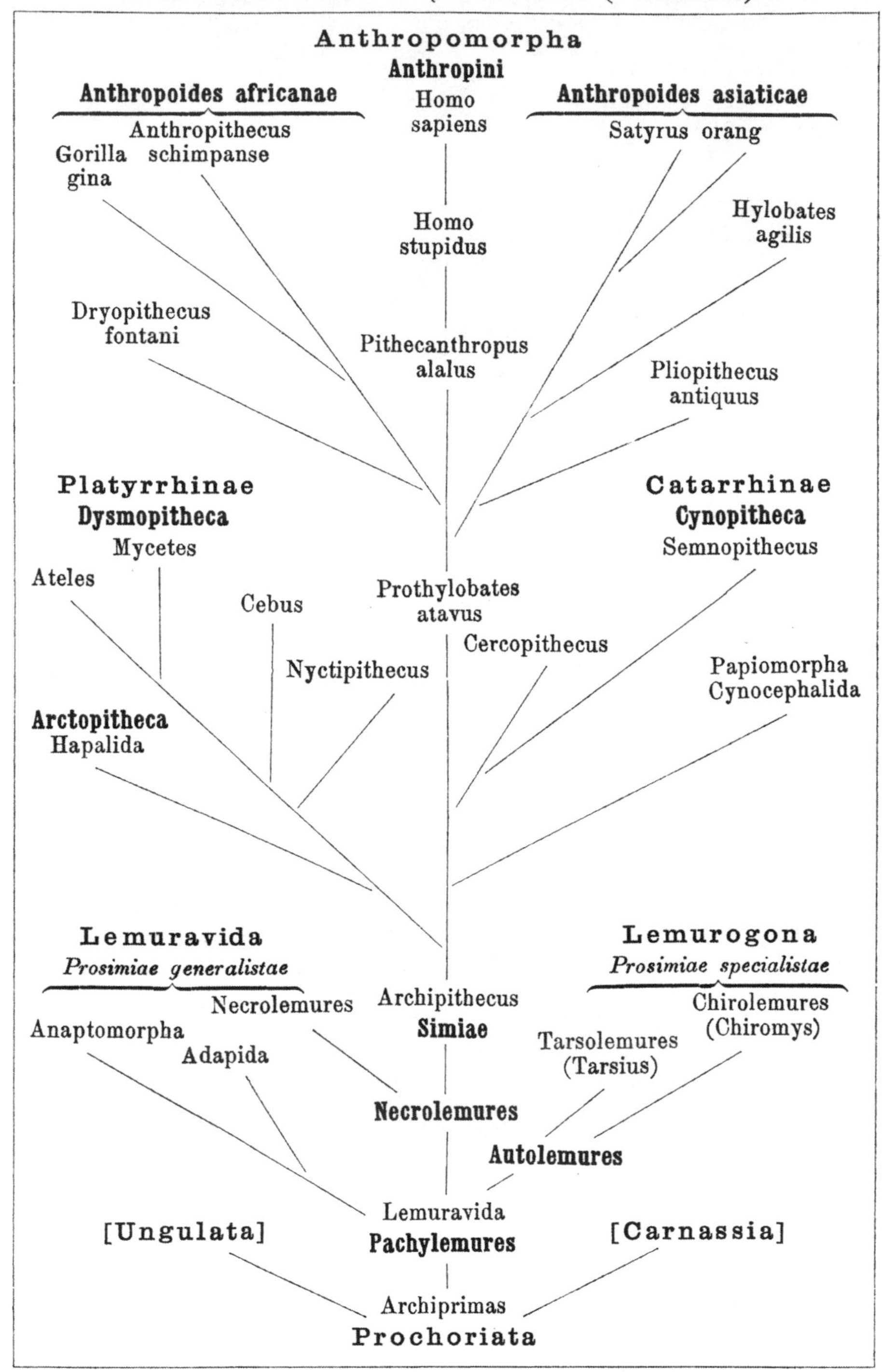

FIGURE 85. A family tree of the vertebrates, showing *Pithecanthropus alalus,* Ernst Haeckel's hypothetical speechless ape-man, ancestral to *Homo stupidus* (stupid man) and *Homo sapiens* (wise or knowing man), 1905.

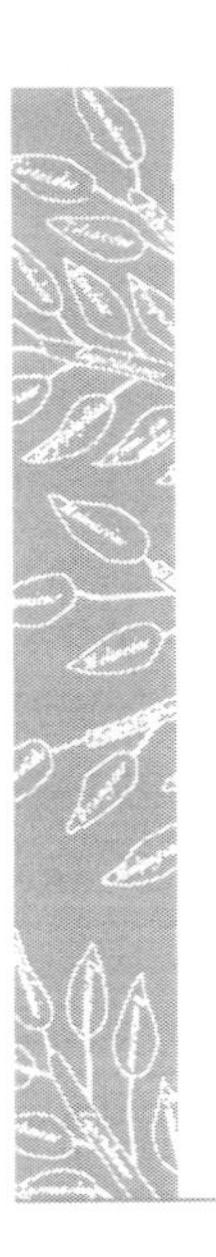

Post-Darwinian Nonconformists

1868–1896

By the early 1860s, and especially with evolutionary theory taking hold among all but a few vocal critics, most notions of maps and networks and geometric and symmetrical graphic displays of plant and animal relationships had been abandoned, but there were a few notable exceptions. One of these is Graceanna Lewis's (1821–1912) triangular view—or is it a cone?—of the animal kingdom published in 1868 (Figure 86). Lewis may well be the first woman to have constructed and published a tree of life. Although she accepted the fact of evolution, she was uncomfortable with Darwin's competitive mechanism of survival of the fittest, preferring to think of evolution as a progressive, benevolent unfolding of God's plan.[1] Her highly symmetrical tree—a central stem of ascending coils diverging dichotomously left and right, with "man" at the zenith of animal descent—reflects her firm belief in the pre-Darwinian assumption held by many that plant and animal relationships are susceptible to mathematical analysis: "There can be no doubt that both animal and vegetable life do group according to some law of numbers."[2]

Another holdover from an earlier time is George Bentham's (1800–1884) 1873 circle diagram of the thirteen tribes of the Compositae (or Asteraceae of current taxonomy), the largest family of vascular plants, containing, among other things, asters, daisies, and sunflowers (Figure 87). The "oldest" tribe, the Anthemideae, is placed at the bottom, with progressively "younger" tribes situated above.[3] In a kind of hybridization of the mapping and networking approaches exemplified by the diagrams of Buffon, Rüling, and Batsch (see Figures 17–19), with the circular imagery of Candolle, Newman, and Milne-Edwards (Figures 33,

43, and 50), relative affinity is indicated by juxtaposition of the circles, while a secondary level of relationship is expressed by a network of dotted lines.

Another circle diagram, a spin-off of the kinds of images constructed by the quinarians (see Figures 35–42), is British marine biologist William Saville-Kent's (1845–1908) relationships of the protozoa published in 1880 (Figure 88). Realizing that it is impossible in an "artificial, lineally arranged table to adequately and intelligibly illustrate the innumerable cross-relationships of lines of evolution that undoubtedly connect these various orders and classes with one another . . . a special diagrammatic scheme . . . has therefore been constructed . . . with the following explanation:"[4]

> Referring to the foregoing diagrammatic scheme, it will be observed that the four primary sections of the Pantostomata, Discostomata, Polystomata, and Eustomata, including their more important classes and orders . . . , are circumscribed by a broader and double circular line. Making use of a convenient metaphor, these circular sections with their varied contents may be compared to so many planetary systems or constellations, all derivable from one common centre and indicating at the points where their peripheries are made to intersect, their mutual relationship to, and interdependence on each other. The centre of the entire series and common source from whence, through the process of evolution, all the various types, orders, and classes of the Protozoa have probably through a more or less extensive epoch of time developed, is undoubtedly to be found among the most simply organized Pantostomata, finding there its typical embodiment in the amoeban order, the hypothetic primeval ancestor of which may, for convenience' sake, appropriately receive the generic name of *Protamoeba*.[5]

More reminiscent of the network approach of the likes of Rüling (see Figure 18) is Alfred William Bennett's (1833–1902) "affinities and classification of algae" published in 1887 (Figure 89). Here Bennett employs chains of taxa connected by lines of relationship that indicate levels of advancement.[6] A similar approach was employed in 1889 by Eduard Hackel (1850–1926) in his attempt to show relationships among grasses of the tribe Andropogoneae (Figure 90). Finally, among these examples of pre-Darwinian approaches to tree-making is Russian botanist Nicolai Ivanovich Kusnezov's (1864–1932) 1896 network of affinities of *Gentiana,* a large genus of herbs, nearly cosmopolitan in temperate regions of the world (Figure 91). Kusnezov specifically rejected the notion of depicting relationships as a branching tree,[7] favoring this reticulated, Batsch-like approach instead (compare with Figure 19).

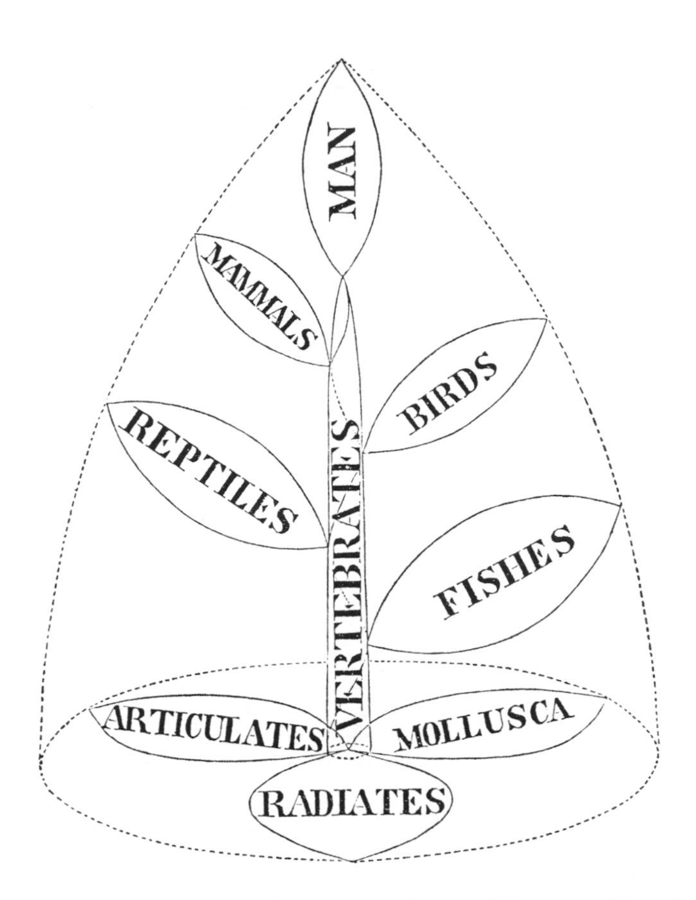

FIGURE 86. A geometric and symmetrical view of the animal kingdom published in 1868 by Graceanna Lewis, perhaps the first woman to construct and publish a tree of life.

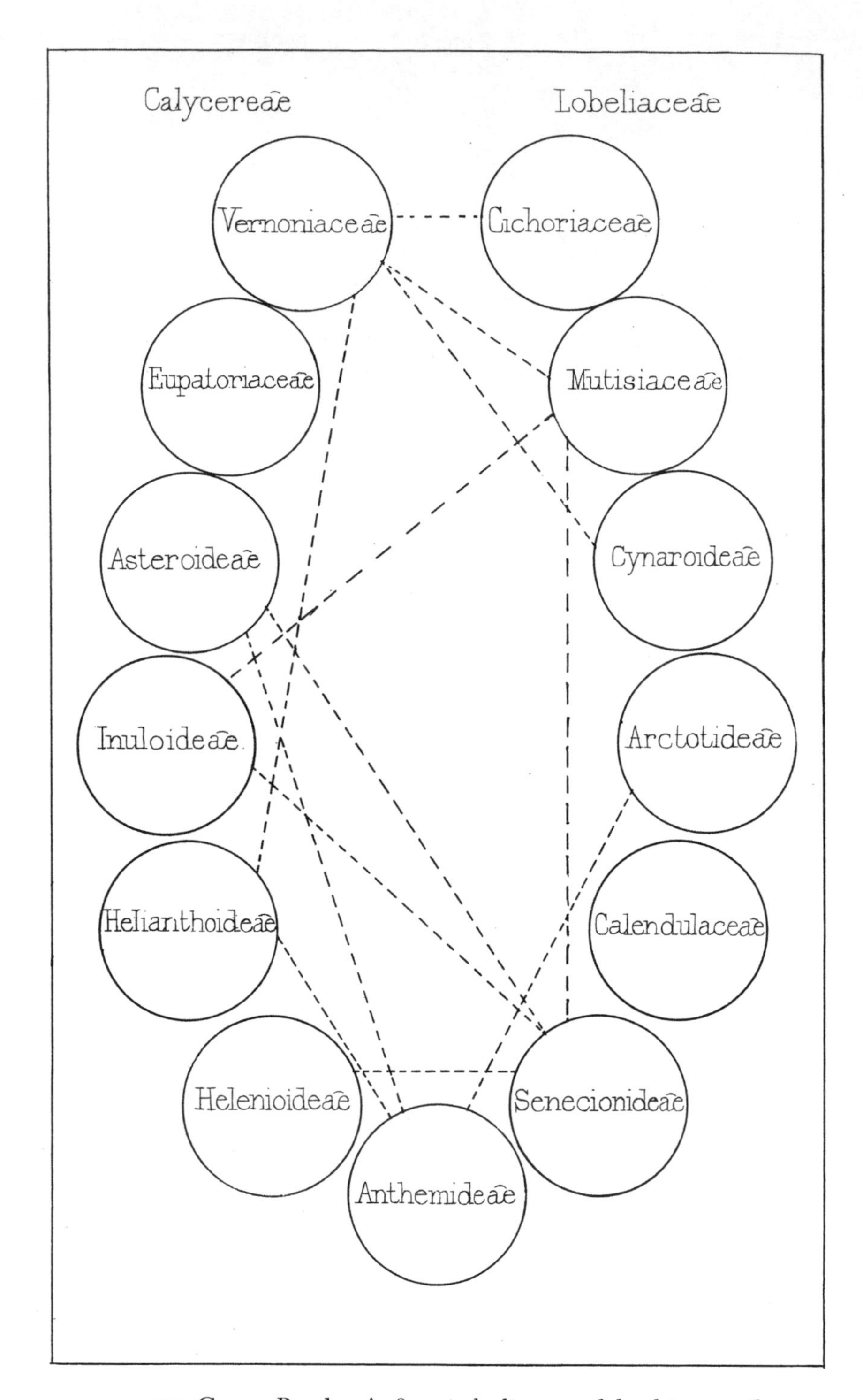

FIGURE 87. George Bentham's 1873 circle diagram of the thirteen tribes of the Compositae, then recognized as the largest family of vascular plants, containing, among other things, asters, daisies, and sunflowers.

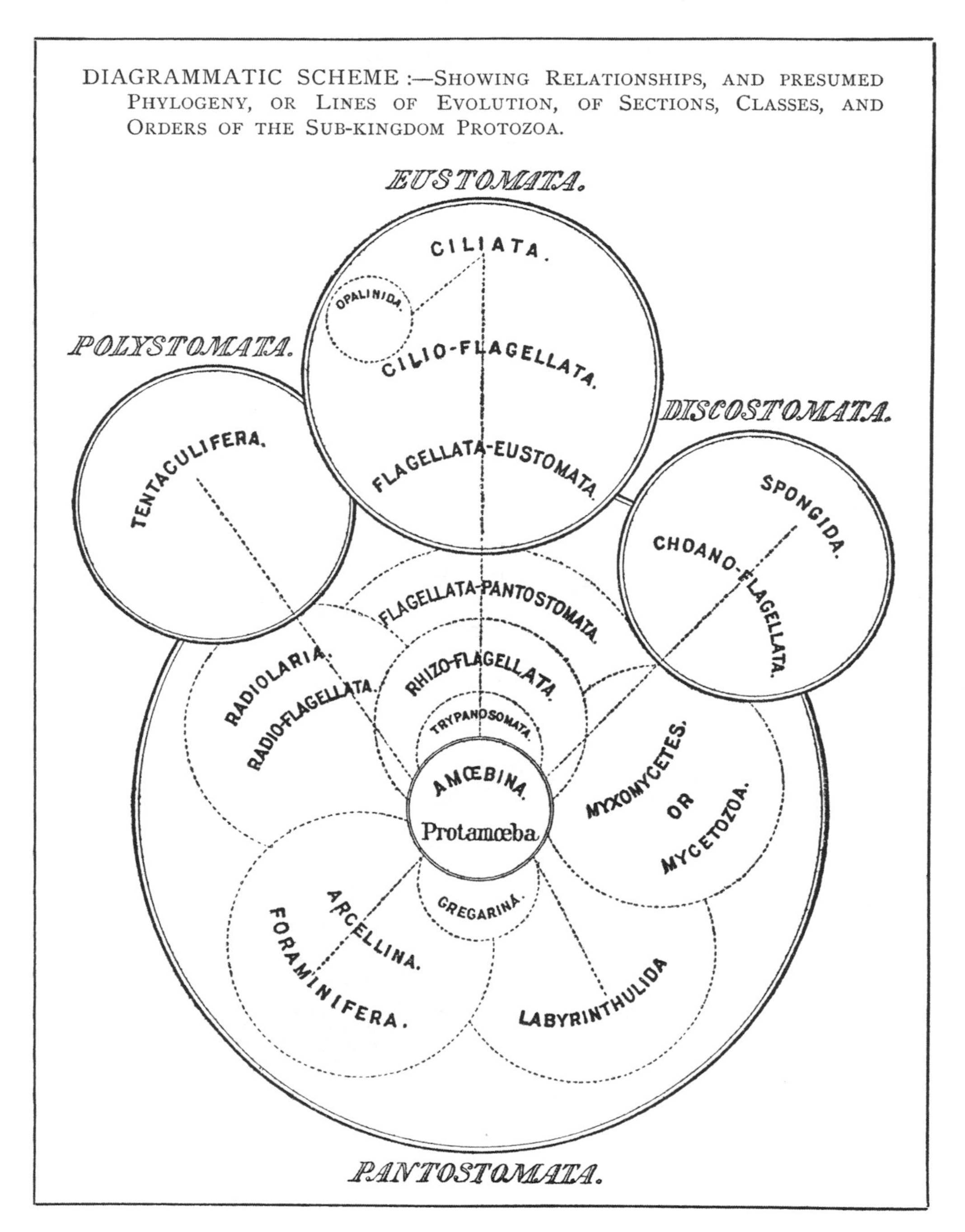

FIGURE 88. William Saville-Kent's relationships of the sections, classes, and orders of the Protozoa, 1880.

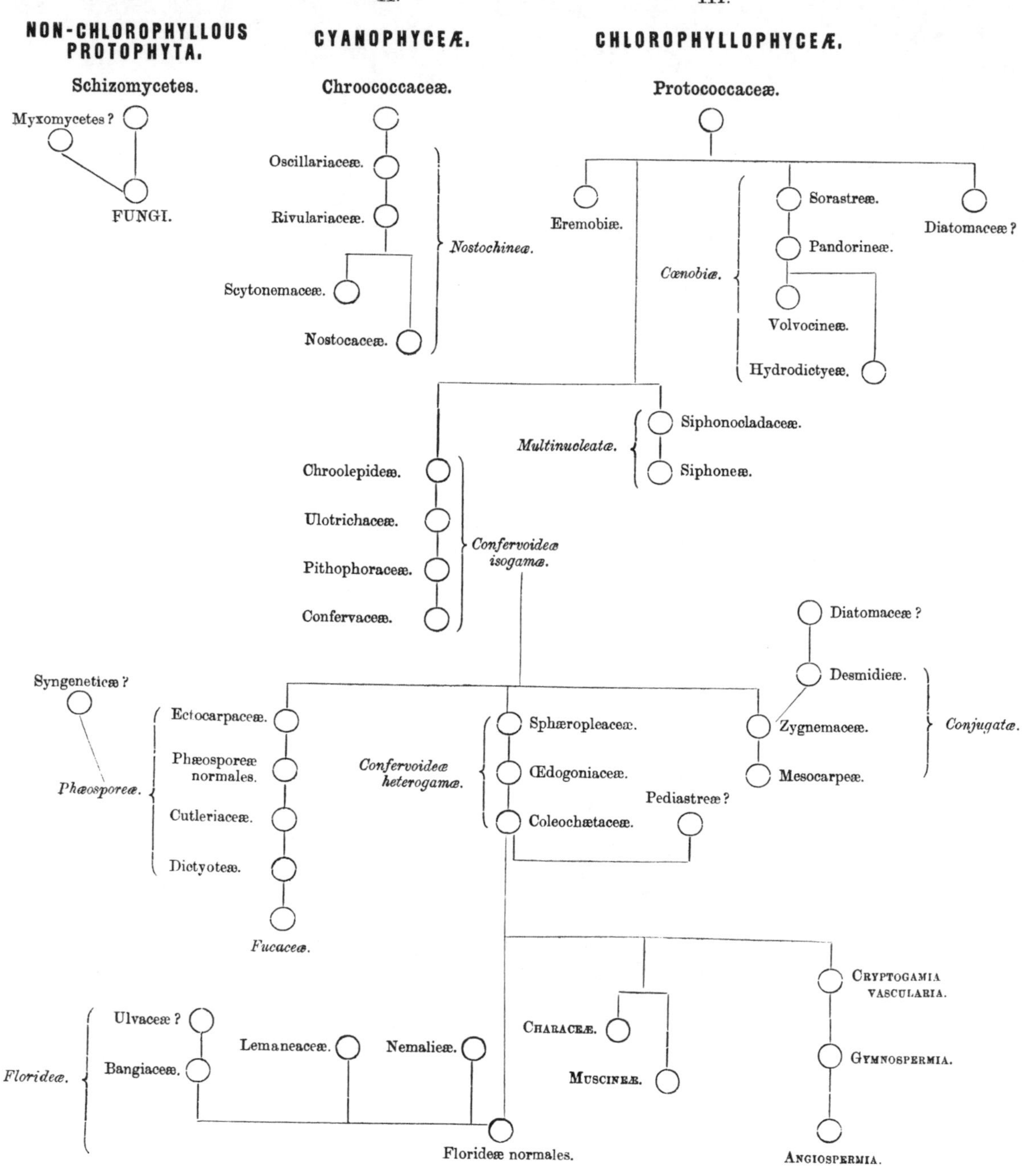

FIGURE 89. A network of the "affinities and classification of algae," by Alfred William Bennett in 1887.

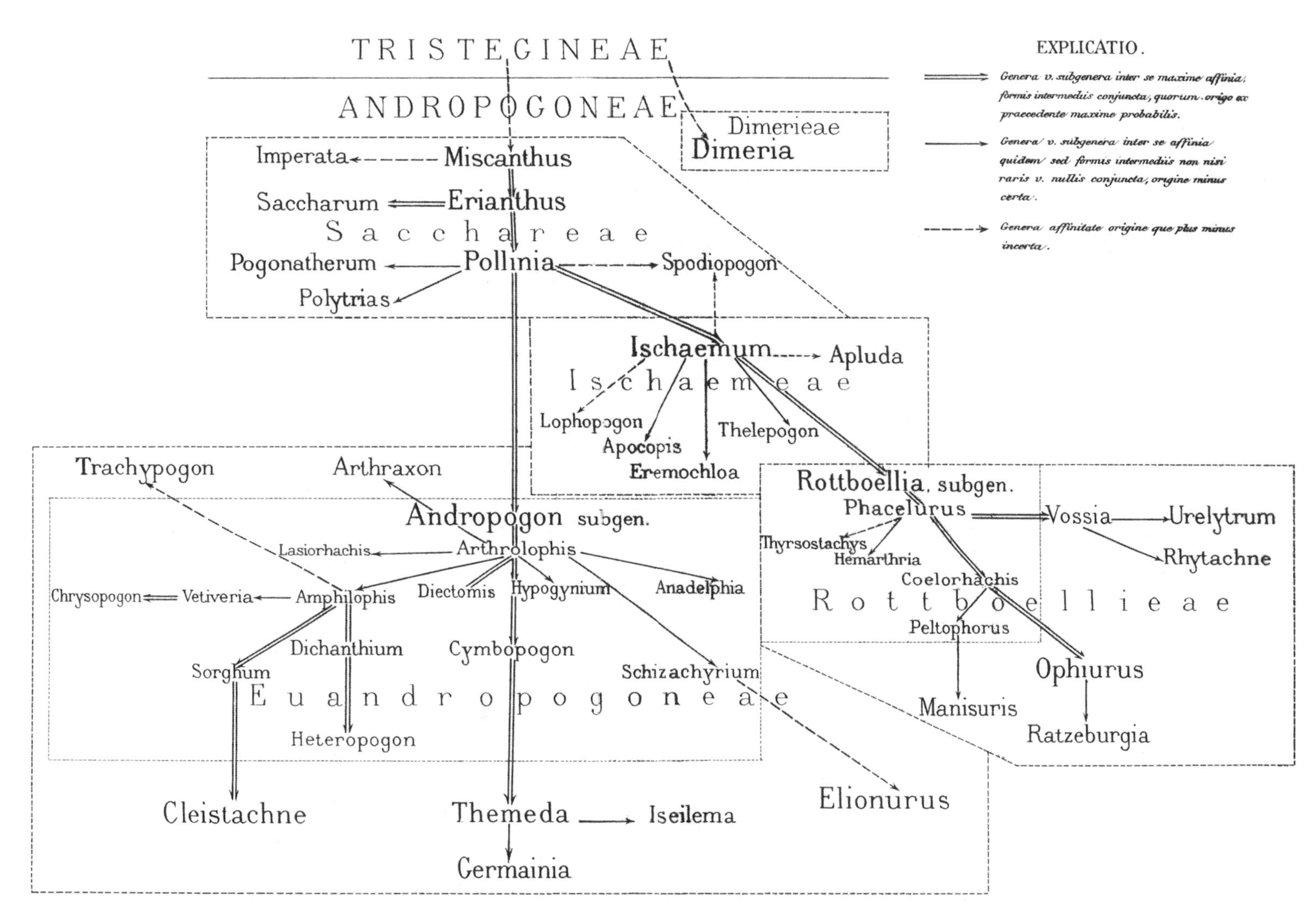

FIGURE 90. Eduard Hackel's 1889 attempt to show the relationships among grasses of the tribe Andropogoneae.

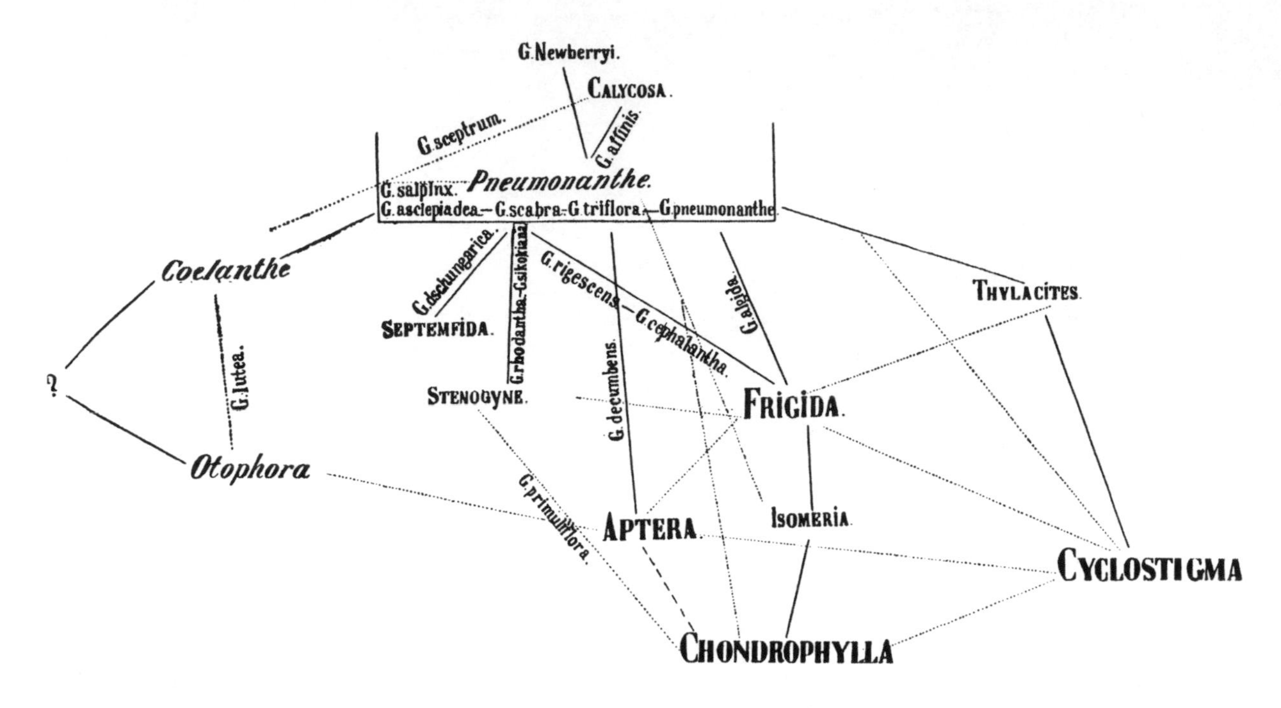

FIGURE 91. A network of affinities of *Gentiana,* a genus of herbs, by Nicolai Ivanovich Kusnezov in 1896.

More Late Nineteenth-Century Trees

1874–1897

German botanist Heinrich Gustav Adolf Engler (1844–1930) is well known for his beautifully illustrated taxonomic works. He is especially known for his plant classification, referred to as the Engler System, which remains the only system that includes all plants in the broad sense (algae to flowering plants) and that is still in use today. In his 1874 review of the Ochnaceae, a family mostly of tropical and subtropical trees and shrubs, he included a very botanical and modern-looking tree, with subfamilies indicated along the length of the branches and the genera lined up in a row along the top (Figure 92). The relative size or number of species contained within each genus is indicated by the diameter of a dot placed at the tip of each twig, the latter terminating at different levels to show their relative advancement.[1]

Engler's 1874 tree is presented in typical lateral view, but later in a stroke of genius he somehow contrived the notion of displaying trees from above, as if looking down on the "canopy" with a bird's-eye view, visualizing the lateral spread of branches from a central axis in what may well be the first three-dimensional tree. In his top-down view of the relationships of plants of the cashew, or sumac, family Anacardiaceae published in 1881 (Figure 93), concentric circles, each corresponding to a morphological feature, provide a measure of relative divergence from a common ancestor.[2] The idea of a tree is further demonstrated by the gradual narrowing of the branches toward their tips.[3]

Perhaps influenced by the innovations brought to tree construction by Heinrich Engler, German anatomist Maximilian Fürbringer (1846–1920) went much farther. His fantastic tree of bird relationships published in 1888—which, by the way, indicates a basal divergence from dinosaurs, a relationship that is commonly understood today—is a triumph of detail and complexity in itself, including all birds, living and fossil.[4] But after displaying the relationships in lateral view (Figure 94), he turned it around 180 degrees and visualized it again from the other side (Figure 95). Dividing each of these lateral projections into lower, middle, and upper "horizons," he then provided corresponding vertical projections or slices through the branches resulting in three map-like diagrams (Figures 96–98), reminiscent of Paul Giseke's free-floating bubble diagram of 1792 (see Figure 16).

Another view of bird relationships, strikingly different from that of Fürbringer, is German ornithologist Anton Reichenow's (1847–1941) "Phylogenetic Tree of the Class Aves" published in 1882 (Figure 99). Best known for his several attempts to classify birds, he proposed, in this early example, four primary lineages or "stems," together encompassing seven series and seventeen orders, all arising from "primaeval birds with teeth."[5] The first "stem" contained only the flightless birds (*Struthiones*); the second gave rise to the swimming birds (*Natatores*) and shorebirds (*Grallatores*); the third, pigeons, doves, and their allies (*Gyrantes*); and the fourth led to gamebirds (*Captatores*), parrots, woodpeckers and their allies (*Fibulatores*), and all the passerines or perching birds (*Arboricolae*). Much later, in his *Handbuch der systematischen Ornithologie* (Handbook of systematic ornithology) of 1913, he reduced the number of series to six, but, unfortunately, despite his efforts and prominence in the ornithological world, none of his proposals were ever adopted by any of his colleagues.[6]

English zoologist Alfred Henry Garrod (1846–1879) is remembered for his work in ornithology, but perhaps more so for his contributions to systematic methodology. In 1874, he published a "pedigree tree" of the parrots that is remarkable for showing character-state distributions for the first time (Figure 100).[7] In a surprisingly modern approach, Garrod reconstructed a hypothetical ancestral parrot by examining morphological variation in parrots and related bird groups and then applying a kind of out-group comparison. Once the characters of the parrot ancestor were established, he identified the groups (Arinae and Palaeornithinae) that differ the least from the ancestor and recognized them as the two primary branches of his tree. The other groups then branch off from these based on additional differences from the ancestor.[8] To further illustrate his classification, he constructed a nested set of circles that correspond with the hierarchy shown in his tree (Figure 101). Character states and a list of corresponding taxa are indicated within each circle. According to Harvard evolutionary biologist Robert O'Hara,

> Garrod's work was novel in that he specifically tried to reconstruct the characters of ancestral taxa and chart the course of character change through evolution. Indeed, Garrod's whole emphasis on characters and character change is relatively novel; previous workers had tended to speak of taxa being in their entirety "close to" or "far from" one another, rather than differing and being similar in particular respects.[9]

Johann Adam Otto Bütschli (1848–1920) was a German zoologist and professor at the University of Heidelberg, well known for his work on invertebrate animals, especially protozoans.[10] In a detailed study of the relationships of arthropods and various groups of "worms," published in 1876, he carefully analyzed various bodily parts, describing homologous versus nonhomologous characters based on criteria such as ontogenetic development and anatomical position within the body.[11] In placing the Annulata—nearly equivalent to the Annelida, including earthworms and leeches—and Arthropoda on distantly related lineages (Figure 102), he argued that segmentation, and related similarities in the nervous system, evolved independently in the two groups.[12] His tree is one of the few ever published with the root placed at the top.

The next two examples are not very pretty or exciting in themselves, but they are included here because they each signify a landmark occurrence or discovery. Two plain-looking "genealogical trees" published by British zoologist Edwin Ray Lankester (1847–1929) in 1881 summarize the relationships of arachnids (spiders and their allies) on the left and those of the larger assemblage of arthropods (arachnids, insects, and crustaceans) on the right (Figure 103). Of great interest in the arachnid genealogy is the inclusion of the Xiphosura, an order containing *Limulus,* the horseshoe crab. Thus, Lankester, primarily a comparative anatomist working mostly on invertebrates, was the first to reject the notion that the horseshoe crab (genus *Limulus*) is a crustacean, providing evidence instead of a close relationship to spiders and their allies, the Arachnida: "*Limulus* is no Crustacean, but simply and unreservedly an Arachnid . . . best understood as an aquatic scorpion."[13]

Again, seemingly uninteresting is Louis Antoine Marie Joseph Dollo's (1857–1931) phylogeny of jawed vertebrates published in 1896 (Figure 104). Dollo, a French-born Belgian paleontologist, best known for formulating Dollo's law—which states that evolution is not reversible[14]—produced the first detailed systematic work on lungfishes, summarizing all earlier research on the interrelationships of coelacanths, lungfishes, and tetrapods.[15] His tree is the first to show lungfishes (*Dipneustes*) as most closely related to tetrapods (*vie terrestre*), with coelacanths (*Crossoptérygiens*) as ancestral to both, a hypothesis that is general accepted today based on both morphological and molecular evidence.[16] He also clearly separated lungfishes from amphibians (*Batraciens*) placing them among the fishes, a hypothesis that was not widely accepted at the time.

A strange-looking tree in a style not seen before was published in 1897 by American botanist Charles Edwin Bessey (1845–1915) to illustrate his work on the relationships of dicotyledonous plants (Figure 105). Based in part on data provided by Wilhelm Philipp Schimper (1808–1880) in his three-volume *Traité de paléontologie végétale* (Treatise of plant paleontology) 1869–1874, Bessey attempted to show the increase in diversity of plant groups through time. This increase is indicated for each taxon by the area of an acute triangle applied to three consecutive geological time periods, Cretaceous, Eocene, and Miocene. The absence of geologically younger groups in older periods and the increase in diversity of all groups are indicated.[17] Bessey is often thought of as the first botanist to develop the concept of angiosperm origins that places flowering plants like *Magnolia* at the base. This view prevailed from the early 1900s through the end of the twentieth century, when molecular phylogenies displaced it.[18]

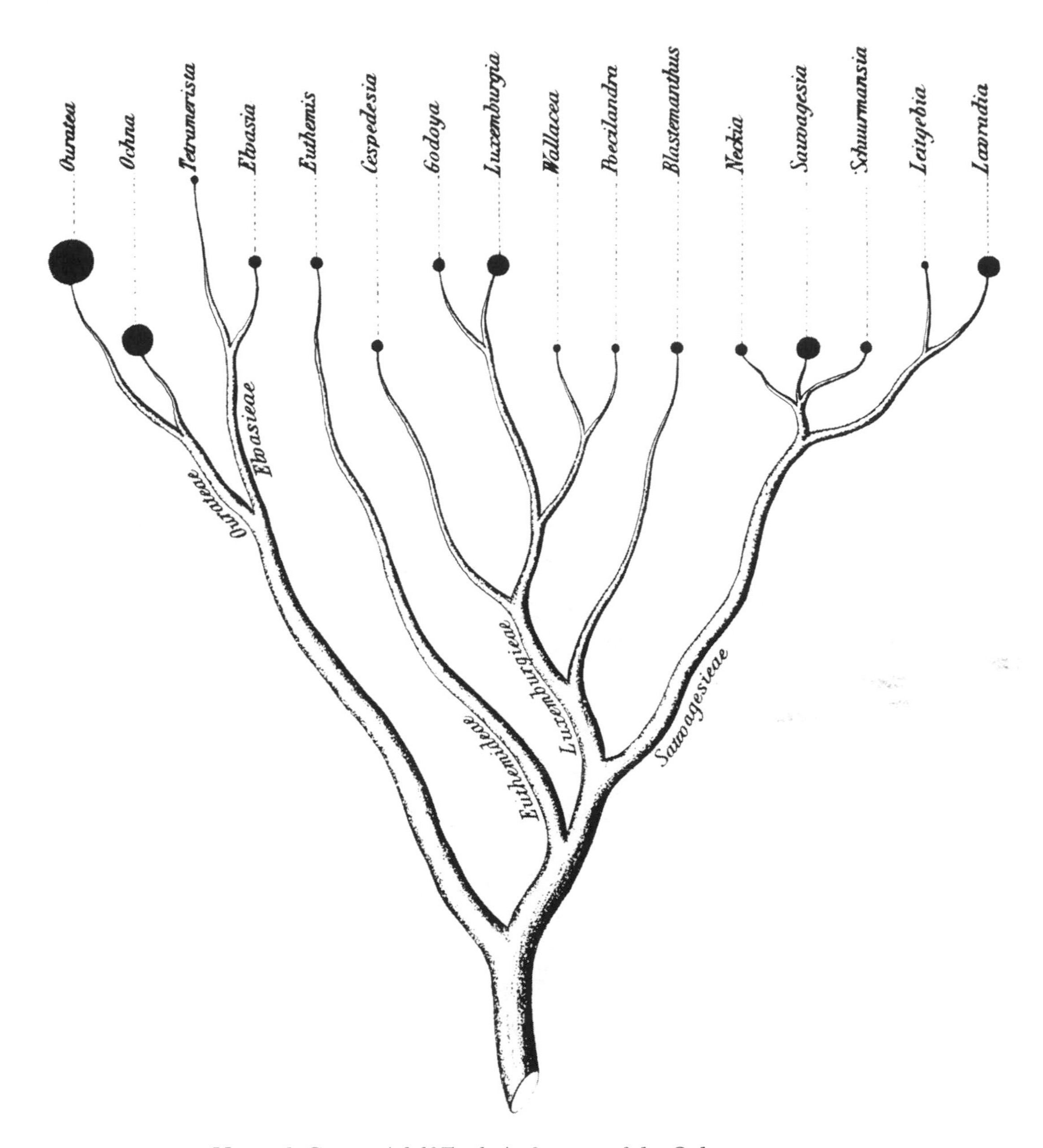

FIGURE 92. Heinrich Gustav Adolf Engler's 1874 tree of the Ochnaceae, a widespread family of mostly trees and shrubs. The relative size of each genus is indicated by the diameter of a dot placed at the tip of each twig, the latter terminating at different levels to show their relative divergence from a basal type.

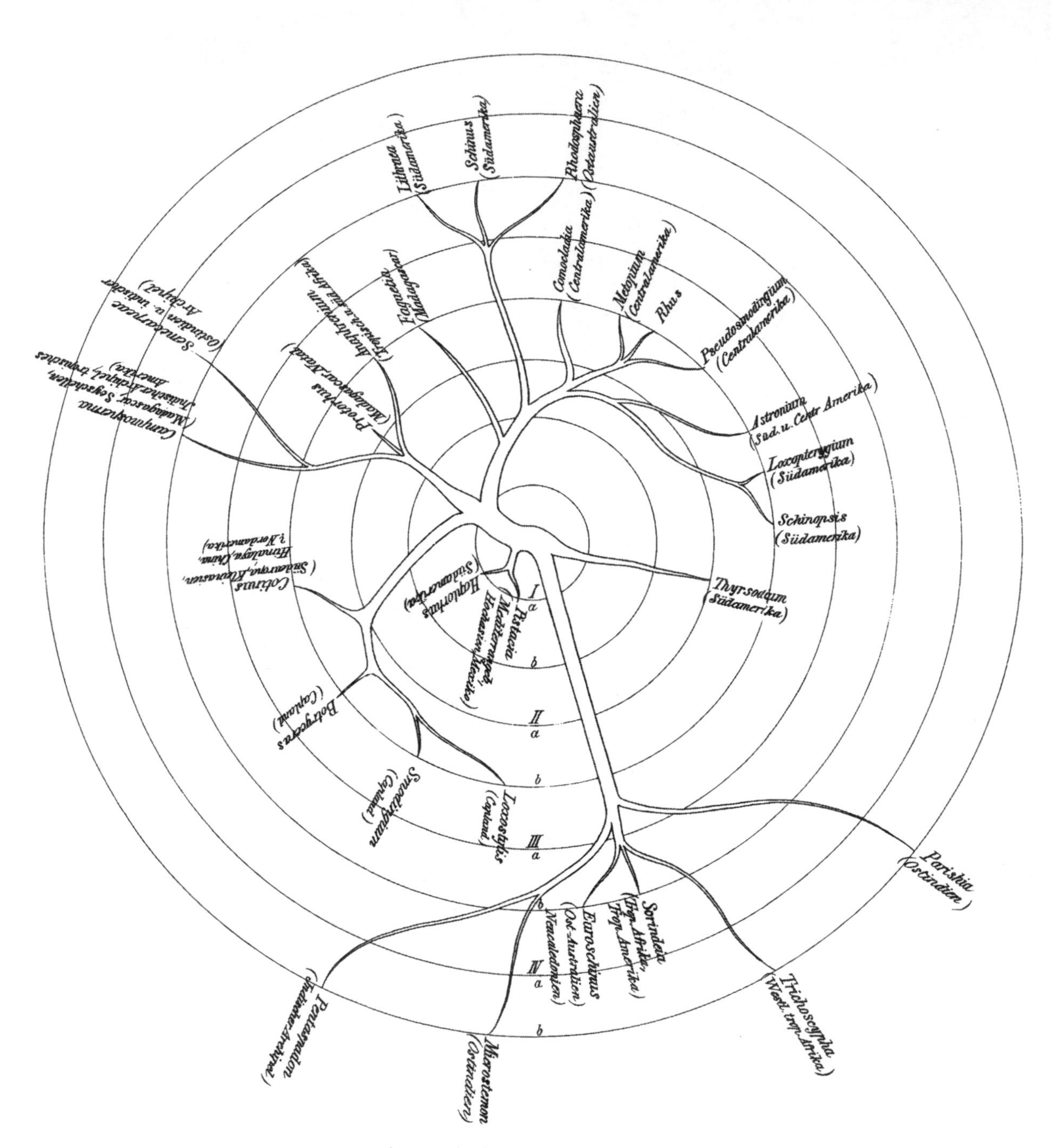

FIGURE 93. A tree of relationships of plants of the cashew or sumac family Anacardiaceae as viewed from above, published by Heinrich Gustav Adolf Engler in 1881.

---

facing page

FIGURE 94. Maximilian Fürbringer's 1888 tree of bird relationships, shown from the side of the Struthiornithes, Rheornithes, Pelargornithes, Hippalectryornithes, Gruiformes, and Ralliformes.

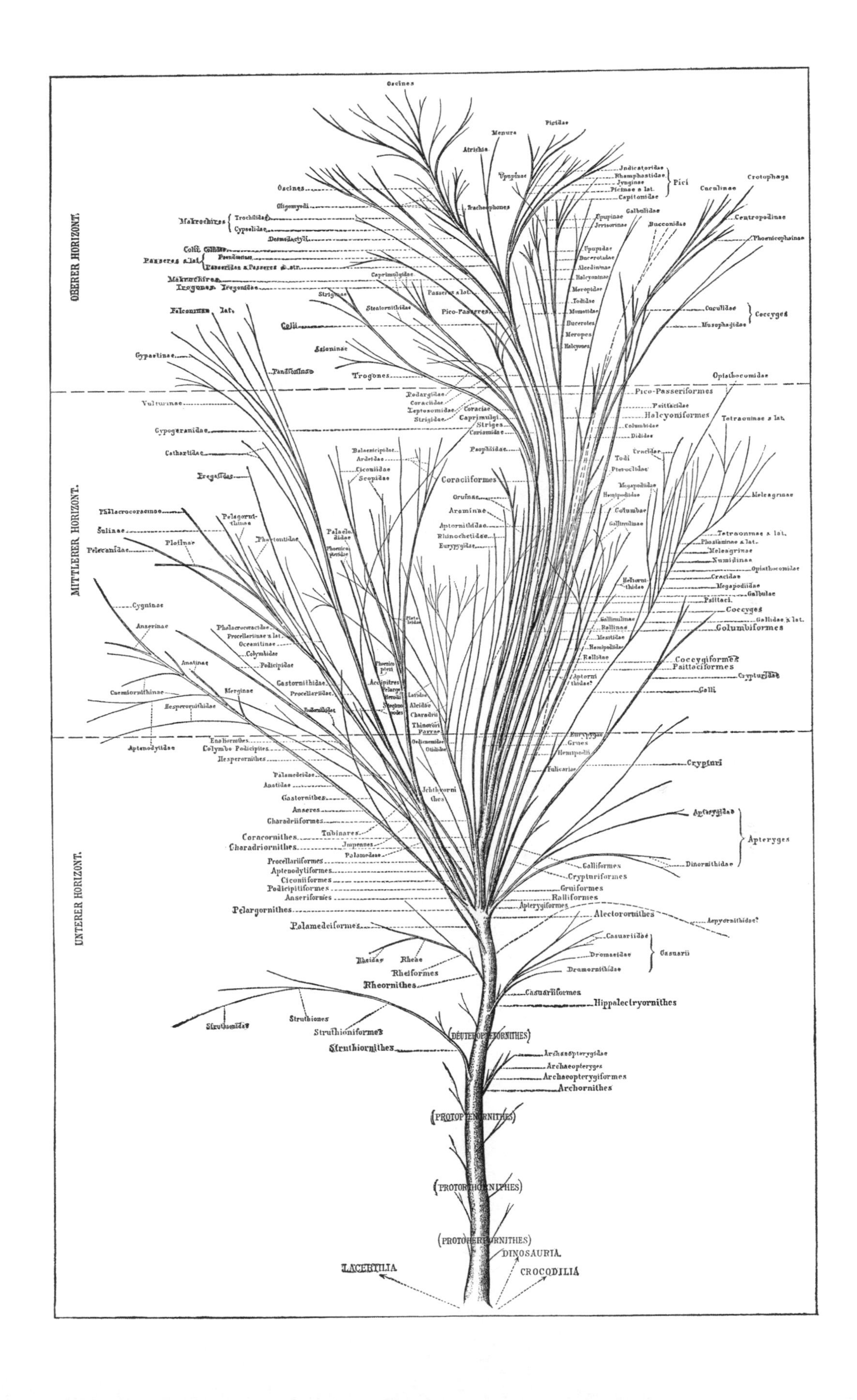

OBERER HORIZONT.
MITTLERER HORIZONT.
UNTERER HORIZONT.
Oscines
Menura
Pittidae
Atrichia
Upupinae
Oscines
Oligomyodi
Tracheophones
Trochilidae
Cypselidae
Makrochires
Desmodactyli
Passeres s. lat.
Caprimulgidae
Passeres s. lat.
Makrochires
Trogones
Strigidae
Steatornithidae
Pico-Passeres
Falconinae s. lat.
Colii
Aesioninae
Gypaetinae
Pandioninae
Trogones
Jndicatoridae
Rhamphastidae
Jynginae
Picinae s. lat.
Capitonidae
Pici
Upupinae
Jrrisorinae
Galbulidae
Bucconidae
Crotophaga
Cuculinae
Centropodinae
Phoenicophainae
Upupidae
Bucerotidae
Alcedininae
Halcyoninae
Meropidae
Todidae
Momotidae
Bucerotes
Meropes
Halcyones
Cuculidae
Musophagidae
Coccyges
Opisthocomidae
Pico-Passeriformes
Psittacidae
Halcyoniformes
Columbidae
Dididae
Tetraoninae s. lat.
Vulturinae
Podargidae
Coraciidae
Leptosomidae
Strigidae
Coraciae
Caprimulgi
Striges
Cariamidae
Gypogeranidae
Cathartidae
Psophiidae
Cracidae
Todi
Pteroclidae
Megapodiidae
Hemipodiidae
Meleagrinae
Fregatidae
Balaenicipidae
Ardeidae
Ciconiidae
Scopidae
Coraciiformes
Gruinae
Aramineae
Columbae
Gallinulinae
Phalacrocoracinae
Sulinae
Pelecanidae
Plotinae
Phaetontidae
Aptornithidae
Rhinochetidae
Eurypygidae
Tetraoninae s. lat.
Phasianinae s. lat.
Meleagrinae
Numidinae
Opisthocomidae
Cracidae
Megapodiidae
Galbulae
Psittaci
Coccyges
Gallidae s. lat.
Columbiformes
Cygninae
Anserinae
Anatinae
Colymbidae
Podicipidae
Gastornithidae
Procellariidae
Merginae
Hesperornithidae
Aptenodytidae
Gallinulinae
Rallinae
Mesitidae
Hemipodiidae
Rallidae
Coccygiformes
Psittaciformes
Crypturidae
Colii
Accipitres
Laridae
Alcidae
Charadrii
Eurypygae
Grues
Hemipodii
Fulicariae
Crypturi
Enaliornithes
Colymbo Podicipites
Hesperornithes
Palamedeidae
Anatidae
Gastornithes
Anseres
Charadriiformes
Coracornithes
Charadriornithes
Tubinares
Impennes
Palamedeae
Jchthyornithes
Apterygidae
Apteryges
Dinornithidae
Galliformes
Crypturiformes
Procellariiformes
Aptenodytiformes
Ciconiiformes
Podicipitiformes
Anseriformes
Gruiformes
Ralliformes
Pelargornithes
Apterygiformes
Alectorornithes
Aepyornithidae?
Palamedeiformes
Casuariidae
Dromaeidae
Dromornithidae
Casuarii
Rheidae
Rheae
Rheiformes
Rheornithes
Casuariiformes
Hippalectryornithes
Struthionidae
Struthiones
Struthioniformes
Struthiornithes
(DEUTEROPTERORNITHES)
Archaeopterygidae
Archaeopteryges
Archaeopterygiformes
Archornithes
(PROTOPTERORNITHES)
(PROTORTHORNITHES)
(PROTOHERPORNITHES)
DINOSAURIA
LACERTILIA
CROCODILIA

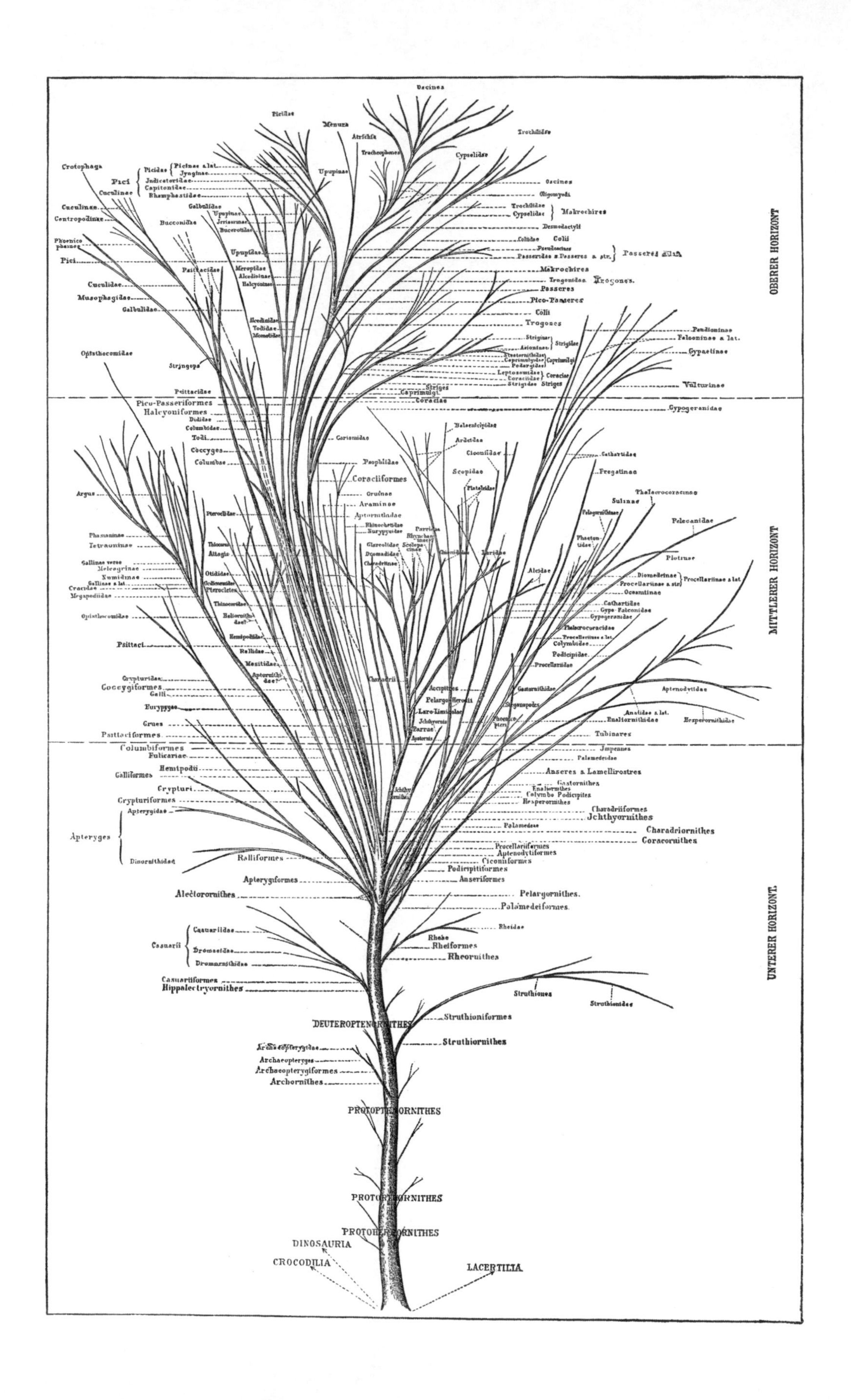

OBERER HORIZONT
MITTLERER HORIZONT
UNTERER HORIZONT
Oscines
Menura
Upupinae
Trochilidae
Cypselidae
Crotophaga
Pici
Cuculinae
Centropodinae
Bucconidae
Psittacidae
Upupidae
Cuculidae
Musophagidae
Galbulidae
Oscines
Oligomyodi
Makrochires
Desmodactyli
Colii
Passeres s. str.
Makrochires
Trogones
Passeres
Pico-Passeres
Colii
Trogones
Opisthocomidae
Stringops
Psittacidae
Pandioninae
Falconinae s. lat.
Gypaetinae
Vulturinae
Striges
Caprimulgi
Coraciae
Gypogeranidae
Pico-Passeriformes
Halcyoniformes
Columbidae
Todi
Coccyges
Columbae
Coriomidae
Psophiidae
Coraciiformes
Gruinae
Araminae
Aptornithidae
Ardeidae
Ciconiidae
Scopidae
Cathartidae
Fregatinae
Thalassoceracinae
Sulinae
Pelecanidae
Argus
Phasianinae
Tetraoninae
Laridae
Alcidae
Phaeton-tidae
Plotinae
Diomedeinae
Procellariinae s. str.
Procellariinae s. lat.
Oceanitinae
Opisthocomidae
Psittaci
Rallidae
Mesitidae
Cathartidae
Gypogeranidae
Colymbidae
Podicipidae
Crypturidae
Coccygiformes
Galli
Eurypygae
Grues
Psittaciformes
Charadrii
Accipitres
Pelargo-Herodii
Laro-Limicolae
Jchthyornis
Barrae
Aptenodytidae
Anatidae s. lat.
Enaliornithidae
Hesperornithidae
Tubinares
Columbiformes
Fulicariae
Galliformes
Hemipodii
Crypturi
Crypturiformes
Apterygidae
Apteryges
Dinornithidae
Palamedeidae
Anseres s. Lamellirostres
Gastornithes
Enaliornithes
Colymbo Podicipites
Hesperornithes
Charadriiformes
Jchthyornithes
Palamedeae
Charadriornithes
Coracornithes
Ralliformes
Procellariiformes
Aptenodytiformes
Ciconiiformes
Podicipitiformes
Anseriformes
Apterygiformes
Alectorornithes
Pelargornithes.
Palamedeiformes.
Casuarii
Casuariidae
Dromaeidae
Dromornithidae
Rheidae
Rhea
Rheiformes
Rheornithes
Casuariiformes
Hippalectryornithes
Struthiones
Struthionidae
Struthioniformes
DEUTEROPTENORNITHES
Struthiornithes
Archaeopterygidae
Archaeopteryges
Archaeopterygiformes
Archornithes
PROTOPTENORNITHES
PROTORTHORNITHES
PROTOHERORNITHES
DINOSAURIA
CROCODILIA
LACERTILIA

facing page

FIGURE 95. *Maximilian Fürbringer's 1888 tree of bird relationships shown from the side of the Aptenodytiformes, Procellariiformes, Charadriiformes, Columbiformes, and Galliformes.*

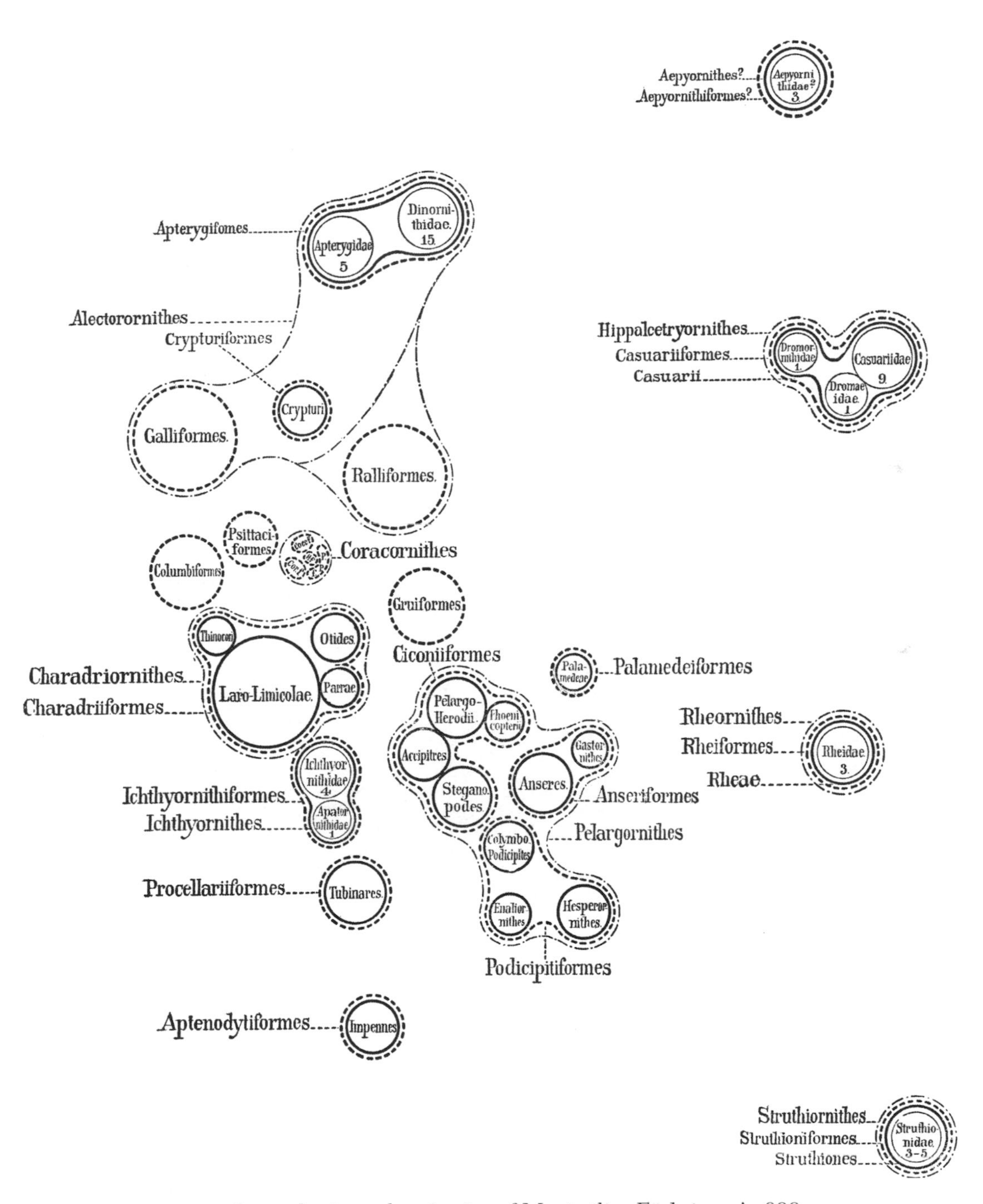

FIGURE 96. Lower horizontal projection of Maximilian Fürbringer's 1888 phylogenetic tree of birds. The number of species in each group is indicated by the diameter of each circle.

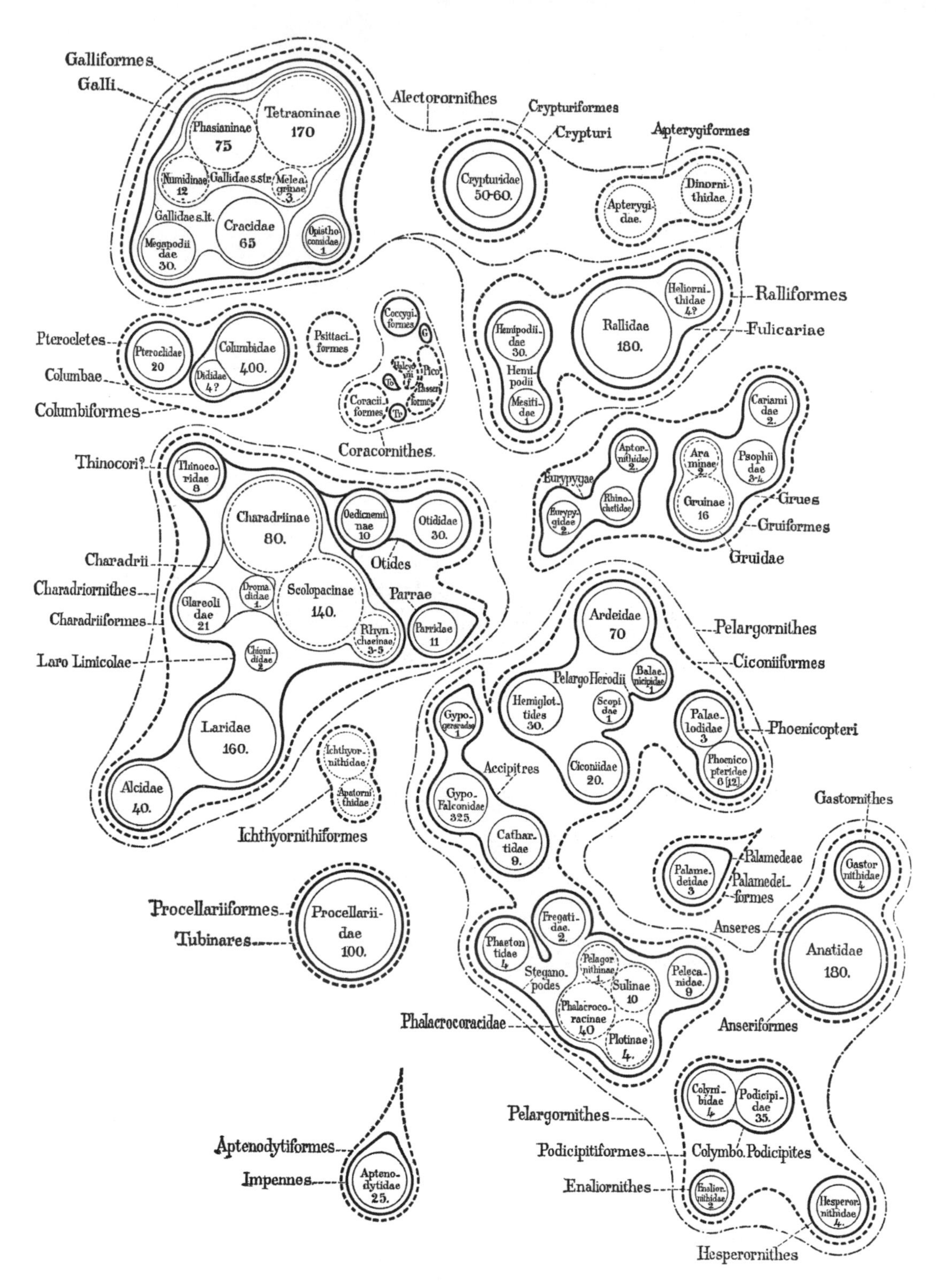

FIGURE 97. Middle horizontal projection of Maximilian Fürbringer's 1888 phylogenetic tree of birds. The number of species in each group is indicated by the diameter of each circle.

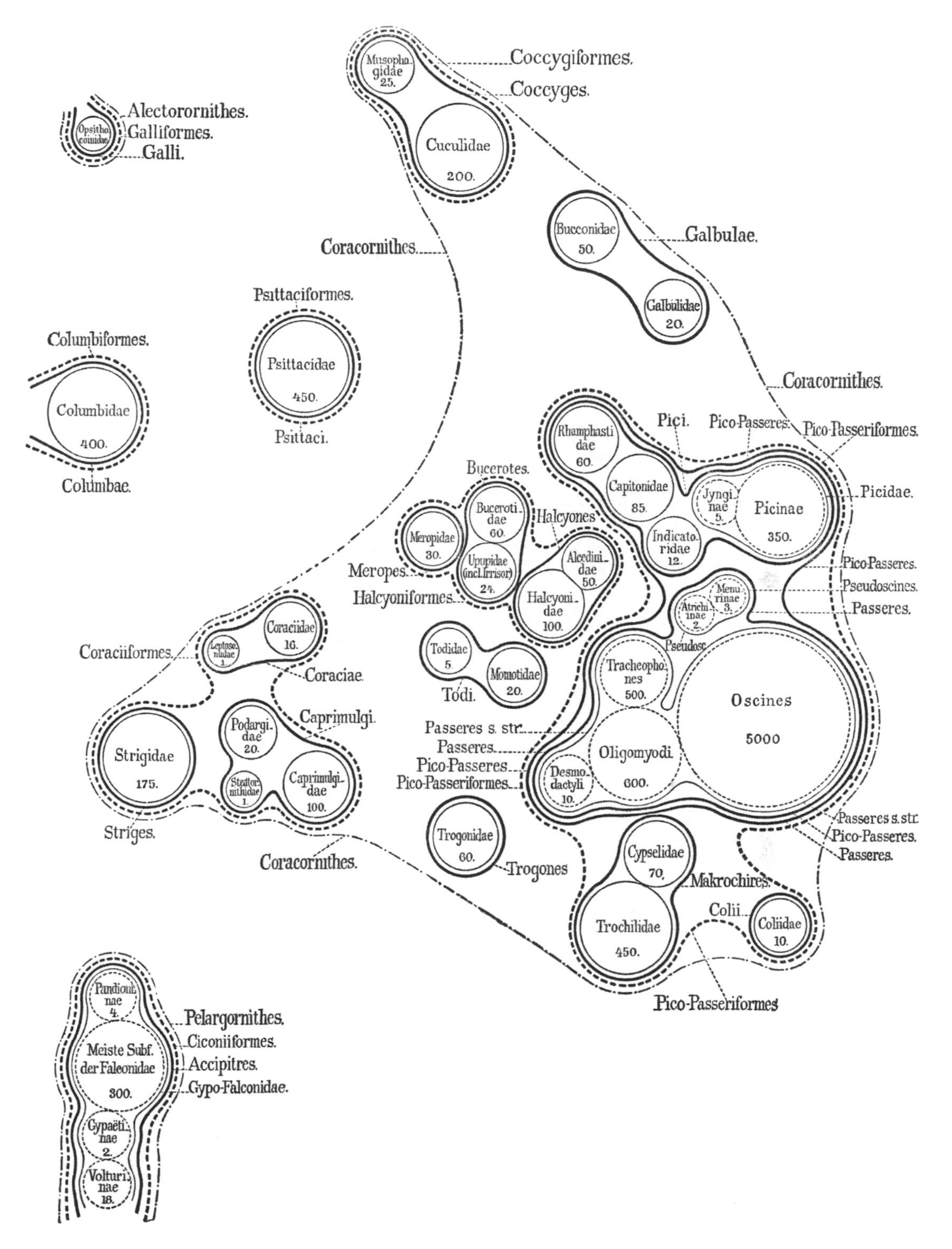

FIGURE 98. Upper horizontal projection of Maximilian Fürbringer's 1888 phylogenetic tree of birds. The number of species in each group is indicated by the diameter of each circle.

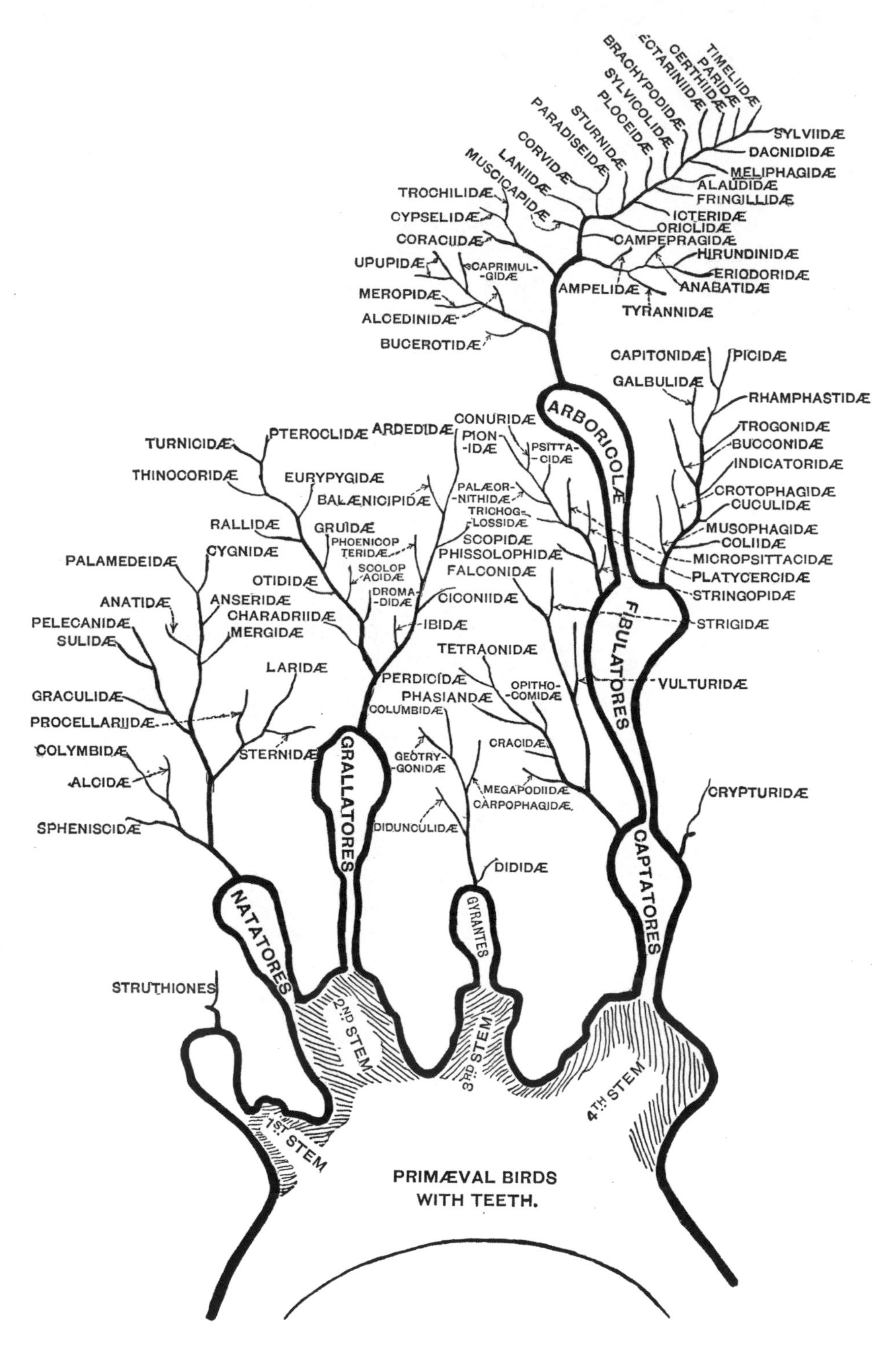
TIMELIIDÆ
PARIDÆ
CERTHIIDÆ
ECTARINIIDÆ
BRACHYPODIDÆ
SYLVICOLIDÆ
PLOCEIDÆ
STURNIDÆ
PARADISEIDÆ
CORVIDÆ
LANIIDÆ
MUSCICAPIDÆ
SYLVIIDÆ
DACNIDIDÆ
MELIPHAGIDÆ
ALAUDIDÆ
FRINGILLIDÆ
TROCHILIDÆ
CYPSELIDÆ
ICTERIDÆ
ORIOLIDÆ
CORACIIDÆ
CAMPEPHAGIDÆ
HIRUNDINIDÆ
UPUPIDÆ
CAPRIMUL-GIDÆ
ERIODORIDÆ
ANABATIDÆ
AMPELIDÆ
MEROPIDÆ
TYRANNIDÆ
ALCEDINIDÆ
BUCEROTIDÆ
CAPITONIDÆ
PICIDÆ
GALBULIDÆ
RHAMPHASTIDÆ
ARBORICOLÆ
CONURIDÆ
PTEROCLIDÆ
ARDEIDÆ
TURNICIDÆ
PION-IDÆ
PSITTA-CIDÆ
TROGONIDÆ
BUCCONIDÆ
INDICATORIDÆ
THINOCORIDÆ
EURYPYGIDÆ
PALÆOR-NITHIDÆ
BALÆNICIPIDÆ
CROTOPHAGIDÆ
CUCULIDÆ
TRICHOG-LOSSIDÆ
RALLIDÆ
GRUIDÆ
MUSOPHAGIDÆ
PHOENICOP TERIDÆ
SCOPIDÆ
COLIIDÆ
CYGNIDÆ
PALAMEDEIDÆ
PHISSOLOPHIDÆ
MICROPSITTACIDÆ
SCOLOP ACIDÆ
FALCONIDÆ
PLATYCERCIDÆ
OTIDIDÆ
DROMA-DIDÆ
STRINGOPIDÆ
ANATIDÆ
ANSERIDÆ
CICONIIDÆ
CHARADRIIDÆ
PELECANIDÆ
STRIGIDÆ
IBIDÆ
SULIDÆ
MERGIDÆ
TETRAONIDÆ
FIBULATORES
LARIDÆ
PERDICIDÆ
VULTURIDÆ
GRACULIDÆ
OPITHO-COMIDÆ
PHASIANDÆ
COLUMBIDÆ
PROCELLARIIDÆ
COLYMBIDÆ
CRACIDÆ
STERNIDÆ
GEOTRY-GONIDÆ
ALCIDÆ
MEGAPODIIDÆ
CRYPTURIDÆ
CARPOPHAGIDÆ
SPHENISCIDÆ
DIDUNCULIDÆ
GRALLATORES
DIDIDÆ
NATATORES
GYRANTES
CAPTATORES
STRUTHIONES
2ND STEM
3RD STEM
4TH STEM
1ST STEM
PRIMÆVAL BIRDS
WITH TEETH.

FIGURE 99. Anton Reichenow's polyphyletic family tree of birds published in 1882, as redrawn by Richard Bowdler Sharpe, 1891.

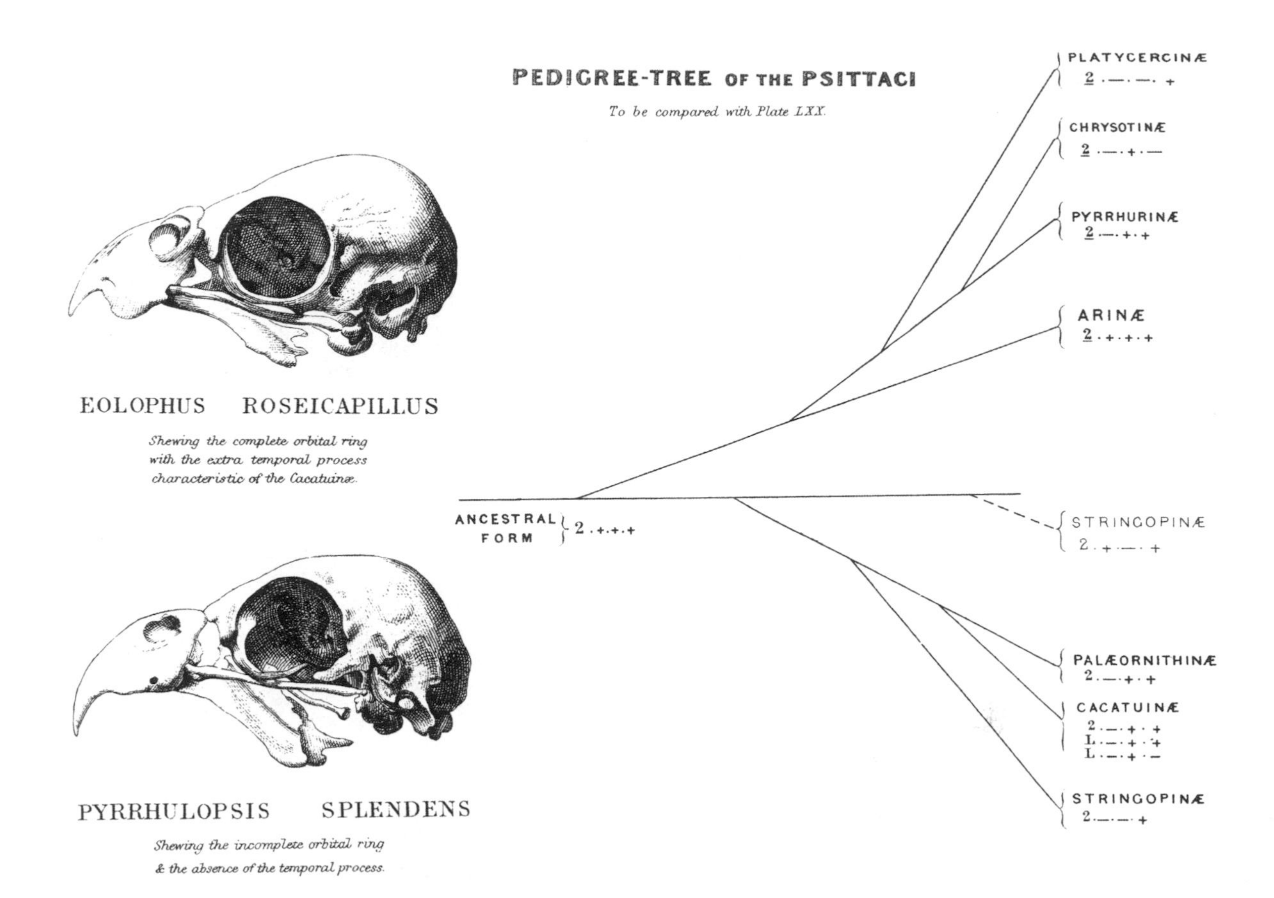

FIGURE 100. A "pedigree tree" of the parrots by Alfred Henry Garrod in 1874, the first known branching diagram to show character-state distributions. One of Garrod's characters, the presence or absence of a complete orbital ring and temporal process, is illustrated on the left. The numeral 2 and the plus and minus signs signify the states of the four characters listed inside the small circle in the center of figure 101 below.

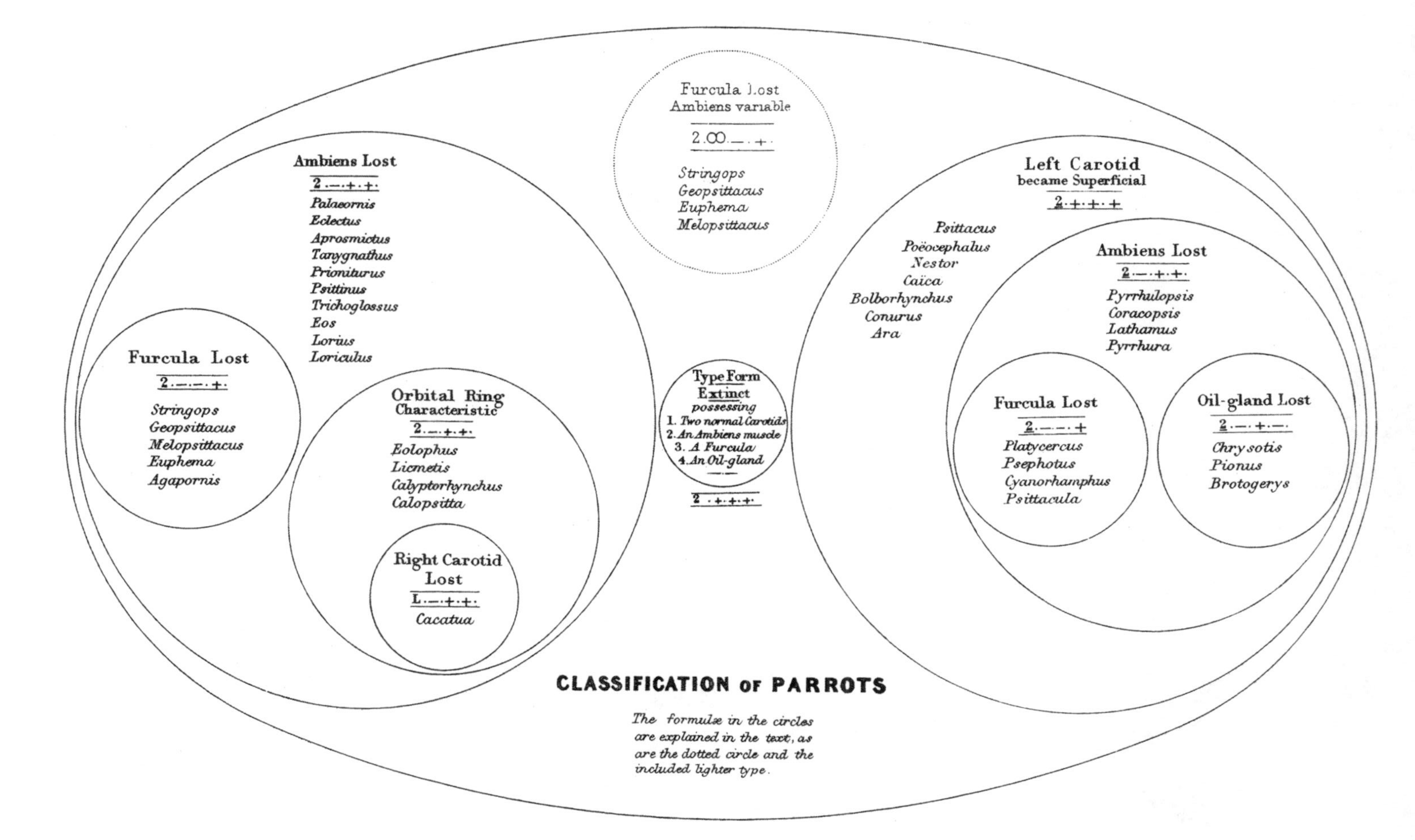

FIGURE 101. Alfred Henry Garrod's 1874 "classification of parrots," showing the distribution of character states among the different genera, constructed to further explain the methodology employed to produce his "pedigree tree" (see figure 100).

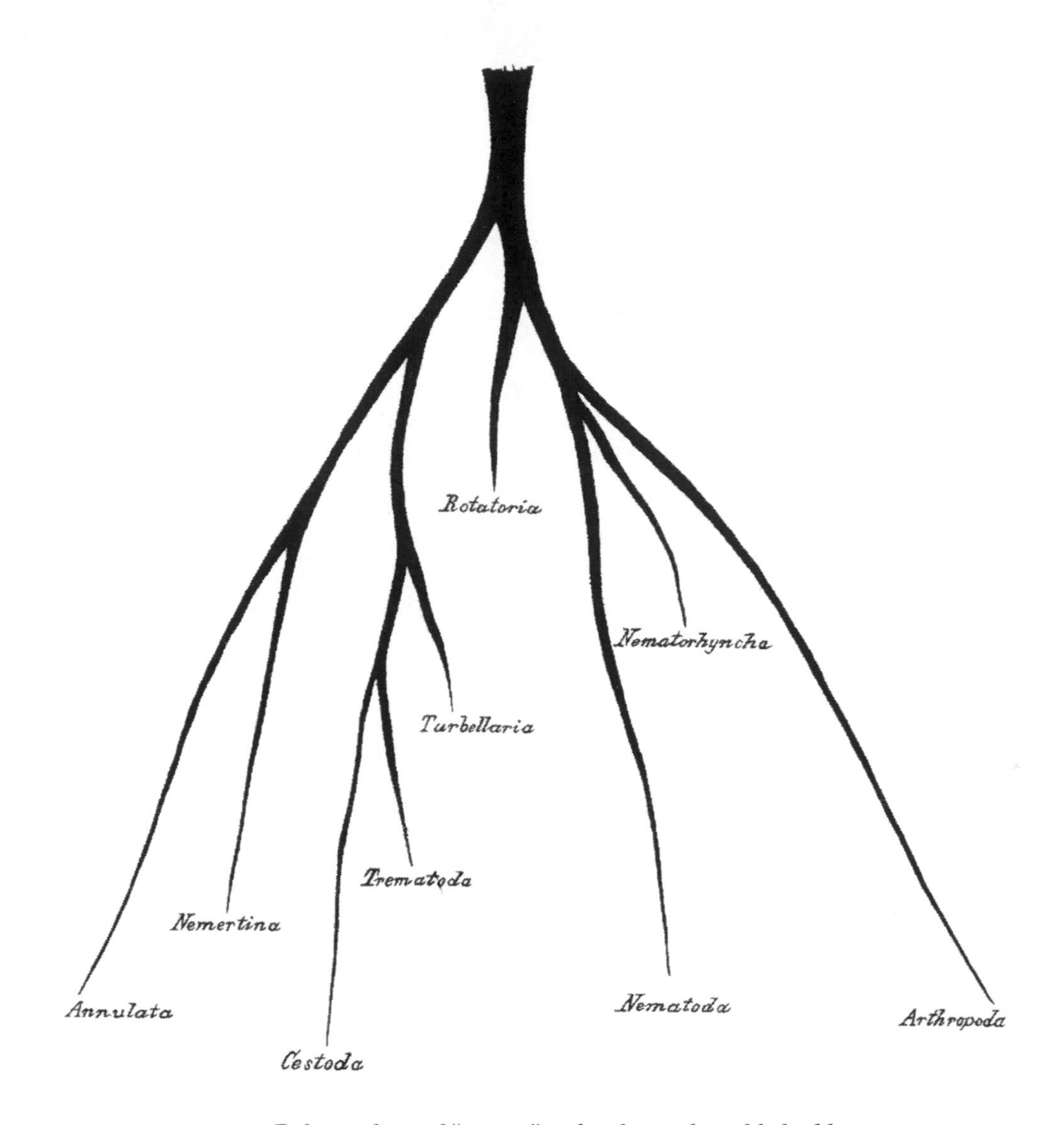

FIGURE 102. Relationships of "worms" and arthropods, published by Johann Adam Otto Bütschli in 1876. One of the few trees ever published with the root placed at the top.

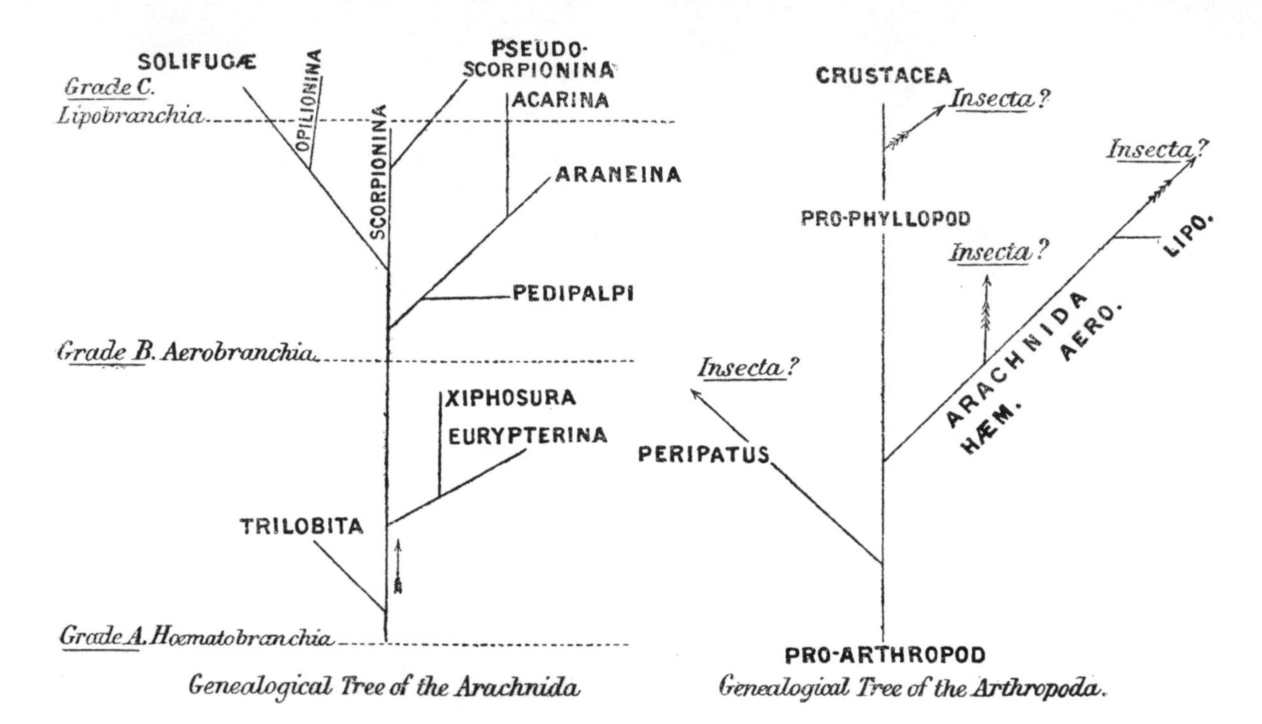

FIGURE 103. A pair of trees published by Edwin Ray Lankester in his famous 1881 paper titled "*Limulus* an arachnid," in which he showed that the horseshoe crab is related to spiders and their allies rather than to crustaceans. The horseshoe crab (*Limulus*) is contained within the order Xiphosura shown on the left; the relationship of spiders and their allies (Arachnida) to crustaceans is on the right.

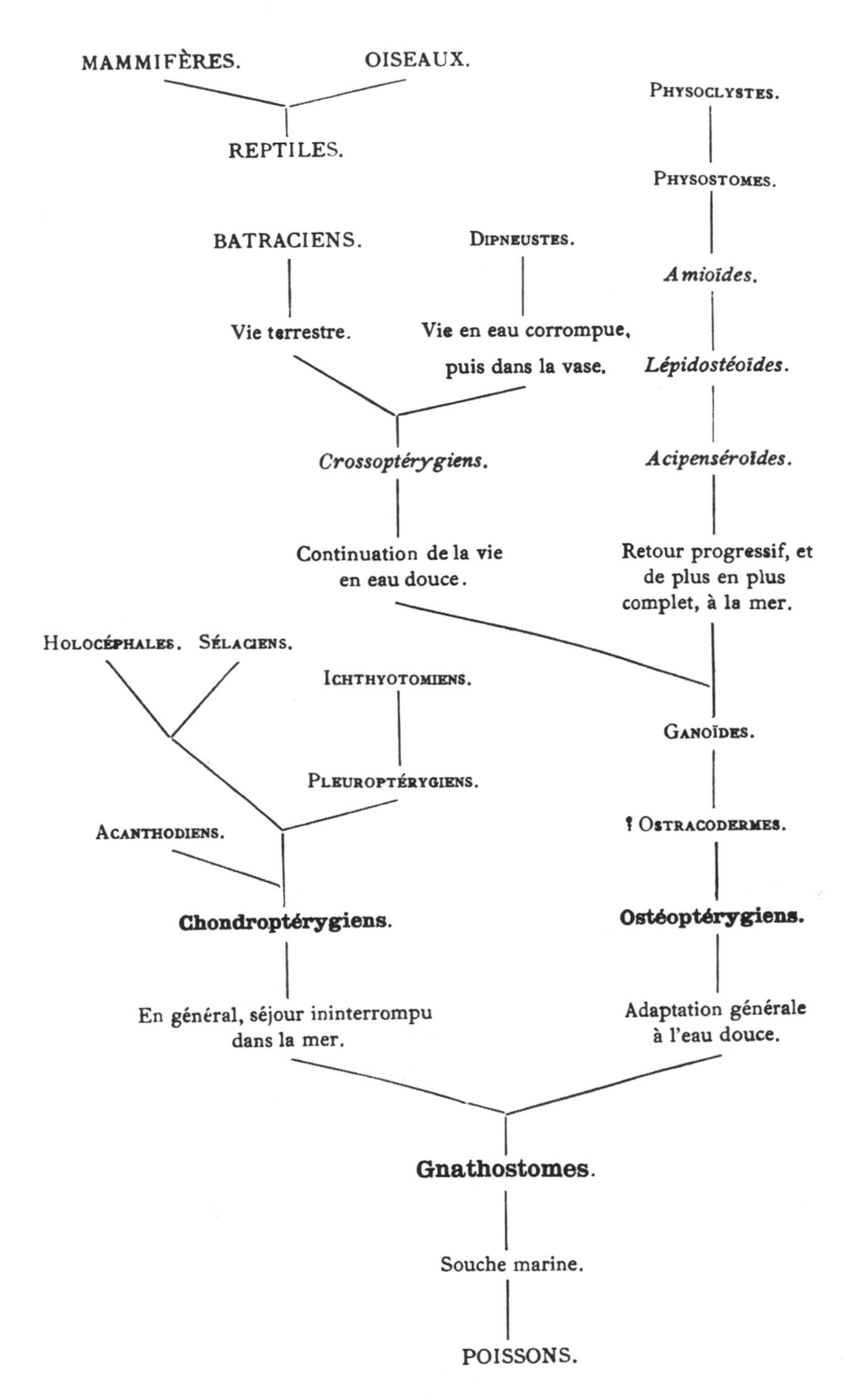

FIGURE 104. A phylogeny of jawed vertebrates by Louis Antoine Marie Joseph Dollo in 1896 showing lungfishes (*Dipneustes*) as most closely related to tetrapods (*vie terrestre*), with coelacanths (*Crossoptérygiens*) ancestral to both.

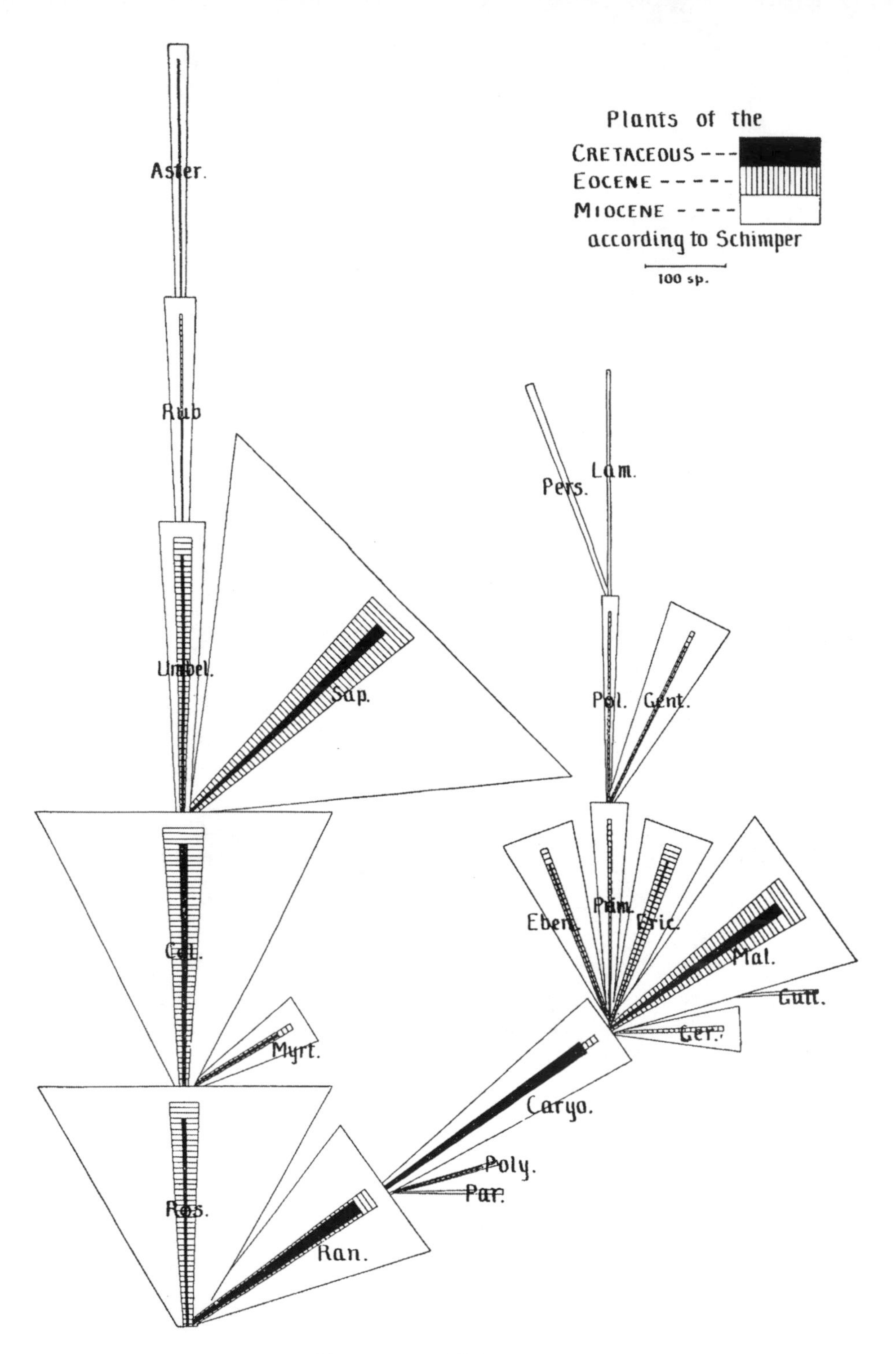

FIGURE 105. Charles Edwin Bessey's 1897 classification of dicotyledonous plants, showing the increase in diversity of each group through time by the relative area of a triangle, applied to three consecutive geological time periods.

Trees of the Early Twentieth Century

1901–1930

With a few prior exceptions—for example, the trees of Nicolas Seringe (1815) and Arthur Garrod (1874) described earlier—branching diagrams of plant and animal relationships were constructed to demonstrate affinities among the organisms themselves; that is, the terminal twigs of trees were labeled only with the names of taxa. No one had put much thought into plotting characters on trees. That changed with the work of Peter Chalmers Mitchell (1864–1945), long-time secretary of the Zoological Society of London (1903–1935) and creator of the world's first open-air zoological park. In a remarkably detailed study of the intestinal tract of birds from 1901, Mitchell produced a series of diagrams that today would be called character-state trees (Figure 106). A half century before Willi Hennig (1913–1976) published his *Theorie der phylogenetischen Systematik* (Theory of phylogenetic systematics, 1950), Mitchell clearly distinguished between primitive and derived characters, which he called "archecentric" and "apocentric" (plesiomorphic and apomorphic in today's terminology). He distinguished also between uniquely derived characters and convergent characters, recognizing that shared primitive characters could not be used as evidence to establish evolutionary relationships.[1] In presenting his trees, Mitchell was concerned that his intent might be misinterpreted:

> In the systematic descriptive part, my task was to treat the characters of the patterns displayed by different birds as nearly as possible as if the gut were the whole animal, and the various . . . Plates . . . I take to be the relations of the intestinal tracts, and not necessarily the relations of the possessors of these tracts. I have been taking, in fact, the anatomical structure as the unit,

> and not the individual or the species. . . . Granting that the Plates attached to this paper represent with approximate accuracy the phylogeny of the intestinal tract in birds, we have yet to learn the relation of the phylogenetic tree of this structure to the phylogenetic trees of other structures, and the relation of all these to the phylogenetic trees of these impermanent combinations of characters that we call species. Although the coincidence of such trees is frequently assumed, there is no *a priori* reason to support such a proposition; and there is much recent work on the nature of characters and of their inheritance to throw doubt on the proposition.[2]

The most explicit and influential phylogenetic tree of Recent and fossil turtles produced during the nineteenth and early twentieth century was published in 1908 by Oliver Perry Hay (1846–1930) in a style not often seen before (Figure 107).[3] Hay hypothesized the genus *Dermochelys* as the most primitive living turtle, diverging from a prochelonid ancestor in the Late Permian, and elaborated on the idea of a diverse ancestral group of extinct Triassic forms called the Amphichelydia, a group characterized by having a nonretractable neck. The use of dotted lines in paleontological trees to indicate hypothesized lineages, in contrast to solid lines indicating actual known fossil material, was new at the time.

Based on his studies of the genus *Ilex,* a group of trees and shrubs commonly referred to as holly, German botanist Ludwig Eduard Theodor Loesener (1865–1941) published an imaginative, but odd-looking, tree in 1908—a cluster of intertwined, finger-like branches emerging from a broad trunk, "mysteriously arising from what seems to be a more or less transparent liquid" (Figure 108).[4] The horizontal dotted line indicates the boundary between past and present; thus, everything below the line is provided with names of hypothesized or known extinct groups. The distinct three-dimensional quality of the diagram is unique.

Russian botanist Constantin Merezhkowsky (1855–1921) is celebrated today for his theory of symbiogenesis. Based primarily on his studies of lichens, he hypothesized soon after the turn of the century that chloroplasts originated from symbiotic blue-green algae. He also proposed a theory for the origin of the nucleus and cytoplasm—and, by extension, all forms of life—from two kinds of organisms and two kinds of protoplasm, which he called "mycoplasm" and "amoeboplasm," detailed in a branching diagram published in 1910 (Figure 109). The chromatin of the nucleus, chloroplasts, and bacteria was supposed to be of the nature of mycoplasm; the cytoplasm, of amoeboplasm. Each kind of protoplasm had an origin in different epochs of Earth's history.[5]

Convinced that natural selection could not adequately explain biological novelty, Merezhkovsky rejected Darwinian evolutionary theory, believing instead that the acquisition and inheritance of microbes were central to the history of life. Merezhovsky's ideas of symbiogenesis are reflected in the modern endosymbiotic theory developed and popularized by Lynn Margulis in the 1970s (see below), which argues that cellular organelles such as chloroplasts and mitochondria are the descendants of bacteria that evolved in an intracellular symbiotic association with early eukaryotic cells.[6]

William Patten (1861–1932), long-time professor of zoology and head of the Department of Biology at Dartmouth, was a champion of evolution, stressing the great importance of its role in the college curriculum.[7] As early as 1898, he offered a course in "Comparative Anatomy of the Vertebrates," and later, in 1920, he organized the first required freshman evolution course in the country.[8] Uniquely designed for his popular textbook *The Evolution of the Vertebrates and Their Kin,* published in 1912, Patten summarized his teaching in a "phylogeny of the principal subdivisions of the animal kingdom" (Figure 110).

Despite his firm belief in evolution, Patten never lost his faith: "I teach evolution because I am confident, and it is a confidence fully justified by long experience, that the teaching of evolution is the only way to bring back a living God into those fields of human thought and experience from which the teachings of "high-brow" philosophy and "low-brow" religion are excluding Him with extraordinary thoroughness and rapidity."[9]

Herbert Fuller Wernham's (1879–1941) 1914 branching diagram of plants of the rubiaceous genus *Sabicea* (commonly known as woodvine) is an early example of phylogeography in which geographic distribution is plotted on a tree of evolutionary relationships (Figure 111). The assemblage is divided initially into four "sections," the Laxae, Sessiles, Capitatae, and Floribundae, based on differences in flower anatomy. The names of African species, found within all three major clusters, are cast in capital letters, while those of American species, clustered primarily on the right between the two curved, dotted lines, are shown in lower case. Species restricted to Madagascar, on the far right, are indicated by italics. Note that some groups, indicated by placement within circles and ovals, share certain morphological characters, in several cases, by convergence. Dutch botanist Herman Johannes Lam (1892–1977) complained that the diagram would have been more successful if designed to show the distribution of the lineages in accordance with the real positions of the continents: America on the left, Africa next, and Madagascar on the right.[10]

Very different from his earlier 1897 tree of dicotyledonous plants (see Figure 105), Charles Bessey's 1915 image of evolutionary relationships among the Recent orders of the phanerogams (or spermatophytes, plants that produce seeds) consists of a series of odd-shaped lobes connected in chain-like fashion—one of the most famous and stylized tree diagrams in flowering plant systematics, referred to as "Bessey's cactus" by botanists in the know (Figure 112).[11] Relationships among the different groups are indicated by position. The area within each lobe is approximately proportional to the number of species contained within each order. The primary structure of the Bessey System was laid down earlier (1897) but revised over the years with respect to some details. The convenience of his approach made it amenable for student use. Deane Bret Swingle (1879–1944), for example, made good use of it in his *Textbook of Systematic Botany* first published in 1928.

Carl Eigenmann (1863–1927) was an American ichthyologist who, along with his wife, Rosa Smith Eigenmann (1858–1947), the first notable female ichthyologist, collected and described many new genera and hundreds of new species of North and South American fishes. On one famous expedition, the Carnegie British Guiana Expedition of 1908, they returned with over 25,000 specimens, resulting in descriptions of 128 new species and 28 new genera. In the first of his five-part series on American characid fishes published in 1917, he published a diagram that he hoped might make sense out of the bewildering morphological variation that he found within the subfamily Tetragonopterinae (Figure 113). Instead of a tree, he invented a strange, oval-shaped affair, placing what he thought were the most primitive genera in the center and arranging the remaining genera around the periphery, fanning out in all directions.[12] Some of the genera are connected to the inner oval by solid lines indicating his confidence in their descent, while others are connected only by dotted lines or by no lines at all. The outer oval is used to segregate the genera into those with and those without a complete lateral line. Eigenmann did not hide the fact that he struggled long and hard with the group:

> It must be quite evident from the foregoing that the subfamily is a paradise for the student of divergent evolution. But the very conditions that make it of interest to the student of evolution make it the despair of the systematist whose object is to express relationship by grouping the species in an orderly array of genera and the individuals in an orderly array of species, always, if possible, in the form of the conventional phylogenetic tree. . . . The Tetragonopterinae seem to form an interlacing fabric rather than a branching tree.[13]

As evolutionary theory became more universally accepted and taught as a critical part of the curriculum of schools and colleges, trees of life were designed for biology textbooks and for popular consumption. Some of the best of these were published by American science educator Benjamin Charles Gruenberg (1875–1965). Notable are the "genealogical trees" of plant and animal life that first appeared in his *Elementary Biology* of 1919 (Figures 114 and 115). His figures are very much like trees in the botanical sense, reminiscent of those of Ernst Haeckel that appeared a half-century before (see Figures 66–75). In an effort to augment interest among his readers, Gruenberg decorated his trees with small images of actual plants and animals placed appropriately at the tips of the branches, an innovation at the time that has become standard today.

Unlike the linked circle diagram of the Compositae published by George Bentham in 1873 (see Figure 87), a tree of the same taxon constructed by James Small (1889–?) in 1919 is typically tree-like (Figure 116).[14] The diagram is based primarily on Small's detailed analysis of the morphology of the styles and stamens of the flowers.[15] Multiple branches representing tribes of the Compositae diverge broadly from both sides of a gradually tapering central axis and pass through horizontal layers of geological time as indicated along the left margin. In yet another early example of phylogeography, numbers placed at the nodes or branching points of the tree correspond to the "geographical centre of origin of each tribe."[16]

Henry Fairfield Osborn (1857–1935), an American geologist, paleontologist, and eugenicist, is best remembered for his massive, two-volume work, *The Proboscidea: A Monograph of the Discovery, Evolution, Migration and Extinction of the Mastodonts and Elephants of the World,* published posthumously in 1936 and 1942, in which he discussed at great length the fossil history and evolution of elephants and their relatives.[17] Throughout the course of his forty years of study, primarily while at the American Museum of Natural History in New York, he published numerous trees, mostly focusing on mammals. One of his earliest (Figure 117), prepared by his student William King Gregory (1876–1970), was designed for Osborn's popular book *The Origin and Evolution of Life* published in 1917. Attempting to demonstrate the adaptive radiation of the mammals, it is reminiscent of the 1844 "spindle" diagram of Agassiz (see Figure 55) and the trees of William Diller Matthew (1871–1930) and Alfred Sherwood Romer (1894–1973), which appeared much later, in the late 1920s and mid-1940s.[18]

A 1921 Osborn diagram showing the "present theory as to the adaptive radiation of the Proboscidea"[19] is unusual and perhaps unique for its time in demonstrating a correspondence between habitat (amphibious and palustral; forests and savannas; forests, grassy plains, savannas, and steppes), feeding behavior (browsers and grazers), and phylogeny (Figure 118). Another Osborn tree, constructed in 1926,[20] again showing the relationships of elephants and their close allies, is nicely decorated with images of the animals themselves (Figures 119), following the precedent set by Benjamin Gruenberg and others only a few years before (see Figures 114 and 115). Yet another proboscidean tree by Osborn, this one of the Mastodontoidea drawn in 1933,[21] focuses more on adaptive radiation or "lines of descent," rather than on phylogenetic relationships (Figure 120). Osborn was extremely prolific and influential, but he was not the best taxonomist: altogether he identified some 352 proboscidean species and subspecies, but only about 164 are recognized today, placed within some forty genera and eight families.[22]

Charles Lewis Camp's (1893–1975) phylogeny of the "families of lizards" published in 1923 (Figure 121) is surprisingly contemporary, incorporating all the basic tenets of phylogenetic systematics (cladistics), suggested earlier by Peter Chalmers Mitchell in 1901 (see Figure 106), and laid down in precise detail by Willi Hennig a quarter of a century later.[23] Camp clearly understood the need for discrete character states to establish monophyly of taxa, recognizing precise criteria by which a transformation series could be polarized. He also made it clear that only derived character states could be used to establish relationships. Unfortunately, Camp failed to develop his ideas in later publications, unlike Hennig who thoroughly promoted his own principles and methods.[24] The black columns placed within "Recent" time indicate the relative specialization of each family; the retention of primitive states is greatest in those groups having the shortest columns. Black sections within the fossil record indicate well-authenticated records; cross-lined sections indicate fairly certain relationships; the dotted lines denote uncertain relationships. The downward directed arrows indicate limbless taxa, a loss that has occurred independently within five lineages.

In a complete makeover of his earlier tree of the "animal kingdom" (see Figure 110), William Patten produced a diagram in 1923 (Figure 122), intended to show "the approximate time when the great classes of animals arose, . . . their probable genetic relations to one another [and] the relative position of each class in the scale of life."[25] It consists of four nested bubbles marked by solid black lines.[26] Geological time, usually set along the vertical axis, is replaced with "progress in organization, brain size, parental care and adaptability." Time is indicated along the horizontal, extending laterally and symmetrically in both directions from the origin of life at the center. Lineages radiate from a twisting central axis, emerging, expanding and, in some cases, diminishing toward extinction, as they ascend and stretch laterally across time. The approximate number of species then known is indicated within each section. Note how arthropods progress directly to vertebrates through ostracoderm fishes.[27]

In 1924, Charles William Theodore Penland (1899–1982) published a rather plain-looking, dichotomous tree, as a result of his systematic studies of *Scutellaria,* a genus of the mint family Labiatae (or Lamiaceae). What makes his diagram interesting is a circle enclosing taxa of three lineages that share the trait of fruits bearing "winged nutlets" (Figure 123). Penland used the tree to demonstrate that this character "has been secondarily acquired [i.e., derived], appearing as it does in several different groups."[28] In another early example of what is now called outgroup comparison,[29] he described his logic for concluding that the trait is derived: "in a hasty review of nutlet-characters of the other genera of the Labiatae, it appears that the character referred to is absent. This is another justification for believing it is of more recent development."[30]

Scottish anatomist and anthropologist Arthur Keith (1866–1955) was a leading figure in the study of human fossils, probably best known in the anthropological world for his studies of primate skulls, summarized in his *Introduction to the Study of Anthropoid Apes* that appeared in 1897. Beyond the realm of science, however, Keith is much better known for his involvement in the discovery of Piltdown Man, the biggest paleontological hoax of all time, which concerned the discovery of the remains of a previously unknown early human. The find consisted of fragments of a skull and jawbone collected in 1912 from a gravel pit at Piltdown, in East Sussex, England. The fragments were thought by many experts of the day to be the fossilized remains of a hitherto unknown form of early man. The significance of the specimens remained the subject of controversy until it was exposed in 1953 as a forgery, consisting of the lower jawbone of an orangutan that had been deliberately but ingeniously combined with the skull of a fully developed modern human.[31] Keith was accused in 1990, long after his death, of preparing the fake specimens that were subsequently planted at Piltdown, but no convincing evidence exists and the case remains open to this day.[32] Always a vociferous advocate for the validity of the fossil, he provided a place for Piltdown Man in a tree of primate relationships published in 1925 (Figure 124).

Arthur Keith was not alone in his enthusiastic support for Piltdown Man. Henry Osborn, whose trees of proboscideans were described in the previous chapter (see Figures 118–120), did not confine his systematic studies to elephants and their allies, but spent considerable time investigating other mammal groups as well, including primates. In a paper titled "Recent discoveries relating to the origin and antiquity of man" (1927a), he published a phylogeny of the Anthropoidea, with images of crania representing the different taxa (Figure 125). He recognized two primary lineages that diverge from one another in the Middle Oligocene, a "Family of Apes" and a "Family of Man." The latter assemblage splits again during the Pliocene to produce Neanderthals and what he called "modern racial stocks." Note the prominent position of Piltdown Man, placed ancestral to all other hominids. Like Keith, Osborn was thoroughly convinced of the validity of the fossil and of its extreme importance in understanding human evolution. He went too far, however, in devoting an entire book to the subject: *Man Rises to Parnassus: Critical Epochs in the Prehistory of Man* (1927b), in which he described the discovery of the faked fossil as "almost a miracle." A product of his time, he was not immune to racial tendencies, comparing, for example, the size of the braincase of Piltdown Man to that of a modern Aboriginal Australian. Luckily, Osborn was saved the embarrassment—the truth of Piltdown Man was not revealed until 1953, long after Osborn's death in 1935.

William Diller Matthew (1871–1930), a vertebrate paleontologist who worked primarily on mammal fossils—first as a curator at the American Museum of Natural History in New York (1895–1927) and later as director of the University of California Museum of Paleontology at Berkeley (1927–1930)—published a number of notable trees. The first of special interest is a "distribution of the major animal groups through geological time" (Figure 126),[33] constructed sometime in the late 1920s (but apparently not published until 1943 by Tracy Storer). The design of this tree undoubtedly influenced Alfred Sherwood Romer, who drew very similar looking diagrams in the mid-1940s (see the chapter devoted to Romer that follows). Two quite different sorts of trees drawn by Matthew appeared in a 1930 paper titled "The Phylogeny of Dogs." The first is a phylogeny of the carnivorous mammals—a branching array of lobular shapes containing the various families, the width of each lobe waxing and waning to provide an approximation of the diversity of each group as it ascends through geological time (Figure 127). A dashed line separates a primitive, relatively unrelated and poorly known assemblage of families that dominated during the Paleocene and Eocene, but which all became extinct by the middle of the Miocene. A resurgence of more modern forms is evident in the Oligocene and Miocene, resulting in seven families that ascend to Recent times.

Focusing then on one of these families, the dog family, Canidae, Matthew provided what might appear at first glance to be an early example of a character-state tree of canid skull evolution. This tree shows evolutionary trends in the shape of the skulls of dogs, all drawn to the same relative scale and compares them with those of the closely related Ursidae and Procyonidae, families that contain the bears and the raccoons and their allies, respectively (Figure 128). But Matthew instead presented his tree as a phylogeny of the families; the images of skulls were used to simply represent the various taxa, not to illustrate the evolutionary progression of cranial characteristics.

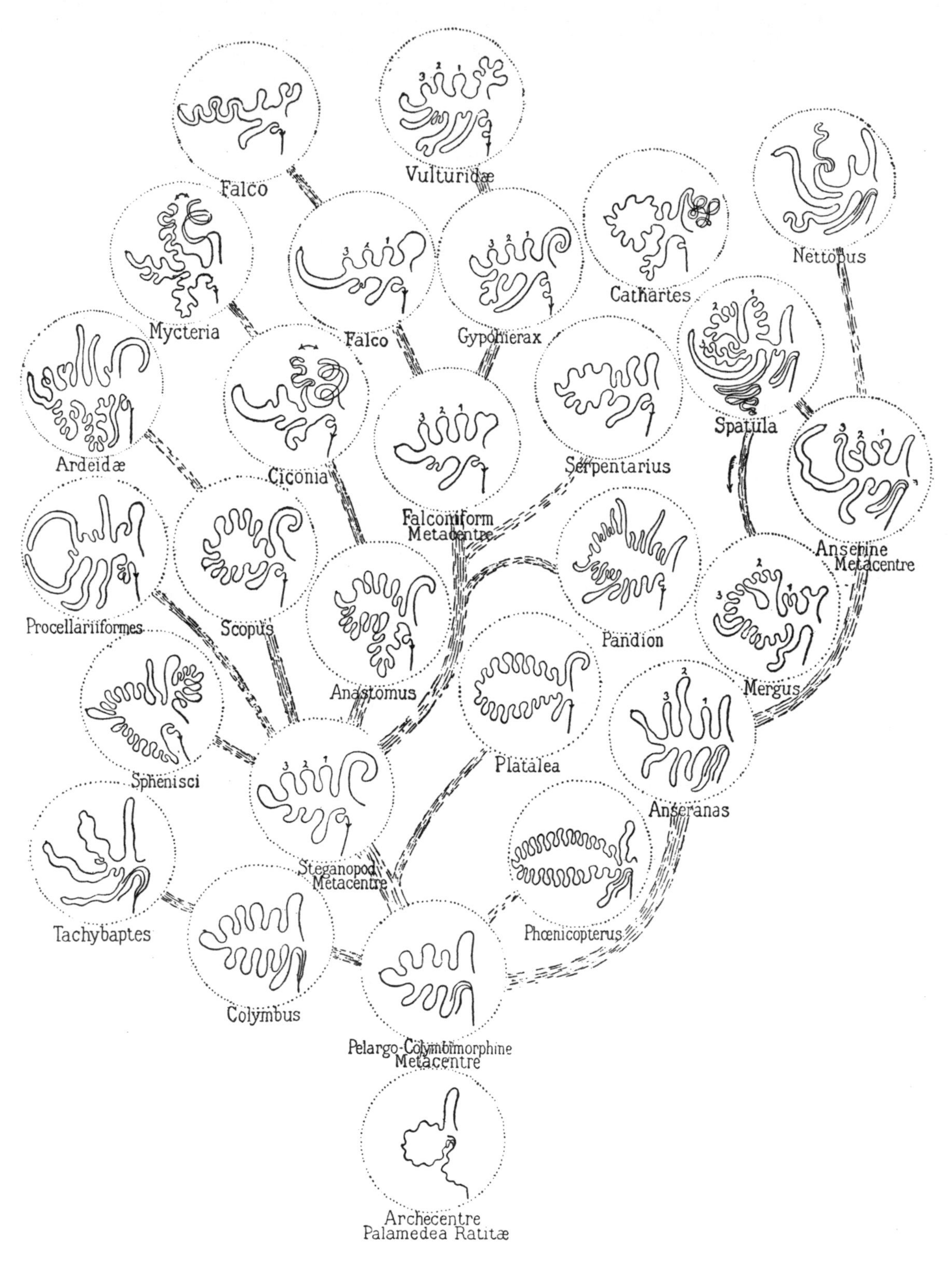

FIGURE 106. Peter Chalmers Mitchell's character-state tree of the intestinal tract of birds from 1901.

Recent	Chelyidae
Pleistocene	Chelydridae
Pliocene	Dermatemydidae, Carettochelyidae, Testudinidae, Dermochelyidae
Miocene	Pelomedusidae, Emydidae, Cheloniidae, Trionychidae
Oligocene	Desmatochelyidae
Eocene	Toxochelyidae, Protostegidae, Plastomenidae
U. Cretaceous	Bothremydidae, Plesiochelyidae, Baënidae, Pleurosternidae, Tretosternidae, Thalassemydidae
L. Cretaceous	Trionychoidea
Jurassic	
Triassic	Primitive Pleurodira, Amphichelydia, Primitive Cryptodira, Primitive Athecae
Permian	Primitive Thecophora, Prochelonida

FIGURE 107. Oliver Perry Hay's 1908 phylogenetic tree of Recent and fossil turtles.

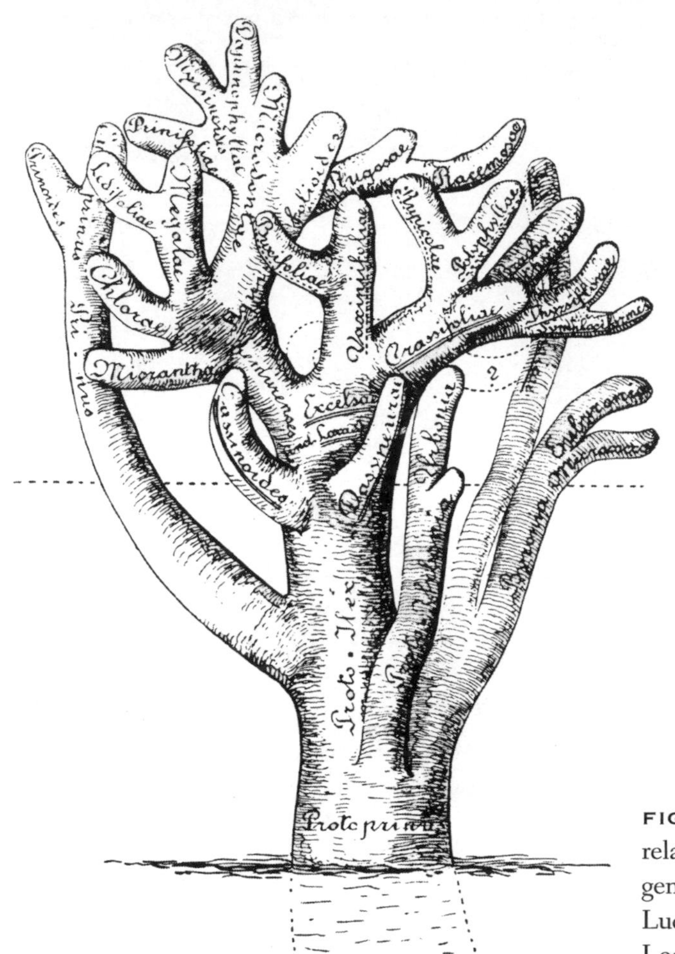

FIGURE 108. A tree of relationships of the plant genus *Ilex,* published by Ludwig Eduard Theodor Loesener in 1908.

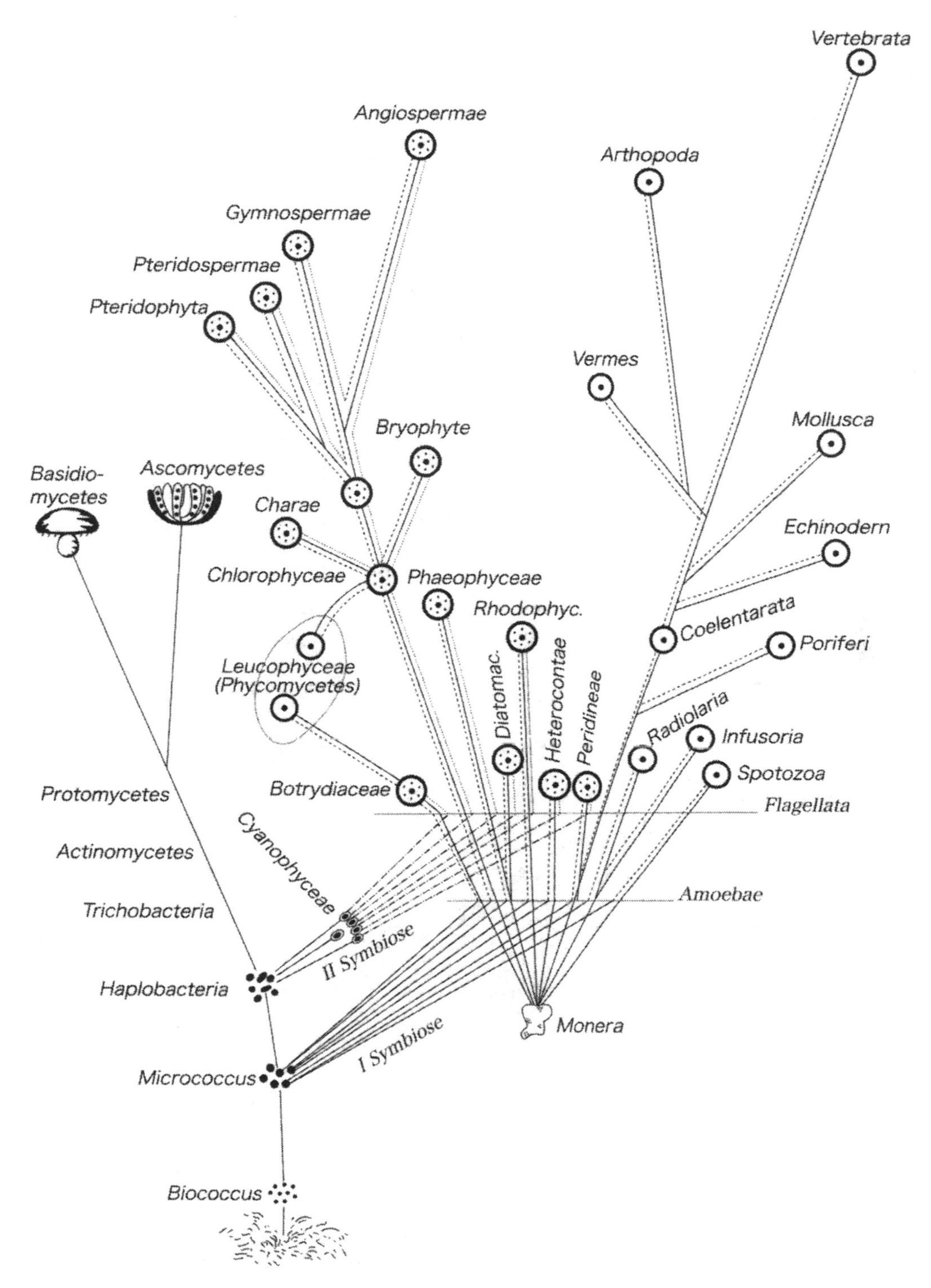

FIGURE 109. Constantin Merezhkowsky's tree of life based on his theory of symbiogenesis, published in 1910.

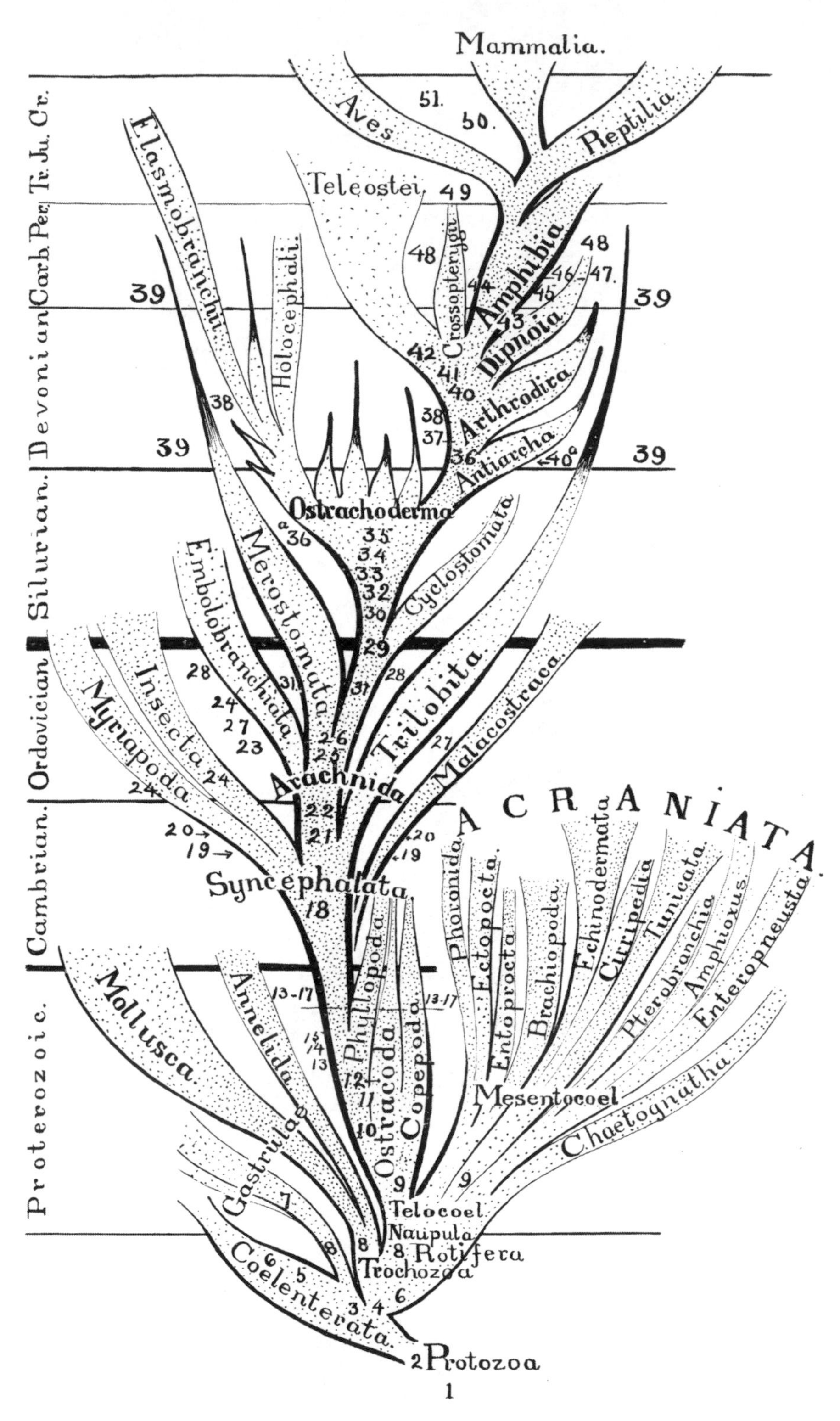

FIGURE 110. William Patten's phylogeny of the "principal subdivisions of the animal kingdom," from 1912. The numbers indicate the approximate periods (defined in his text) in which some of the more important events in the evolution of structures and functions took place.

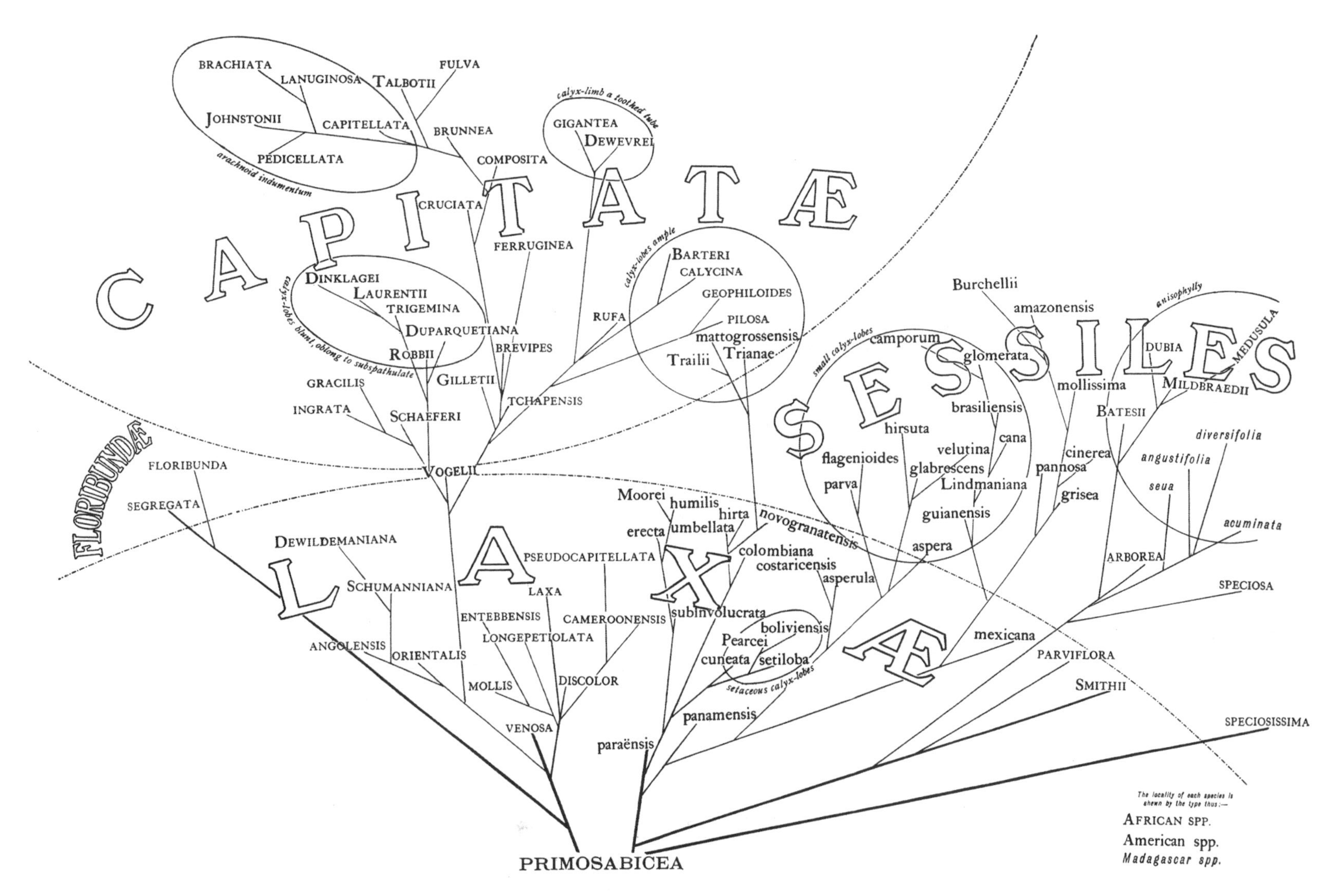

FIGURE 111. Relationships of plants of the rubiaceous genus *Sabicea* constructed by Herbert Fuller Wernham in 1914, an early example of phylogeography.

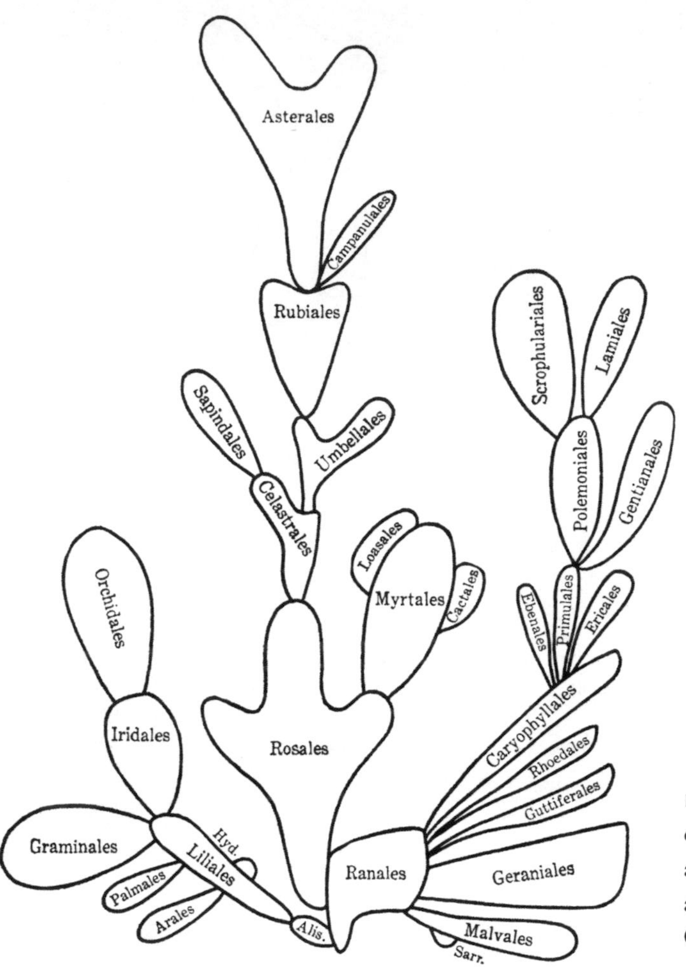

FIGURE 112. "Bessey's cactus": a tree of relationships among the Recent orders of angiosperms, published by Charles Bessey in 1915.

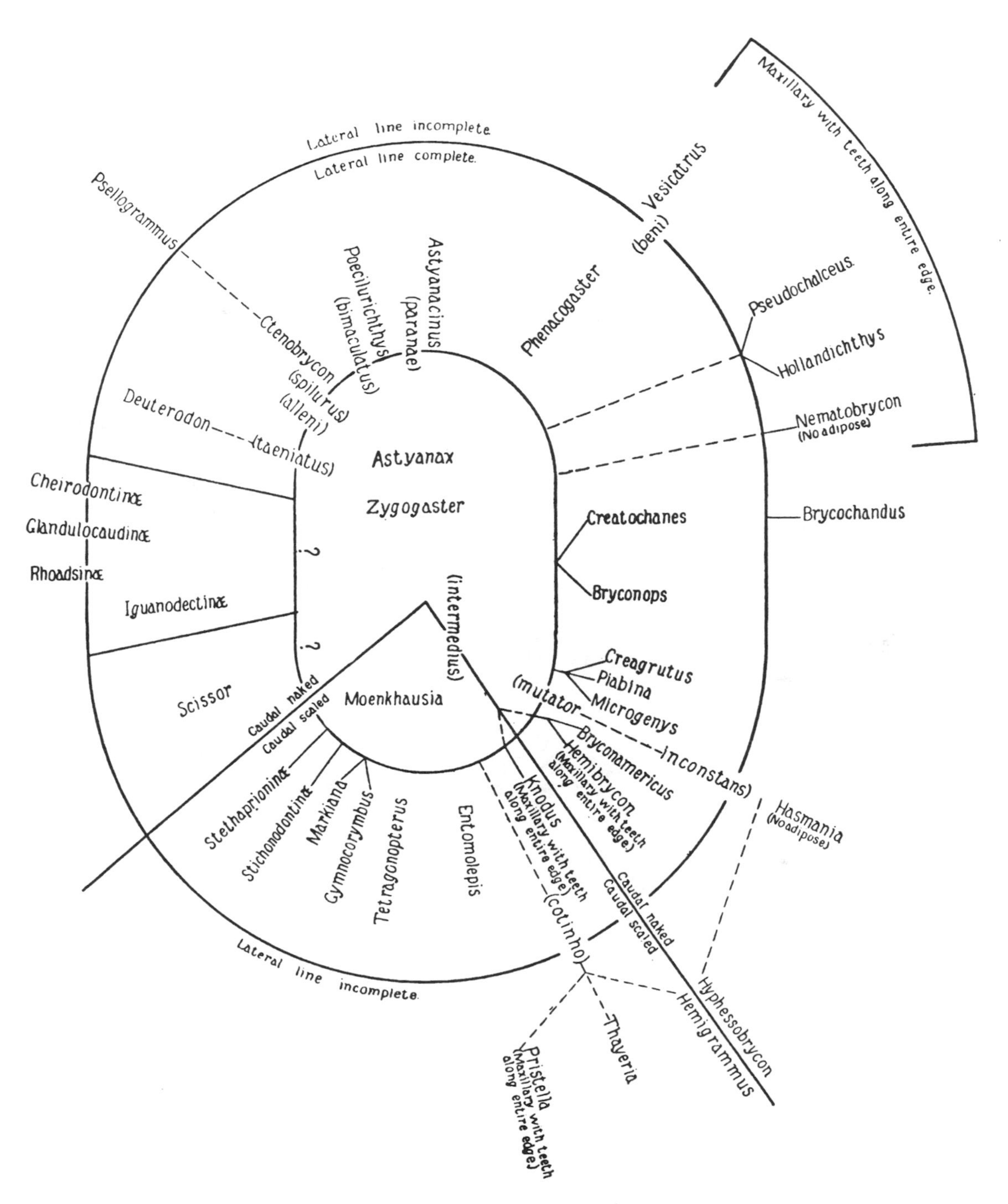

FIGURE 113. A branching diagram intended to show affinities among South American fishes of the characid subfamily Tetragonopterinae, published by Carl Eigenmann in 1917.

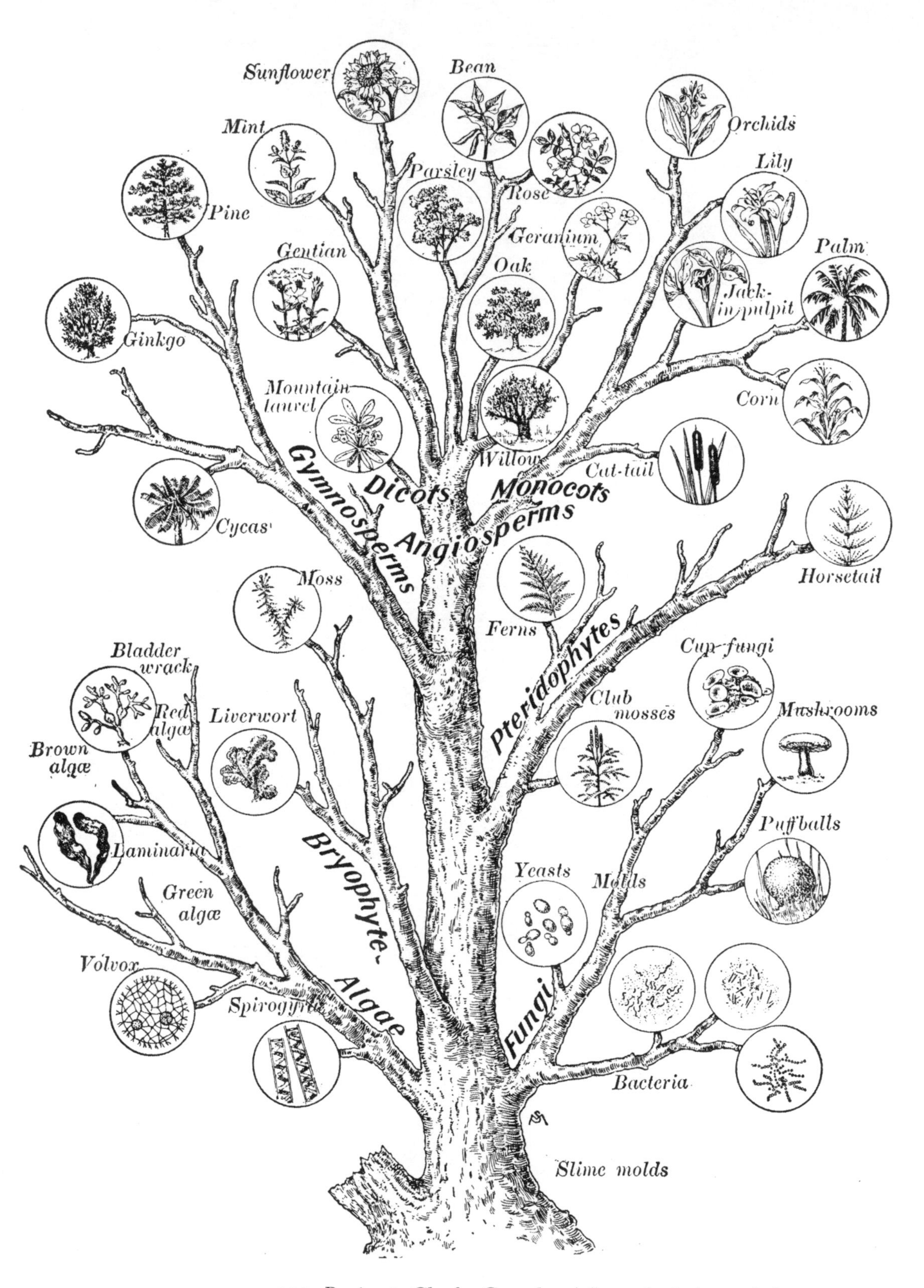

FIGURE 114. Benjamin Charles Gruenberg's "genealogical tree of plant life" constructed for his *Elementary Biology* of 1919.

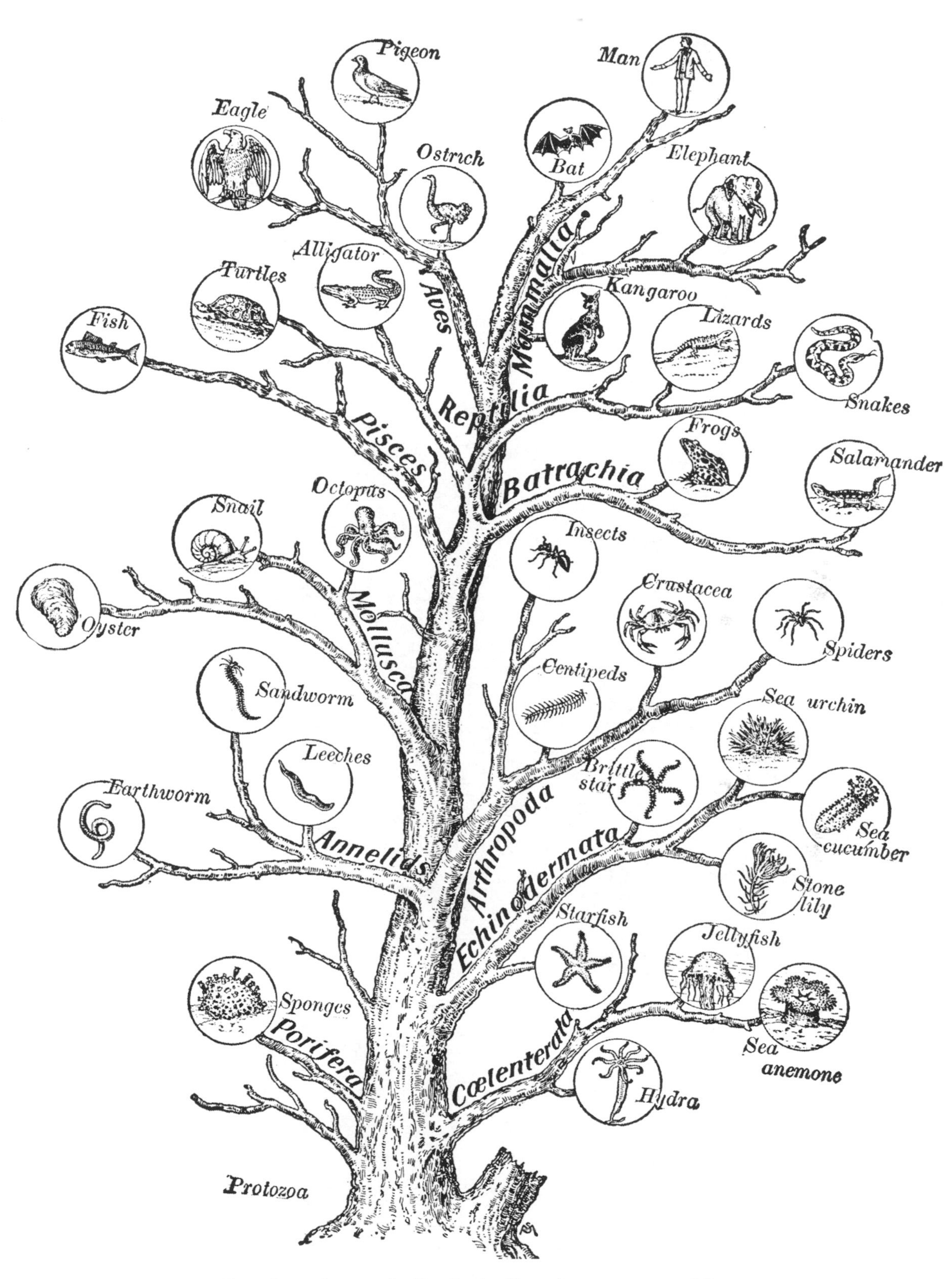

FIGURE 115. Branching diagram by Benjamin Gruenberg constructed for his *Elementary Biology* of 1919, this one a "genealogical tree of animal life."

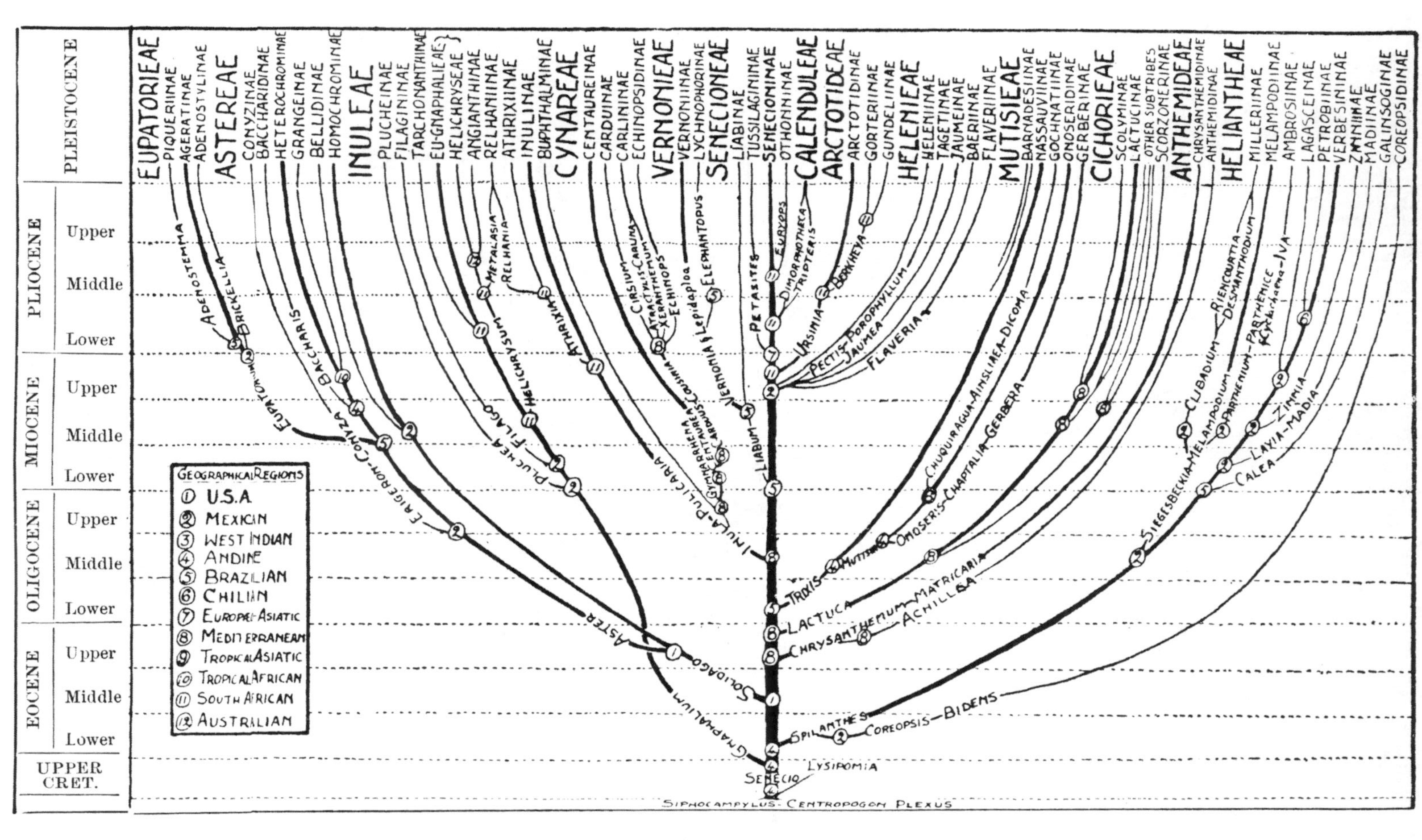

FIGURE 116. "The evolution of the Compositae in time and space" by James Small, 1919. In an early example of phylogeography, the numbers placed at the branching points of the tree correspond to geographic regions.

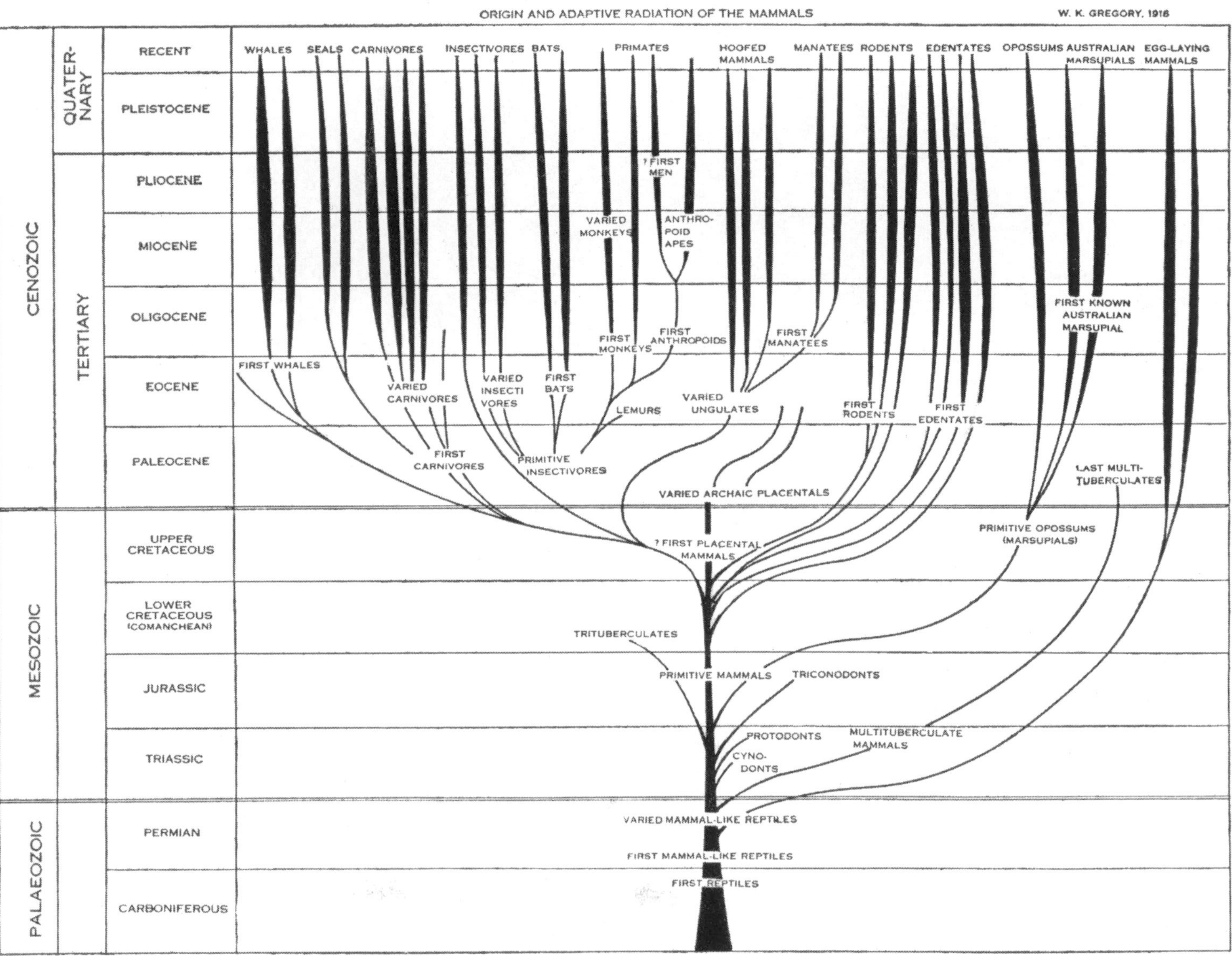

FIGURE 117. A tree of mammal relationships prepared by William King Gregory for Henry Fairfield Osborn's popular book *The Origin and Evolution of Life*, 1917.

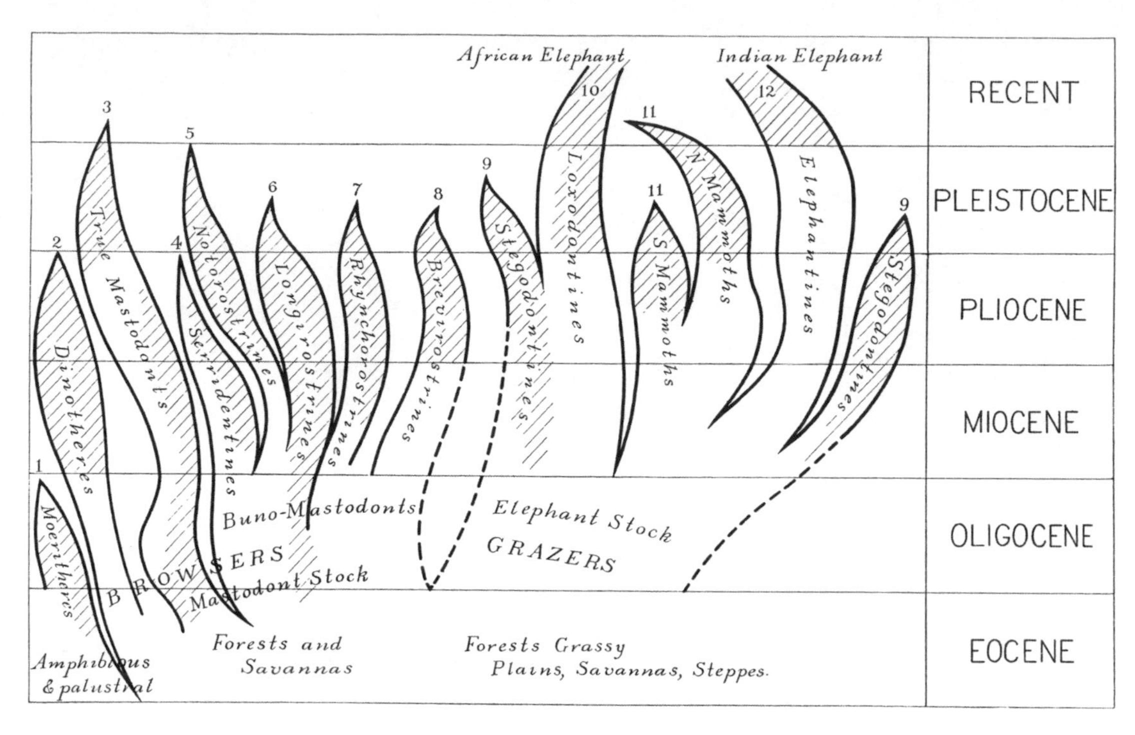

FIGURE 118. Adaptive radiation of the elephants and their allies published in 1921 by Henry Fairfield Osborn, unique for its time in demonstrating a correspondence between habitat, feeding behavior, and phylogeny.

---

facing page

FIGURE 119. A phylogeny of elephants constructed by Henry Fairfield Osborn in 1926, this one a "pictogram," showing relationships, but also an indication of relative size, and the remarkable convergence of general body shape among modern forms descended from markedly different ancestors.
Osborn, 1934a, *American Naturalist,* 68(716):212, fig. 3; courtesy of the University of Chicago Press. Used with permission.

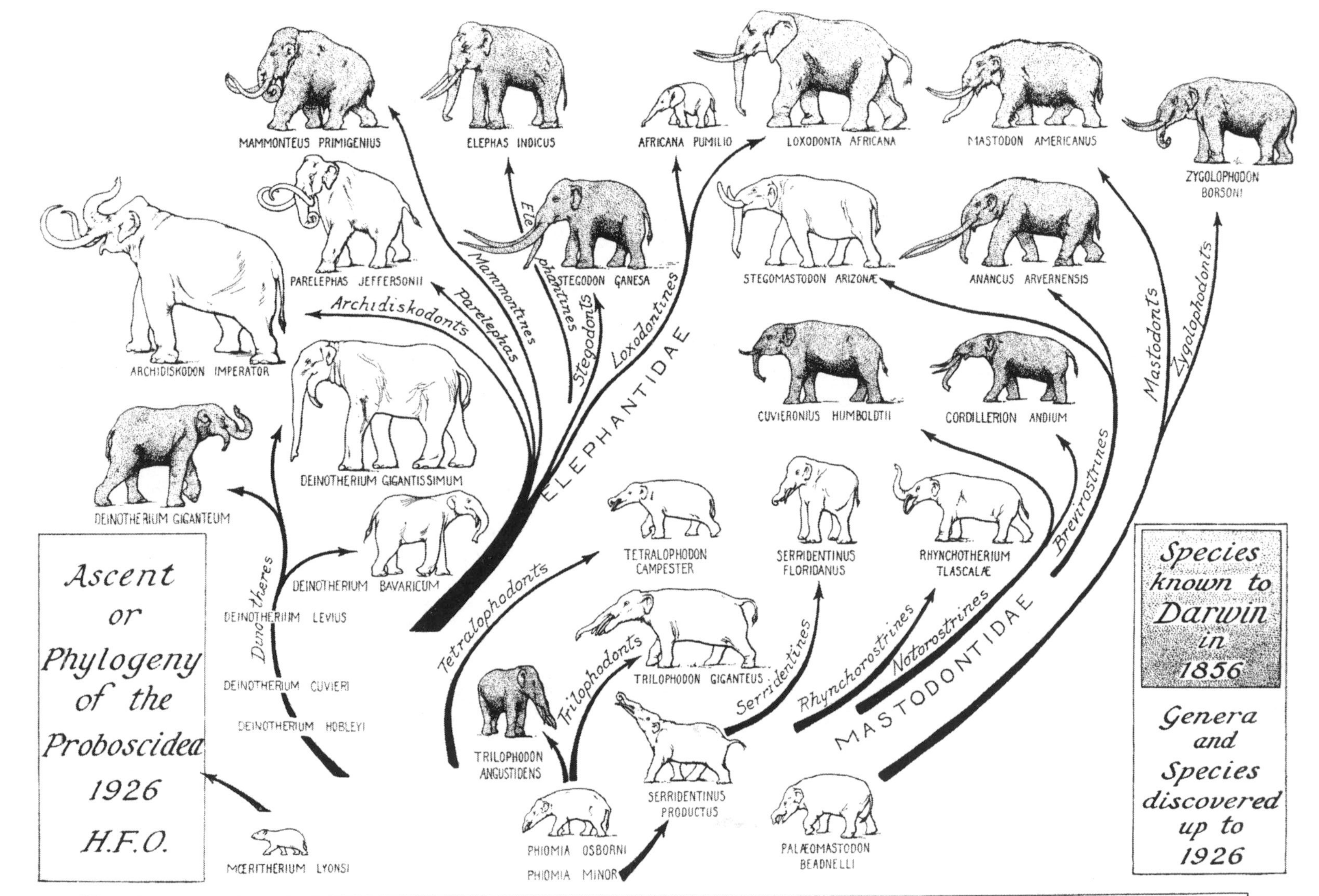
MAMMONTEUS PRIMIGENIUS
ELEPHAS INDICUS
AFRICANA PUMILIO
LOXODONTA AFRICANA
MASTODON AMERICANUS
ZYGOLOPHODON BORSONI
PARELEPHAS JEFFERSONII
STEGODON GANESA
STEGOMASTODON ARIZONÆ
ANANCUS ARVERNENSIS
ARCHIDISKODON IMPERATOR
Archidiskodonts
Parelephas
Mammontines
Elephantines
Stegodonts
Loxodontines
ELEPHANTIDAE
Mastodonts
Zygolophodonts
CUVIERONIUS HUMBOLDTII
CORDILLERION ANDIUM
DEINOTHERIUM GIGANTISSIMUM
DEINOTHERIUM GIGANTEUM
DEINOTHERIUM BAVARICUM
DEINOTHERIUM LEVIUS
DEINOTHERIUM CUVIERI
DEINOTHERIUM HOBLEYI
Dinotheres
TETRALOPHODON CAMPESTER
SERRIDENTINUS FLORIDANUS
RHYNCHOTHERIUM TLASCALÆ
Brevirostrines
Tetralophodonts
TRILOPHODON GIGANTEUS
Trilophodonts
Serridentines
Rhynchorostrines
Notorostrines
MASTODONTIDAE
TRILOPHODON ANGUSTIDENS
SERRIDENTINUS PRODUCTUS
PALÆOMASTODON BEADNELLI
PHIOMIA OSBORNI
PHIOMIA MINOR
MŒRITHERIUM LYONSI
Ascent or Phylogeny of the Proboscidea 1926 H.F.O.
Species known to Darwin in 1856
Genera and Species discovered up to 1926

Ascent phyla of Elephants and Mastodonts. After Osborn in the year 1926

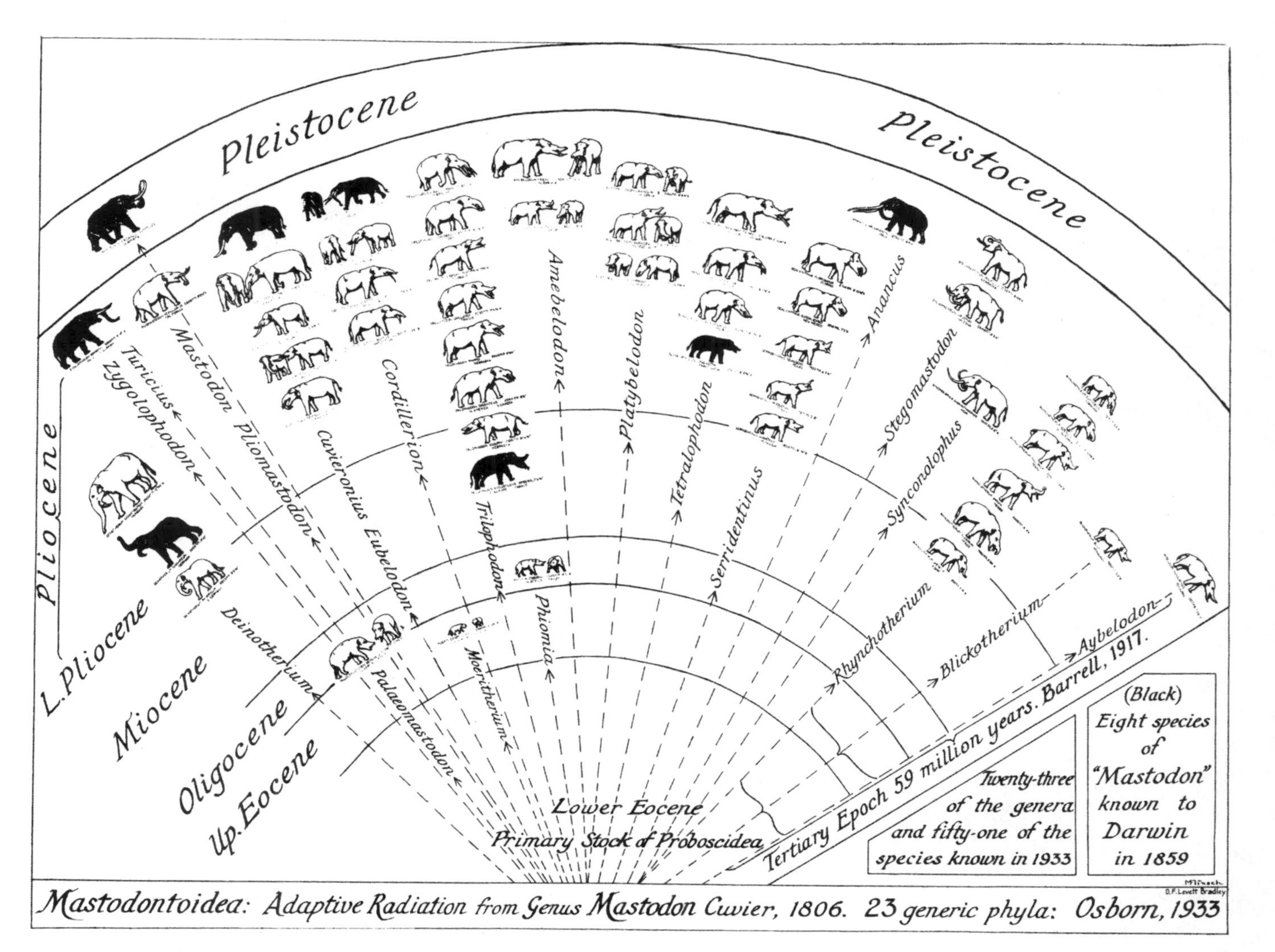

FIGURE 120. Lines of descent of the Mastodontoidea drawn by Henry Osborn in 1933 and published in 1934.

Osborn, 1934a *American Naturalist,* 68(716):214, fig. 4; courtesy of the University of Chicago Press. Used with permission.

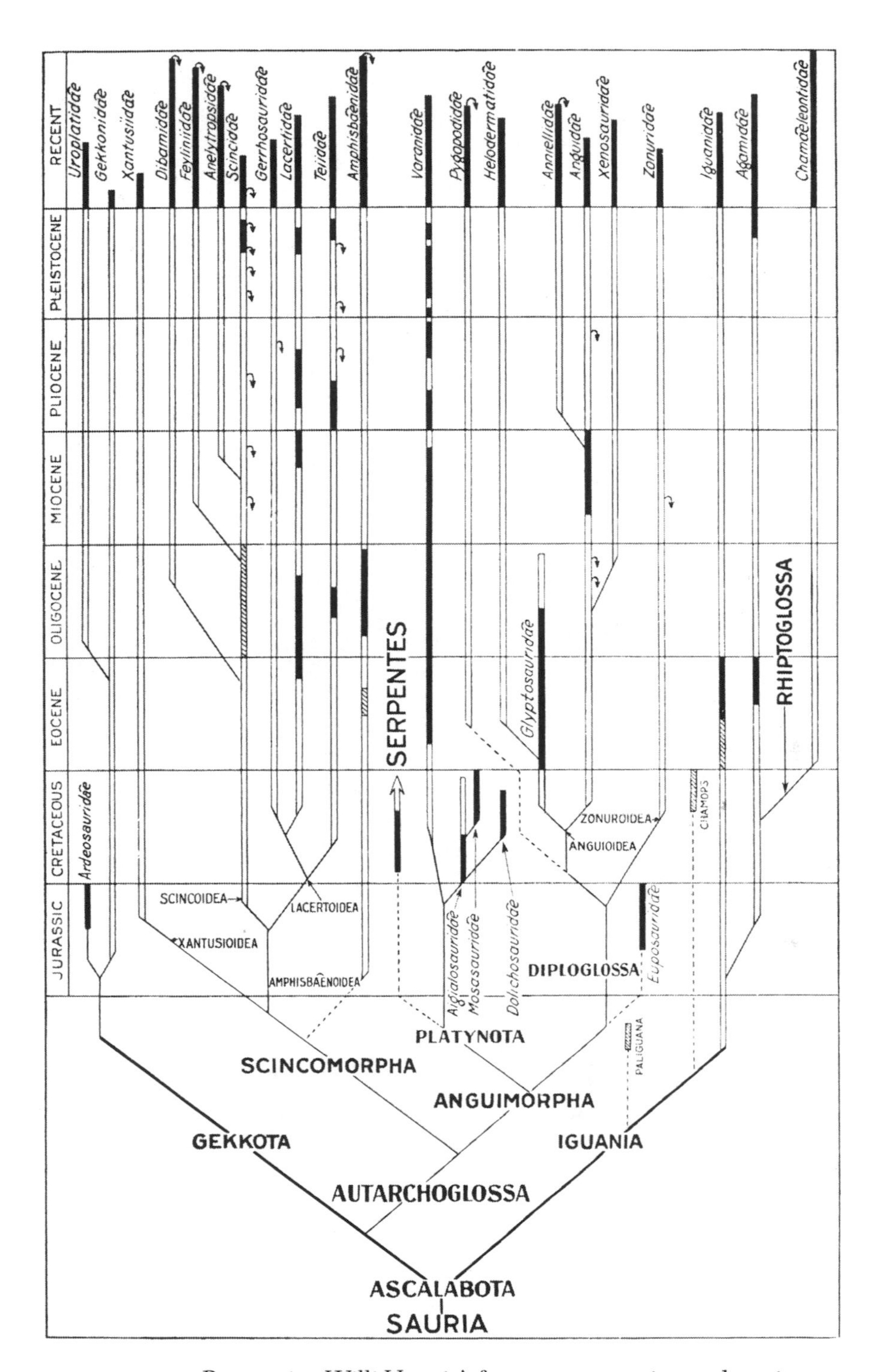

FIGURE 121. Preempting Willi Hennig's famous 1950 treatise on the principles of phylogenetic systematics by a quarter of a century, Charles Lewis Camp's "cladistic" phylogeny of the families of lizards recognizes only monophyletic groups. It also clusters groups on the basis of shared-derived characters. Camp, 1923, *Bulletin of the American Museum of Natural History*, 48(11):333, unnumbered figure; courtesy of the American Museum of Natural History. Used with permission.

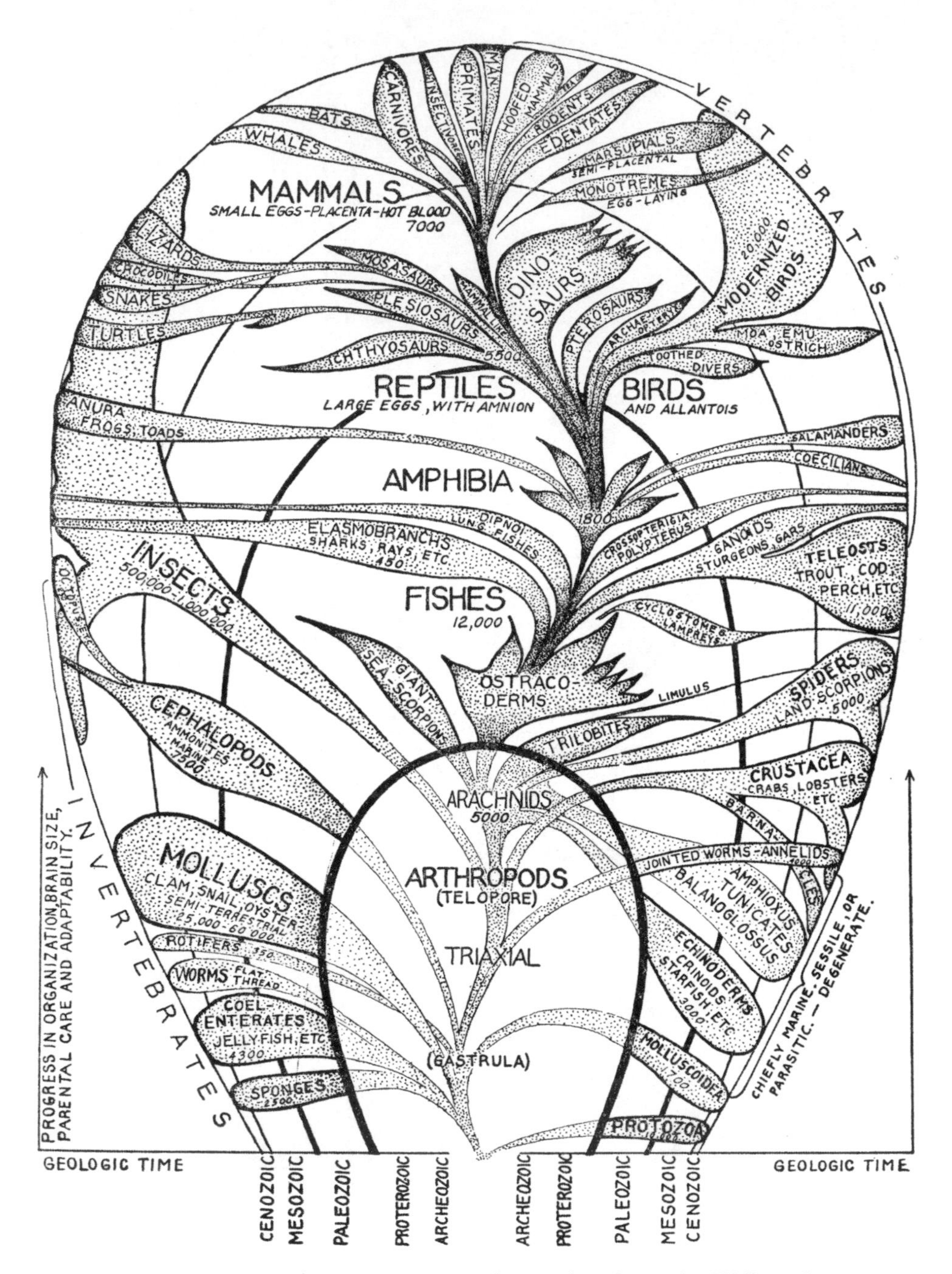

FIGURE 122. An ingenious tree of animal evolution by William Patten in 1923, showing "progress in organization, brain size, parental care, and adaptability" along the vertical axis, with geological time indicated along the horizontal, extending laterally in both directions from the origin of life at the center.

Patten, 1923, *Evolution, Part Two: The Evolution of Plant and Animal Life,* pl. 1, following page 50; courtesy of Deborah J. Gray and Dartmouth Press. Used with permission.

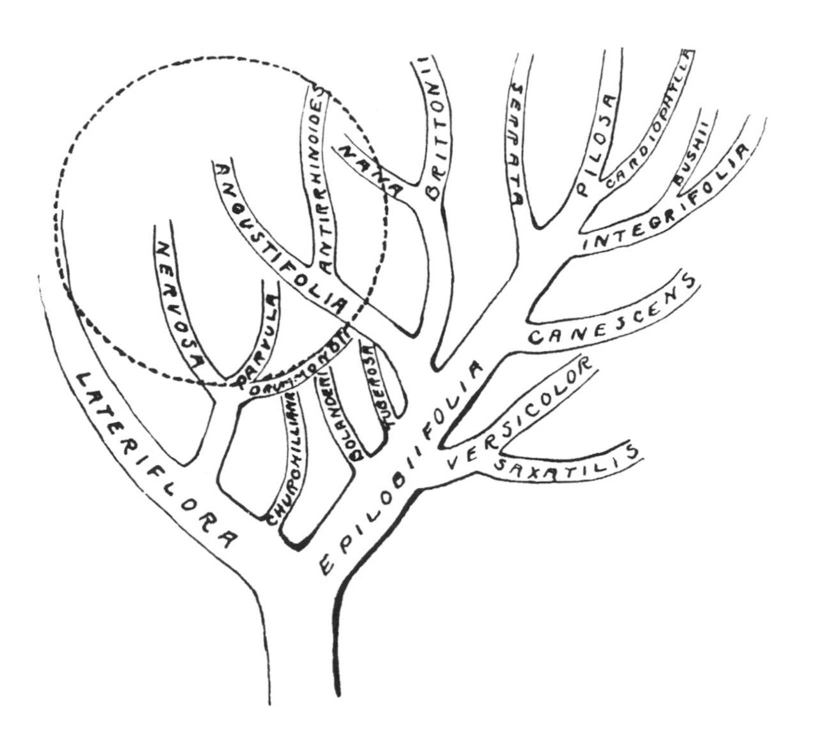

FIGURE 123. Charles William Theodore Penland's 1924 tree of relationships within *Scutellaria,* a genus of the mint family. The circle encloses taxa of three independent lineages that share derived features of the fruiting bodies.

Penland, 1924, *Rhodora, Journal of the New England Botanical Club,* 26(304):63, diagram 1; courtesy of Elizabeth Farnsworth and the New England Botanical Club. Used with permission.

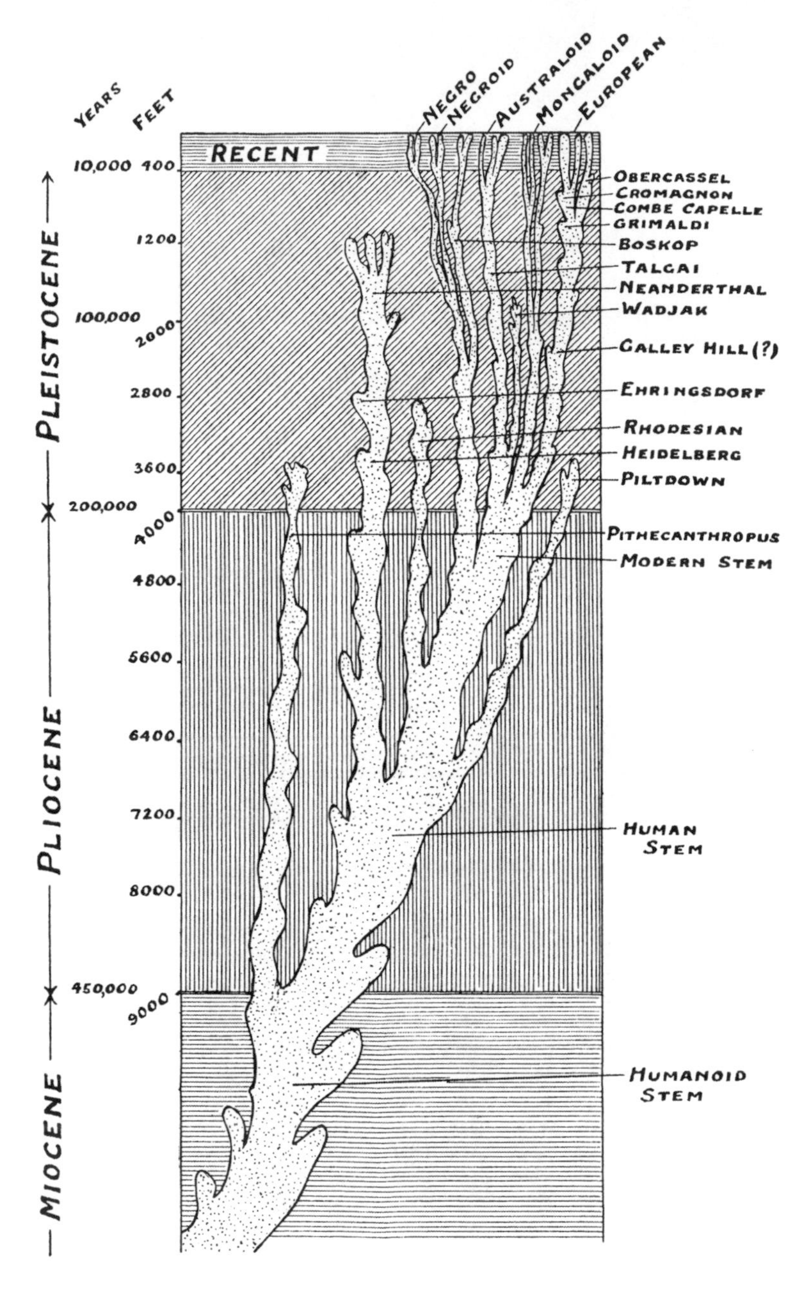

FIGURE 124. Arthur Keith's 1925 "genealogical tree of man's ancestry," showing Piltdown Man, the subject of the famous paleontological hoax perpetrated in 1912, emerging from the "human stem" during the Middle Pliocene. Note also the inclusion of Ernst Haeckel's *Pithecanthropus,* now included within the genus *Homo.* Geological time periods, represented by the relative depth of their deposits, are shown on the left.
Keith, 1925, *The Antiquity of Man,* vol. 2, p. 714, fig. 263.

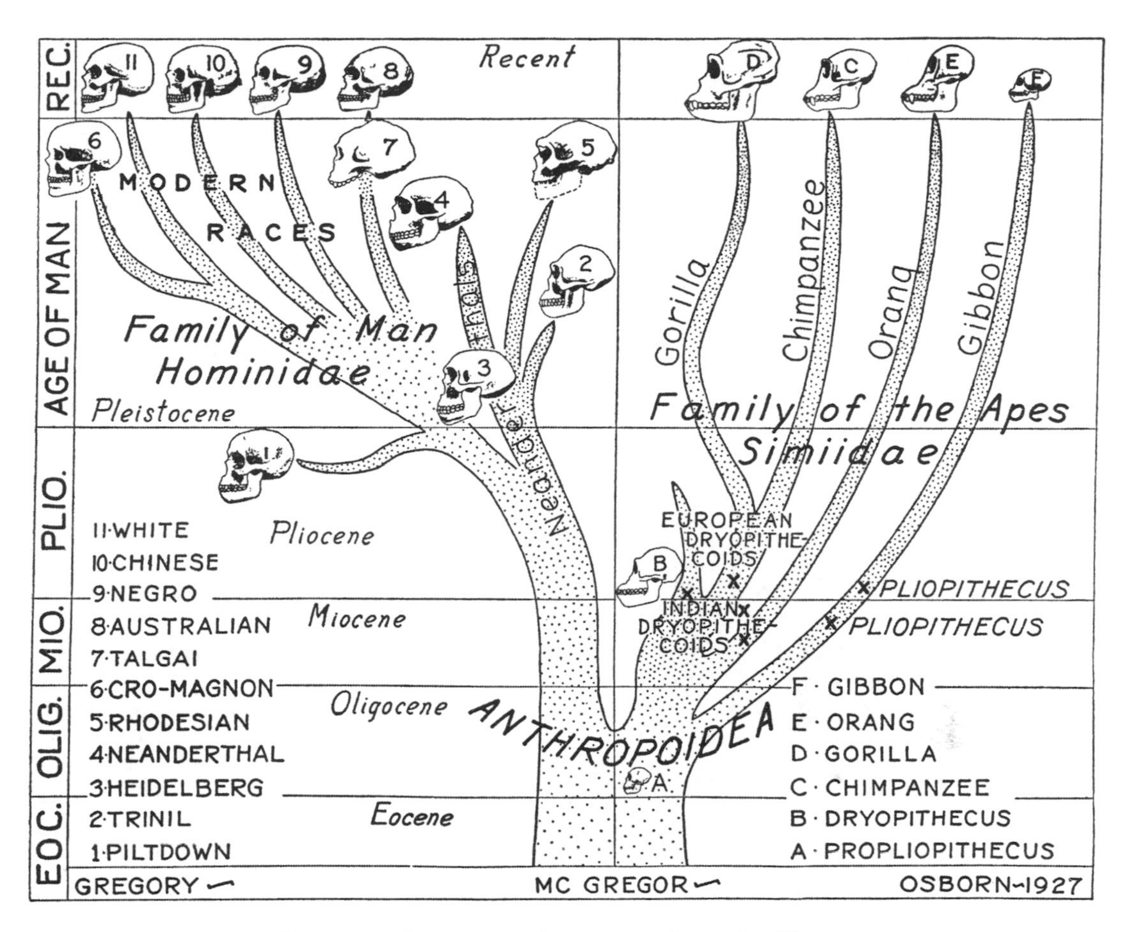

FIGURE 125. A tree illustrating the origin and antiquity of man, by Henry Fairfield Osborn in 1927, showing Piltdown Man, diverging during Late Pliocene, as ancestral to all other hominids.

Osborn, 1927a, *Science,* 65:486, fig. 2, based on an address presented to the American Philosophical Society on 28 April 1927; courtesy of Elizabeth Sandler and Mary McDonald. Reprinted with permission of the American Association for the Advancement of Science and the American Philosophical Society.

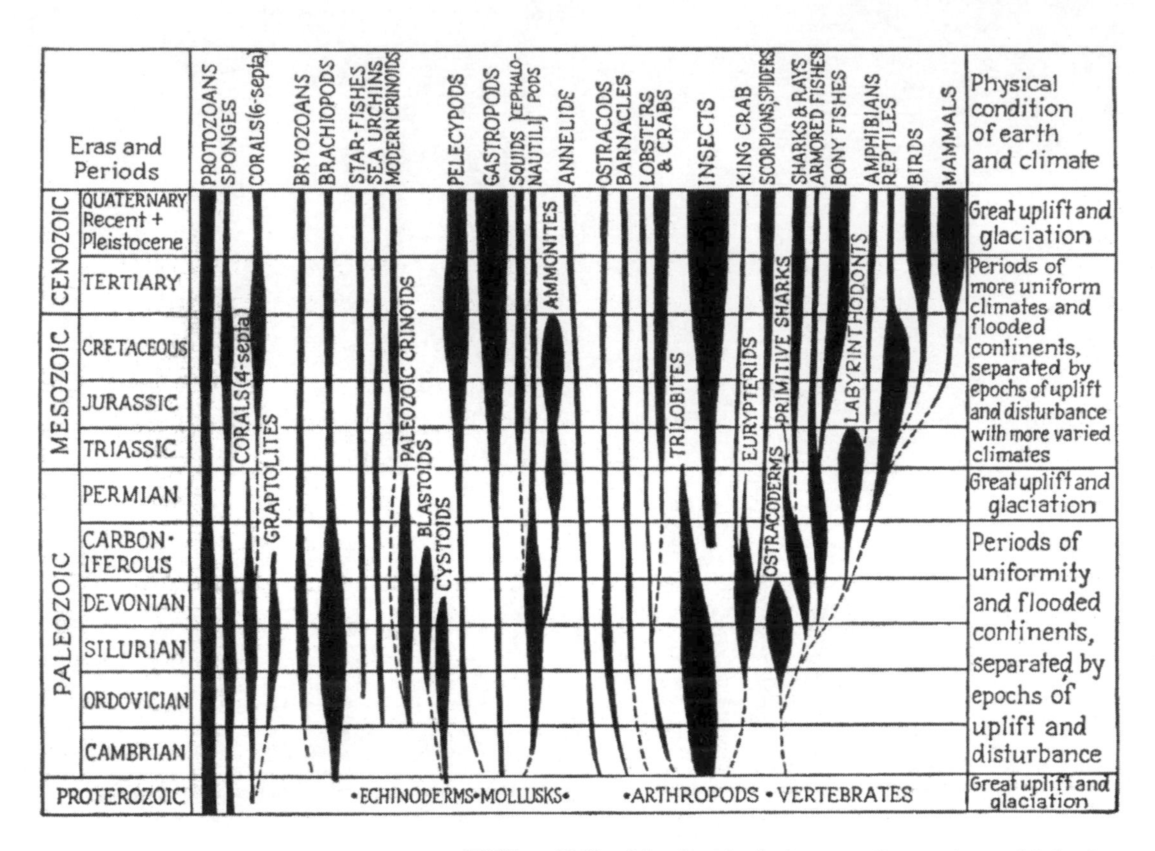

FIGURE 126. William Diller Matthew's phylogeny of animals—published for the first time in Tracy Storer's 1943 *General Zoology*—showing eras and periods of geological time along the left and the physical conditions of earth and climate along the right, reminiscent of the 1844 "spindle" diagram of Louis Agassiz.

Storer, 1943, *General Zoology*, p. 2, fig. 1-1.

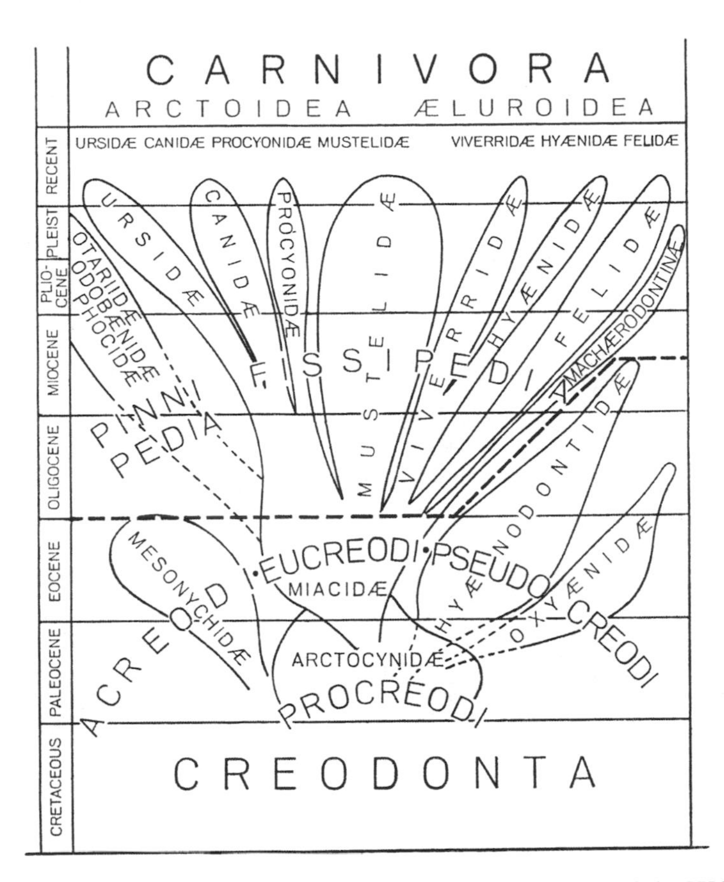

FIGURE 127. A phylogeny of the carnivorous mammals by William Diller Matthew, 1930.

Matthew, 1930, *Journal of Mammalogy,* 11(2):119, fig. 1; courtesy of Daniela Bone, the American Society of Mammalogists, and Allen Press Publishing Services. Used with permission.

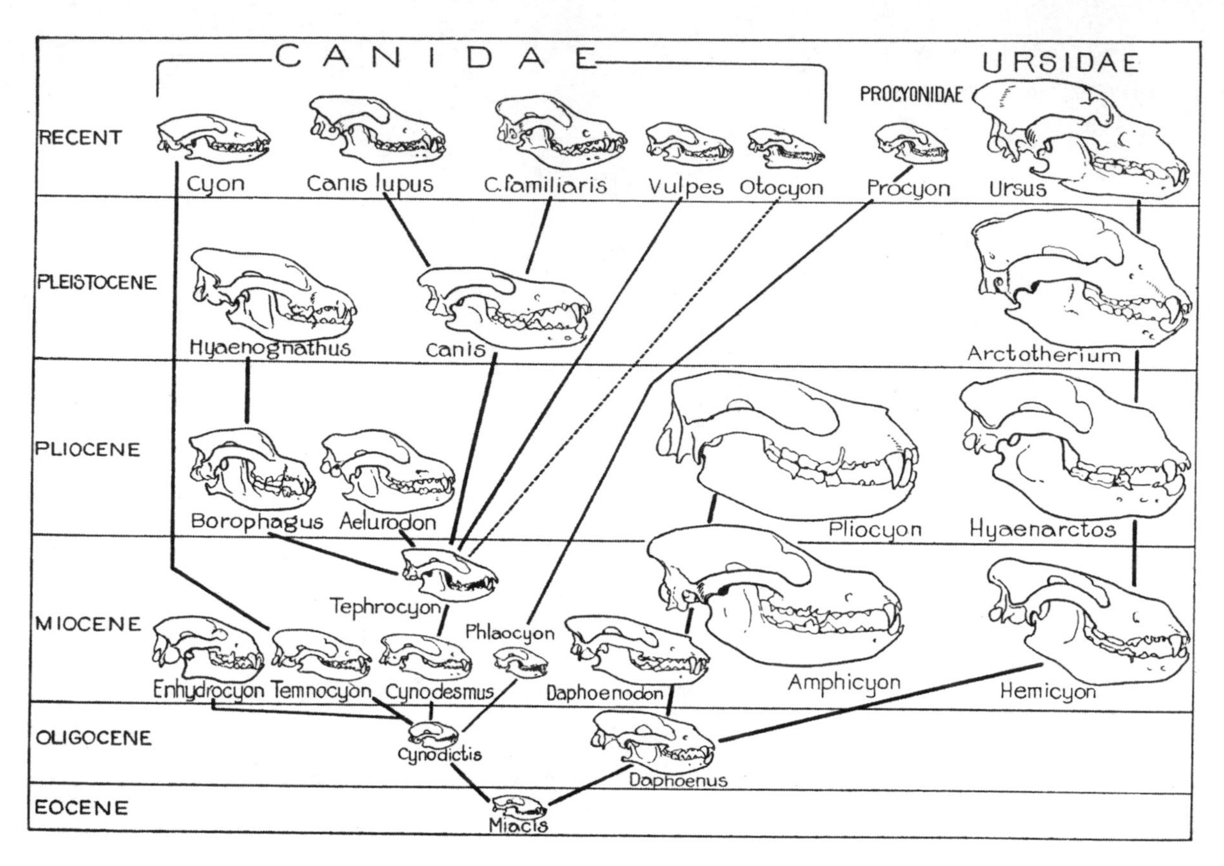

FIGURE 128. Tree by William Diller Matthew in 1930, focusing on the relationships of dogs and their close affinities with bears and with raccoons and their allies. The various taxa are represented by images of their skulls all drawn to the same scale.

Matthew, 1930, *Journal of Mammalogy,* 11(2):132, fig. 3; courtesy of Daniela Bone, the American Society of Mammalogists, and Allen Press Publishing Services. Used with permission.

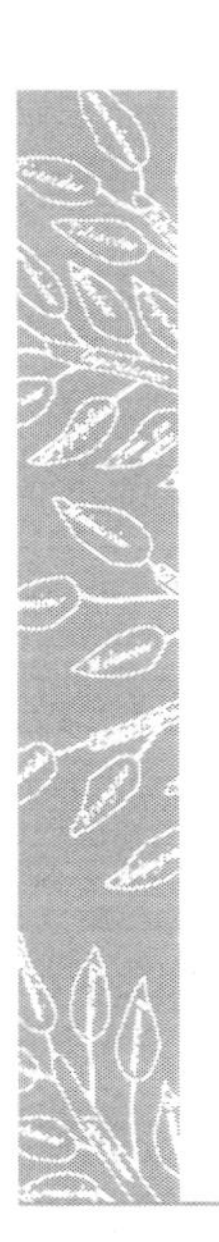

The Trees of Alfred Sherwood Romer

1933–1966

After Ernst Haeckel, Alfred Sherwood Romer (1894–1973) might be called the next great tree builder. Romer studied at Amherst College and Columbia University before joining the department of paleontology at the University of Chicago in 1923. From there, in 1934, he became professor at Harvard University, eventually taking on the directorship of the Museum of Comparative Zoology in 1946. A paleontologist and comparative anatomist, he specialized in vertebrate evolution, making numerous important contributions to the scientific literature and building massive collections of fossil material. However, Romer is remembered most fondly for his brilliant teaching and for his many well-written and beautifully illustrated books. Anyone who majored in biology or took comparative anatomy or paleontology during the 1950s, 60s, and 70s will be familiar with the distinctive style of his now classic texts: *Man and the Vertebrates* and *Vertebrate Paleontology,* both first published in 1933, and *The Vertebrate Body* and *Vertebrate Story,* both in 1949. In *Man and the Vertebrates,* as well as the several subsequent editions of *The Vertebrate Body* and *Vertebrate Story,* Romer included a series of generalized trees, intended primarily to provide advanced undergraduates with the basic underpinnings of evolutionary relationships. Three of these diagrams are shown here, the first, "a simplified family tree of the invertebrates to show the probable descent of vertebrates"[1] (Figure 129); the second, taken from the same edition, is a family tree of the reptiles (Figure 130); and the third, which appeared for the first time in the 1955 second edition of *The Vertebrate Body,* a tree of the orders of the eutherian or placental mammals (Figure 131).

Strikingly different, more detailed, and considerably more appealing are Romer's trees designed originally for the second edition of *Vertebrate Paleontology* published in 1945, many of which were revised and updated, and a few new ones added, in later editions.[2] Although, in some ways these paleontological diagrams are original, they quite obviously hark back to Louis Agassiz's 1844 paleontological tree of fishes (see Figure 55) and also to some of the early trees of William King Gregory (Figure 117) and William Diller Matthew (Figure 126). The best of them are reproduced here (Figures 132–140). Of course, it should not be forgotten that Romer also constructed numerous additional trees to illustrate his many journal articles, one example of which is shown below (Figure 141).

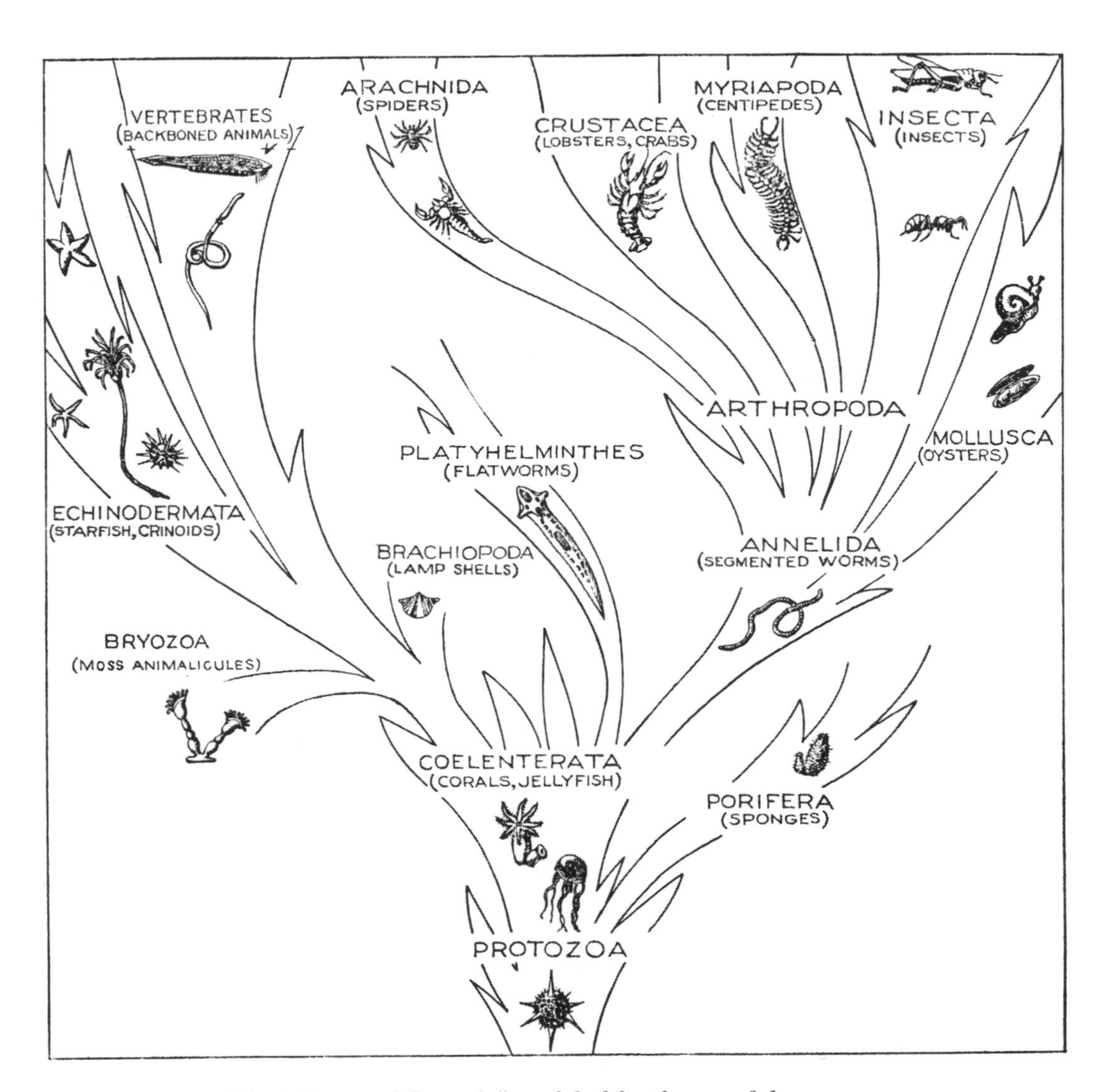

FIGURE 129. Alfred Sherwood Romer's "simplified family tree of the invertebrates to show the probable descent of vertebrates," from 1933.

Romer, 1933a, *Man and the Vertebrates,* 1st ed., p. 16, fig. 7; courtesy of Perry Cartwright and the University of Chicago Press. Used with permission.

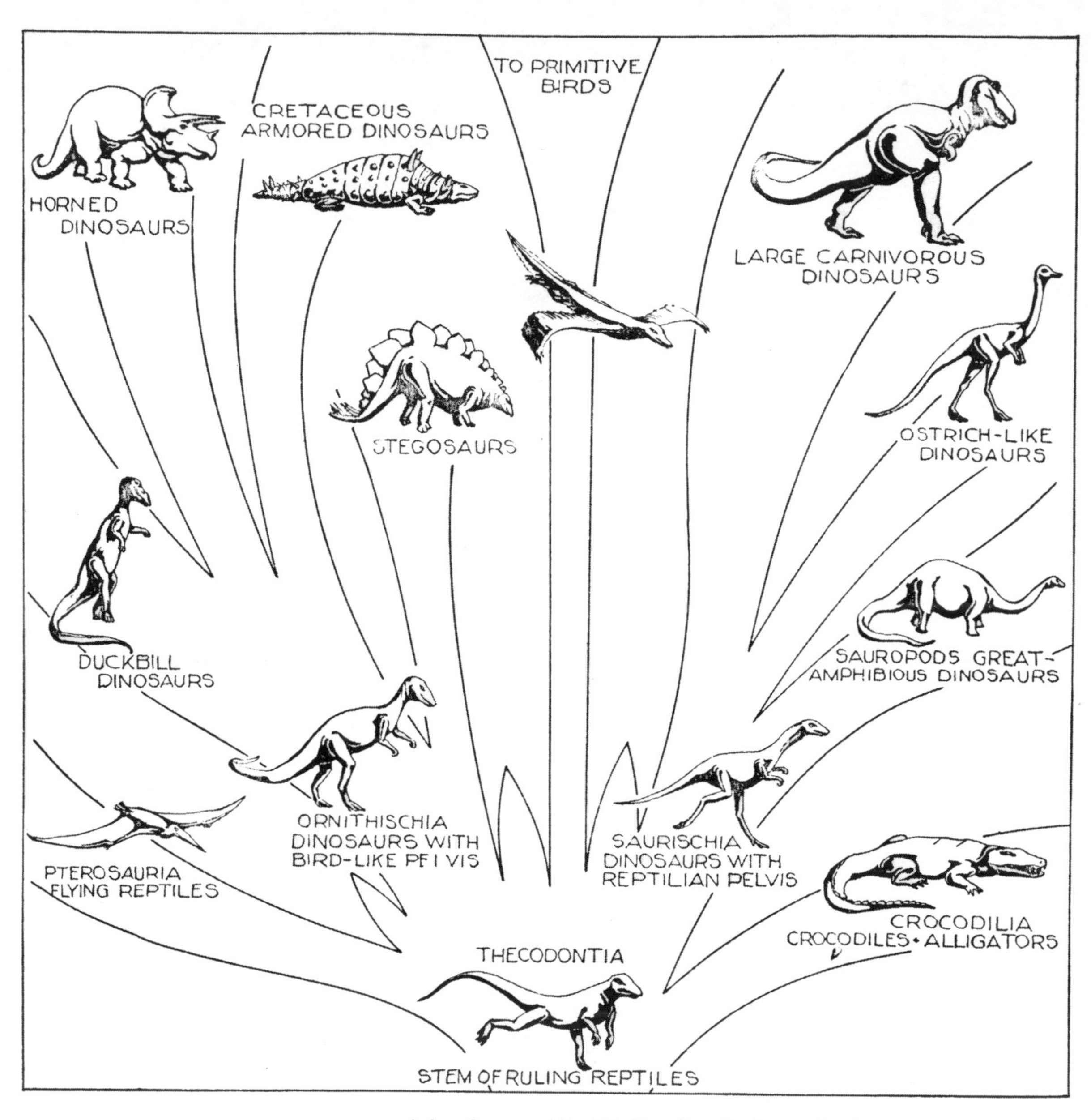

FIGURE 130. A family tree of the "Ruling Reptiles"—an obsolete term originally used to describe a diverse array of dinosaurs, pterosaurs, crocodilians, and birds.

Romer, 1933a, *Man and the Vertebrates,* 1st ed., p. 63, fig. 43; courtesy of Perry Cartwright and the University of Chicago Press. Used with permission.

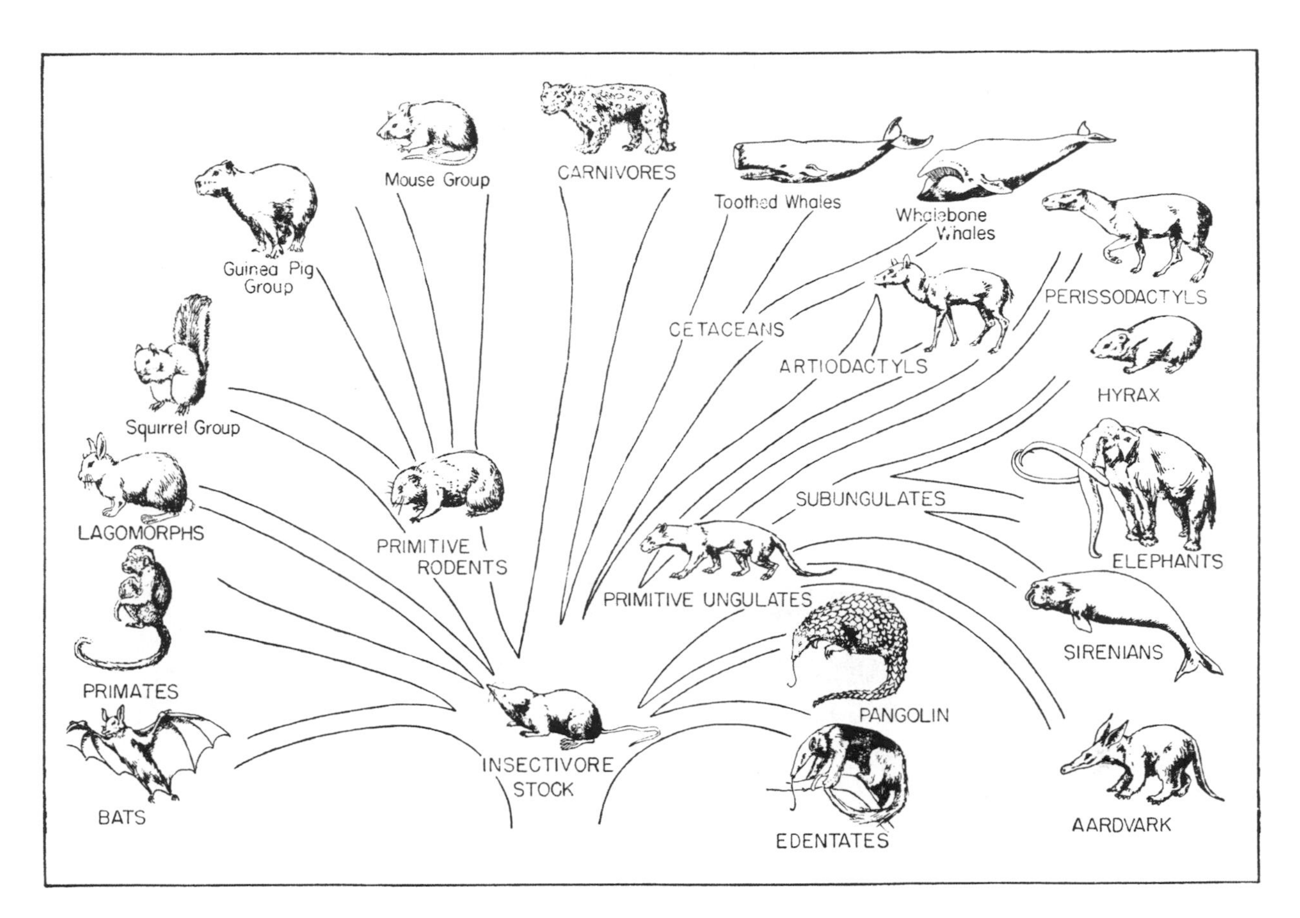

FIGURE 131. A tree of placental mammals.

Romer, 1955, *The Vertebrate Body,* 2nd ed., p. 72, fig. 41; © 1955, 1986 Brooks/Cole, a part of Cengage Learning, Inc., www.cengage.com/permissions. Used with permission.

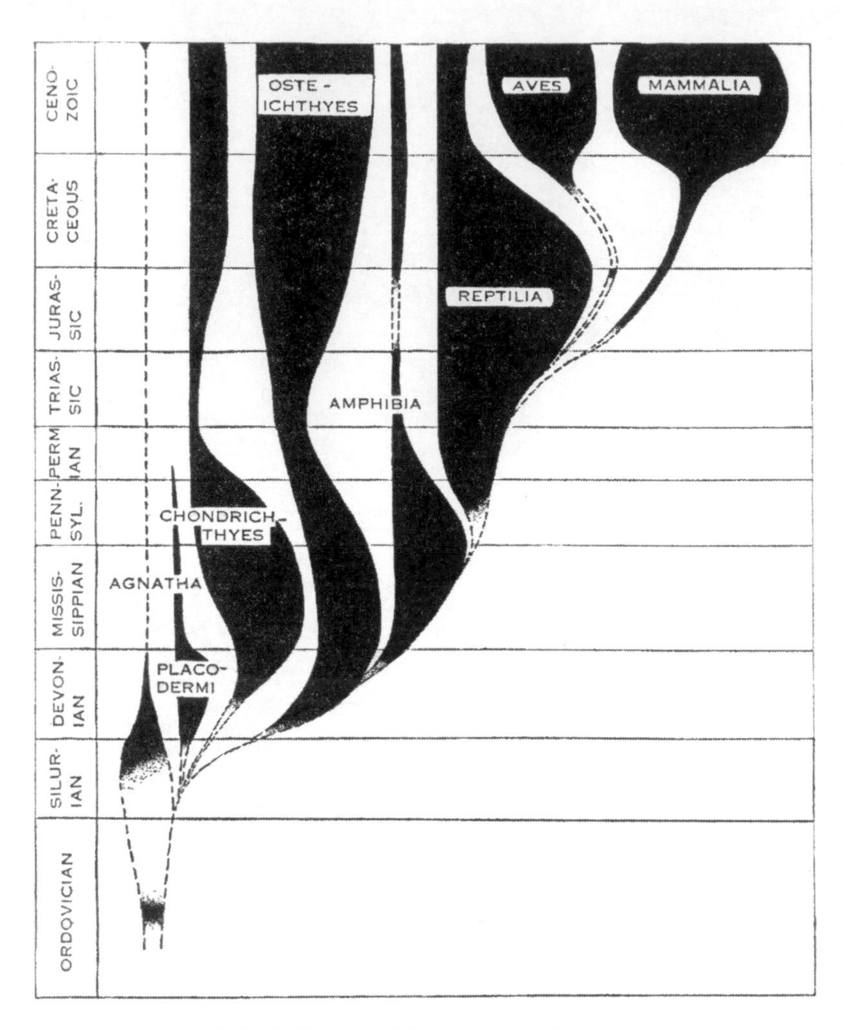

FIGURE 132. Alfred Sherwood Romer's 1945 family tree of the vertebrates. In this tree and in the Romer trees that follow, a rough indication of the comparative abundance of the various groups is furnished by the thickness of the various branches.

Romer, 1945, *Vertebrate Paleontology,* 2nd ed., p. 22, fig. 13; courtesy of Perry Cartwright and the University of Chicago Press. Used with permission.

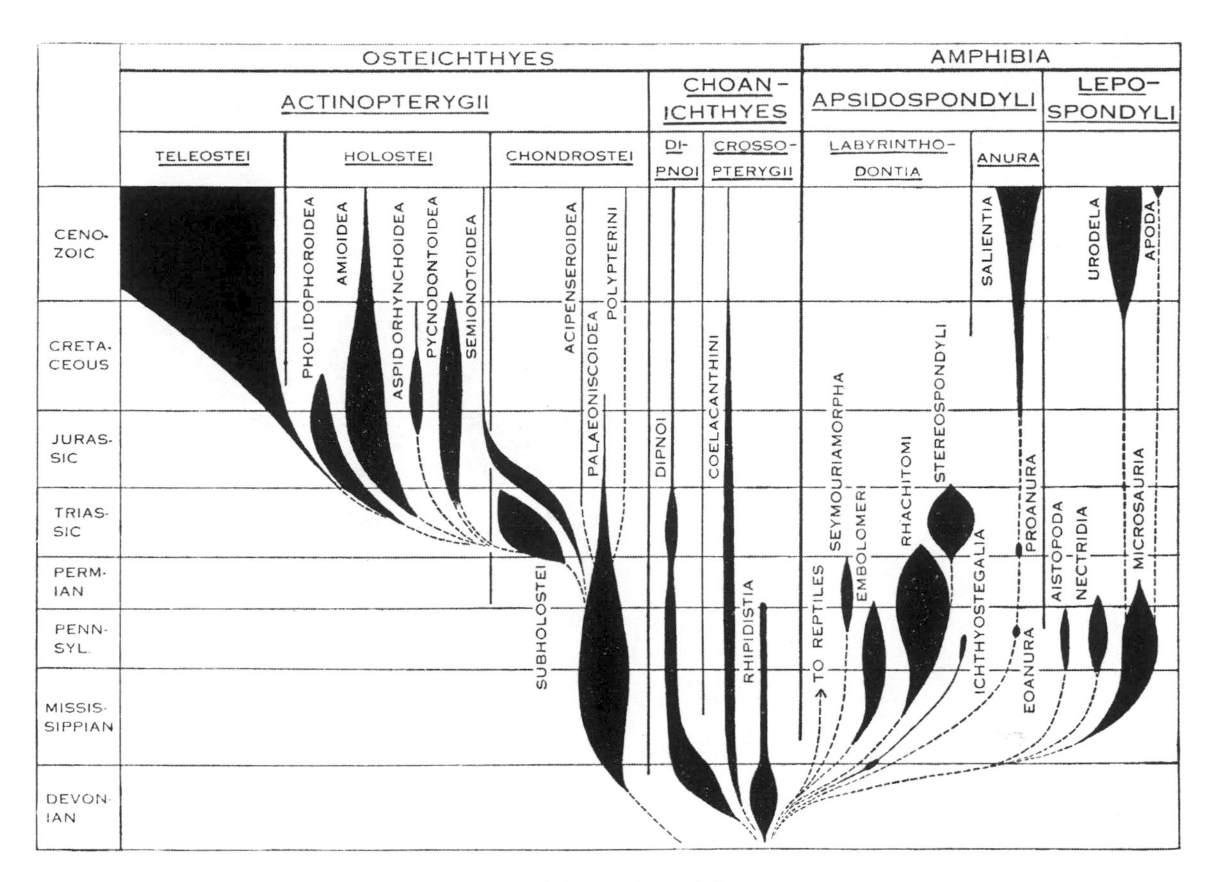

FIGURE 133. A phylogeny of the bony fishes and amphibians.

Romer, 1945, *Vertebrate Paleontology,* 2nd ed., p. 76, fig. 60; courtesy of Perry Cartwright and the University of Chicago Press. Used with permission.

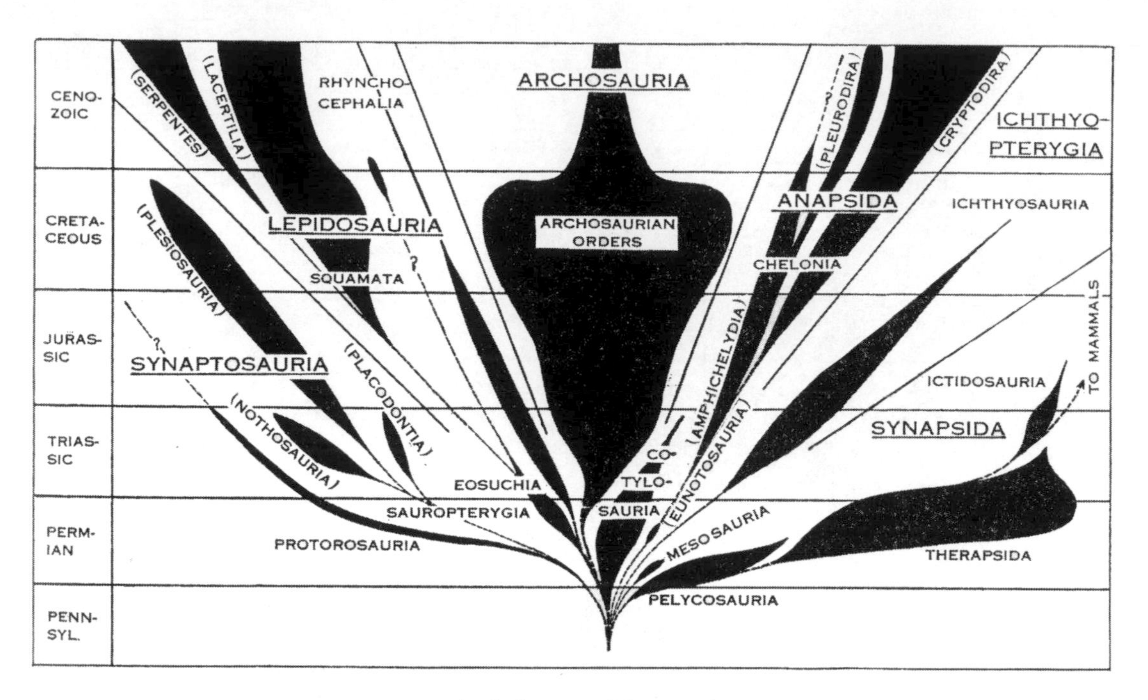

FIGURE 134. A phylogeny of the reptiles.

Romer, 1945, *Vertebrate Paleontology,* 2nd ed., p. 171, fig. 133; courtesy of Perry Cartwright and the University of Chicago Press. Used with permission.

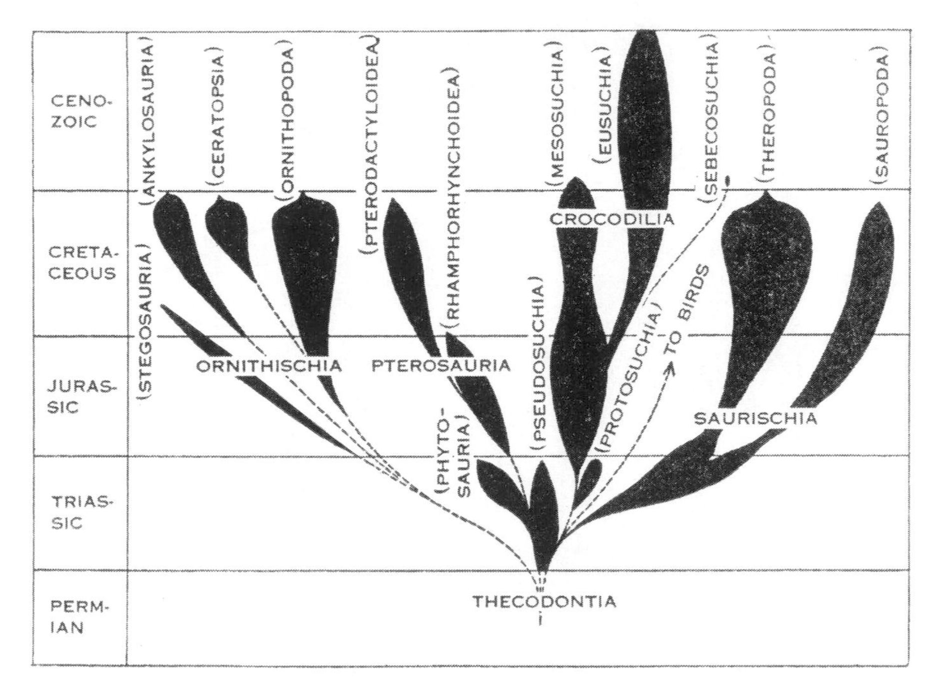

FIGURE 135. A phylogeny of the Thecodontia or "Ruling Reptiles" (see figure 130).

Romer, 1945, *Vertebrate Paleontology,* 2nd ed., p. 210, fig. 173; courtesy of Perry Cartwright and the University of Chicago Press. Used with permission.

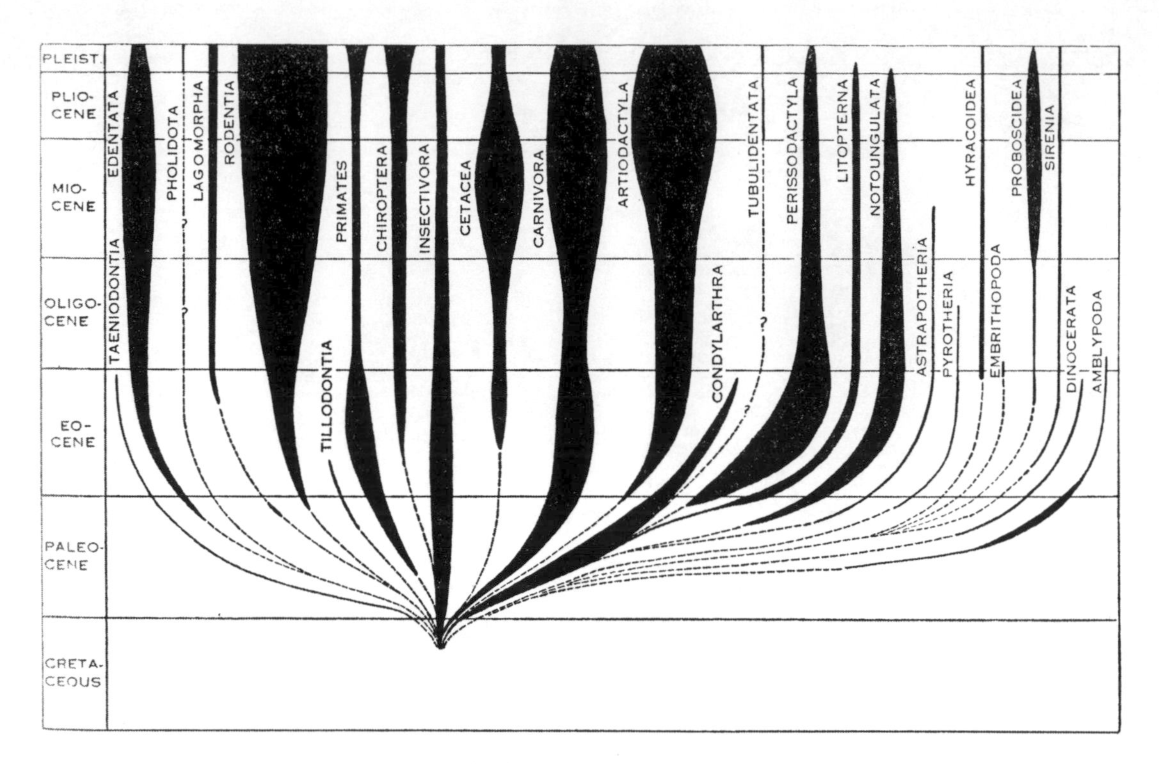

FIGURE 136. A phylogeny of the placental mammals.

Romer, 1945, *Vertebrate Paleontology,* 2nd ed., p. 325, fig. 257; courtesy of Perry Cartwright and the University of Chicago Press. Used with permission.

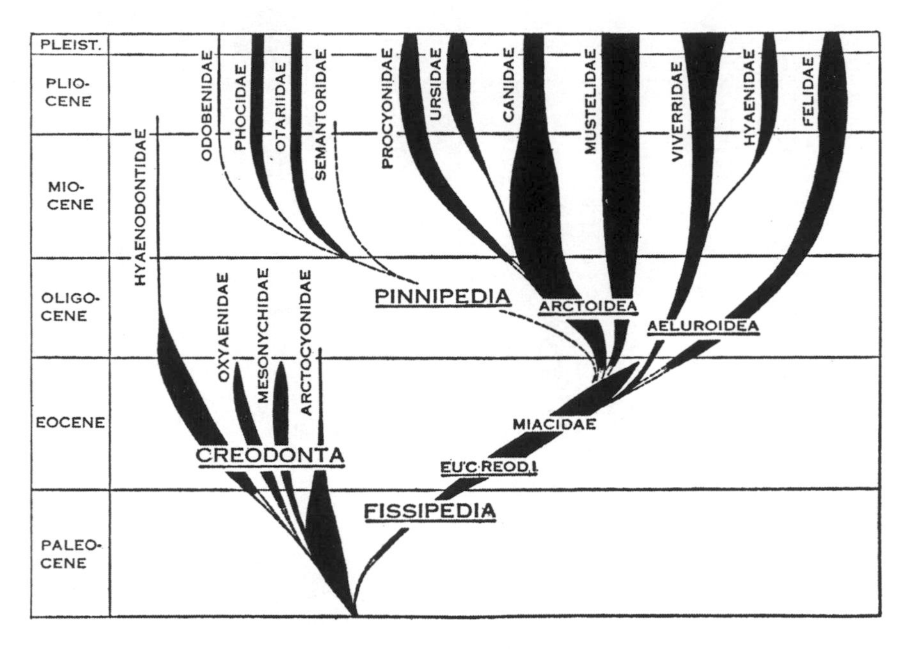

FIGURE 137. A phylogeny of the carnivorous mammals.

Romer, 1945, *Vertebrate Paleontology,* 2nd ed., p. 363, fig. 274; courtesy of Perry Cartwright and the University of Chicago Press. Used with permission.

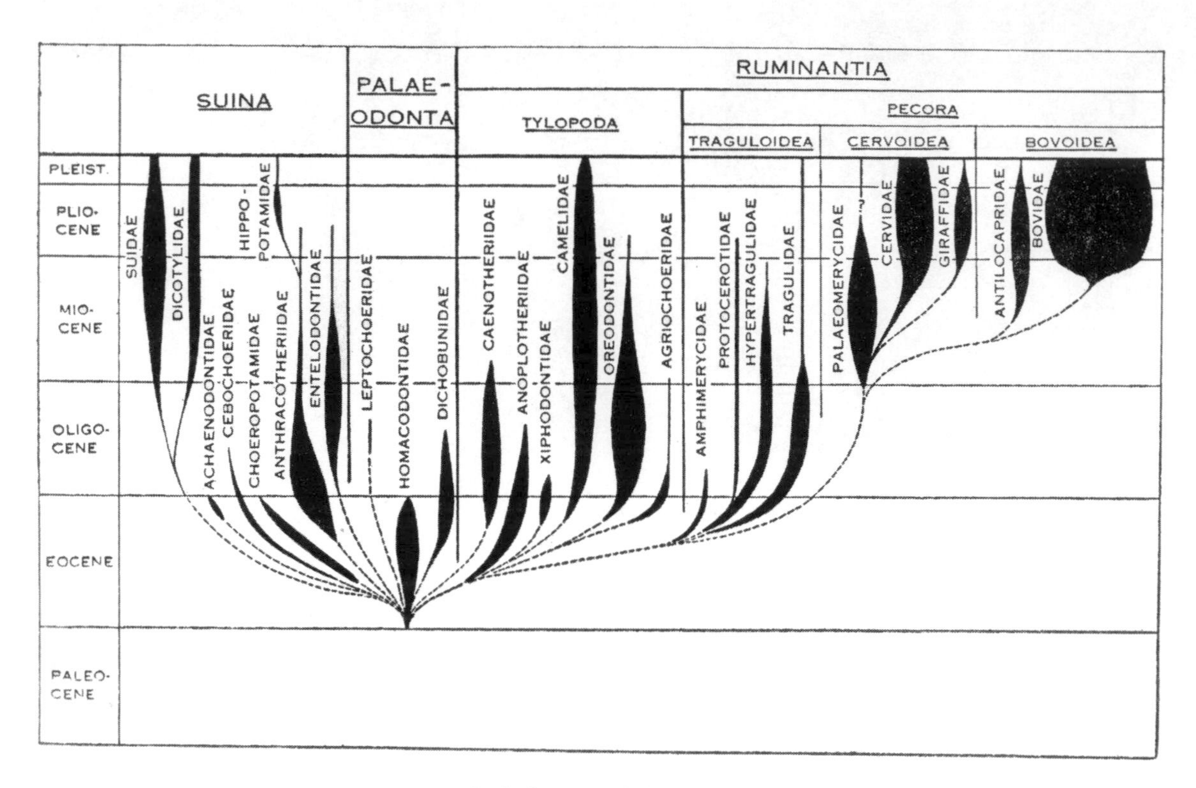

FIGURE 138. A phylogeny of the Artiodactyla, the even-toed hoofed mammals.

Romer, 1945, *Vertebrate Paleontology,* 2nd ed., p. 443, fig. 334; courtesy of Perry Cartwright and the University of Chicago Press. Used with permission.

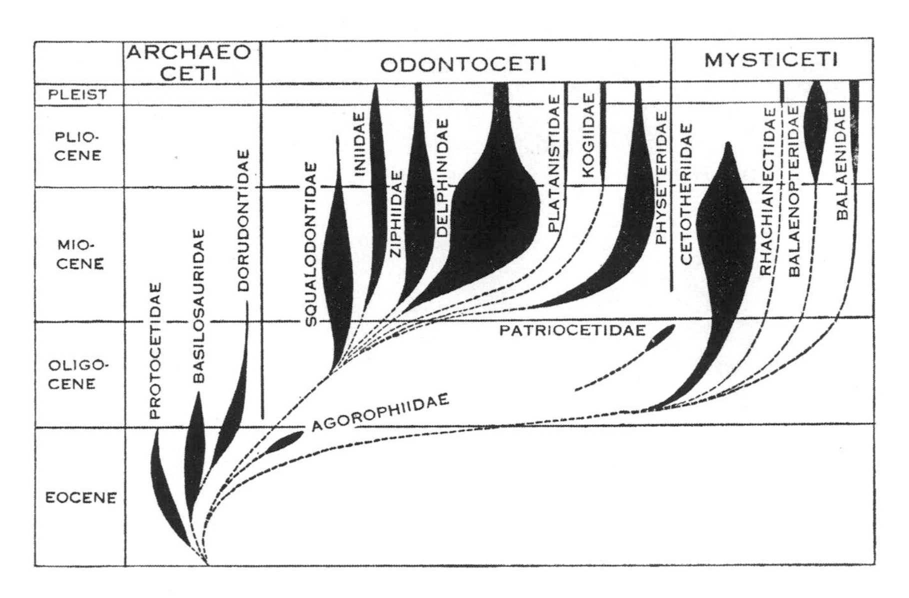

FIGURE 139. A phylogeny of the whales.

Romer, 1945, *Vertebrate Paleontology,* 2nd ed., p. 492, fig. 368; courtesy of Perry Cartwright and the University of Chicago Press. Used with permission.

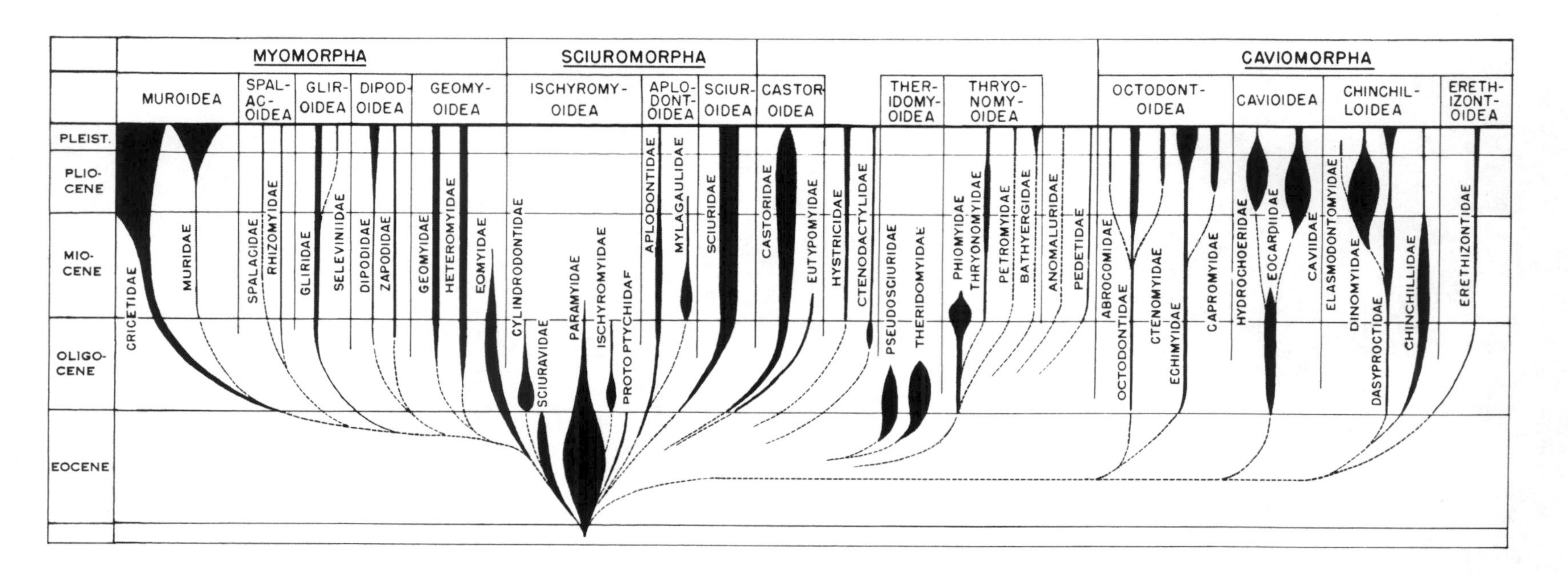

FIGURE 140. A phylogeny of the rodents.

Romer, 1966, *Vertebrate Paleontology*, 3rd ed., p. 303, fig. 435; courtesy of Perry Cartwright and the University of Chicago Press. Used with permission.

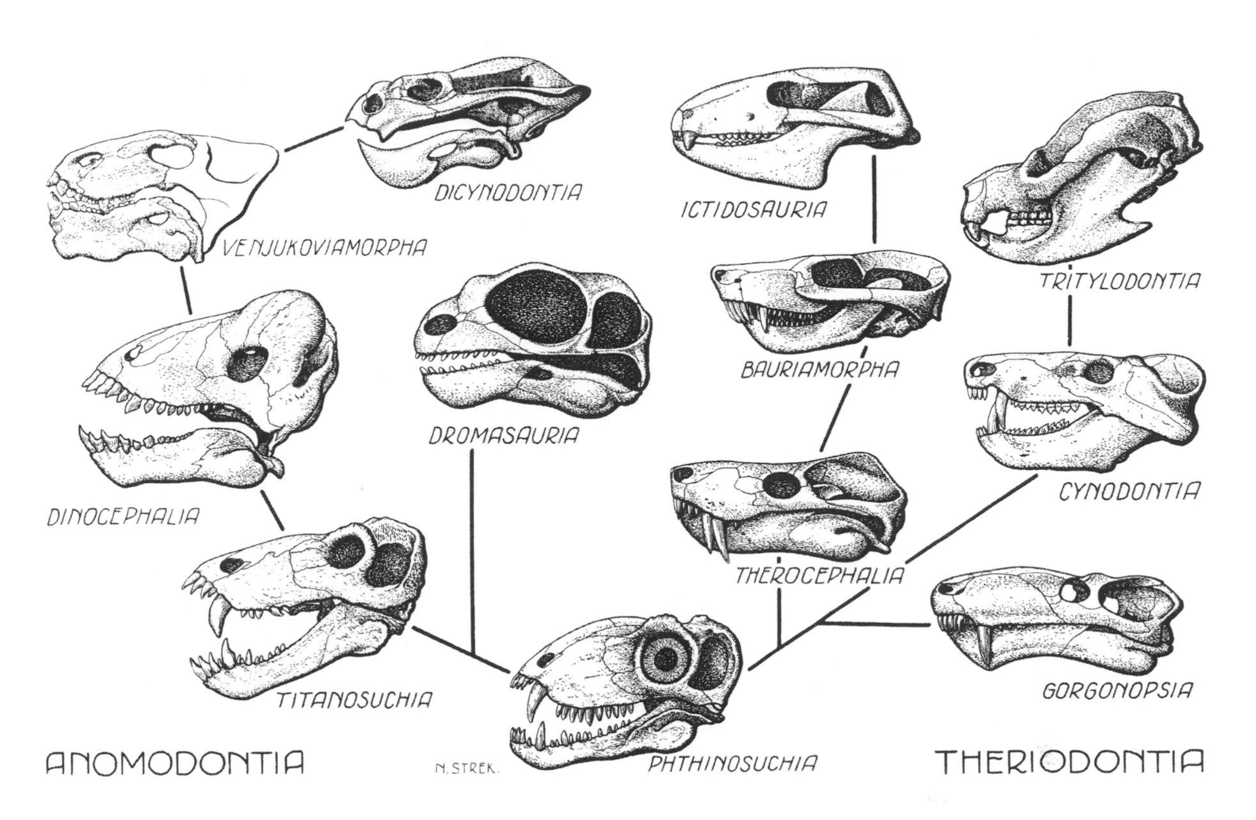

FIGURE 141. A phylogeny of the mammal-like reptiles, the various groups represented by images of their skulls.

Romer, 1961, "Synapsid evolution and dentition," p. 27, fig. 10; courtesy of Sophie Dejaegher and the Koninklijke Vlaamse Academie Voor Wetenschappen. Used with permission.

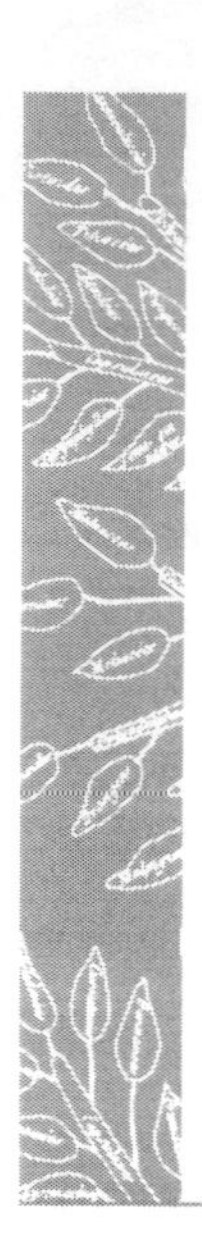

Additional Trees of the Mid-Twentieth Century

1931–1943

A decidedly contemporary, theoretical tree, published by German botanist Walter Max Zimmermann (1892–1980) in 1931, was constructed to illustrate the concept of "phylogenetic relationship" (Figure 142), coming two decades before Willi Hennig's 1950 monograph on the subject (see Figures 190–192). Clearly defined in terms of recency of common descent, Zimmermann stated in no uncertain terms that two species are more closely related to one another than either is to a third species if and only if they share a more recent common ancestor.[1] Like Peter Mitchell and Charles Camp before him (see Figures 106 and 121), Zimmermann failed to adequately develop his ideas in later publications. Otherwise, his name might now be in everyday use to describe the almost universally accepted approach to resolving plant and animal relationships championed a generation later by Hennig.

British zoologist Walter Garstang (1868–1949) is best known for *Larval Forms, and Other Zoological Verses,* a book of collected poems published posthumously in 1951, which describes the form and function of various marine invertebrate larvae and illustrates some of the significant controversies of evolutionary biology of the time. But he also produced a large body of serious scientific work, including a remarkable but largely forgotten classification of teleost fishes that is surprisingly close to current thought (Figure 143). Garstang offered his classification as a replacement for the artificial scheme of relationships prevalent at the time, which he described as "an array of detached and isolated orders, which conveys no explicit outline of the evolutional succession [and whose] very flatness and lack of relief is indeed a misrepresentation of nature."[2] Instead of drawing osteological characters from the skull and jaws, as was the approach then in vogue, he suggested that the primary subdivision of the Teleostei should be based on differences in the otophysic connection between the swim bladder and the ear. Largely ignored at the time, Garstang's approach was revitalized some thirty-five years later by Peter Humphry Greenwood (1927–1995) and colleagues (1966) and stands today as the primary basis for our understanding of teleost fish evolution (see, for example, Figure 182).

A semi-circular array of branches gracefully spreading from a short thick trunk was used by British entomologist William Edward China (1895–1979) in 1933 to display the relationships among insects of the order Hemiptera (Figure 144).[3] Dedicated during his entire career to the taxonomy of this one group, he was prolific, eventually publishing 265 scientific papers, describing 98 new genera and 248 species.[4]

Two trees published by botanist John Henry Schaffner (1866–1939), in a 1934 paper on the "phylogenetic taxonomy of plants" are, in his own words, intended to demonstrate a "phyletic system of classification" based on a list of twenty-one fundamental principles or "dicta" of evolutionary theory. Rigorous application of these principles reveals in turn a succession of ten fundamental stages or "subkingdoms" of plants. While Schaffner's reasoning behind this differentiation is not entirely clear, the final result is not radically different from his contemporaries. Plant relationships as a whole are shown in a somewhat three-dimensional diagram (Figure 145), in which the classes and subclasses emerge as short branches from both sides of a thick stalk. His second diagram focuses on relationships among the orders of the phylum Anthophyta (Figure 146), the flowering plants, containing the twelve classes that form the crown of his first tree. In both diagrams, the solid black portion of the branches indicates the relative diversity of living forms; the dotted portion represents known fossil material. Note in Figure 145 a short, abruptly truncated stalk labeled "Animalia," emerging from the base of the Archeophyta, the latter defined by Schaffner as the "genesis of life or the transition of the first living things from the nonliving"[5]

American phycologist Josephine Elizabeth Tilden (1869–1957) published a branching diagram in her well-known treatment of *Algae and Their Life Relations* of 1935, intended primarily to show the "evolutionary levels" of the various groups of algae (Figure 147).[6] In its extreme linear verticality, it is reminiscent of some of Ernst Haeckel's trees of the 1860s and 70s (see Figures 79 and 80). But not seen before in any tree is a measure along the horizontal, extending from the center of the diagram and spreading out in both directions, of "mutational activity" among the different groups. The main vertical axis of the tree serves to separate groups "submerged in salt water" from those exposed to air during low tide, submerged in fresh water, and adapted to a terrestrial existence. At the midpoint along the central axis, the ancestors of animals emerge, ascending and spreading to the left in polyphyletic fashion, leading to various groups of marine algae and eventually to invertebrate and vertebrate animals.

In a classic paper on "phylogenetic symbols, past and present," published in 1936 by Dutch botanist Herman Johannes Lam (1892–1977), various kinds of branching diagrams are presented, along with a discussion of their details and the methodology behind their construction. Among the various images described, he includes several abstract constructions of his own. One of these (Figure 148) is a "schematical diagram [in which] the system of the organic world is represented, with [an estimate of] the relative diversity of the groups (in thousands of species), recent and extinct, aquatic and terrestrial."[7] Another Lam diagram (Figure 149) taken from the same publication is an extraordinary "spherical system of the microcosm, consisting of an infinite number of concentric time spheres,"[8] that attempts to express time, differentiation and relationship, and biological diversity. The system is "imagined as a sort of translucent sphere, the ever-growing radius of which [corresponds] to the advancing time, while the surface contains, in a proper arrangement, the amount of non-living and living matter at any moment."[9]

In 1938, following the lead set in 1866 by Ernst Haeckel, with his proposal to place microorganisms within a new category called the "Protista" (see Figure 66) American biologist Herbert Faulkner Copeland (1902–1968) argued persuasively that the two-kingdom arrangement of Plantae and Animalia that had been around since antiquity was inadequate to explain biological diversity. Instead he provided evidence to the effect that "organisms can be arranged, naturally and more conveniently than in the past, in four kingdoms": the Monera, organisms without nuclei, the cells solitary or physiologically independent; Protista, largely unicellular organisms, but containing nuclei; Plantae, organisms with plastids, containing starch and cellulose; and Animalia, multicellular organisms, passing through blastula and gastrula stages in development.[10] Copeland's 1938 tree, designed to illustrate his new paradigm, consists of a series of tapering cylinders (Figure 150). The Monera are represented by a tight bundle of six, column-like cylinders on the far left. The evolutionary origin of the nucleus, occurring sometime in the late Archaeozoic, is responsible for the rise of everything else: the Plant Kingdom on the left, emerging with the origin of starch and cellulose; the Animal Kingdom on the right, with the origin of blastula and gastrula; and the numerous subgroups of the Protista clustered in between.

Margery Joan Milne (1914–) and Lorus Johnson Milne (1910–1982), both experts in the biology of caddisflies, order Trichoptera, published in 1939 an extraordinary three-dimensional picture of caddisfly evolution based on differences in caddis worm-case construction (Figure 151). The diagram is remarkable for several reasons, not least of which, images like it did not exist before.[11] Time, which equates to probable relationship, is indicated on the vertical axis; differentiation, in terms of morphological similarity, is indicated along the horizontal from right to left; while habitat differentiation from ponds and slow streams to fast streams is shown in the third dimension, trending from front to back. Letters and numbers indicate character states; the first major dichotomy, for example, labeled 1 and 2, separates herbivores and scavengers, with movable cases on the left; from carnivores, building either fixed shelters or none at all, on the right. Lineages labeled 3 are characterized by having short cases, while those labeled 4, long cases coiled in a flat spiral; 5 and 6 indicate fixed silken tubes versus fixed nets and snares, and so on. According to Milne and Milne, the analysis illustrated in the tree nicely separates the five families and twenty-five subfamilies of the Trichoptera into groups most apparent to contemporary systematists.[12]

Percy Edward Raymond (1879–1952) published in 1939 a branching diagram highly unusual for the time (Figure 152), showing the probable relationships of the important groups of animals, in "progress upward from the Protozoa."[13] The radial lines emanating from the various names of taxa indicate that each group has evolved in all possible directions, but only a few lines have led to progress toward a higher group. Raymond was not a fan of traditional "family trees," stating that a truer idea of the paths of evolutionary change may be gained from his diagram, "in which is expressed the idea that although on the whole there has been progress in an upward direction, it has not been straight forward but is the result of the response of animals to all conditions of the environment. In other words, progress has been the more or less fortuitous result of radiative adaptation."[14]

Ruben Arthur Stirton (1901–1966), well-known curator of fossil mammals at the Museum of Paleontology at the University of California at Berkeley, was for a time the foremost authority on horse evolution. The fossil record of horses is more complete than that of any other family of mammals, thus phylogenies can be based on actual specimens at numerous points within each lineage.[15] Stirton took good advantage of this fact in constructing a number of trees, two of which are shown here. The first is one of his earliest, dating from 1940 (Figure 153). While showing the apparent relationships of the various genera, it also indicates their approximate geological distribution throughout the Tertiary. A much revised version, revealing considerably greater resolution, was published in 1959 (Figure 154). Of all the past diversity, especially during the Late Miocene and Pliocene, only the genus *Equus,* the modern horse, with eight or nine living species, has survived to recent times.

Konrad Zacharias Lorenz (1903–1989) was an Austrian zoologist, ethologist, ornithologist, and one of the founders of the modern study of animal behavior. Lorenz—who shared the 1973 Nobel Prize for Physiology or Medicine with Nikolaas "Niko" Tinbergen (1907–1988) and Karl Ritter von Frisch (1886–1982)—published a remarkable branching diagram in 1941, showing relationships among twenty species of ducks, based primarily on behavioral characters (Figure 155).[16] The species are represented by ascending lines, parallel from their base, but diverging to the left and right as they rise as limbs, emerging from the trunk of the tree. The ascending lines are connected at various points along their length with horizontal lines that symbolize characters, many of which represent derived characters, as expressly indicated by Lorenz. Thus, Lorenz clearly distinguished between primitive and derived, and, in true cladistic fashion, he united nearly all assemblages of two or more species in his tree with shared derived characters.[17] Also thoroughly modern was his thinking about evolutionary relationships: "one must rid oneself of the idea once and for all that a linear arrangement of kinds would ever represent true phylogenetic relationships. . . . All extant animals are living branch tips in the 'tree of life' and hence cannot stem from one another."[18]

Tracy Irwin Storer (1889–1973), long-time professor of zoology at the University of California at Davis, published over two hundred scientific articles and books, but he is probably best remembered for his textbooks. The first edition of his *General Zoology,* which appeared in 1943, contains a "genealogical tree" of the animal kingdom (Figure 156), indicating the "probable relationships and relative position of the major groups (named in boldface)."[19] In modern cladistic fashion, Storer defined all the major groups of animals in his tree with derived character states indicated directly on the diagram, stating that "all groups above a given characteristic (named in italics) possess that character."[20]

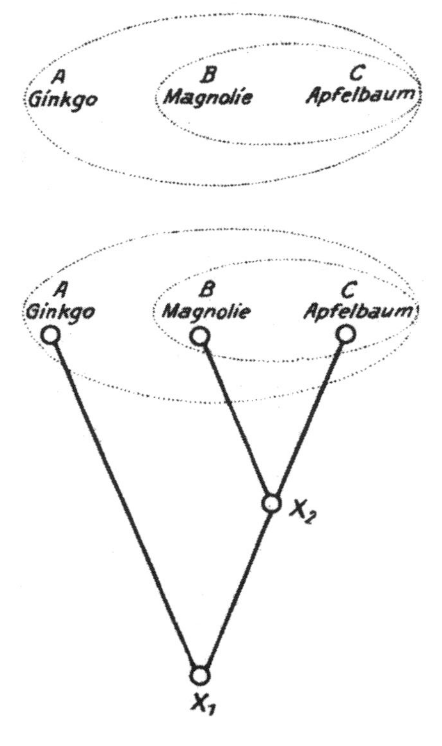

FIGURE 142. Well ahead of his time, Walter Max Zimmermann published this tree in 1931, to illustrate the concept of "phylogenetic relationship," nearly twenty years before publication of Willi Hennig's 1950 monograph on phylogenetic systematics (see figures 190–192).

Zimmermann, 1931, "Arbeitsweise der botanischen Phylogenetik und anderer Gruppierubgswissenschaften," p. 1004, fig. 179; Urban & Schwarzenberg, Berlin.

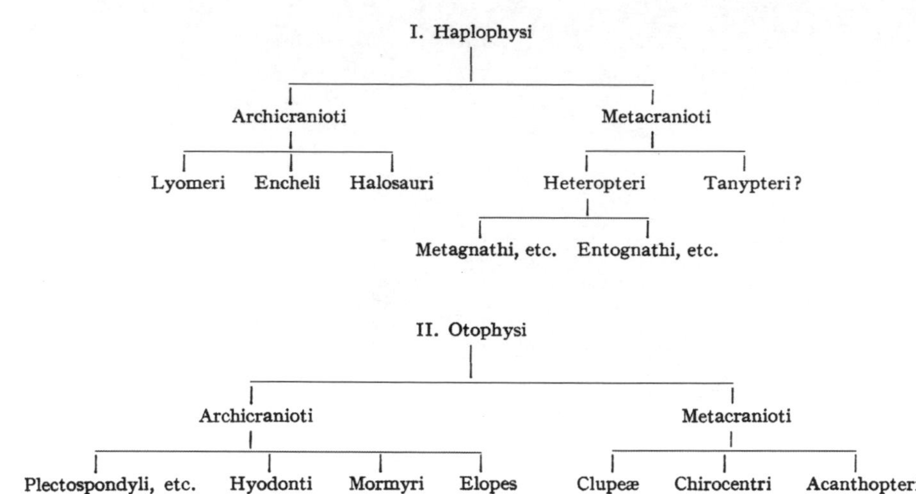

FIGURE 143. A phylogeny of teleostean fishes proposed by Walter Garstang in 1931, based on differences in the otophysic connection between the swim bladder and the ear, a hypothesis that was resurrected thirty-five years later (see figure 182) and now stands as the primary basis for our understanding of teleost fish evolution.

Garstang, 1931, *Proceedings of the Leeds Philosophical Society (Scientific Section)*, 2(5):253, unnumbered figure; courtesy of Anthony C. T. North and the Leeds Philosophical and Literary Society. Used with permission.

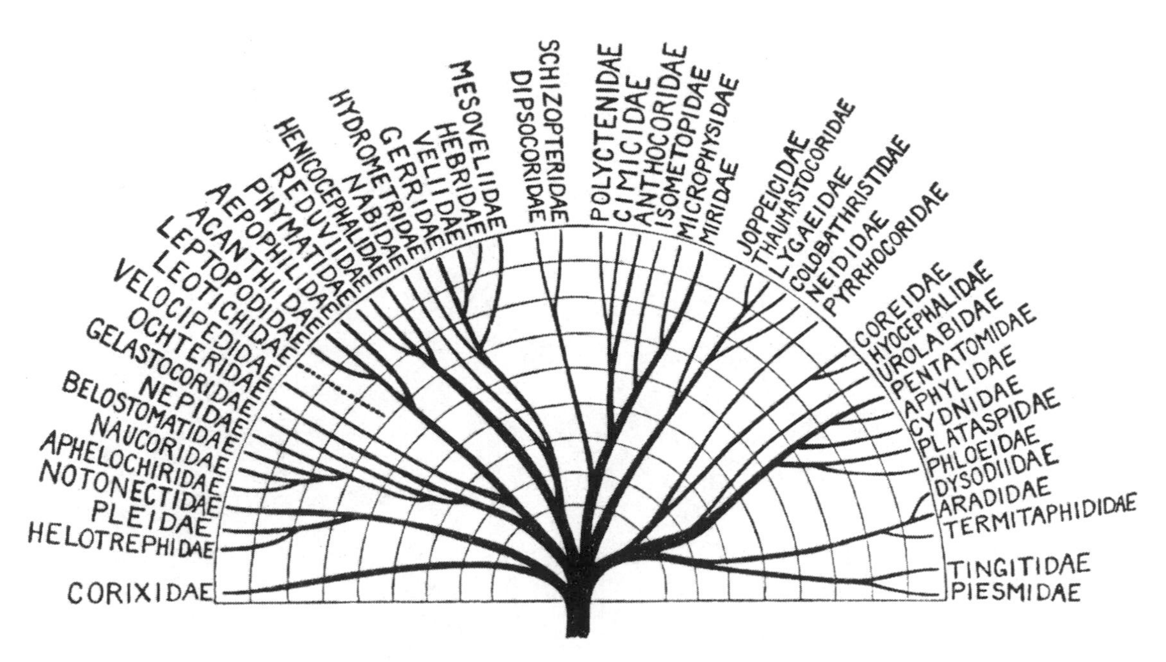

FIGURE 144. William Edward China's 1933 diagram showing the origin and relationships of insects of the order Hemiptera.

China, 1933, *Annals and Magazine of Natural History,* Series 1, 12:195, fig. 4.

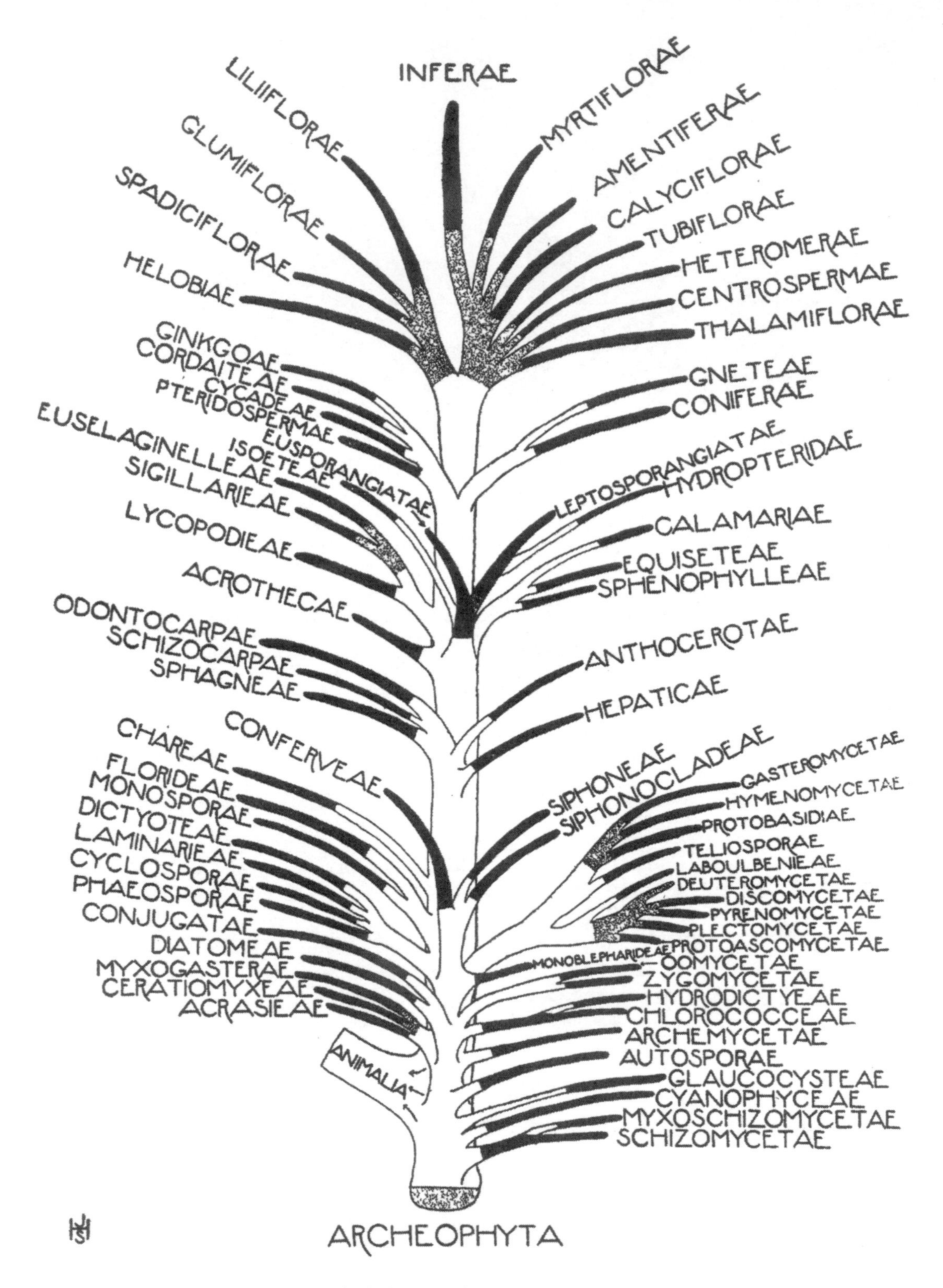

FIGURE 145. A phylogeny of plants, by John Henry Schaffner.

Schaffner, 1934, *Quarterly Review of Biology*, 9(2):142, fig. 1; courtesy of Perry Cartwright and the University of Chicago Press. Used with permission.

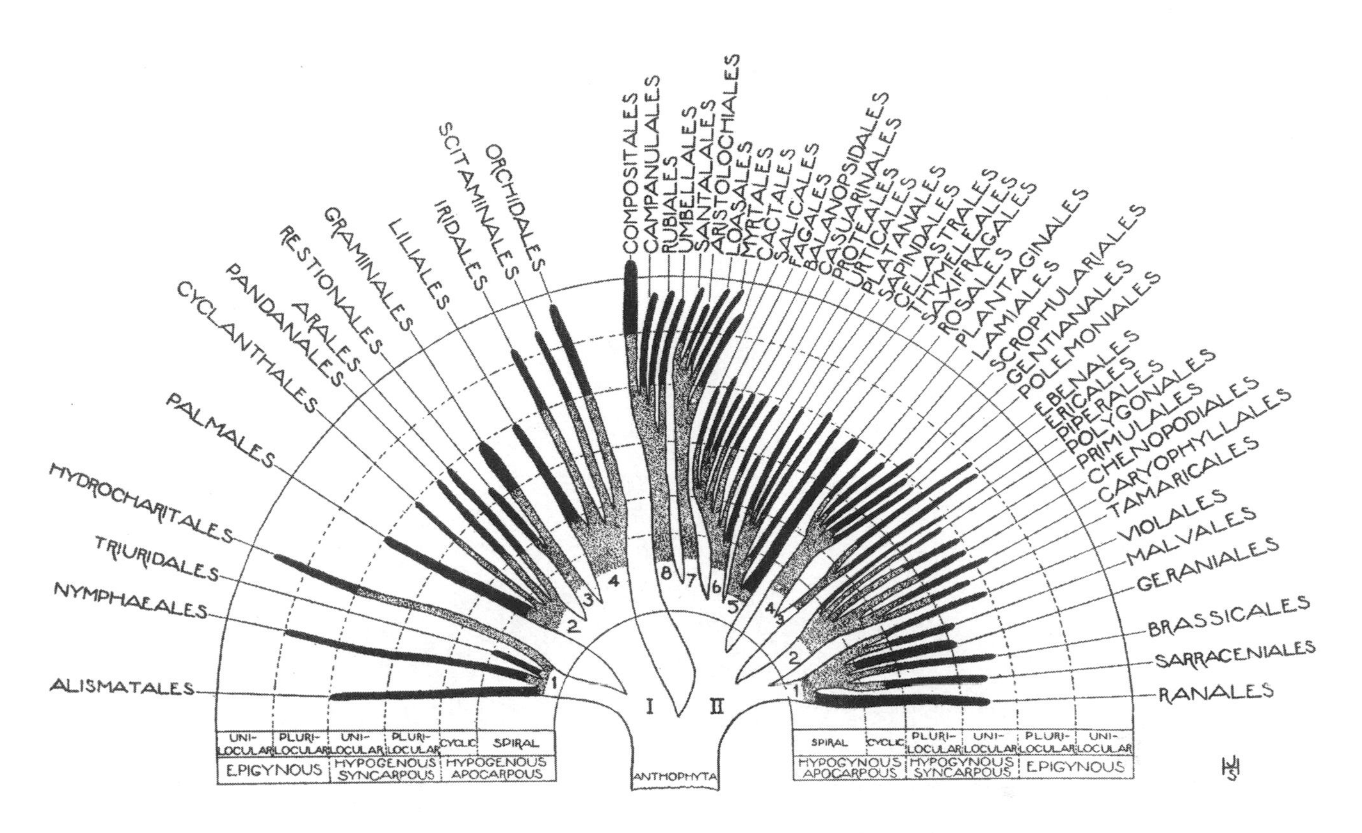

FIGURE 146. Tree by John Henry Schaffner showing the relationships of the flowering plants. The early split at the base of the tree leads to the monocotyledonous plants on the left and the dicotyledons on the right. Schaffner, 1934, *Quarterly Review of Biology*, 9(2):150, fig. 2; courtesy of Perry Cartwright and the University of Chicago Press. Used with permission.

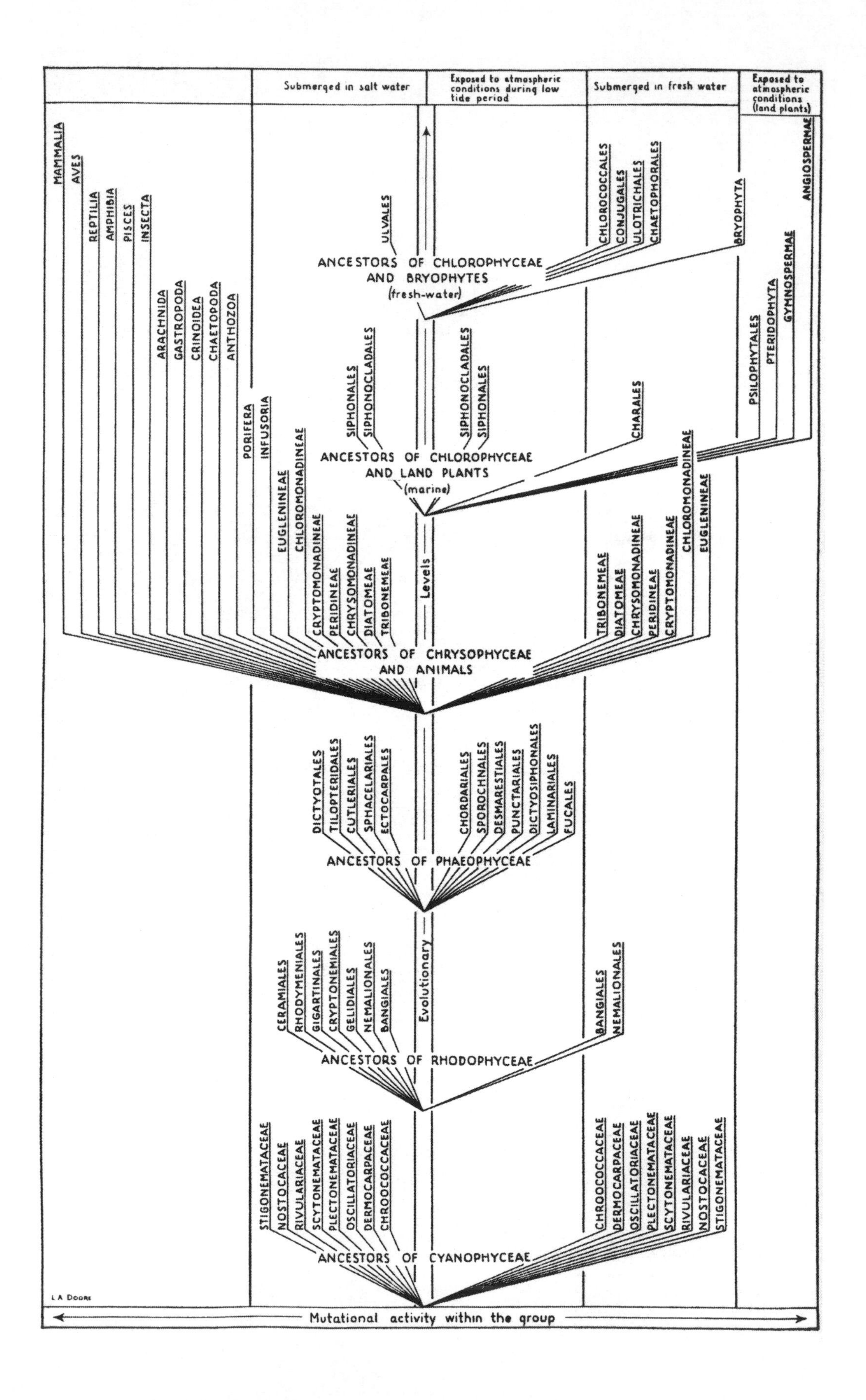

Submerged in salt water
Exposed to atmospheric conditions during low tide period
Submerged in fresh water
Exposed to atmospheric conditions (land plants)
MAMMALIA
AVES
REPTILIA
AMPHIBIA
PISCES
INSECTA
ARACHNIDA
GASTROPODA
CRINOIDEA
CHAETOPODA
ANTHOZOA
PORIFERA
INFUSORIA
ULVALES
ANCESTORS OF CHLOROPHYCEAE AND BRYOPHYTES (fresh-water)
CHLOROCOCCALES
CONJUGALES
ULOTRICHALES
CHAETOPHORALES
BRYOPHYTA
ANGIOSPERMAE
GYMNOSPERMAE
PTERIDOPHYTA
PSILOPHYTALES
SIPHONALES
SIPHONOCLADALES
ANCESTORS OF CHLOROPHYCEAE AND LAND PLANTS (marine)
CHARALES
EUGLENINEAE
CHLOROMONADINEAE
CRYPTOMONADINEAE
PERIDINEAE
CHRYSOMONADINEAE
DIATOMEAE
TRIBONEMEAE
ANCESTORS OF CHRYSOPHYCEAE AND ANIMALS
DICTYOTALES
TILOPTERIDALES
CUTLERIALES
SPHACELARIALES
ECTOCARPALES
CHORDARIALES
SPOROCHNALES
DESMARESTIALES
PUNCTARIALES
DICTYOSIPHONALES
LAMINARIALES
FUCALES
ANCESTORS OF PHAEOPHYCEAE
CERAMIALES
RHODYMENIALES
GIGARTINALES
CRYPTONEMIALES
GELIDIALES
NEMALIONALES
BANGIALES
ANCESTORS OF RHODOPHYCEAE
STIGONEMATACEAE
NOSTOCACEAE
RIVULARIACEAE
SCYTONEMATACEAE
PLECTONEMATACEAE
OSCILLATORIACEAE
DERMOCARPACEAE
CHROOCOCCACEAE
ANCESTORS OF CYANOPHYCEAE
Evolutionary Levels
L A DOORE
Mutational activity within the group

facing page

FIGURE 147. Josephine Elizabeth Tilden's tree of the relationships of algae.

Tilden, 1935, *The Algae and Their Life Relations: Fundamentals of Phycology,* p. 40, pl. 2; courtesy of Jeff Moen and the University of Minnesota Press. Used with permission.

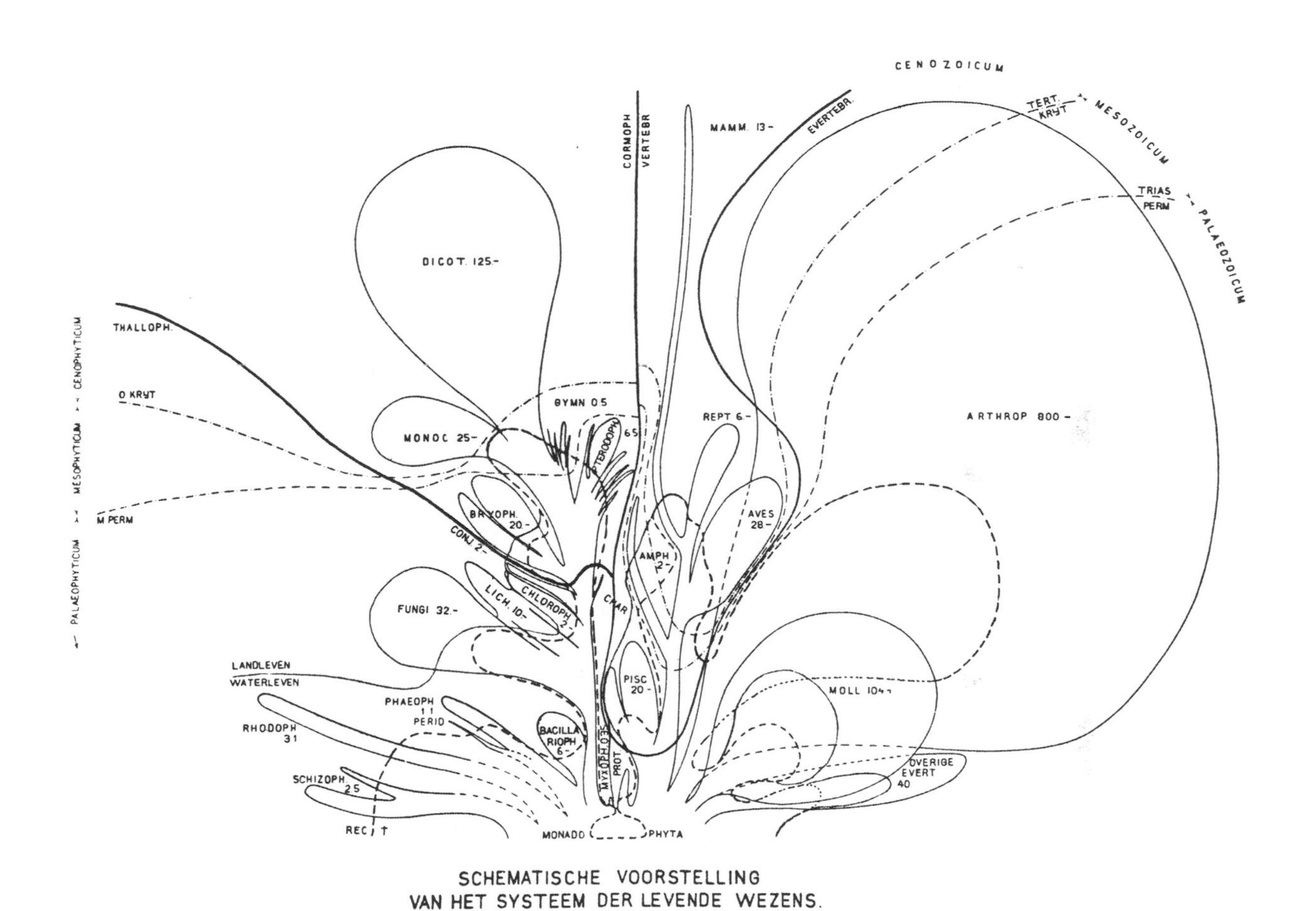

FIGURE 148. A "schematical diagram of the system of organisms" by Herman Johannes Lam.

Lam, 1936, *Acta Biotheoretica,* 2(3):171, fig. 18; courtesy of the Jan van der Hoeven Foundation and Tjard de Cock Buning, Editor in Chief of *Acta Biotheoretica*. Used with permission.

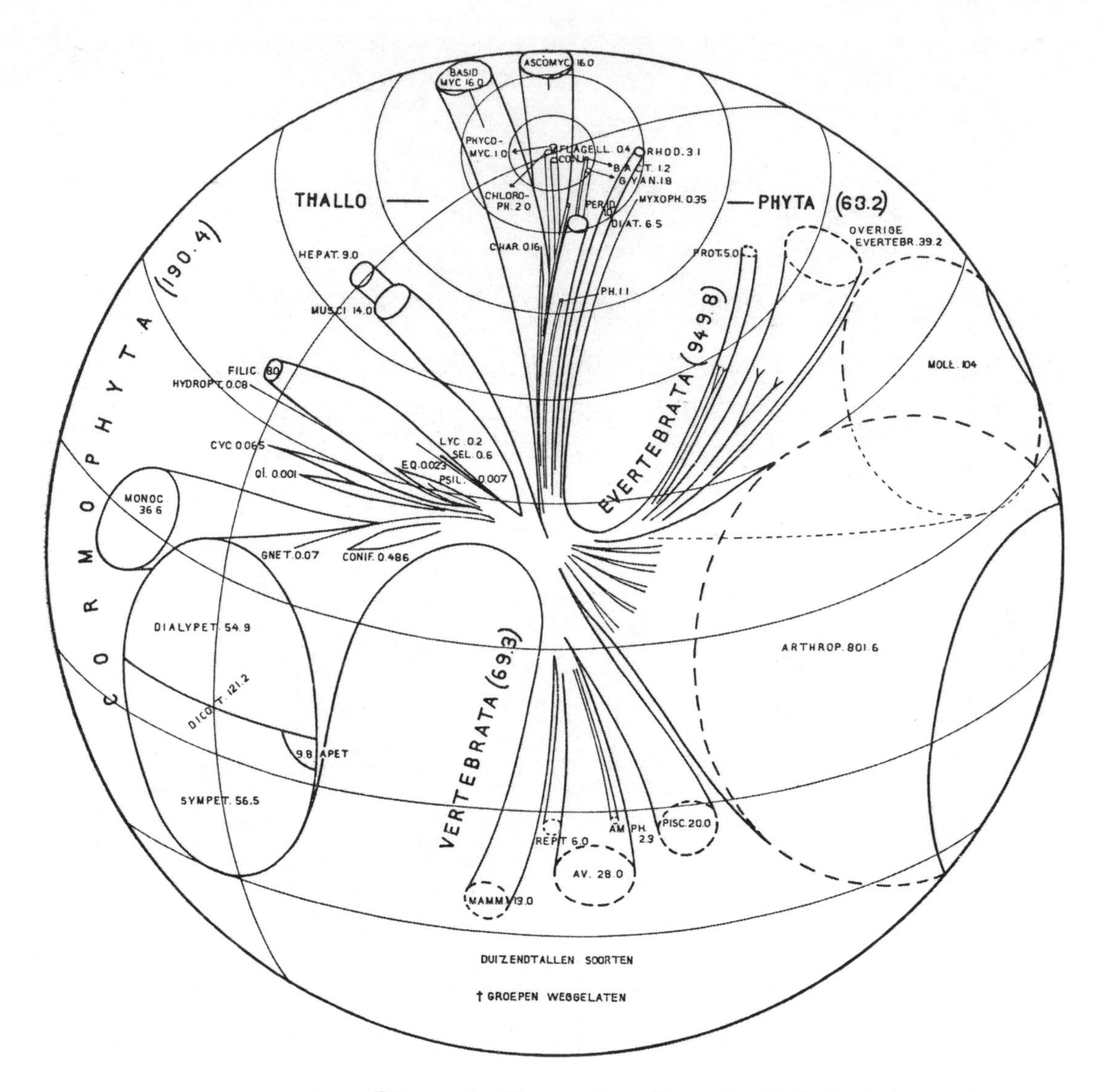

FIGURE 149. Diagram by Herman Lam illustrating his "spherical system of the microcosm, consisting of an infinite number of concentric time spheres," expressing time, differentiation and relationship, and diversity.

Lam, 1936, *Acta Biotheoretica,* 2(3):173, fig. 21; courtesy of the Jan van der Hoeven Foundation and Tjard de Cock Buning, Editor in Chief of *Acta Biotheoretica*. Used with permission.

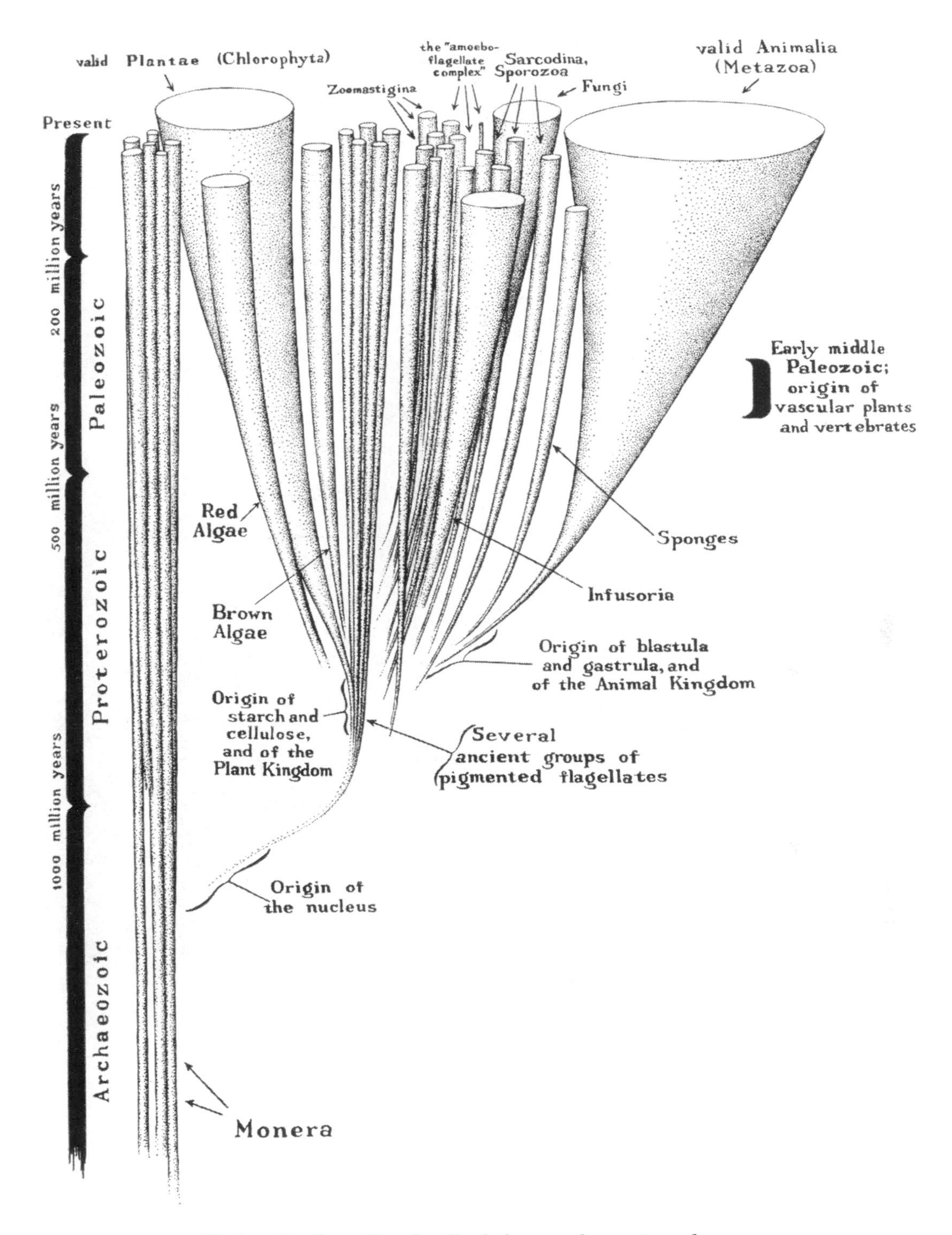

FIGURE 150. Herbert Faulkner Copeland's phylogeny of organisms that proposed four kingdoms to replace the traditional two-kingdom dichotomy between Plantae and Animalia: the Monera along the left margin, the Plantae above on the left and Animalia on the right, and the various small subgroups of the Protista in between.

Copeland, 1938, *Quarterly Review of Biology,* 13(4):410, fig. 8; courtesy of Perry Cartwright and the University of Chicago Press. Used with permission.

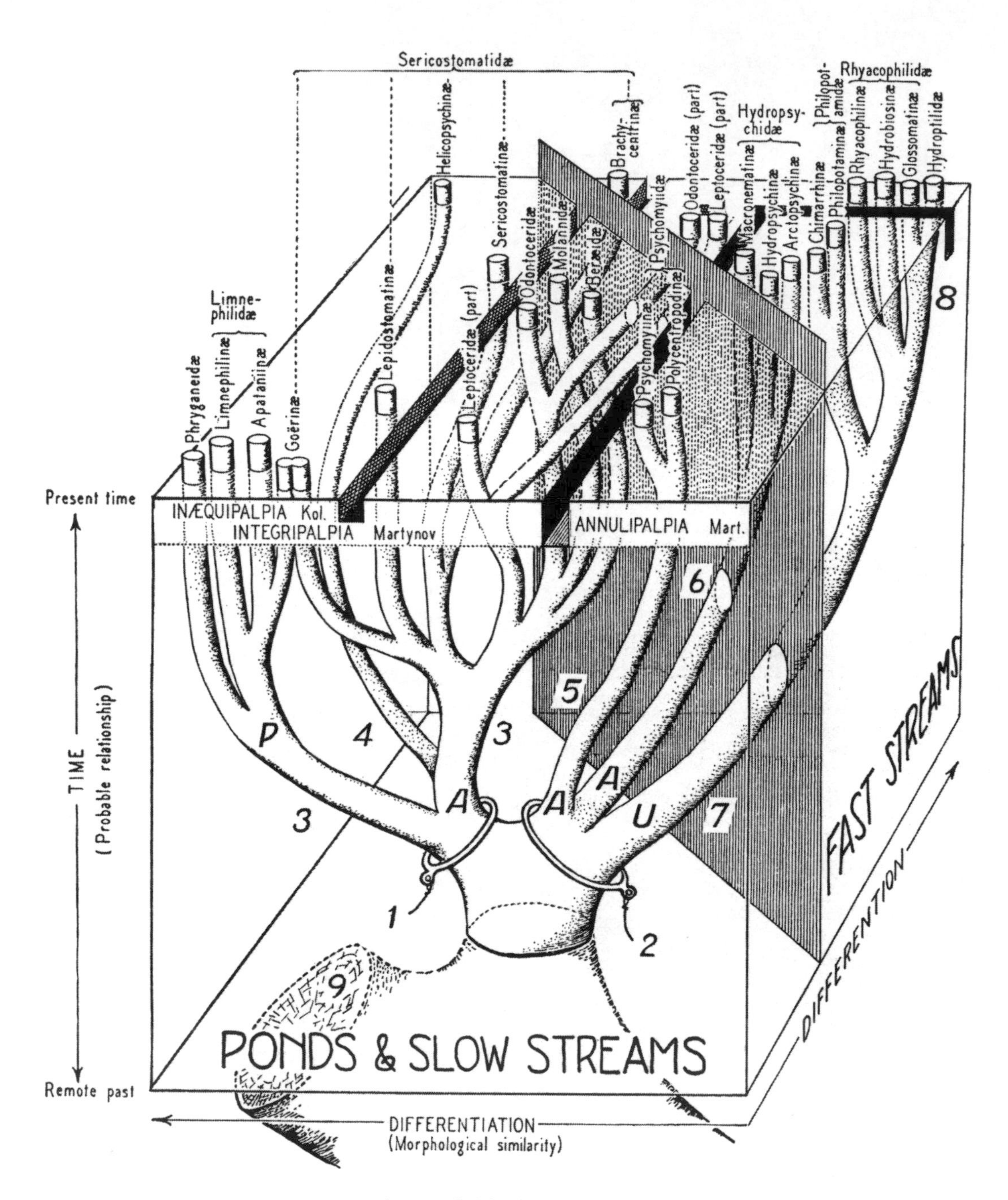

FIGURE 151. A remarkable three-dimensional view of the relationships among caddisflies based largely on caddis worm case construction, by Lorus Johnson Milne and Margery Joan Milne.

Milne and Milne, 1939, *Annals of the Entomological Society of America,* 32(3):540, fig. 1, courtesy of Alan Kahan and the Entomological Society of America. Used with permission.

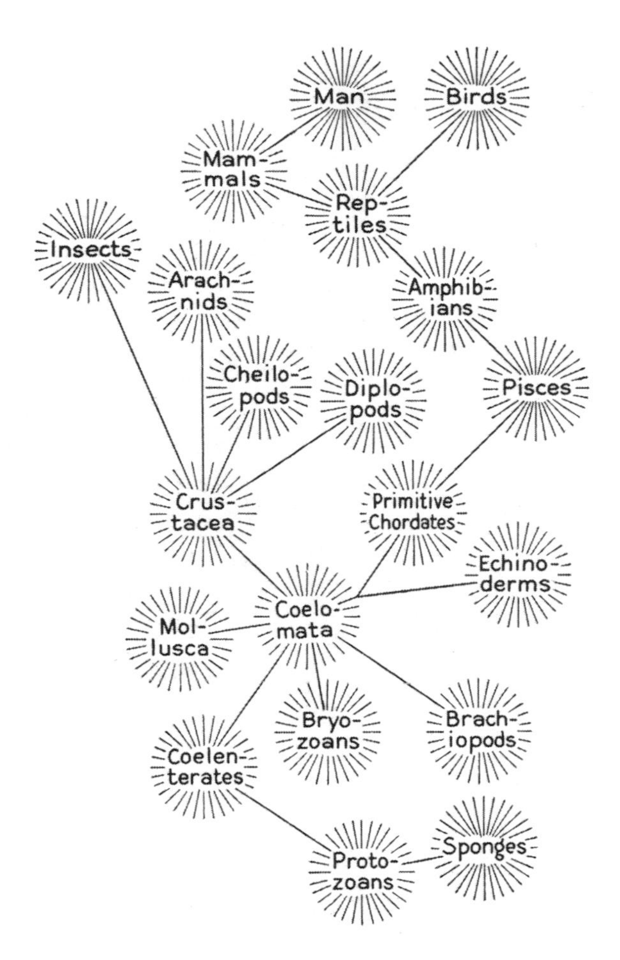

FIGURE 152. Percy Edward Raymond's branching diagram of 1939, showing the probable relationships of the important groups of animals, in "the progress upward from the Protozoa."

Reprinted by permission of the publisher from *Prehistoric Life* by Percy E. Raymond, 309, p. fig. 156, Cambridge, Massachusetts: The Belknap Press of Harvard University Press, Copyright © 1939, 1947, by the President and Fellows of Harvard College. Copyright © renewed 1966, 1975 by Ruth Elspeth Raymond. Courtesy of Scarlett R. Huffman.

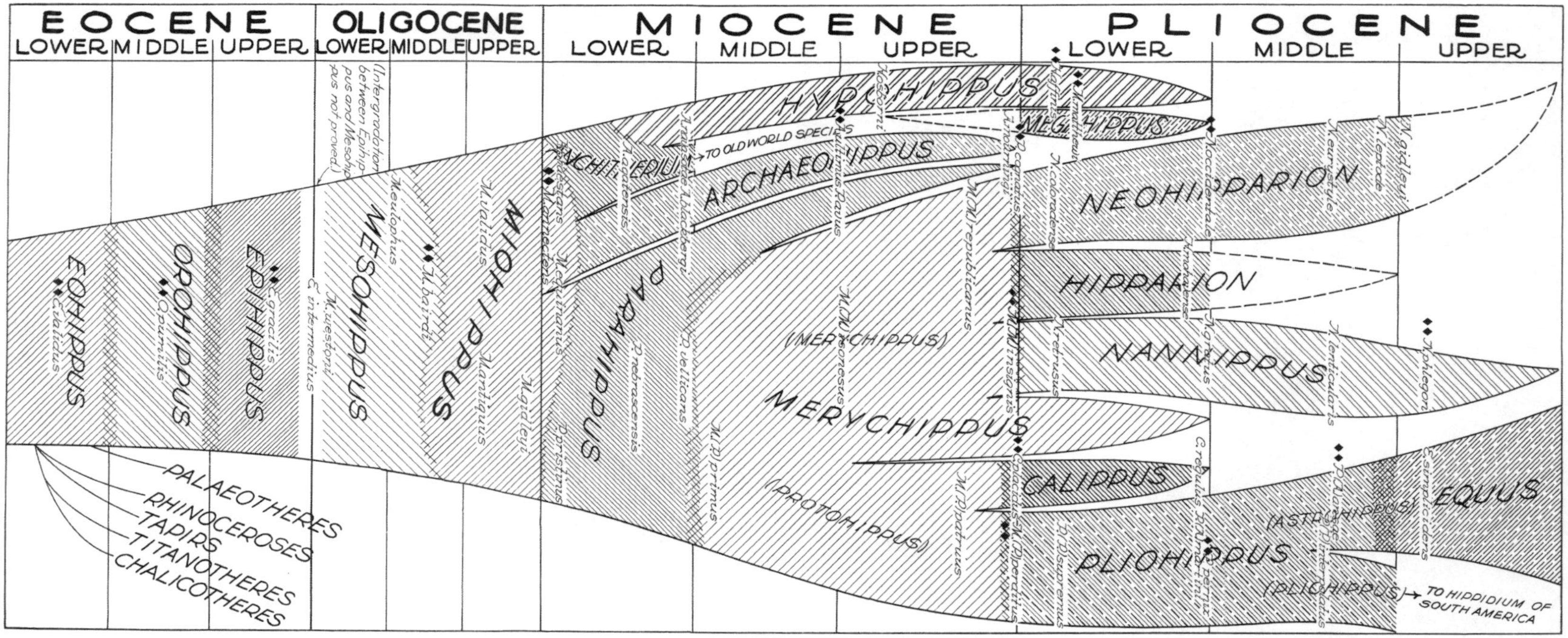

PLIOCENE UPPER
PLIOCENE MIDDLE
PLIOCENE LOWER
MIOCENE UPPER
MIOCENE MIDDLE
MIOCENE LOWER
OLIGOCENE UPPER
OLIGOCENE MIDDLE
OLIGOCENE LOWER
EOCENE UPPER
EOCENE MIDDLE
EOCENE LOWER

EQUUS
E. simplicidens
TO HIPPIDIUM OF SOUTH AMERICA
(ASTROHIPPUS)
P. (A.) ansae
P. interpolatus
PLIOHIPPUS
(PLIOHIPPUS)
P. pernix
P. (A.) martini
P. (P.) supremus
CALIPPUS
C. regulus
C. placidus
M. (P.) perditus
N. phlegon
N. lenticularis
NANNIPPUS
N. gratus
N. retrusus
HIPPARION
H. mohavense
N. gidleyi
N. leptode
N. eurystyle
NEOHIPPARION
N. occidentale
N. coloradense
H. affinis
M. matthewi
MEGAHIPPUS
P. coenatus
A. mourningi
M. (M.) insignis
M. (M.) republicanus
M. (P.) patruus
MERYCHIPPUS
(MERYCHIPPUS)
(PROTOHIPPUS)
H. osborni
HYPOHIPPUS
A. ultimus
P. avus
M. (M.) isonesus
TO OLD WORLD SPECIES
ARCHAEOHIPPUS
M. (P.) primus
A. blackbergi
P. vellicans
ANCHITHERIUM
A. agatensis
P. nebrascensis
PARAHIPPUS
M. equiranus
P. pristinus
M. annectens
M. gidleyi
MIOHIPPUS
M. validus
M. antiquus
M. bairdi
M. eulophus
MESOHIPPUS
(Intergradation between Epihippus and Mesohippus not proved.)
M. westoni
E. intermedius
E. gracilis
EPIHIPPUS
OROHIPPUS
O. pumilis
EOHIPPUS
E. validus
PALAEOTHERES
RHINOCEROSES
TAPIRS
TITANOTHERES
CHALICOTHERES

facing page

FIGURE 153. A phylogeny of horses showing their geological distribution throughout the Tertiary, by Ruben Arthur Stirton.

Stirton, 1940, plate following page 198; courtesy of Rebecca Wells and the University of California Press. Used with permission.

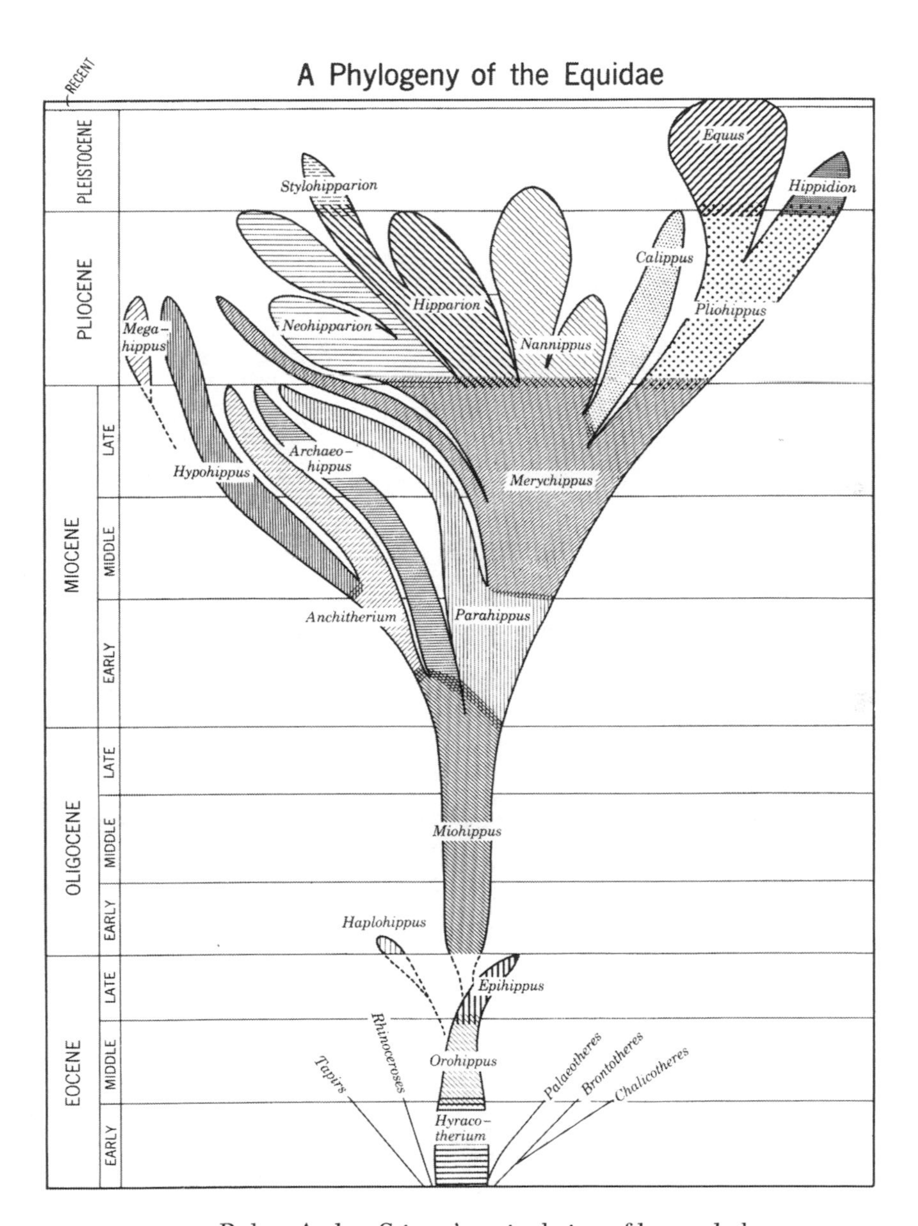

FIGURE 154. Ruben Arthur Stirton's revised view of horse phylogeny.

Stirton, 1959, *Time, Life, and Man: The Fossil Record,* p. 466, fig. 250; courtesy of Sheik Safdar and John Wiley & Sons, Inc. Used with permission.

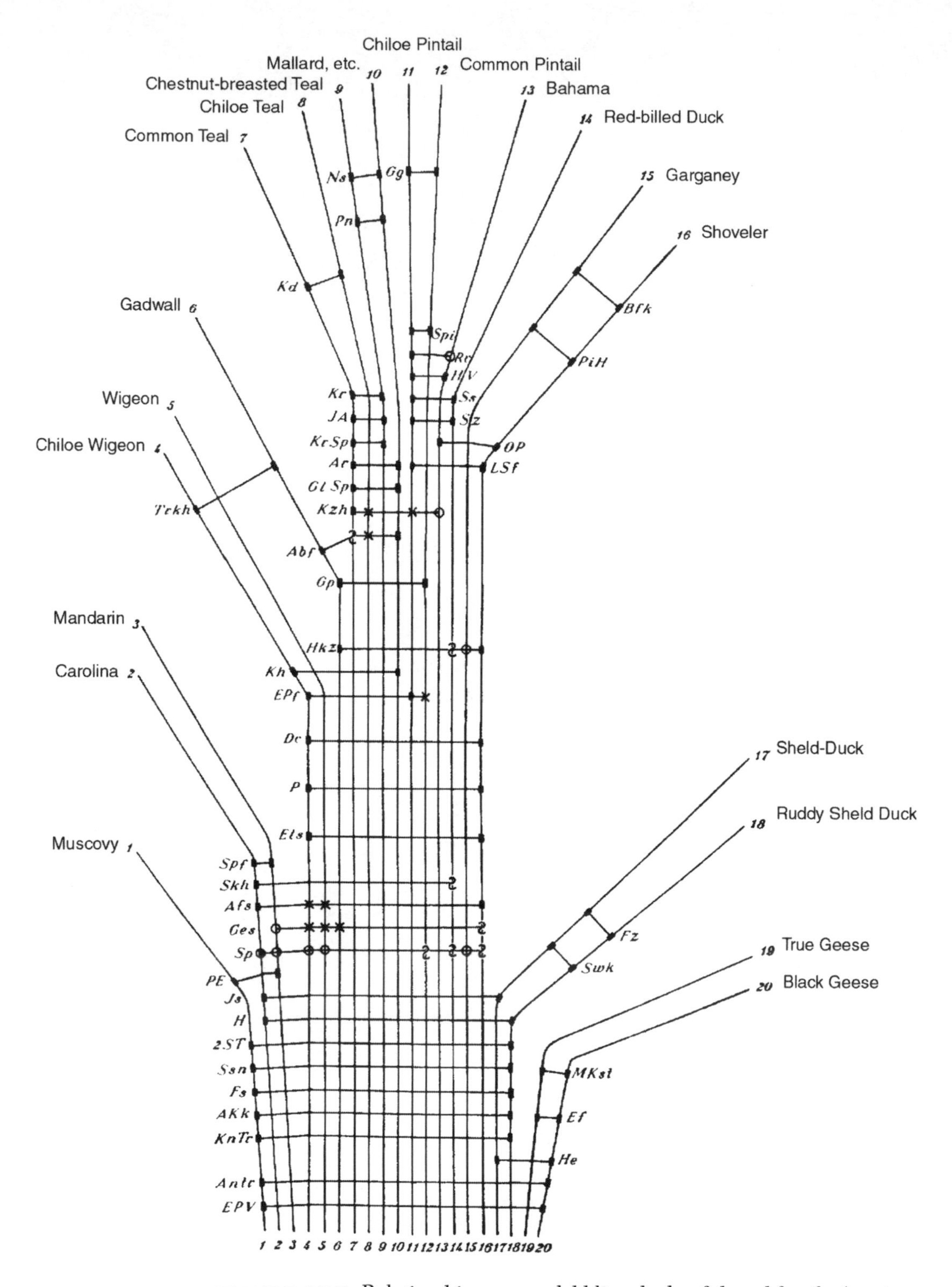

FIGURE 155. Relationships among dabbling ducks of the subfamily Anatinae, based primarily on behavioral characters, by Konrad Lorenz in 1941, as it appears with common names added in an English translation of his work.

Lorenz, 1953, *Avicultural Magazine,* 59:90, unnumbered figure, courtesy of Peter Stocks and the Avicultural Society. Used with permission.

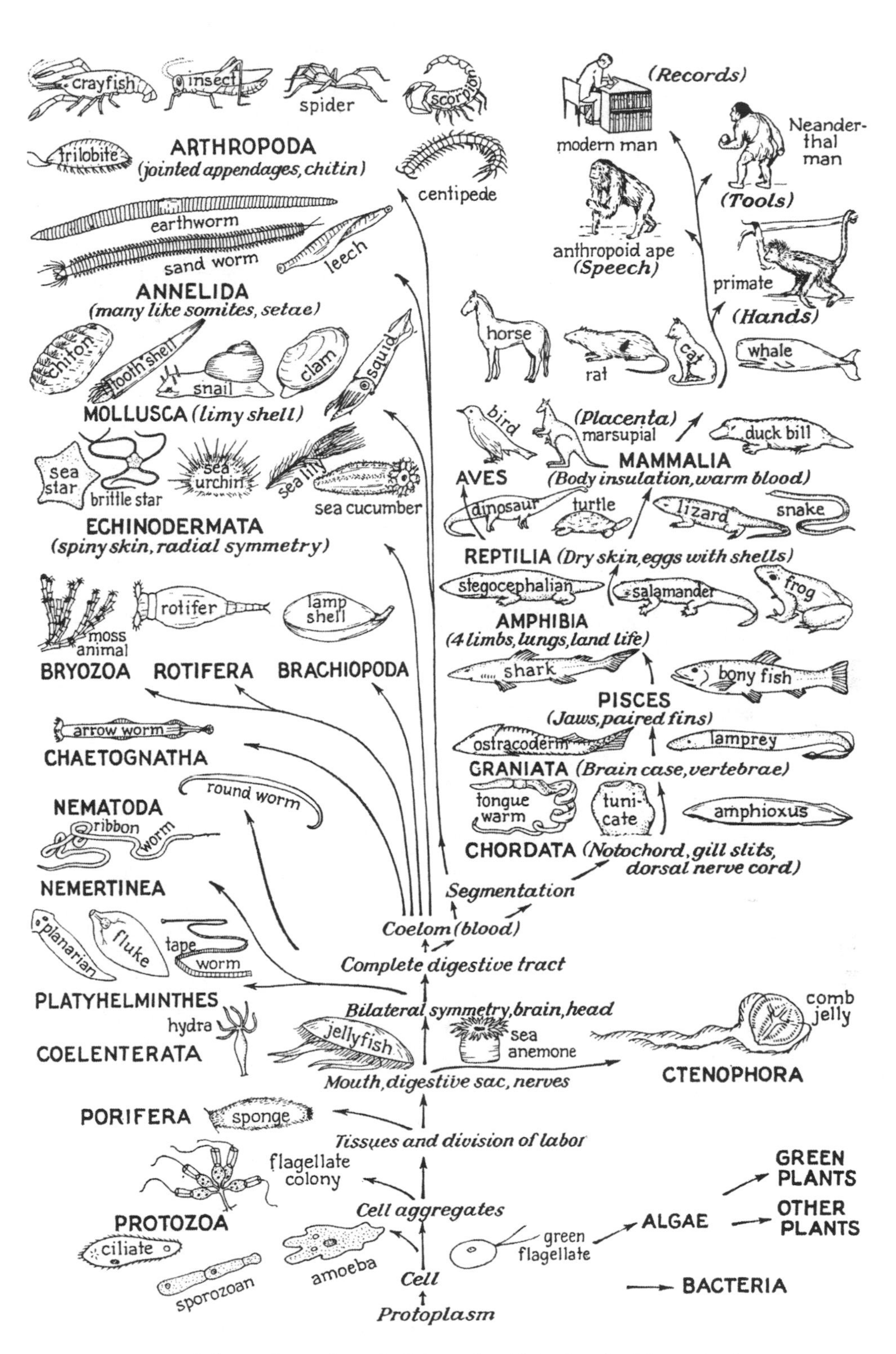

FIGURE 156. A "genealogical tree" of the animal kingdom by Tracy Irwin Storer, indicating the probable relationships and relative position of the major groups (named in boldface). All groups above a given attribute (such as "eggs with shells," in italics) possess that character.

Storer, 1943, *General Zoology*, p. 2, fig. 8-8.

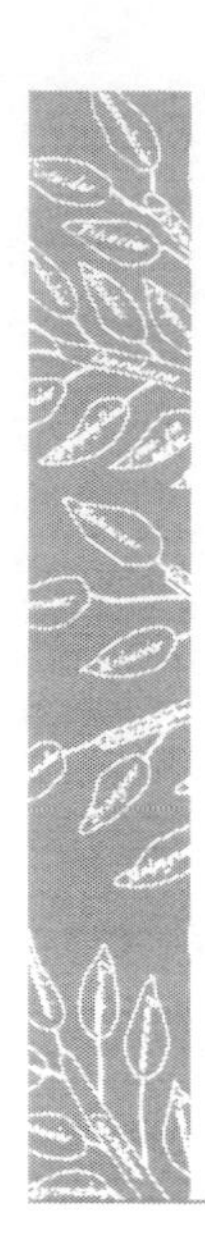

The Trees of William King Gregory

1938–1951

American zoologist William King Gregory (1876–1970) surely published more evolutionary trees than anyone else, by far eclipsing Ernst Haeckel, and Alfred Romer who comes in as a distant third. As an undergraduate at Columbia University, he become a favorite of Henry Osborn (see Figures 118–120, and 125) who took him on as his research assistant. After receiving his doctorate at Columbia in 1910, he was appointed the following year to the scientific staff of the American Museum of Natural History, where he remained for his entire career. He is famous for his work in paleontology, primatology, and functional and comparative morphology but well known also for his dedication to students and for promoting science, particularly evolution, to the general public through books and museum exhibits. One of his earliest trees, designed to illustrate Osborn's *Origin and Evolution of Life* published in 1917, has already been described (see Figure 117).

During the 1920s and 30s, Gregory was constructing trees to accompany research results that appeared in various publications of the New York Zoological Society. Typical of his branching diagrams of that period is a 1935 phylogeny of the sharks and rays, showing hypothesized relationships in a unique way that gives a sense of the animals swimming across the page (Figure 157). In the mid-1930s, he formed an active collaboration with George Miles Conrad (1911–1964), also of the American Museum, with whom he published a number of well-illustrated papers. They jointly proposed a phylogeny of carnivorous characins in 1938 (Figure 158). Gregory alone constructed a "comparative history of the invertebrates and vertebrates" in 1942[1] that was unique in providing information on the estimated age of the various taxa, their definite or probable occurrence, the lack of fossils, and an indication of relationship inferred from comparative anatomy, including embryology (Figure 159). Notable for their complexity are three trees dating from 1945: "branching evolution" in reptiles, birds, and mammals (Figure 160); a unique superimposition of phylogeny, life zones, and locomotion in vertebrates (Figure 161); and a character-state tree of palatal types in birds, both living and fossil (Figure 162).[2]

A comparative osteologist and expert on mammal teeth, Gregory produced a great number of branching diagrams focusing on these elements: a phylogeny of dogs and their allies based on cranial morphology in 1946 (Figure 163);[3] tooth evolution in bears, raccoons, and pandas, also in 1946 (Figure 164);[4] and evolution of the humerus of vertebrates, from crossopterygians to Recent mammals in 1948 (Figure 165).[5] A very different approach was adopted for a family tree of rodents constructed in 1946, showing the animals in various three-dimensional, natural poses, instead of the flat two-dimensional profiles in common use at the time (Figure 166).[6]

Gregory's massive two-volume *Evolution Emerging* that appeared in 1951 is a celebration of evolutionary trees if there ever was one, containing nearly a hundred examples, all beautifully designed and rendered. The first volume opens with an ingenious "Procession of the Vertebrates" as a frontispiece (Figure 167)—a tight array of lineages emerging from the depths of a narrow rocky chasm of geological time, and diverging and ascending to recent times, with major events in the history of life displayed along the right margin.

The remaining figures of *Evolution Emerging* are clustered in the second volume. Only a very few of the best are reproduced here: a phylogeny of deep-sea anglerfishes (Figure 168);[7] evolution of the pineal aperture in the crania of terrestrial vertebrates (Figure 169); a phylogeny of lizards (Figure 170), of antelopes and their relatives based on skulls and horns (Figure 171), and the hands and feet of primates (Figures 172 and 173).

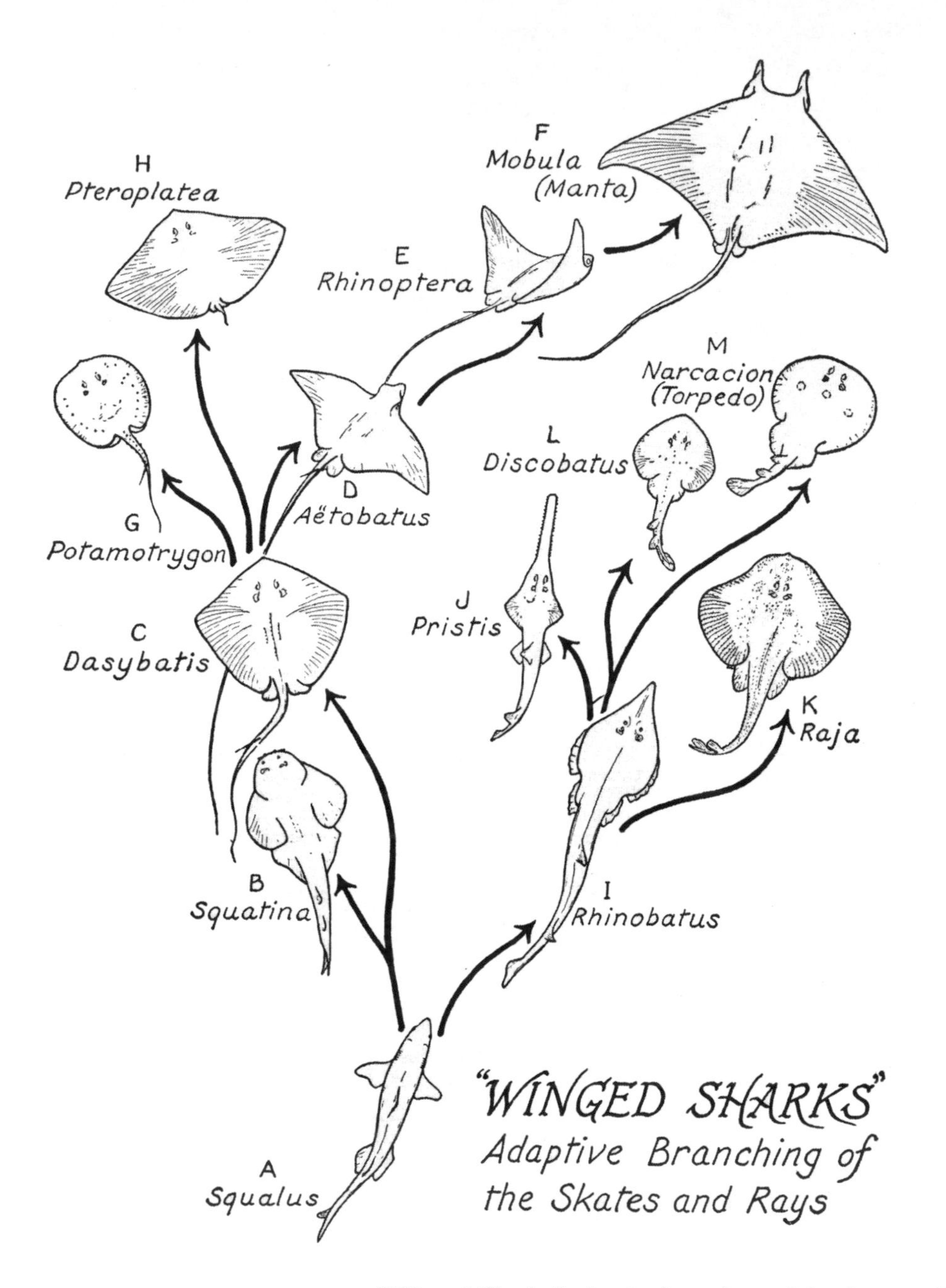

FIGURE 157. "Winged Sharks," adaptive branching of the skates and rays, based on work published by William King Gregory in 1935.

Gregory, 1951, *Evolution Emerging: A Survey of Changing Patterns from Primeval Life to Man,* vol. 2, p. 86; fig. 6.1; courtesy of Mary DeJong, Mai Qaraman, and the American Museum of Natural History. Used with permission.

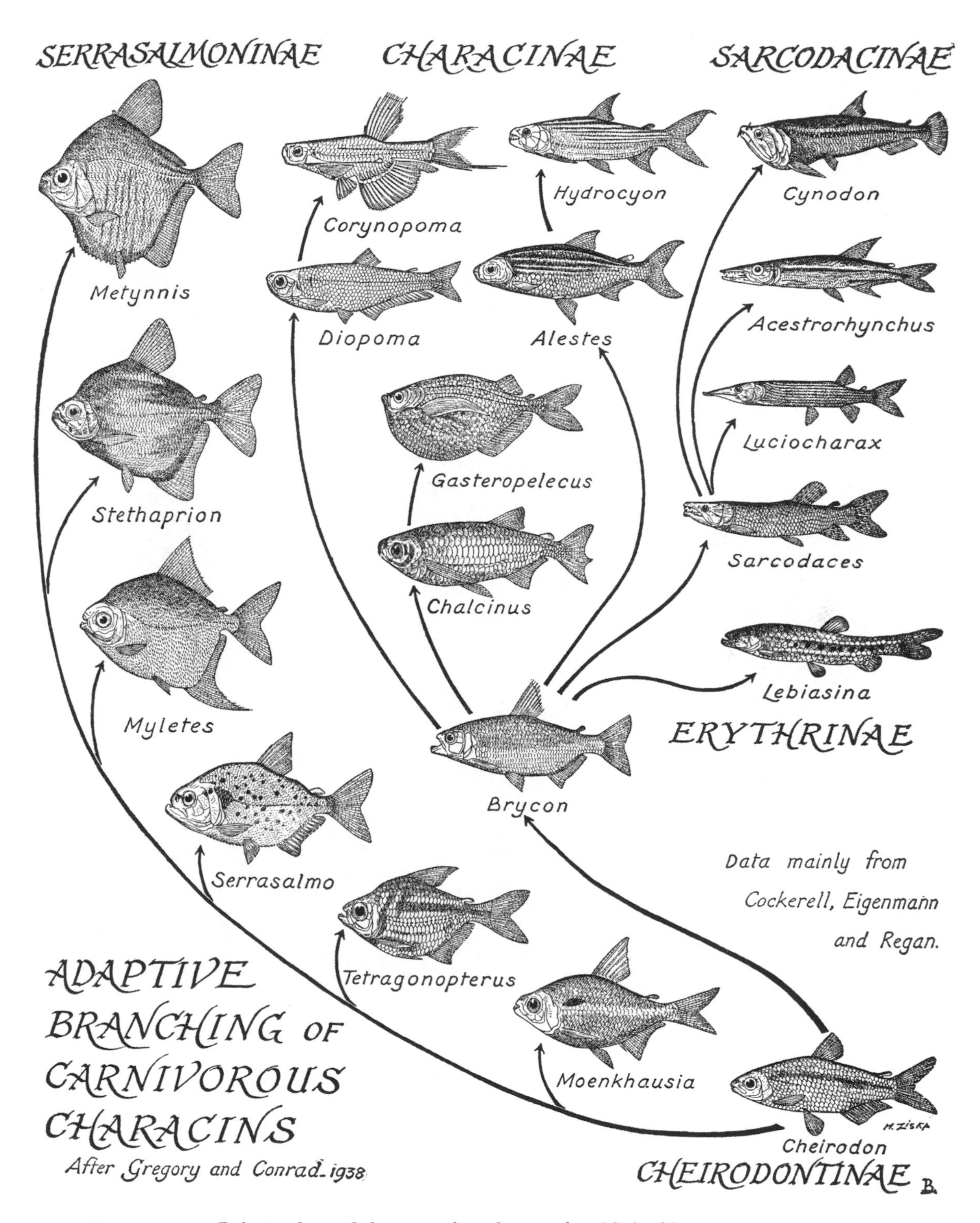

FIGURE 158. Relationships of characins, based on work published by William King Gregory and George Miles Conrad in 1938.

Gregory, 1951, *Evolution Emerging: A Survey of Changing Patterns from Primeval Life to Man*, vol. 2, p. 170; fig. 8.60; courtesy of Mary DeJong, Mai Qaraman, and the American Museum of Natural History. Used with permission.

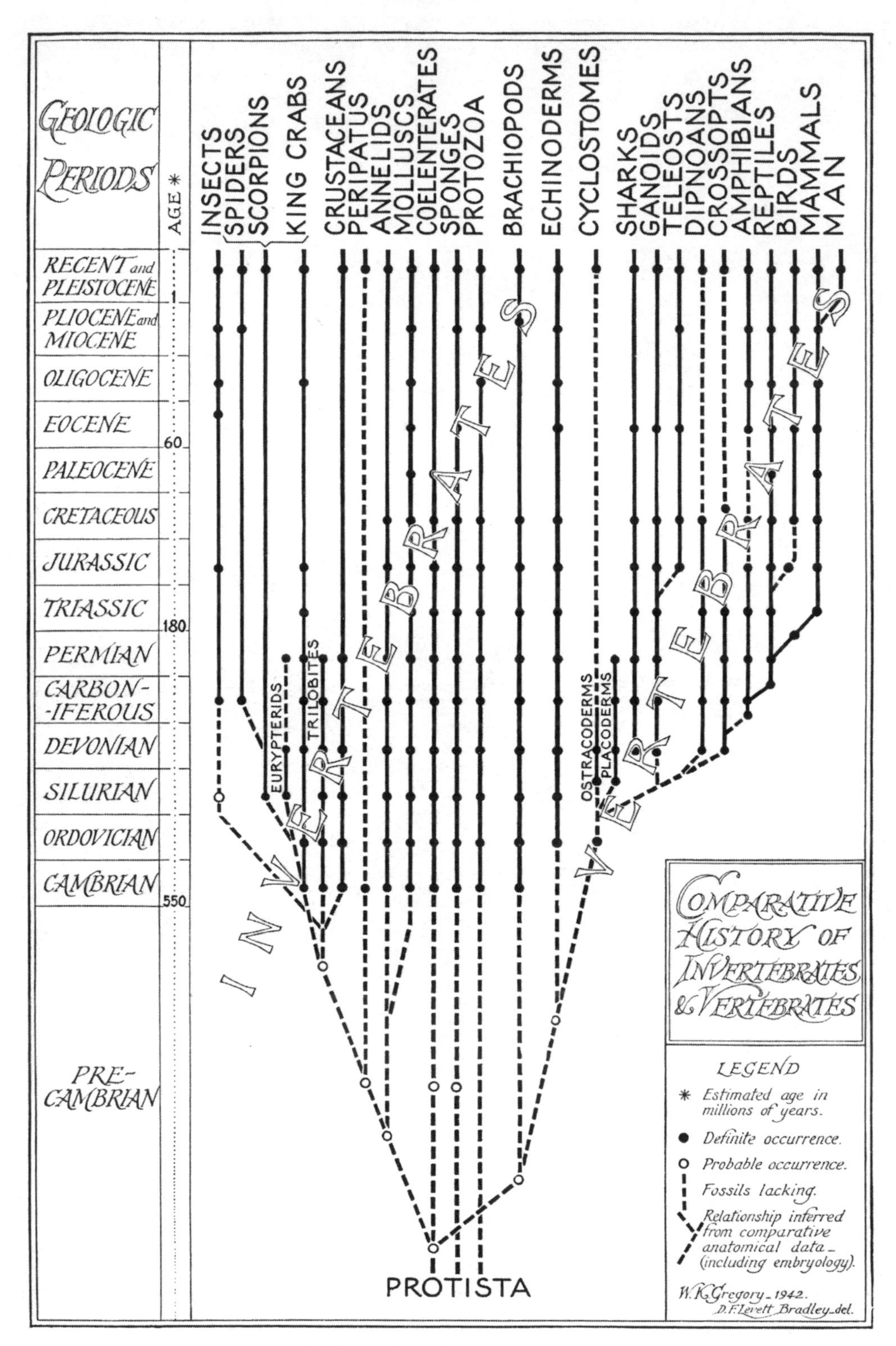

FIGURE 159. William King Gregory's 1942 phylogeny of invertebrates and vertebrates.

Gregory, 1951, *Evolution Emerging: A Survey of Changing Patterns from Primeval Life to Man*, vol. 2, p. 213; fig. 9.155; courtesy of Mary DeJong, Mai Qaraman, and the American Museum of Natural History. Used with permission.

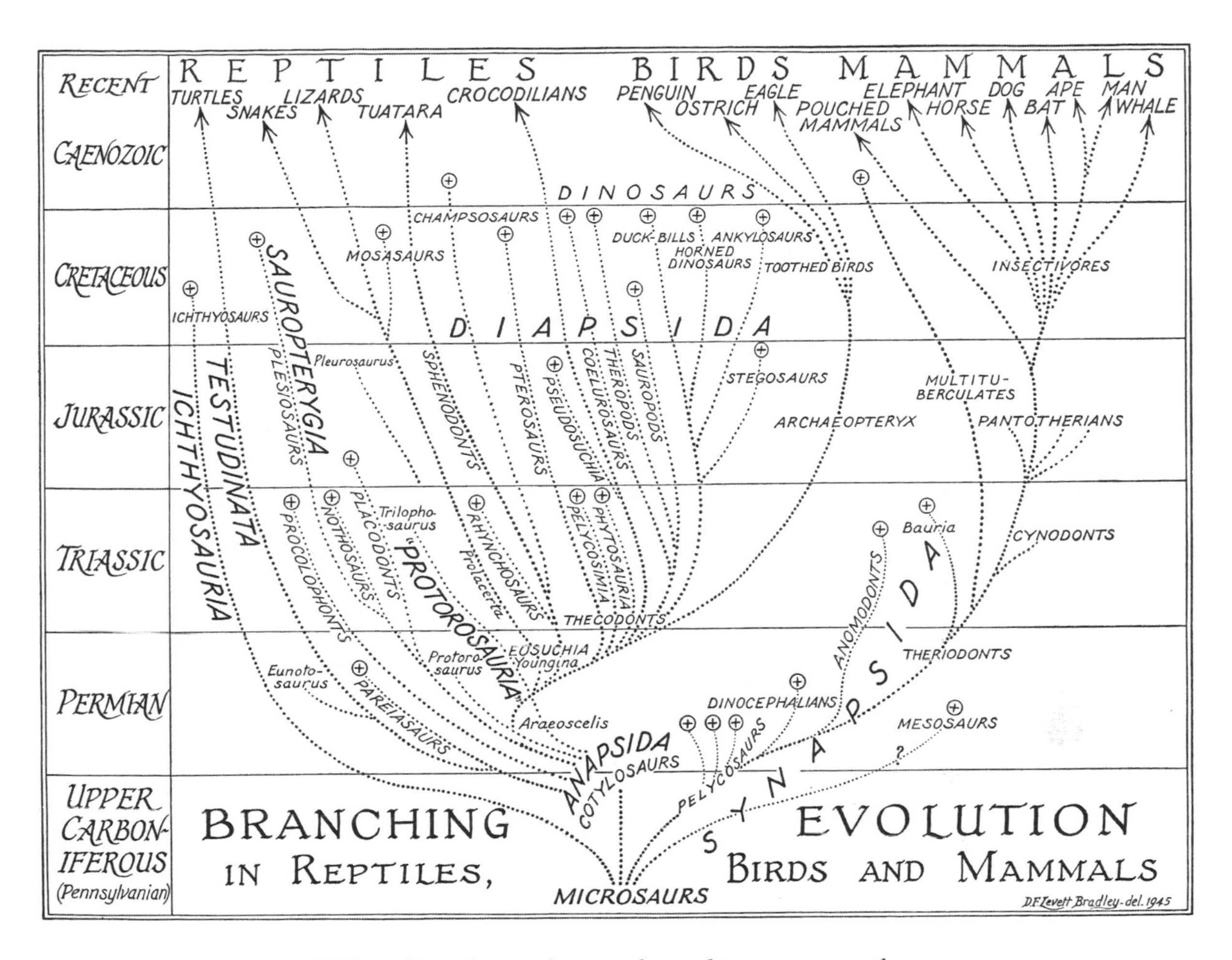

FIGURE 160. William King Gregory's 1945 relationships among reptiles, birds, and mammals.

Gregory, 1951, *Evolution Emerging: A Survey of Changing Patterns from Primeval Life to Man,* vol. 2, p. 312, fig. 9.47; courtesy of Mary DeJong, Mai Qaraman, and the American Museum of Natural History. Used with permission.

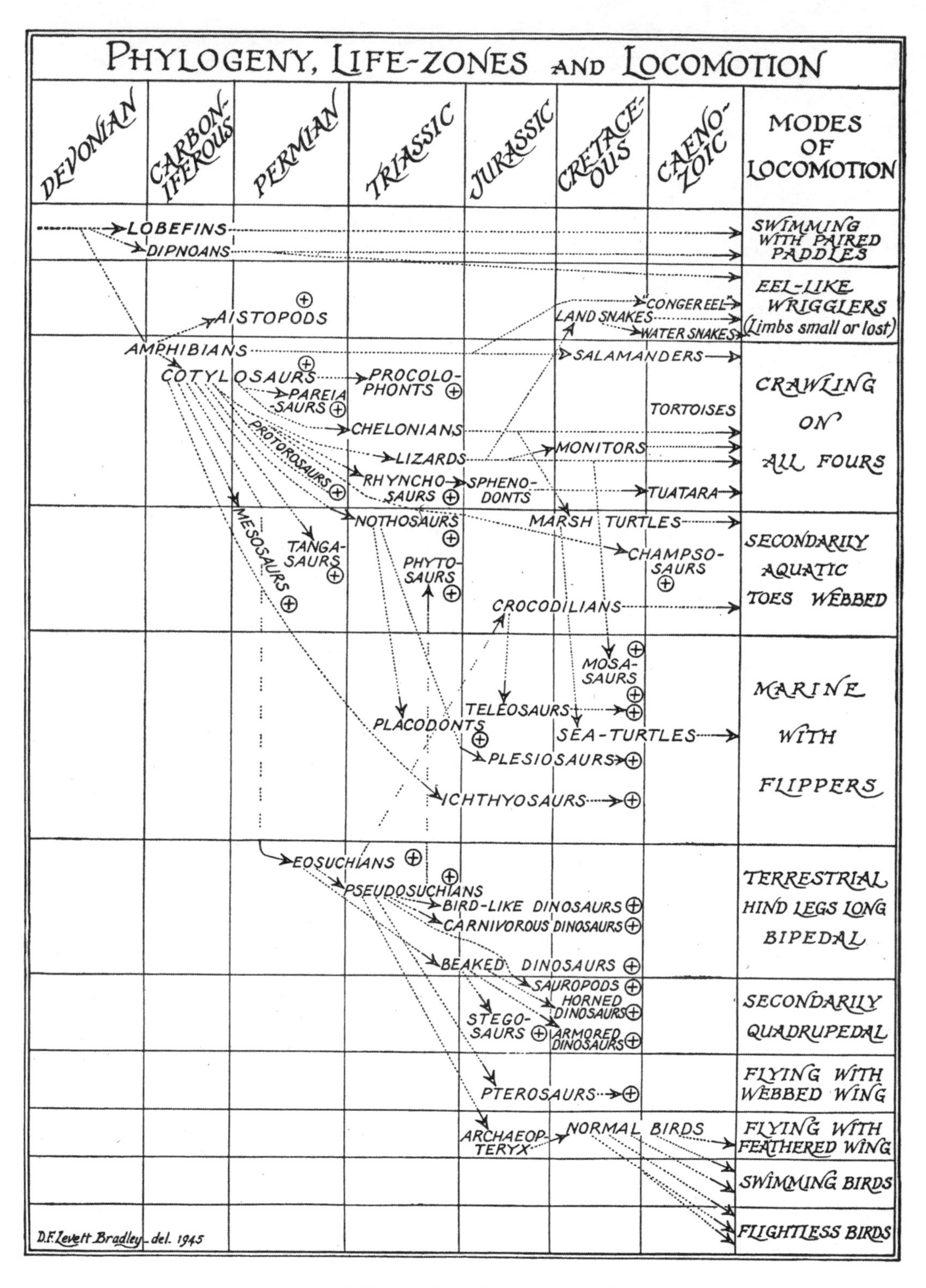

FIGURE 161. William King Gregory's 1945 "phylogeny, life-zones, and locomotion" of vertebrates.

Gregory, 1951, *Evolution Emerging: A Survey of Changing Patterns from Primeval Life to Man*, vol. 2, p. 378, fig. 12.1; courtesy of Mary DeJong, Mai Qaraman, and the American Museum of Natural History. Used with permission.

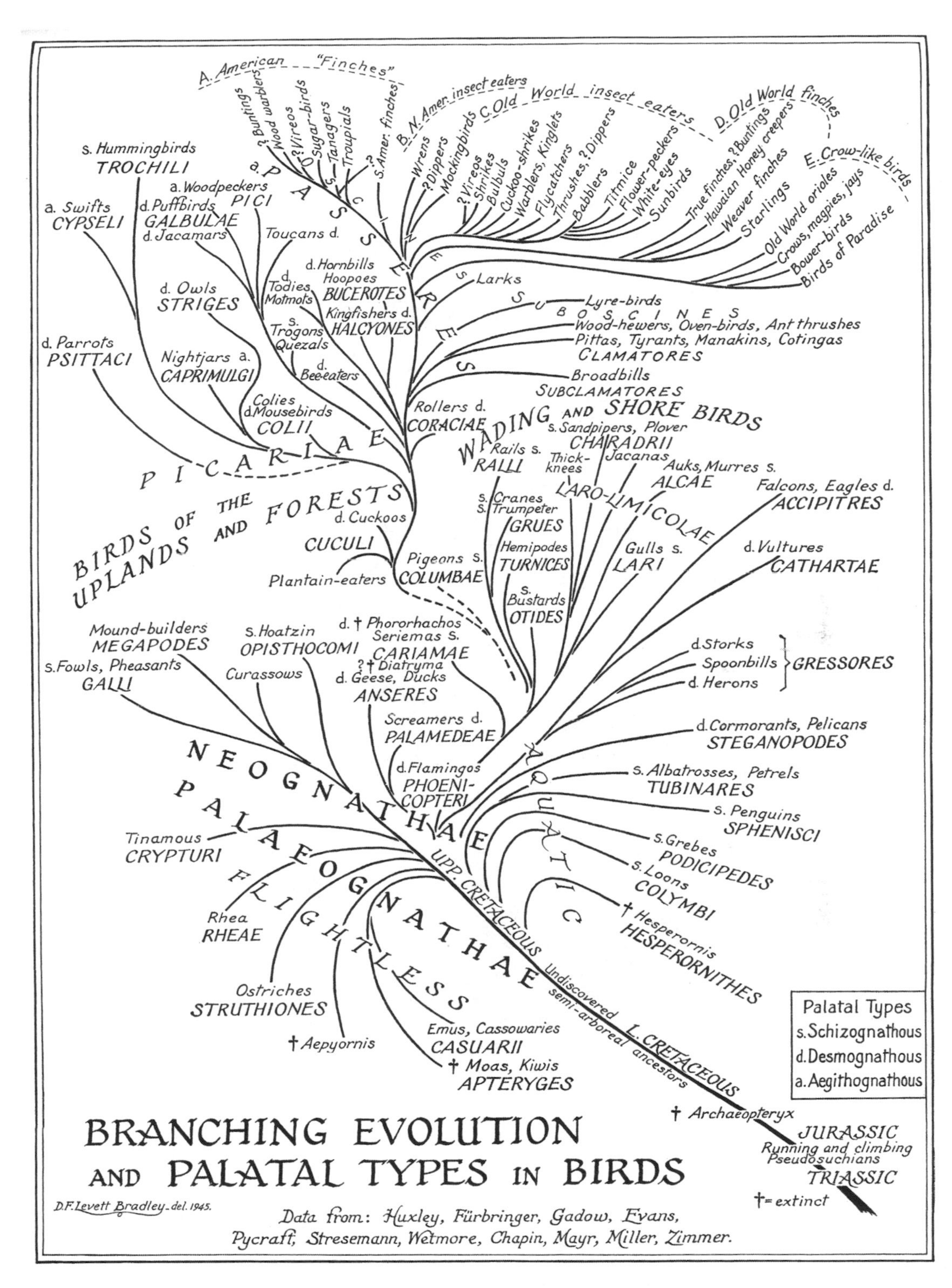

FIGURE 162. A character-state tree of palatal types in birds, constructed by William King Gregory in 1945.

Gregory, 1951, *Evolution Emerging: A Survey of Changing Patterns from Primeval Life to Man*, vol. 2, pp. 440–441, figs. 12.52a and 12.52b; courtesy of Mary DeJong, Mai Qaraman, and the American Museum of Natural History. Used with permission.

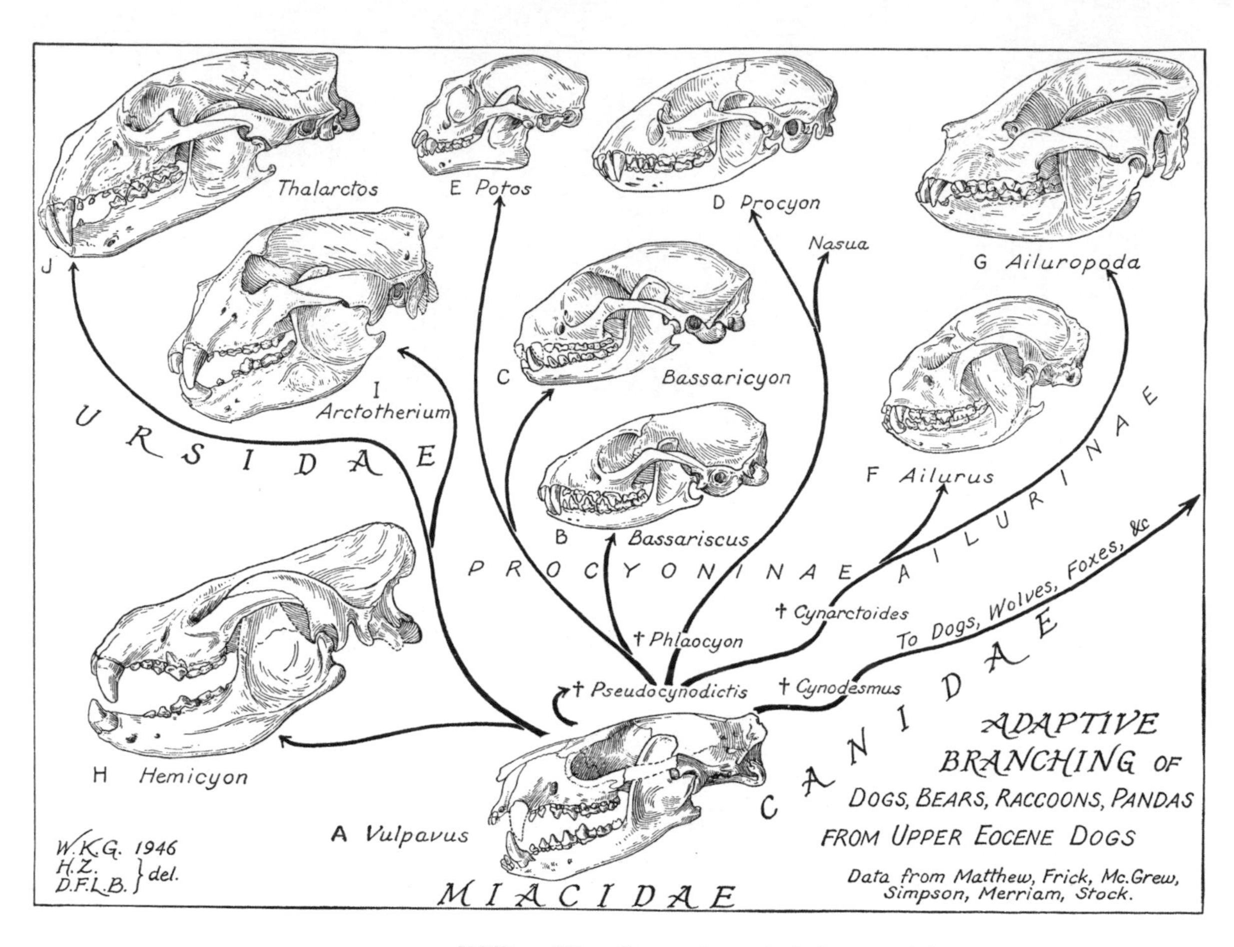

FIGURE 163. William King Gregory's 1946 phylogeny of dogs and their close relatives, based on cranial morphology.

Gregory, 1951, *Evolution Emerging: A Survey of Changing Patterns from Primeval Life to Man*, vol. 2, pp. 442–443, figs. 12.53a and 12.53b; courtesy of Mary DeJong, Mai Qaraman, and the American Museum of Natural History. Used with permission.

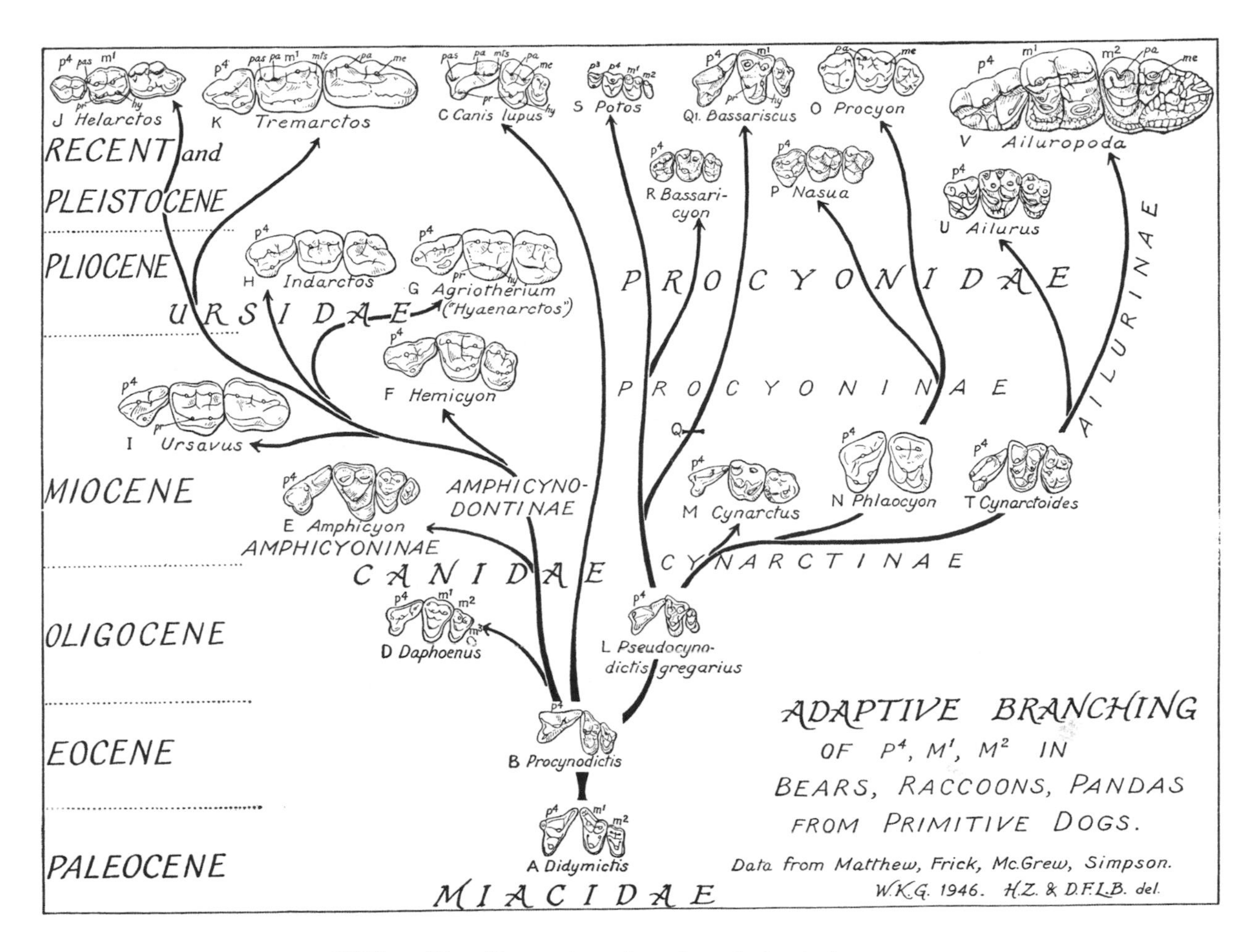

FIGURE 164. William King Gregory's 1946 tooth evolution in bears, raccoons, and pandas.

Gregory, 1951, *Evolution Emerging: A Survey of Changing Patterns from Primeval Life to Man,* vol. 2, p. 468; fig. 14.1; courtesy of Mary DeJong, Mai Qaraman, and the American Museum of Natural History. Used with permission.

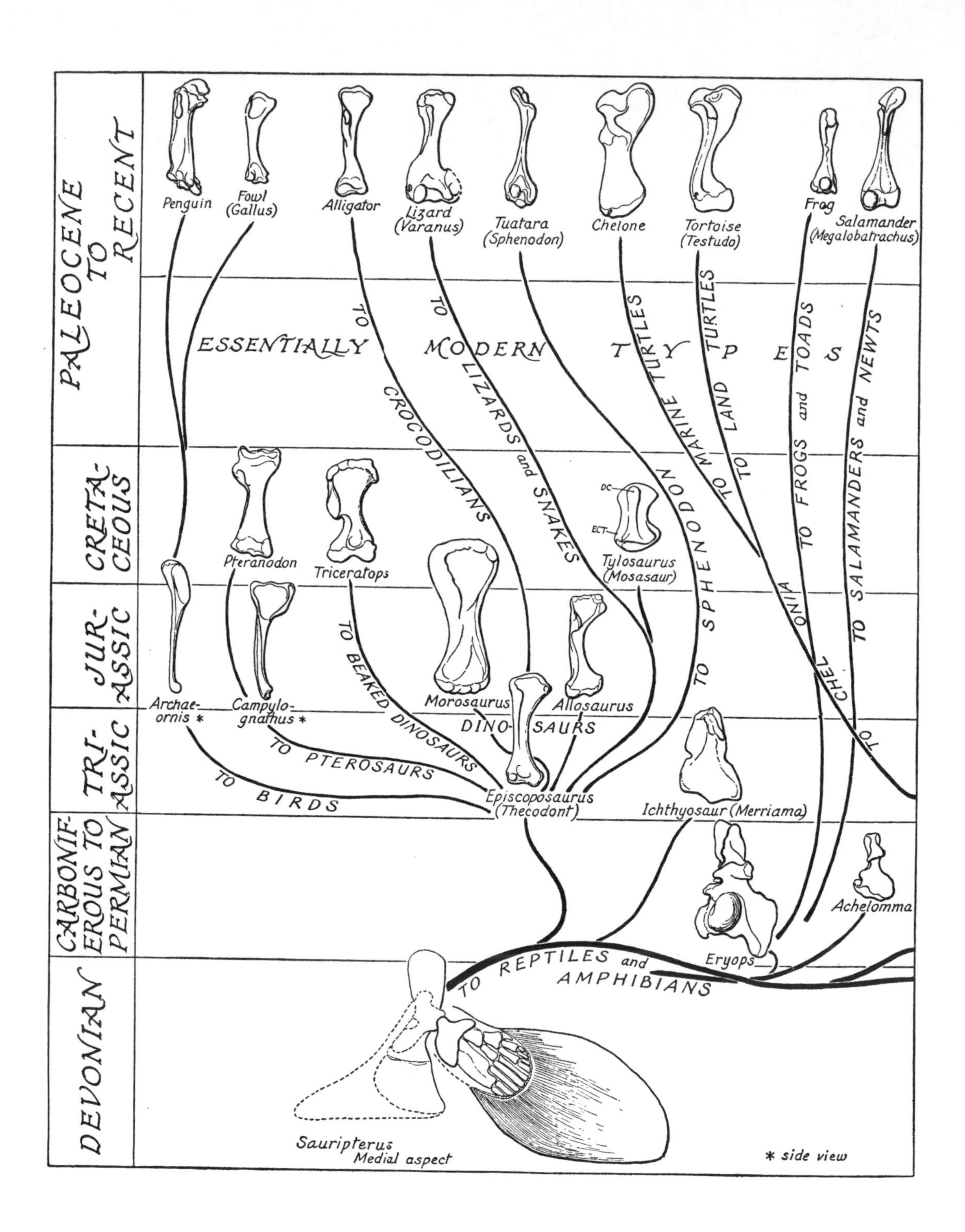
PALEOCENE TO RECENT
CRETACEOUS
JURASSIC
TRIASSIC
CARBONIFEROUS TO PERMIAN
DEVONIAN
Penguin
Fowl (Gallus)
Alligator
Lizard (Varanus)
Tuatara (Sphenodon)
Chelone
Tortoise (Testudo)
Frog
Salamander (Megalobatrachus)
ESSENTIALLY MODERN TYPES
TO CROCODILIANS
TO LIZARDS and SNAKES
TO SPHENODON
TO MARINE TURTLES
TO LAND TURTLES
TO FROGS and TOADS
TO SALAMANDERS and NEWTS
TO CHELONIA
Pteranodon
Triceratops
DC
ECT
Tylosaurus (Mosasaur)
Archaeornis *
Campylognathus *
TO BEAKED DINOSAURS
Morosaurus
Allosaurus
DINOSAURS
TO PTEROSAURS
TO BIRDS
Episcoposaurus (Thecodont)
Ichthyosaur (Merriama)
Achelomma
Eryops
TO REPTILES and AMPHIBIANS
Sauripterus
Medial aspect
* side view

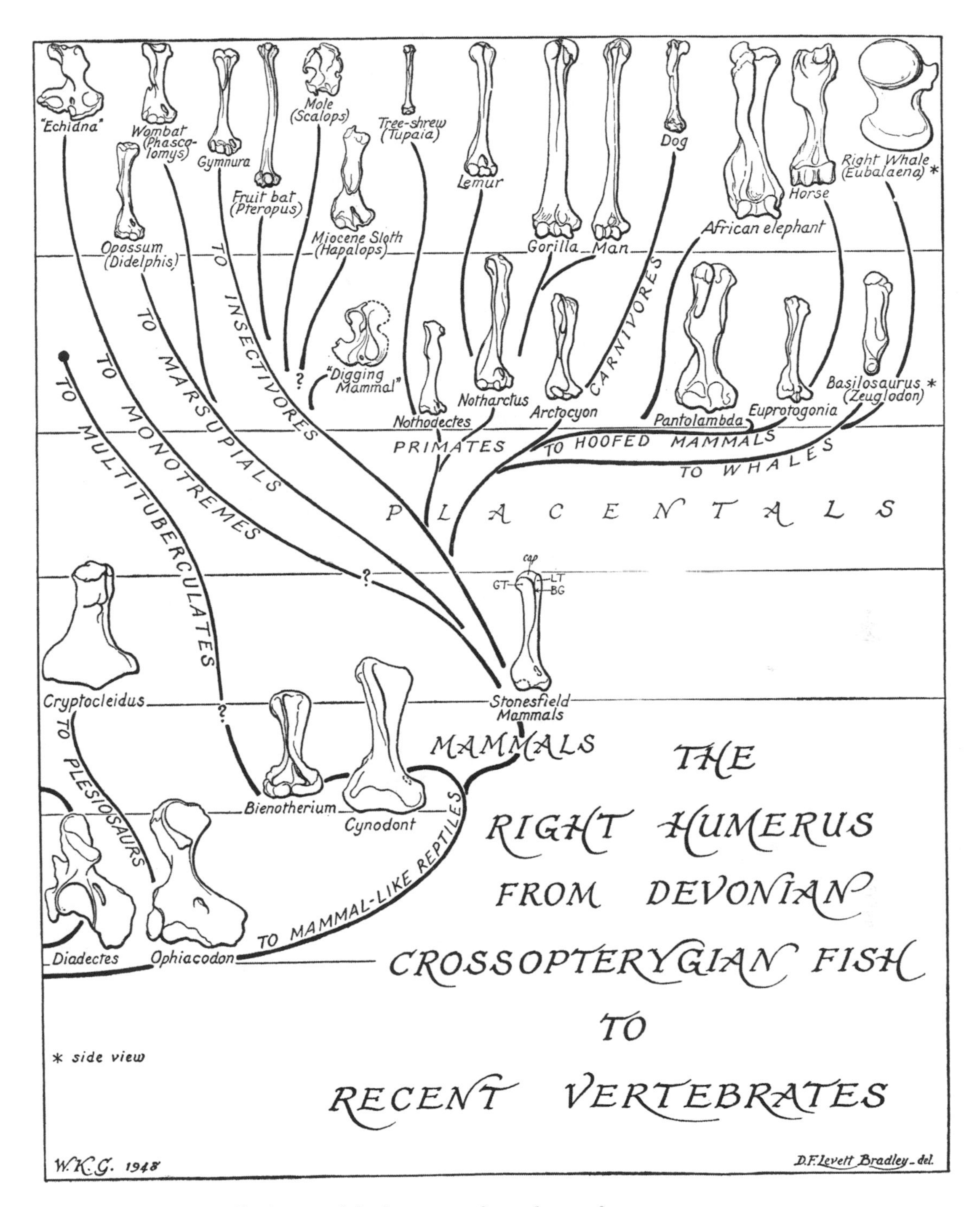

FIGURE 165. Evolution of the humerus of vertebrates, from crossopterygians to recent mammals, by William King Gregory in 1948.

Gregory, 1951, *Evolution Emerging: A Survey of Changing Patterns from Primeval Life to Man,* vol. 2, p. 547, fig. 19.40; courtesy of Mary DeJong, Mai Qaraman, and the American Museum of Natural History. Used with permission.

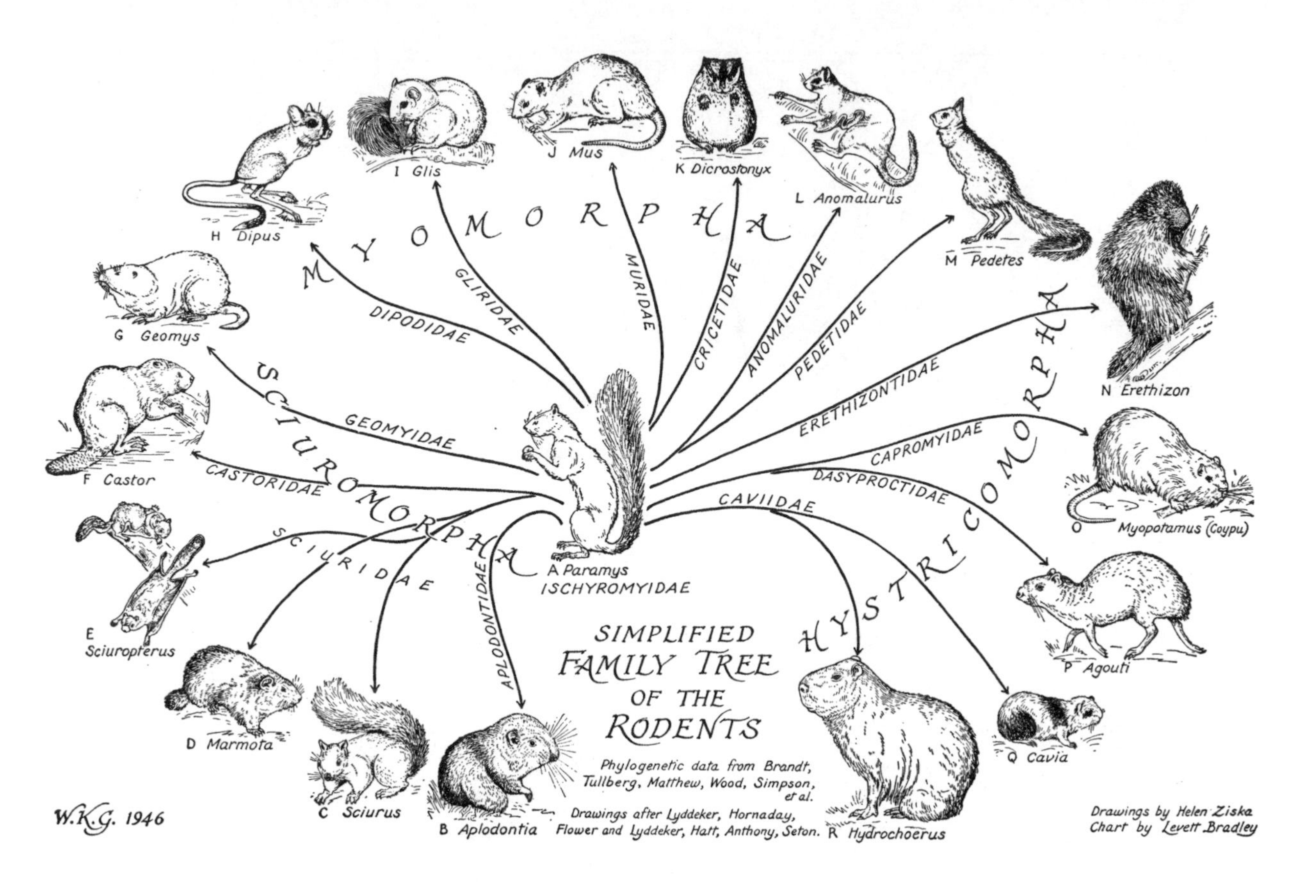

FIGURE 166. William King Gregory's 1946 tree of rodent relationships.

Gregory, 1951, *Evolution Emerging: A Survey of Changing Patterns from Primeval Life to Man,* vol. 2, p. 757; fig. 20.33; courtesy of Mary DeJong, Mai Qaraman, and the American Museum of Natural History. Used with permission.

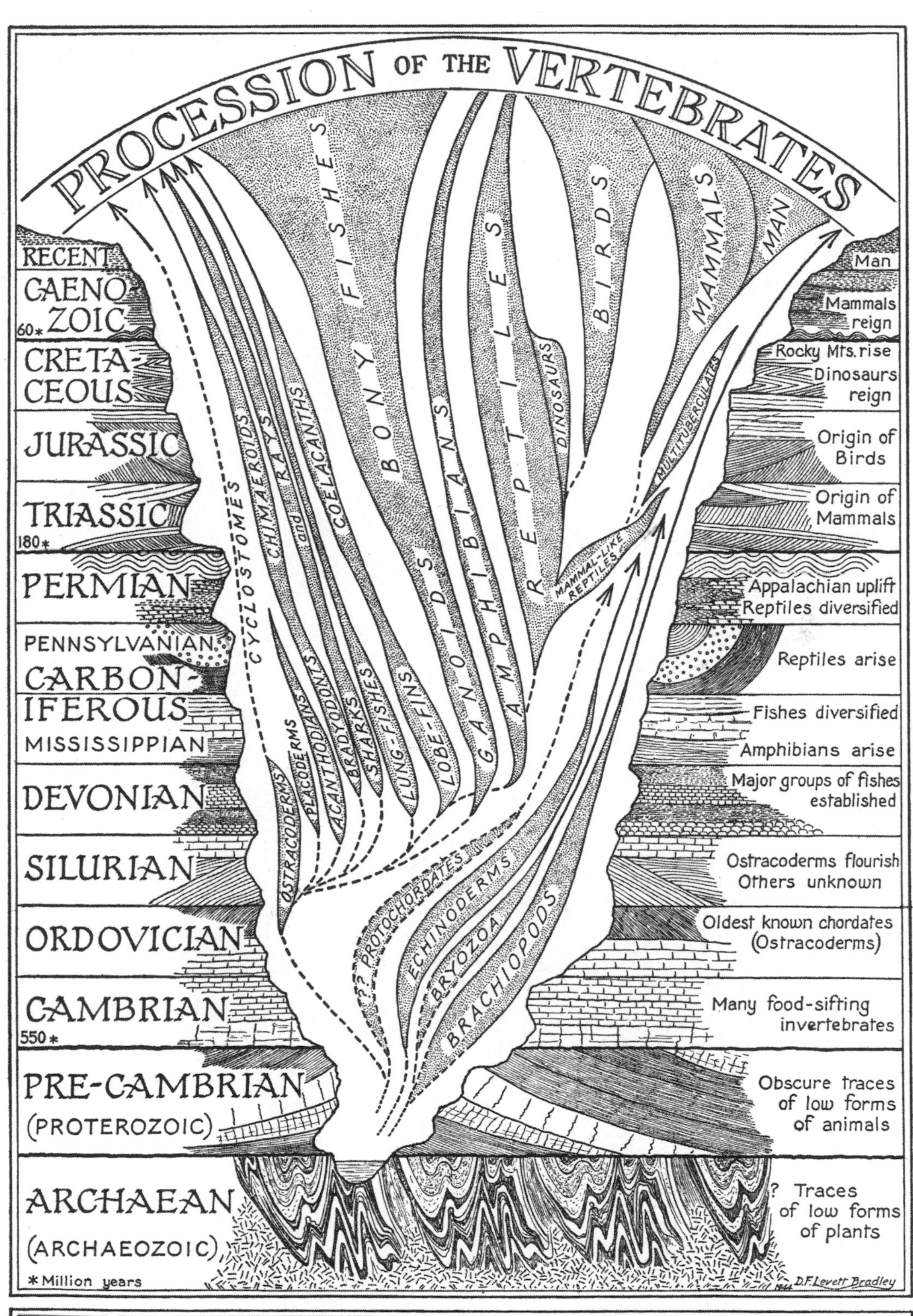
PROCESSION OF THE VERTEBRATES
RECENT
CAENOZOIC
60*
CRETACEOUS
JURASSIC
TRIASSIC
180*
PERMIAN
PENNSYLVANIAN
CARBONIFEROUS
MISSISSIPPIAN
DEVONIAN
SILURIAN
ORDOVICIAN
CAMBRIAN
550*
PRE-CAMBRIAN
(PROTEROZOIC)
ARCHAEAN
(ARCHAEOZOIC)
*Million years
CYCLOSTOMES
CHIMAEROIDS
RAYS
COELACANTHS
BONY FISHES
LOBE-FINS
GANOIDS
AMPHIBIANS
REPTILES
DINOSAURS
BIRDS
MAMMALS
MAN
MULTITUBERCULATES
MAMMAL-LIKE REPTILES
OSTRACODERMS
PLACODERMS
ACANTHODIANS
BRADYODONTS
SHARKS
SHARKS
LUNG-FISHES
?? PROTOCHORDATES
ECHINODERMS
BRYOZOA
BRACHIOPODS
Man
Mammals reign
Rocky Mts. rise
Dinosaurs reign
Origin of Birds
Origin of Mammals
Appalachian uplift
Reptiles diversified
Reptiles arise
Fishes diversified
Amphibians arise
Major groups of fishes established
Ostracoderms flourish
Others unknown
Oldest known chordates
(Ostracoderms)
Many food-sifting invertebrates
Obscure traces of low forms of animals
? Traces of low forms of plants
D.F. Levett Bradley

EVOLUTION EMERGING

(*page* 229)

FIGURE 167. The frontispiece of William King Gregory's two-volume *Evolution Emerging.*

Gregory, 1951, *Evolution Emerging: A Survey of Changing Patterns from Primeval Life to Man,* vol. 1, frontispiece; courtesy of Mary DeJong, Mai Qaraman, and the American Museum of Natural History. Used with permission.

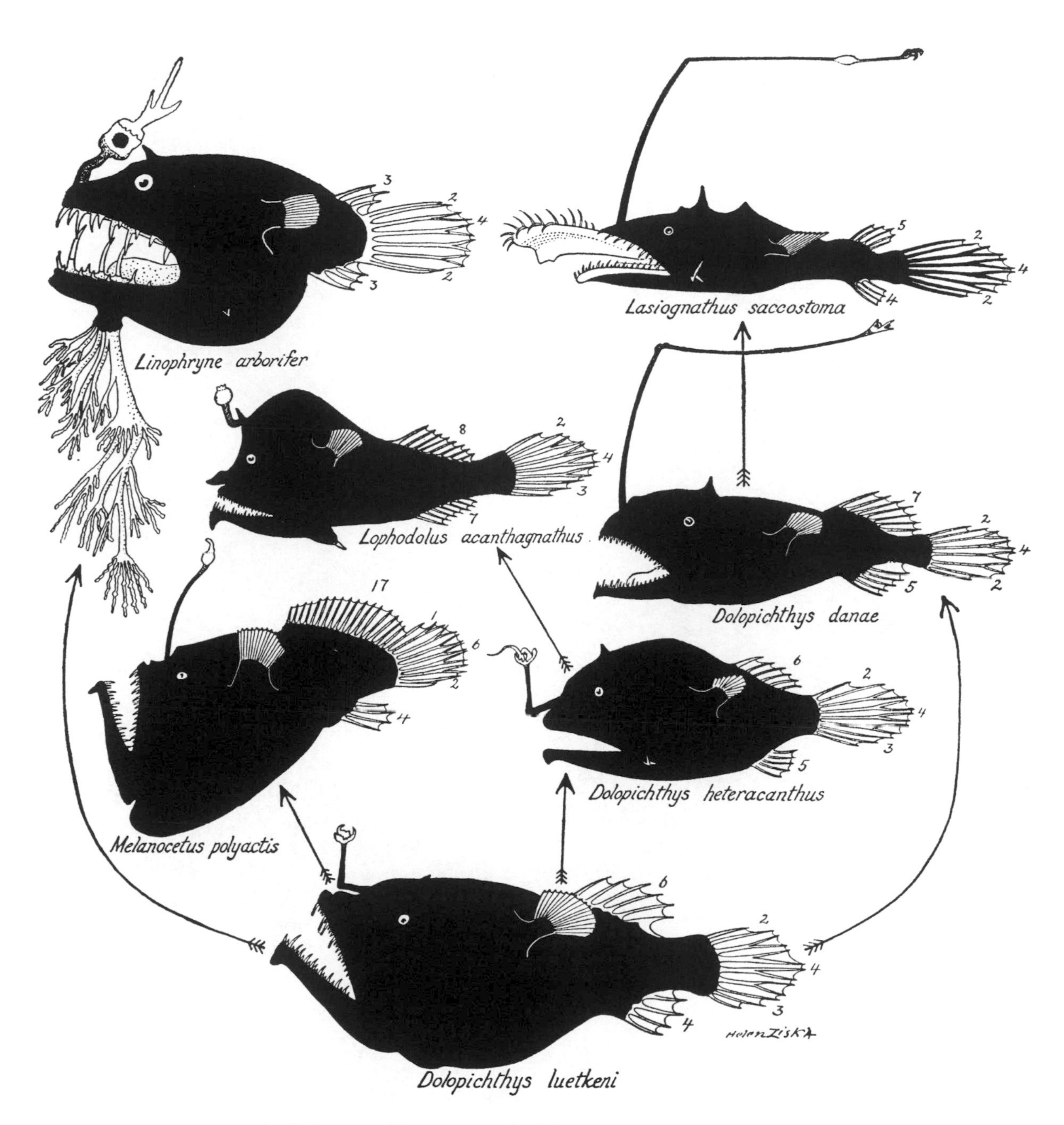

FIGURE 168. A phylogeny of deep-sea anglerfishes.

Gregory, 1951, *Evolution Emerging: A Survey of Changing Patterns from Primeval Life to Man,* vol. 2, p. 758, fig. 20.34; courtesy of Mary DeJong, Mai Qaraman, and the American Museum of Natural History. Used with permission.

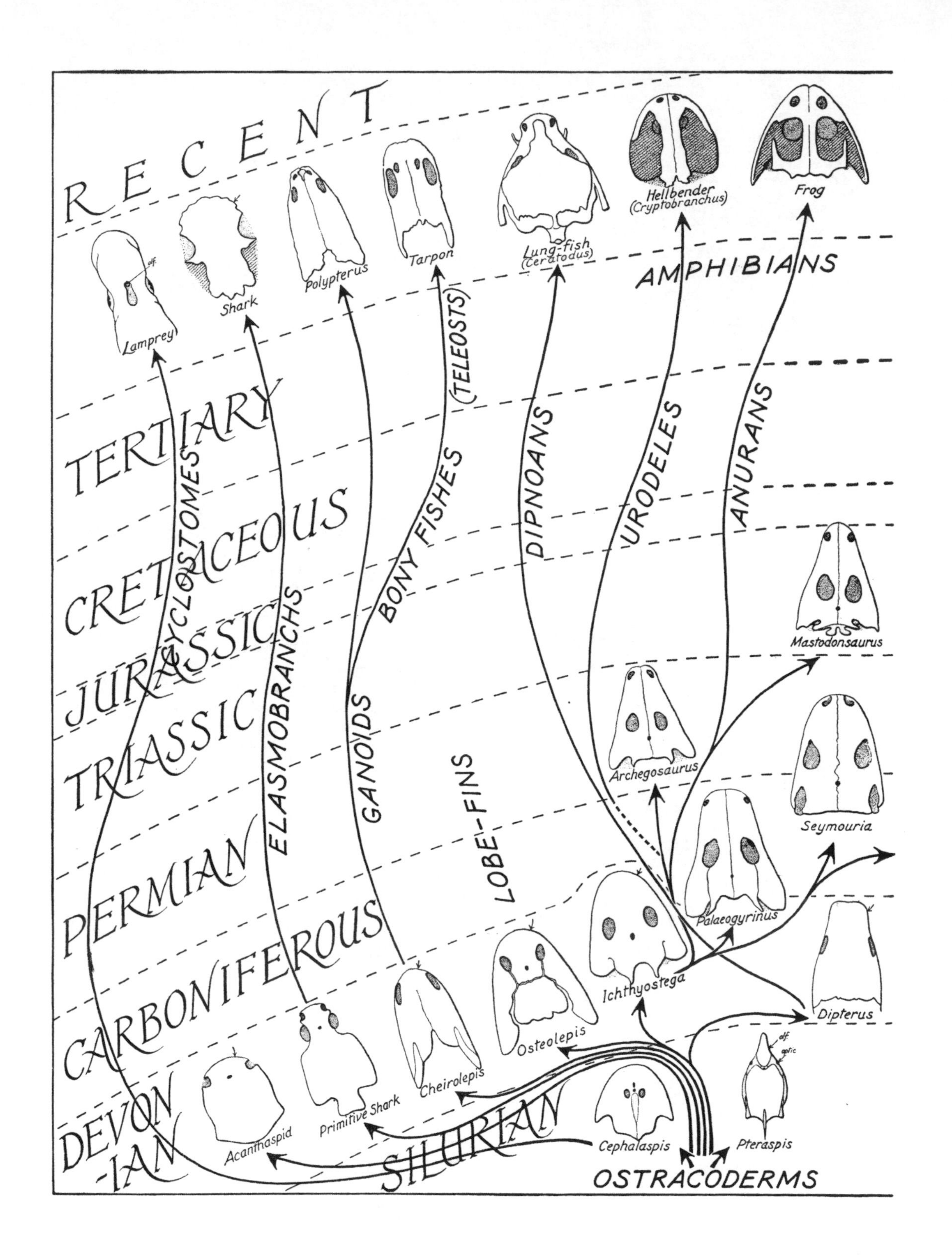

RECENT
TERTIARY
CRETACEOUS
JURASSIC
TRIASSIC
PERMIAN
CARBONIFEROUS
DEVON-IAN
SILURIAN
Lamprey
Shark
Polypterus
Tarpon
Lung-fish
(Ceratodus)
Hellbender
(Cryptobranchus)
Frog
AMPHIBIANS
(TELEOSTS)
CYCLOSTOMES
ELASMOBRANCHS
GANOIDS
BONY FISHES
LOBE-FINS
DIPNOANS
URODELES
ANURANS
Mastodonsaurus
Archegosaurus
Seymouria
Palaeogyrinus
Ichthyostega
Dipterus
Osteolepis
Cheirolepis
Primitive Shark
Acanthaspid
Cephalaspis
Pteraspis
OSTRACODERMS

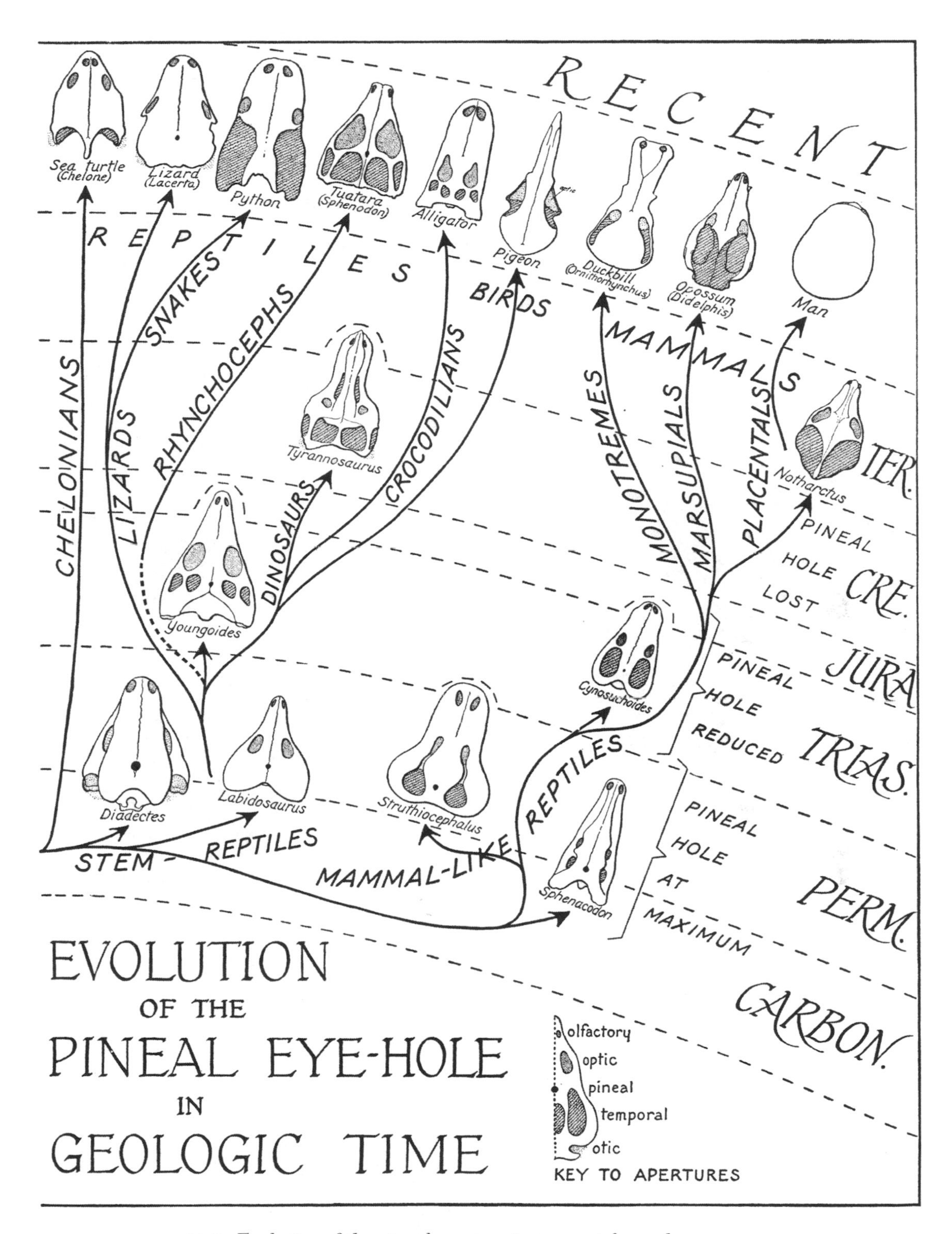

FIGURE 169. Evolution of the pineal aperture in terrestrial vertebrates.

Gregory, 1951, *Evolution Emerging: A Survey of Changing Patterns from Primeval Life to Man*, vol. 2, p. 718, figs. 21.116a and 21.116b; courtesy of Mary DeJong, Mai Qaraman, and the American Museum of Natural History. Used with permission.

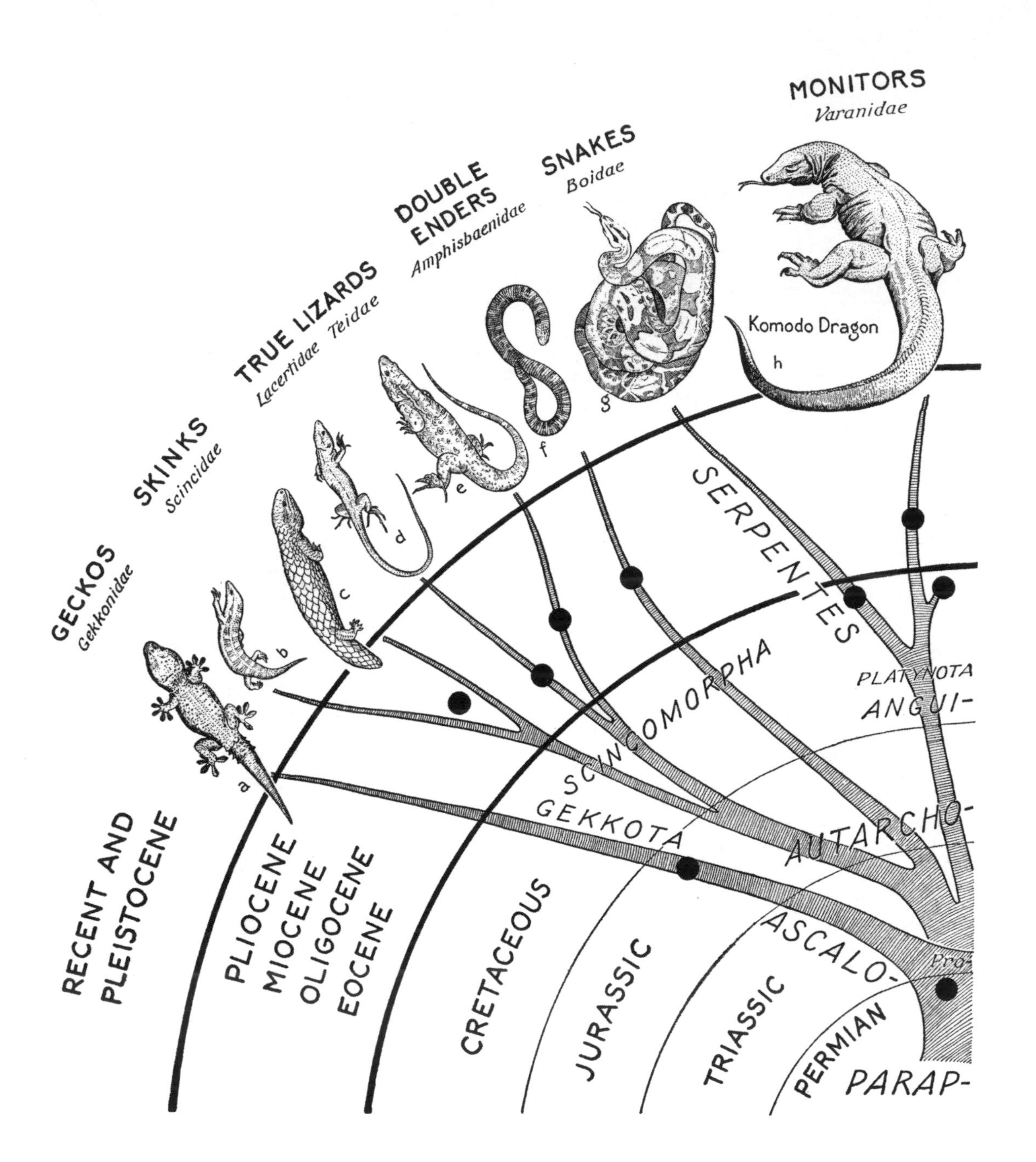

MONITORS
Varanidae
SNAKES
Boidae
DOUBLE ENDERS
Amphisbaenidae
TRUE LIZARDS
Lacertidae Teidae
SKINKS
Scincidae
GECKOS
Gekkonidae
Komodo Dragon
a
b
c
d
e
f
g
h
SERPENTES
PLATYNOTA
ANGUI-
SCINCOMORPHA
GEKKOTA
AUTARCHO-
ASCALO-
Pro-
PARAP-
RECENT AND PLEISTOCENE
PLIOCENE
MIOCENE
OLIGOCENE
EOCENE
CRETACEOUS
JURASSIC
TRIASSIC
PERMIAN

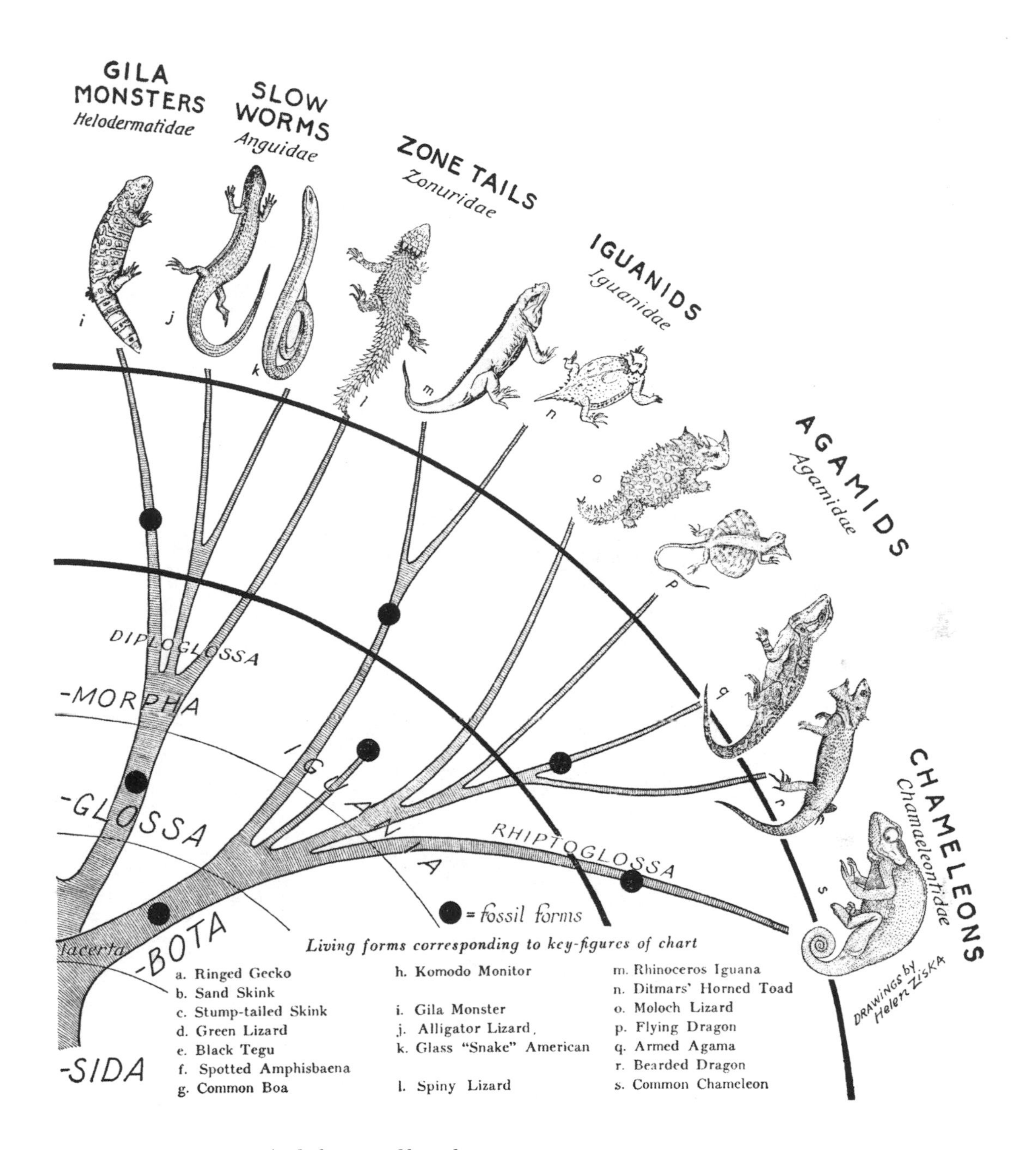

FIGURE 170. A phylogeny of lizards.

Gregory, 1951, *Evolution Emerging: A Survey of Changing Patterns from Primeval Life to Man*, vol. 2, pp. 886–887; figs. 24.5a and 24.5b; courtesy of Mary DeJong, Mai Qaraman, and the American Museum of Natural History. Used with permission.

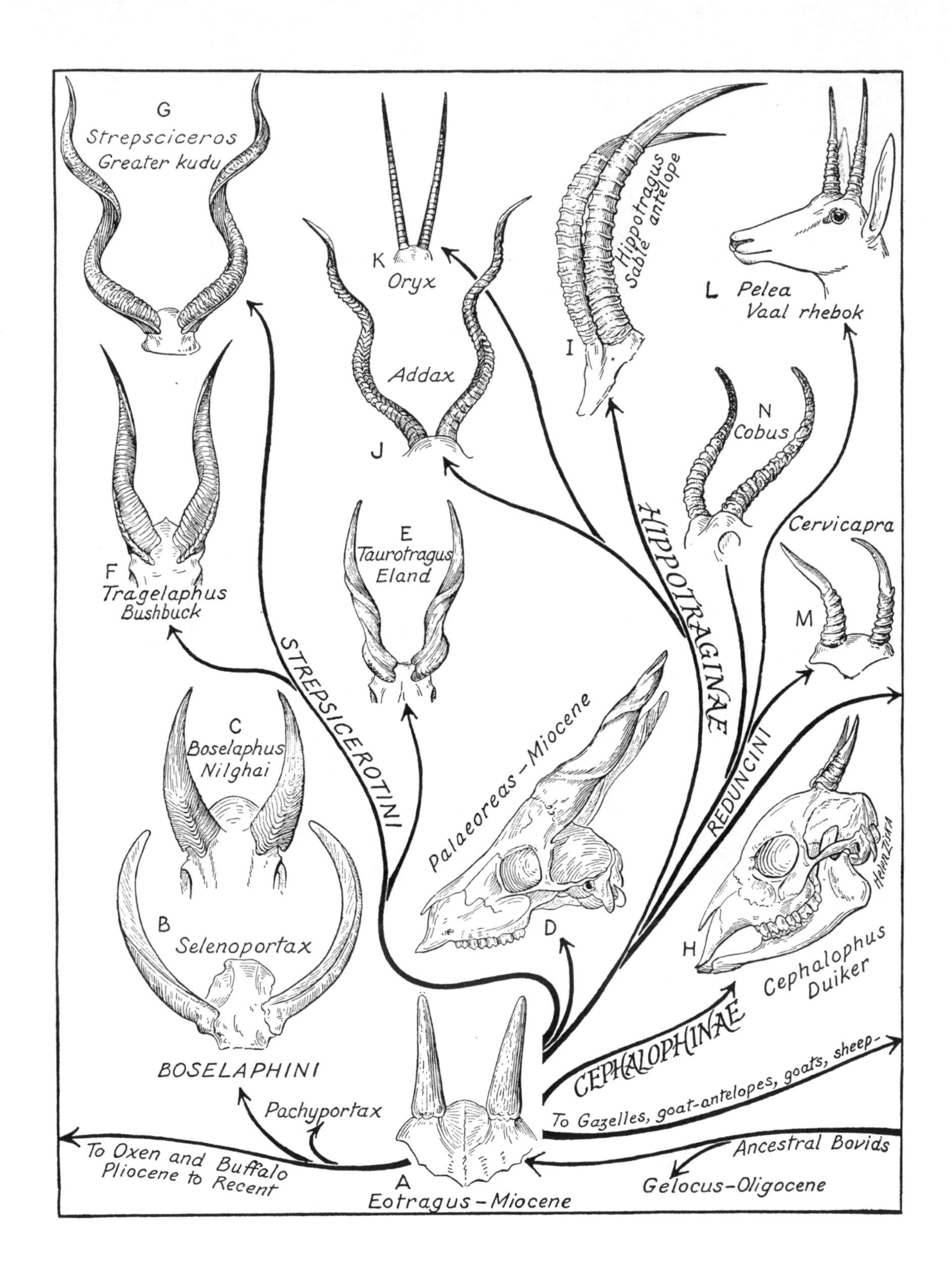

G
Strepsiceros
Greater kudu
K
Oryx
Hippotragus
Sable antelope
I
L Pelea
Vaal rhebok
Addax
J
N
Cobus
E
Taurotragus
Eland
HIPPOTRAGINAE
Cervicapra
F
Tragelaphus
Bushbuck
M
STREPSICEROTINI
C
Boselaphus
Nilghai
Palaeoreas–Miocene
REDUNCINI
B
Selenoportax
D
H
Cephalophus
Duiker
CEPHALOPHINAE
BOSELAPHINI
To Gazelles, goat-antelopes, goats, sheep-
Pachyportax
Ancestral Bovids
To Oxen and Buffalo
Pliocene to Recent
A
Eotragus–Miocene
Gelocus–Oligocene

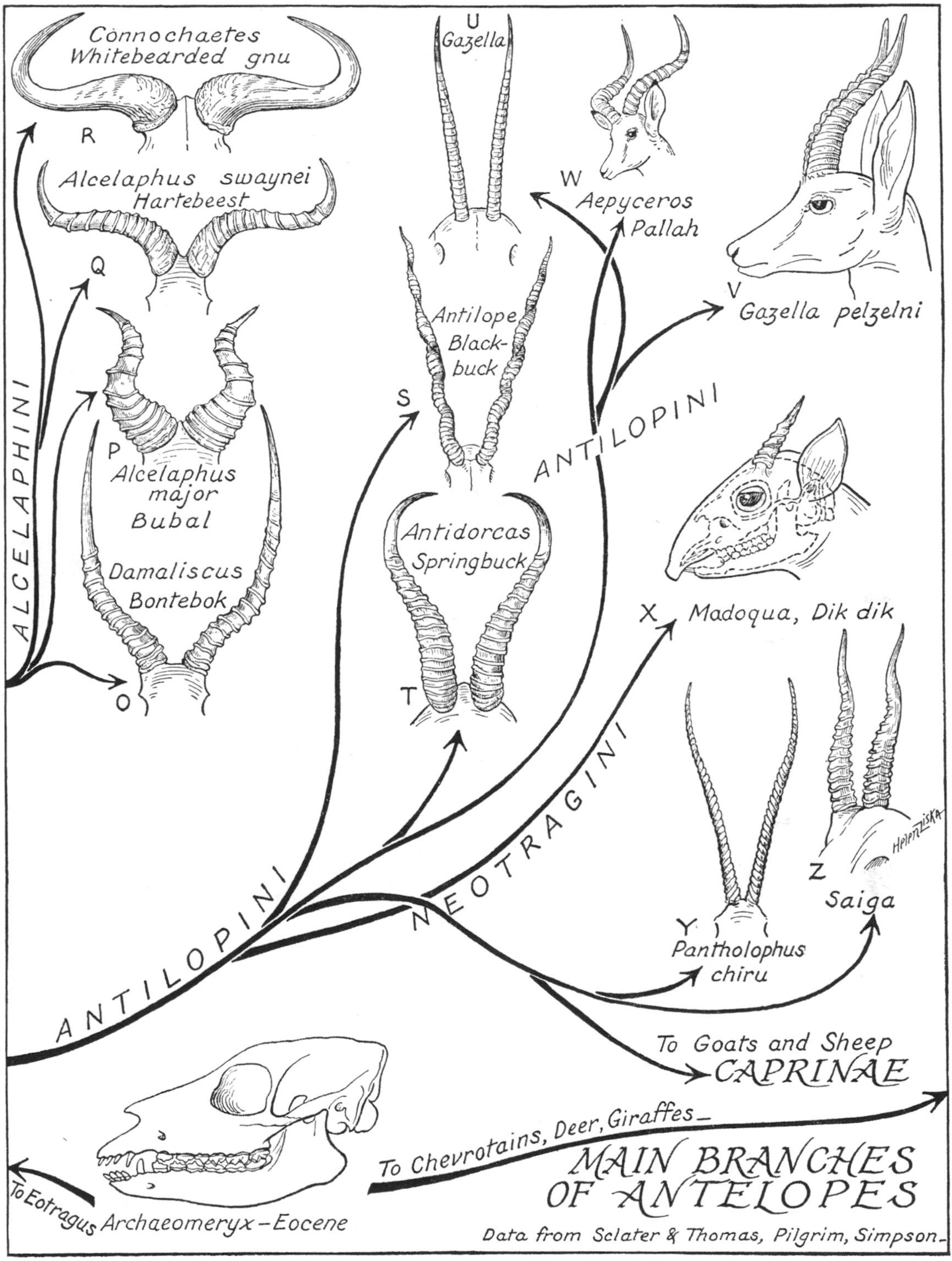

FIGURE 171. A phylogeny of antelopes and their allies based on skulls and horns.

Gregory, 1951, *Evolution Emerging: A Survey of Changing Patterns from Primeval Life to Man*, vol. 2, pp. 998–999, figs. 24.10a and 24.10b; courtesy of Mary DeJong, Mai Qaraman, and the American Museum of Natural History. Used with permission.

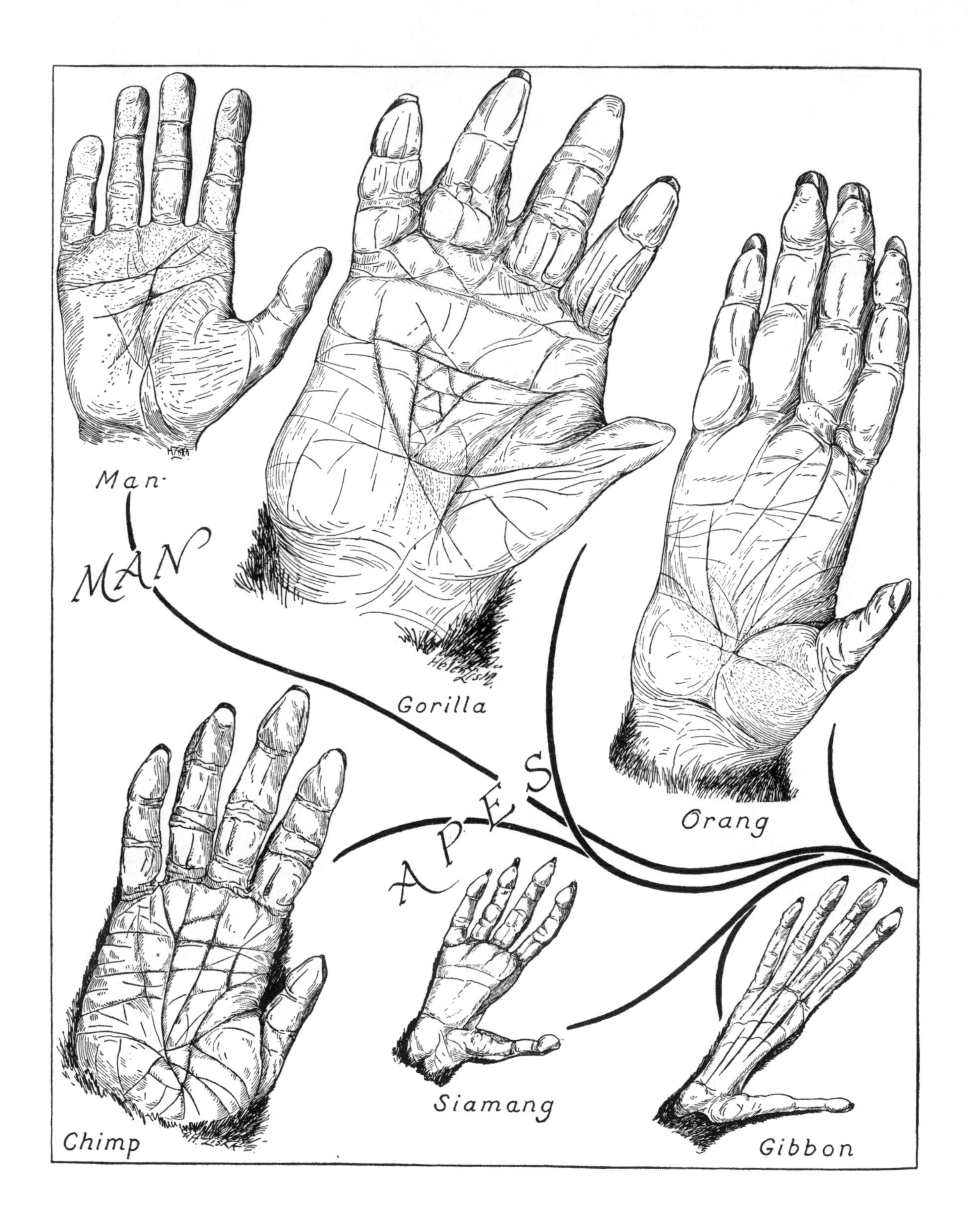
Man
MAN
Gorilla
Orang
APES
Chimp
Siamang
Gibbon

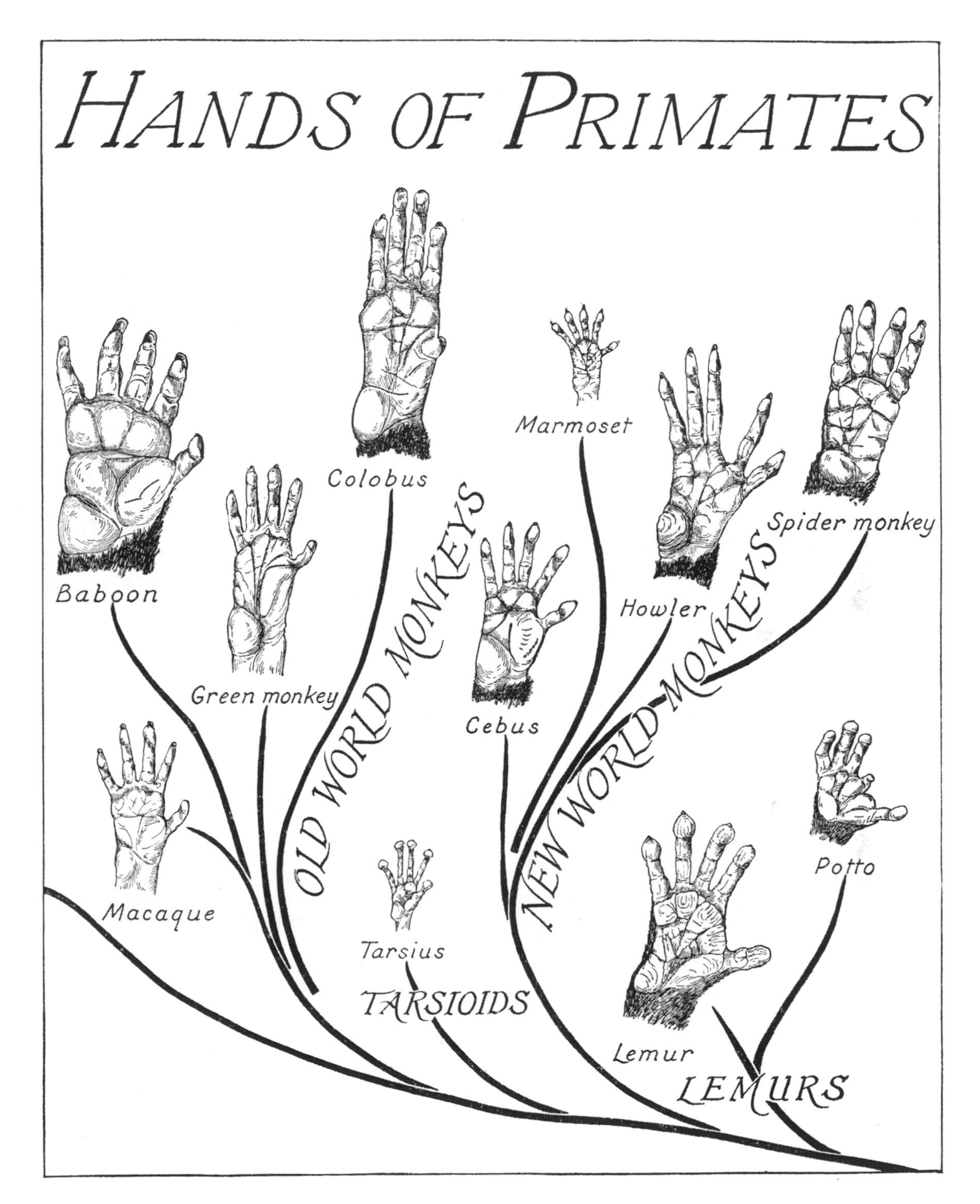

FIGURE 172. The hands of primates.

Gregory, 1951, *Evolution Emerging: A Survey of Changing Patterns from Primeval Life to Man,* vol. 2, pp. 1004–1005; figs. 24.11a and 24.11b; courtesy of Mary DeJong, Mai Qaraman, and the American Museum of Natural History. Used with permission.

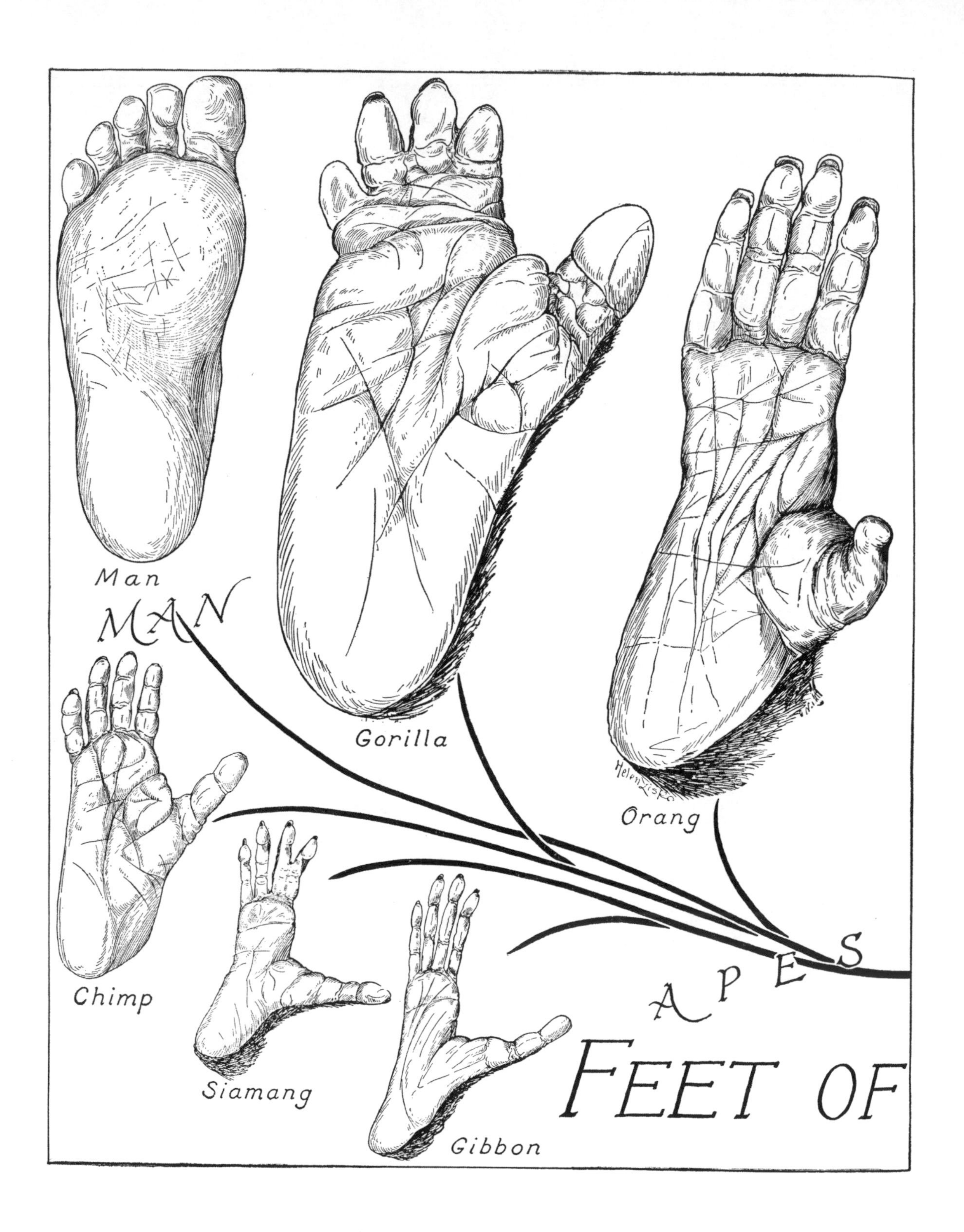
Man
MAN
Gorilla
Orang
Chimp
Siamang
Gibbon
APES
FEET OF

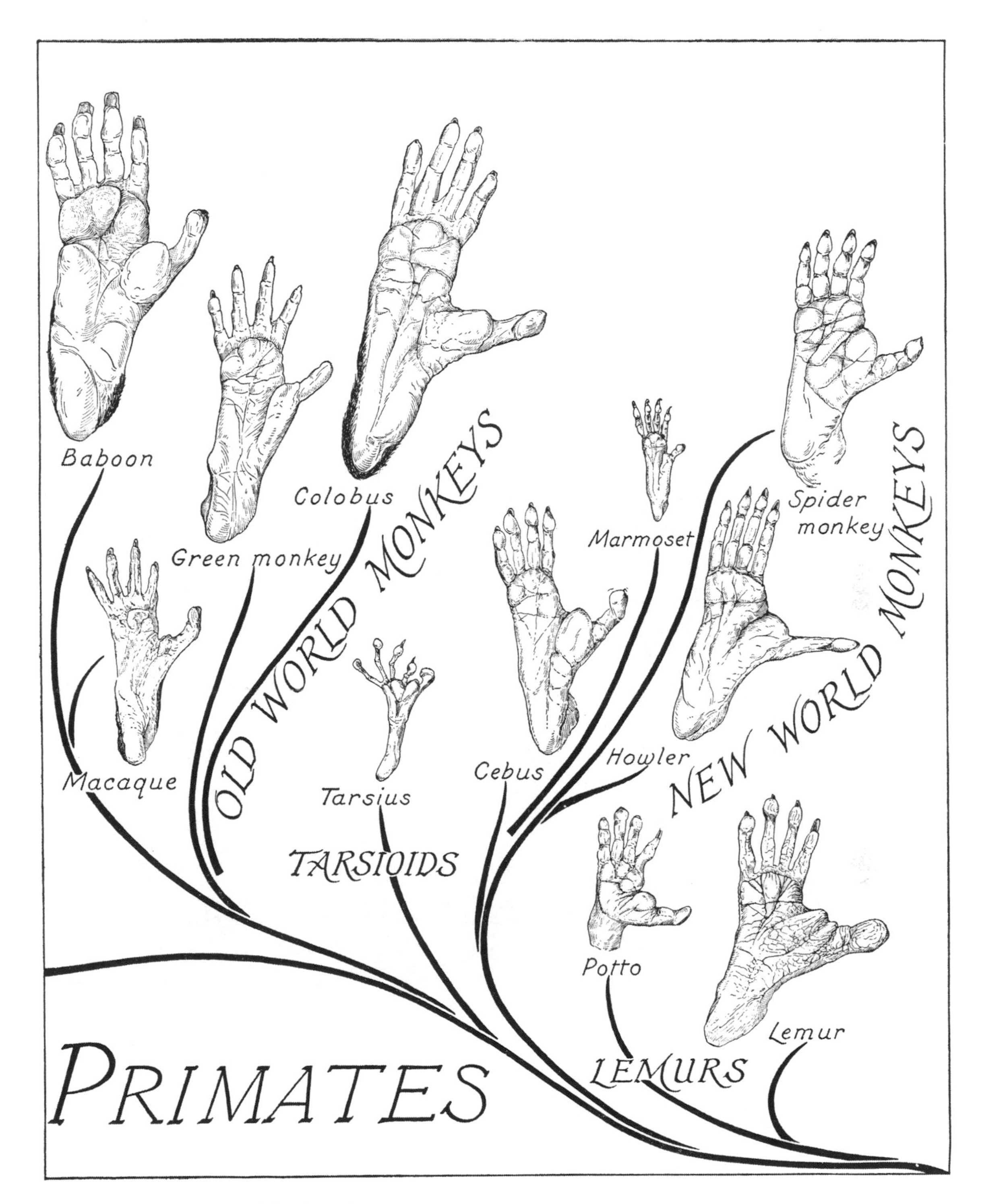

FIGURE 173. The feet of primates.

Gregory, 1951, *Evolution Emerging: A Survey of Changing Patterns from Primeval Life to Man,* vol. 2, pp. 1006–1007; pl. 15.I; courtesy of Mary DeJong, Mai Qaraman, and the American Museum of Natural History. Used with permission.

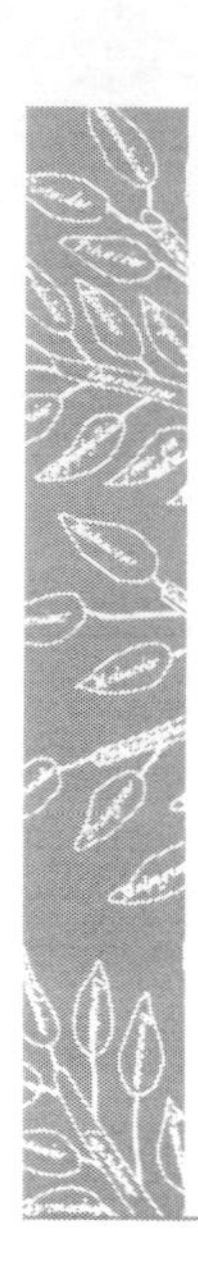

Hints of New Approaches

1954–1969

Early in his career at the University of California at Berkeley, American botanist and cytologist Thomas Harper Goodspeed (1887–1966) became obsessed with the tobacco genus *Nicotiana,* devoting much of his career to tracing the origin (via hybridization) and evolution of the numerous varieties of cultivated tobacco from their wild ancestors and from one another. In a now classic monograph on the subject, published in 1954, Goodspeed summarized his work using several branching diagrams, two of which are reproduced here (Figures 174 and 175). The various taxa, hypothesized and real, are shown distributed within a series of concentric arcs. Numerous connections between the taxa—especially between the diploid (twelve pairs of chromosomes) and tetraploid (twenty-four pairs of chromosomes) groups shown in the outer two arcs—reflect the high frequency of polyploidy (having more than two sets of chromosomes) and hybridization (the latter assumed in many cases, as was the style of the day) for which tobacco is well known.[1] Reading the two trees as one, the first shows the species-level relationships among the diploids. In the second, the diploid clusters are reduced to circles with subgeneric or section names (e.g., the cluster around "longibracteata" corresponds to section Acuminatae and the cluster associated with "noctiflora" is section Noctiflorae), and only includes those diploid groups that Goodspeed believed contributed to the tetraploid taxa.[2]

Not satisfied with traditional branching diagrams to display evolutionary relationships, British botanist Ronald D'Oyley Good (1896–

1992) published in 1956 a diagram intended to show affinities among monocotyledonous plants (Figure 176) that is not unlike Paul Giseke's (1792) circle diagram of plant families that appeared more than 160 years before (see Figure 16). But, instead of circles, Good chose a three-dimensional model—unlabeled spheres, like planets, floating in space, their size, relative position, and distance from one another correlating with taxonomic content and hypothesized affinity. Arguing that a two-dimensional diagram cannot possibly show the correct "mutual position" of taxa, "according to their total morphological resemblances and differences," he found it relative easy to "make a model in which each group is represented by a *sphere* of appropriate size (namely of a diameter roughly the cube root of the species number) which at least will show us a great deal about the monocotyledons, and which will help us to visualize both them and some aspects of their possible evolution in novel fashion."[3]

Another three-dimensional approach, similar to that of Good, but with rod-like interconnections (Figure 177), was employed by Oleg Lysenko (1931–1985) and Peter Henry Andrews Sneath (1923–) in attempting to show the relationships of the Enterobacteriacaea, a large family of bacteria, including many of the more familiar pathogens, such as *Salmonella* and *Escherichia coli*. The various groupings are represented by spheres connected by rods or sticks indicating taxonomic distance.[4]

Following the precedent set by Ernst Haeckel in 1866 and Herbert Copeland in 1938 (see Figures 66, 150), American ecologist Robert Harding Whittaker (1920–1980) summarized new evidence to support yet another alternative hypothesis to visualize the earliest branches of the tree of life (Figure 178). In 1959, citing recent fundamental changes in the classification of algae, the need to accommodate "lower" organisms that cannot easily be divided between the plant and animal kingdoms, and new hypotheses concerning the origin of the fungi, Whittaker proposed a modified four-kingdom system (Figure 179): the Protista, or unicellular organisms; the Plantae, or multicellular plants; the Fungi; and the Animalia, or multicellular animals.[5]

In 1969, Whittaker revised his thinking, again based on new evidence, this time proposing a five-kingdom system emphasizing three levels of organization (Figure 180): the prokaryotic (kingdom Monera), eukaryotic unicellular (kingdom Protista), and eukaryotic multicellular and multinucleate (kingdoms Plantae, Fungi, and Animalia).[6] On each level he proposed divergence in relation to three principal modes of nutrition: photosynthetic, absorptive, and ingestive. Ingestive nutrition is lacking in the Monera, the three modes are continuous along evolutionary lines in the Protista, but, on the multicellular-multinucleate level, the nutritive modes lead to the widely different kinds of organization that characterize the three more derived kingdoms.

Elmar Emil Leppik (1898–1978), the foremost authority on the biology of rust fungi, published a long series of papers in the 1950s and 60s, detailing his ideas about biological specialization as an adaptation of obligate rust parasites that live only on specific plant hosts. Providing early insight into the process of coevolution, which was little explored at the time, he was able to show that the phylogenetic history of plant rusts is strongly correlated with the evolution of their host plants, illustrating this relationship in a tree published in 1965 (Figure 181). The diagram is rather complex and requires some explanation: rust fungi are characterized by a genetically fixed alteration of generations, with an irreversible sequence of spore forms, each generation parasitizing a different host plant. As further described by Leppik,

> one generation must keep its old host until the other generation is well established to a new host. Then the first generation frees itself from the old host and chooses some phylogenetically younger group of plants for this new host. In this way, the rusts, with their two generations, aecidial and telial, climb the "hologenetic ladder."[7]

This so-called climbing is represented in the diagram by the alternation of the telial (small black dots) and aecidial (small open circles) stages of fungal reproduction, switching back and forth between rust families (encircled within thick black lines) and host plant families (within thin black lines), the rust and plant taxa connected by solid lines. Dashed lines represent assumed connections. Thus, "the oldest rusts still live on ferns, the intermediate forms, on gymnosperms, and the modern groups, on angiosperms."[8]

In the early 1960s, Peter Humphry Greenwood (1927–1995) and his colleagues Donn Eric Rosen (1929–1986), Stanley Howard Weitzman (1927–), and George Sprague Myers (1905–1985), realizing that teleost classification was in a state of disarray—that many of the "most generally recognized orders are no more than catch-alls for separate lineages which have attained a comparable stage of specialization or complexity"[9]—resolved to rectify the problem by analyzing "the predominate evolutionary trends" displayed by the group.[10] They did away with the linear view of teleost relationships that had been in place for nearly a century, little modified since of the time of Theodore Gill (see Figure

13), and replaced it with three major divisions (plus a possible fourth) all evolving independently from among extinct holostean-like fishes of the Pholidophoroidei (Figure 182). The divisions are largely defined by differences in the anatomy of the swim bladder relative to the inner ear known as an "otophysic connection," as originally proposed by Walter Garstang in 1931 (see Figure 143).

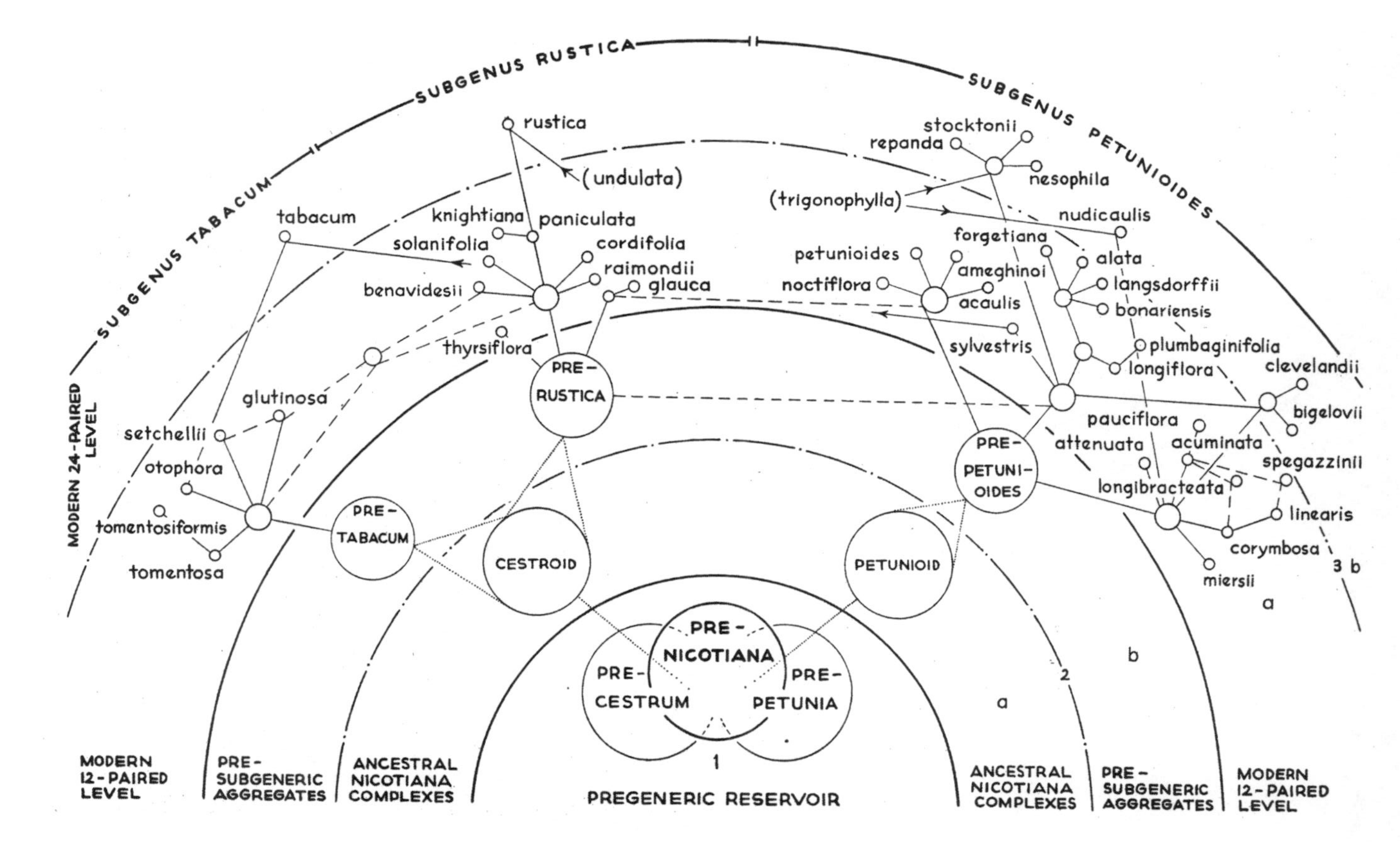

FIGURE 174. Origin, evolution, and relationships of the sixty species of tobacco, genus *Nicotiana*, by Thomas Harper Goodspeed.

Goodspeed, 1954, *The Genus* Nicotiana; *Origins, Relationships and Evolution of its Species in the Light of their Distribution, Morphology and Cytogenetics*, pp. 310, 312, figs. 57 and 58; Chronica Botanica Company, Waltham, Massachusetts.

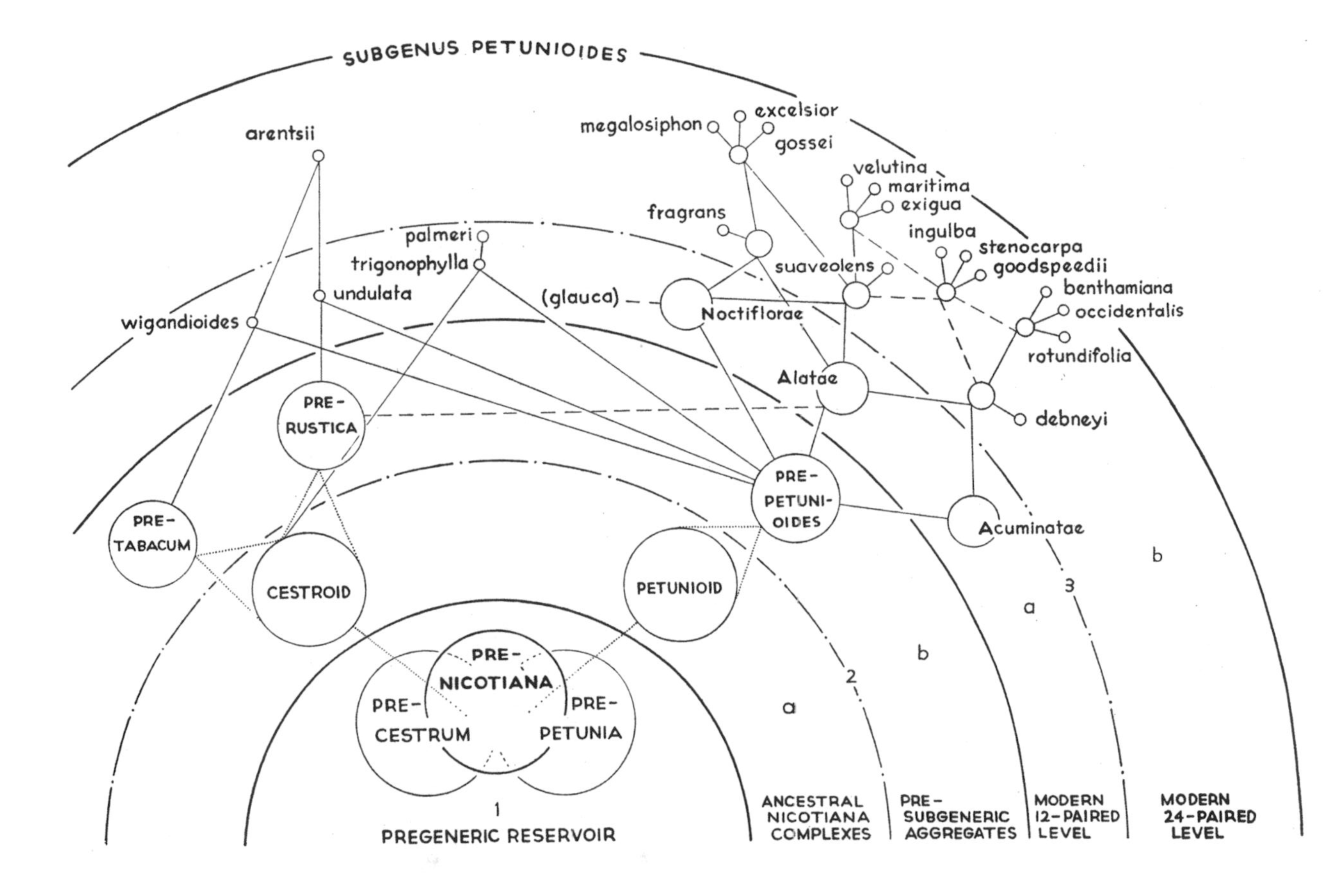

FIGURE 175. Origin, evolution, and relationships of the sixty species of tobacco, genus *Nicotiana* (continued), by Thomas Harper Goodspeed.

Goodspeed, 1954, *The Genus* Nicotiana; *Origins, Relationships and Evolution of its Species in the Light of their Distribution, Morphology and Cytogenetics,* pp. 310, 312, figs. 57 and 58; Chronica Botanica Company, Waltham, Massachusetts.

FIGURE 176. A stellar model of relationships of monocotyledonous plants, by Ronald D'Oyley Good in 1956, in which each group is represented by a circle, the diameter of which is proportionate to the cube root of the number of contained species.

Good, 1956, *Features of Evolution in the Flowering Plants*, p. 117, fig. 31.

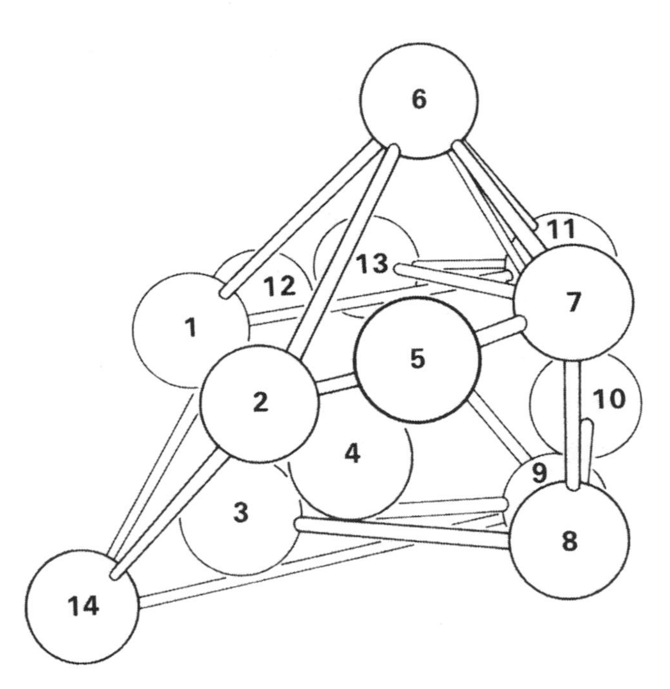

FIGURE 177. A three-dimensional branching diagram of relationships of the Enterobacteriacaea, by Peter Sneath and Robert Sokal, based on a model by Oleg Lysenko and Sneath in 1959. The taxa are represented by spheres connected by rods indicating taxonomic distance.

Sneath and Sokal, 1973, *Numerical Taxonomy: The Principles and Practice of Numerical Classification,* p. 260, figs. 5–15; courtesy of Peter H. A. Sneath and Robert R. Sokal. Used with permission.

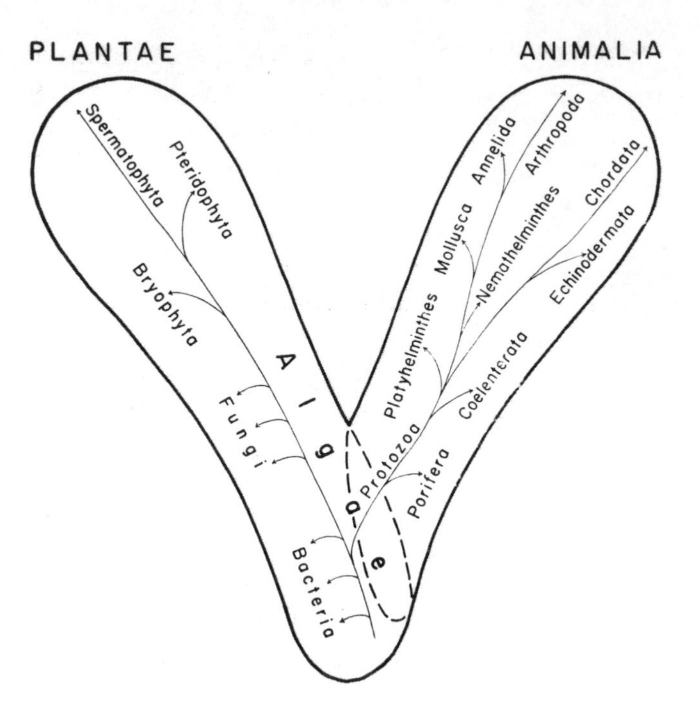

FIGURE 178. Robert Harding Whittaker schematic, two-kingdom view of life, with evolutionary relationships of major groups suggested in simplified form.

Whittaker, 1959, *Quarterly Review of Biology,* 34(3):211, 217, figs. 1 and 2; courtesy of Perry Cartwright and the University of Chicago Press. Used with permission.

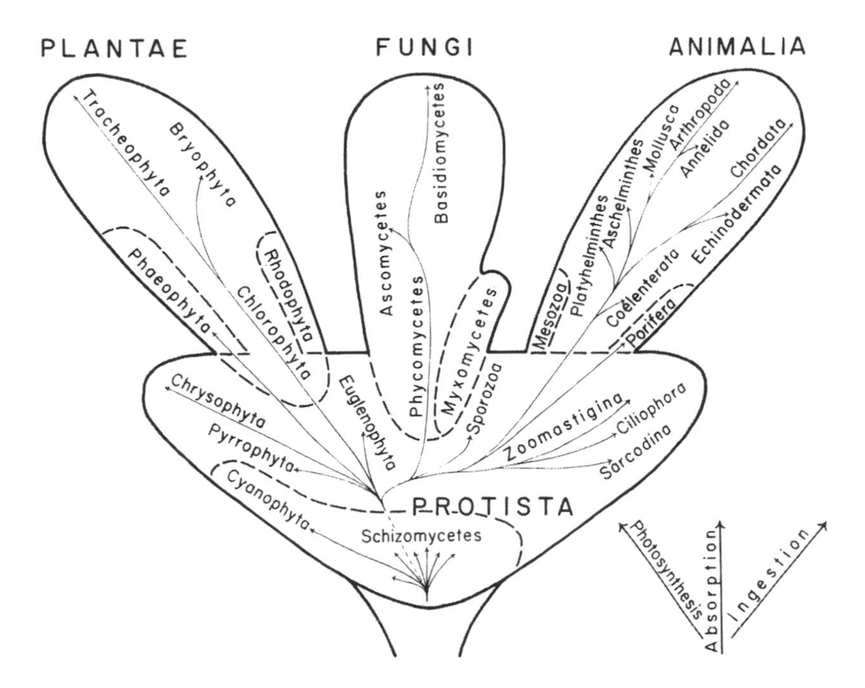

FIGURE 179. A four-kingdom system of classification suggested by Robert Harding Whittaker to replace the old traditional two-kingdom system of Plantae and Animalia, emphasizing evolutionary relationships in terms of three modes of nutrition and directions of evolution—the photosynthetic of the green plants, the ingestive of the animals, and the absorptive of the bacteria and fungi.

Whittaker, 1959, *Quarterly Review of Biology*, 34(3):211, 217, figs. 1 and 2; courtesy of Perry Cartwright and the University of Chicago Press. Used with permission.

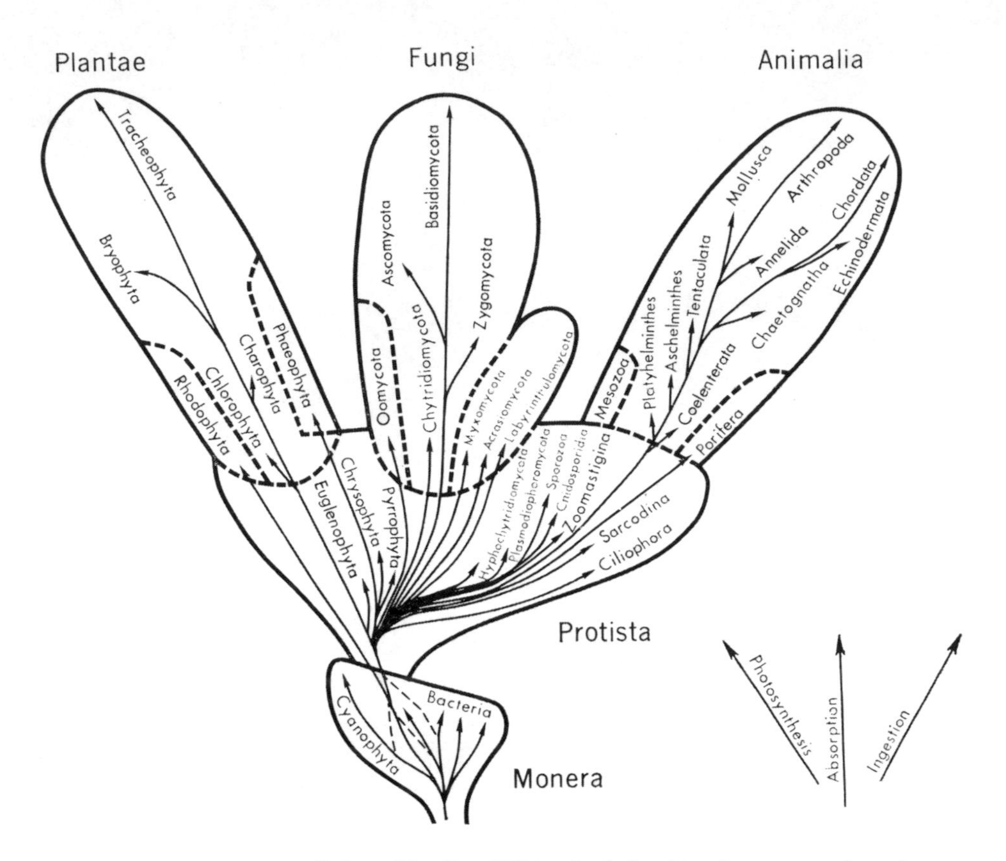

FIGURE 180. Robert Harding Whittaker's five-kingdom system based on three levels of organization: the prokaryotic (kingdom Monera), eukaryotic unicellular (kingdom Protista), and eukaryotic multicellular and multinucleate (kingdoms Plantae, Fungi, and Animalia).

Whittaker, 1969, *Science*, 163(3863):157, fig. 3; courtesy of Elizabeth Sandler and the American Association for the Advancement of Science. Used with permission.

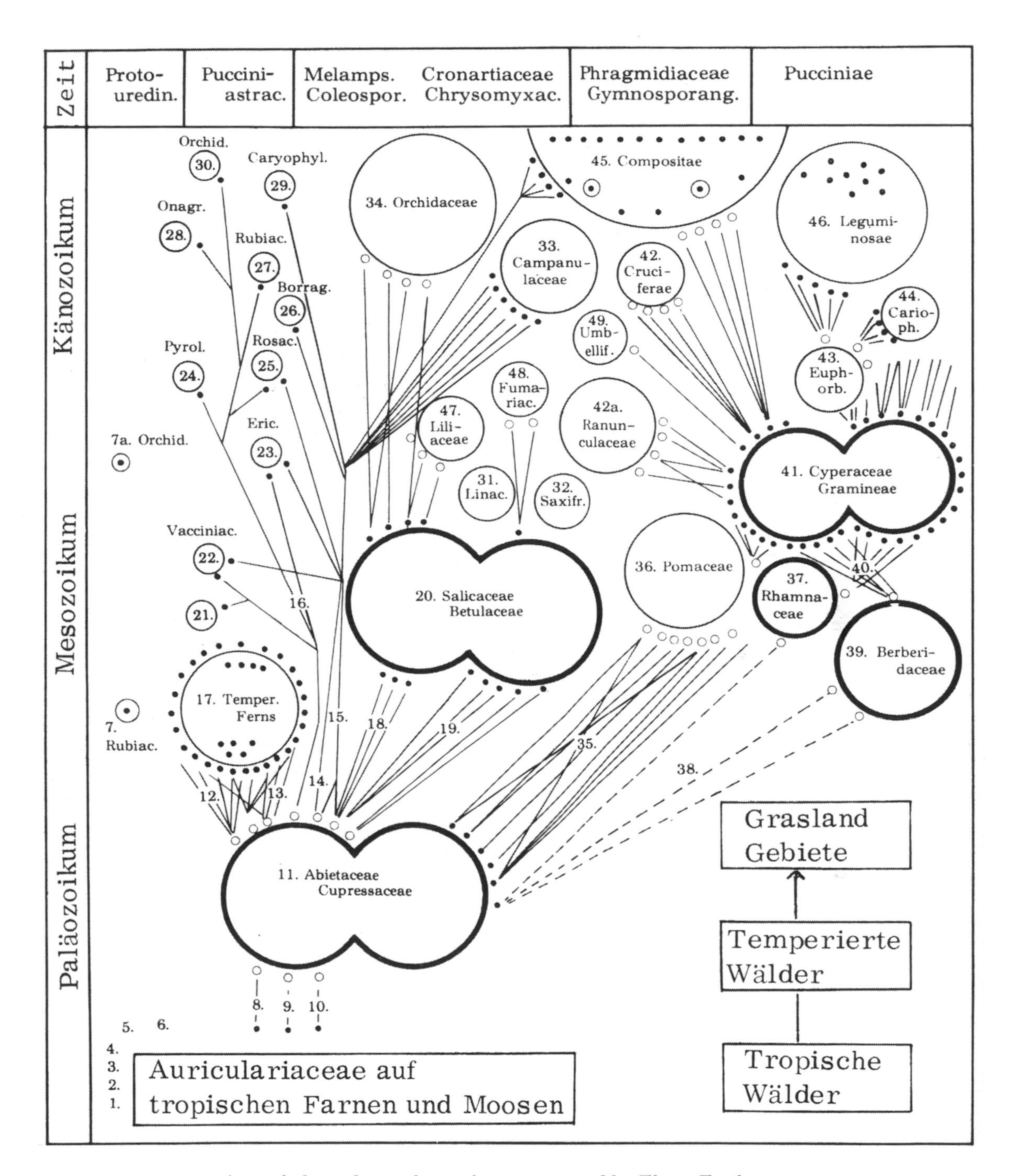

FIGURE 181. An early hypothesis of coevolution proposed by Elmar Emil Leppik between parasitic rust fungi and their plant hosts.

Leppik, 1965, *Mycologia,* 57:14, fig. 6; courtesy of Jeffrey K. Stone and Karen M. Snetselaar. Reprinted with permission from *Mycologia,* © The Mycological Society of America.

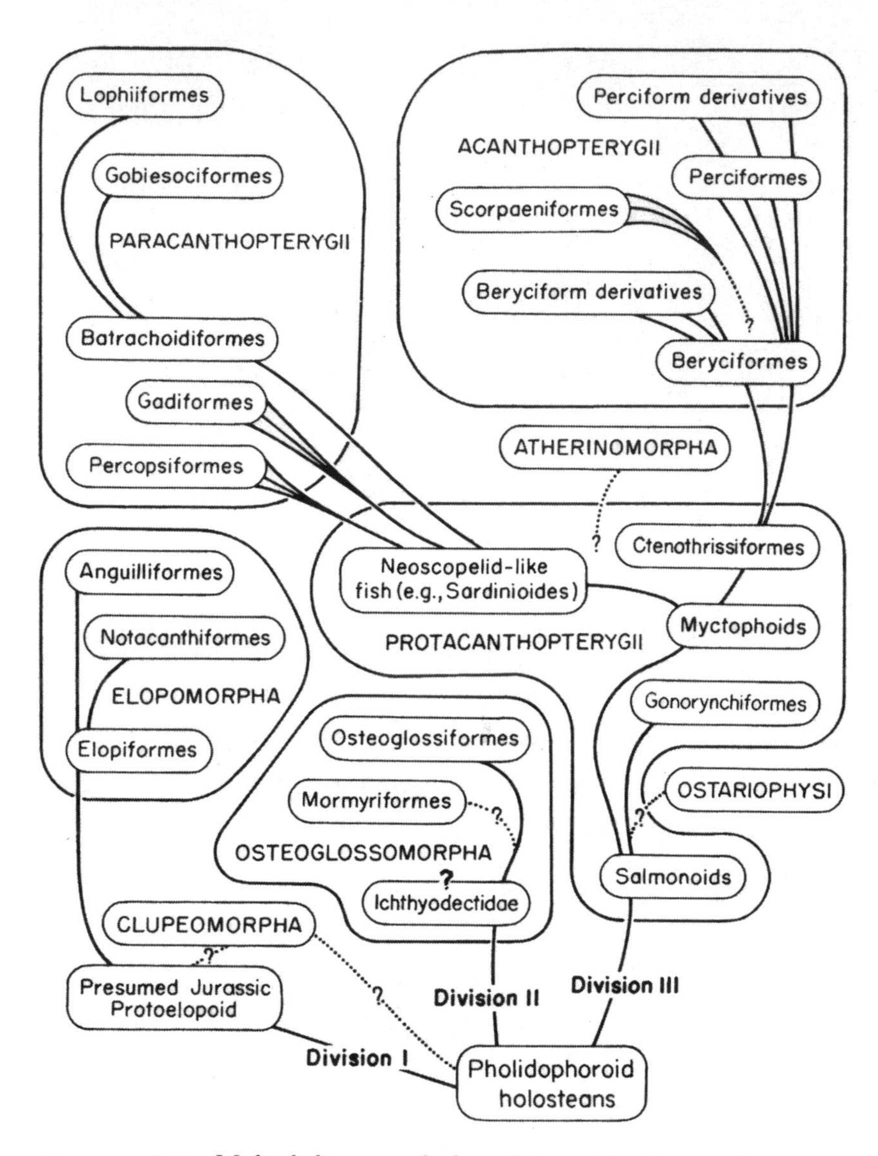

FIGURE 182. Multiple lineages of teleost fishes arising from extinct pholidophoroid holostean ancestors.

Greenwood et al., 1966, *Bulletin of the American Museum of Natural History,* 131(4):349, fig. 1; courtesy of Melanie L. J. Stiassny, Mai Qaraman, and the American Museum of Natural History. Used with permission.

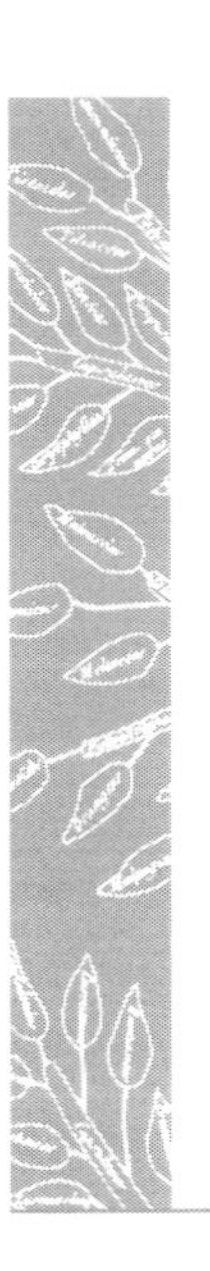

Phenograms and Cladograms

1958–1966

Pheneticism and cladism, two alternative approaches to tree building that developed during the 1960s and 70s, "challenged the entire edifice of evolutionary systematics to its methodological Foundations."[1] These two schools, themselves at philosophical odds, arose out of concern that groupings of taxa in classical evolutionary systematics were based on a combination of measures of similarity and nonempirical hypotheses of relationship. Previous ingrained ideas or hypotheses held by systematists were said to influence the "weighting" of characters in the construction of evolutionary trees. In addition, those opposed to business as usual complained that the construction of classifications required the intervention of subjectivity, often relying to a large extent on the intuitive judgment of the expert and even on artistic and pragmatic criteria.[2]

Phenetics, sometimes called numerical taxonomy, seemed, at least for some, a way to circumvent these problems.[3] The tree of the pheneticist is the phenogram: a branching diagram that depicts taxonomic relationships among organisms based on overall similarity of many characters, without regard to evolutionary history or assumed significance of specific characters. Phenograms are nearly always computer generated and typically not very attractive. An early example was produced by Robert Reuben Sokal (1926–) and Charles Duncan Michener (1918–) in 1958 (Figure 183). Starting with correlation coefficients (based on numerous morphological characters) among species of leaf-cutter bees, the authors developed a method for grouping species, and regrouping the resultant assemblages, to form a classificatory hierarchy that was similar to that previously established by classical systematic methods.[4]

The resulting branching diagram is said to be the first phylogeny ever produced by numerical methods.[5]

More often than not phenograms are placed on their side, the base of the trunk to the left and the tips of the twigs on the far right—appearing not unlike the old bracket diagrams first used by Renaissance naturalists—and usually associated with a scale of phenetic similarity or phenetic distance. Of numerous additional examples of these horizontal phenograms, only three are reproduced here. The first is a diagram published in 1965 by Luigi Luca Cavalli-Sforza (1922–) and Anthony William Fairbank Edwards (1935–) that shows the relationships of human populations inferred from blood-group polymorphism gene frequencies, said to be the earliest numerical phylogeny ever produced by parsimony (Figure 184).[6] The second is a tree of relationships among the species of gulls and terns (Figure 185), based on a much larger assemblage of taxa (a matrix of 81 species and 51 skeletal measurements), published in 1970 by Gary Dean Schnell (1942–).[7] And the third, relationships among seventeen species of mites of the family Dermanyssidae by W. Wayne Moss (1942–) that appeared in 1967 (Figure 186).

A few more interesting-looking diagrams appeared during this time, also based on a computer driven, phenetic approach. For example, Wayne Moss took the data from his 1967 study of mite relationships a step further by calculating the phenetic distance between all possible pairs of taxa, producing a "graph of relationships" (Figure 187, numbered as in Figure 186) that is similar to the old reticulated maps of Morison (1672) and Buffon (1755; see Figures 15 and 17). In yet another step further, Moss then drew concentric circles of arbitrary, but equal, intervals around each taxon location (numbered as in Figures 186 and 187), forming another kind of topographic map (Figure 188). The taxon locations can be thought of as peaks and the spaces between them as valleys, the relative depths of the valleys corresponding to the degree of taxon similarity.[8]

Notwithstanding precedent set by a number of earlier researchers, for example Peter Mitchell (1901), Charles Camp (1923), and Walter Zimmermann (1931), among several others (see Figures 106, 121, and 142), German entomologist Willi Hennig laid the foundation for another alternative to the idiosyncratic methods employed at the time by most systematists. Called phylogenetic systematics or cladistics, Hennig's approach was described in German in 1950, with the publication of *Grundzüge einer Theorie der phylogenetischen Systematik* (Principles of a theory of phylogenetic systematics), but broad attention was drawn to his work only after publication of *Phylogenetic Systematics* in English in 1966.[9]

Definitely at odds with phenetics, as demonstrated nicely by a tree published in 1965 by Robert Sokal and Joseph Harvey Camin (1922–1979; Figure 189), cladistics is defined by recency of common descent. Only monophyletic groups, or "clades," are recognized, and shared derived characters are the critical evidence for the existence of clades. Hennig illustrated his ideas with a number of theoretical diagrams, three of which are reproduced here. The first demonstrates the phylogenetic relationships between species of a monophyletic group, shown in lateral view and from above, looking down (Figure 190), reminiscent of the bird's-eye views published, for example, by Heinrich Engler in 1874 and Maximilian Fürbringer in 1888 (see Figures 93 and 96–98). The second is a now-famous diagram that shows Hennig's "scheme of argumentation of phylogenetic systematics" (Figure 191), in which all groups considered to be monophyletic are distinguished by the possession of derived character states, or stages of expression (black rectangles), of at least one character. The third serves to explain how the concept of similarity can be thought of in at least three ways (Figure 192), by the sharing of primitive character states ("symplesiomorphy"), the sharing of derived states ("synapomorphy"), or through parallel or convergent evolution.

For reasons beyond the scope of the present work, phenetics, or numerical taxonomy, was out of favor by the late 1970s, while cladistics took on a firm hold, so that nearly every credible phylogenetic analysis since then has followed Hennigian methodology, as represented by most all of the remaining trees presented in this volume. Examples of early cladograms are in the thousands, but only a few are shown here. Notable are the diagrams constructed by Winter Patrick Luckett (1937–) in 1975 to illustrate relationships among primates (Figure 193); Edward Orlando Wiley (1944–), 1976, relationships among bony fishes (Figure 194); and Donn Eric Rosen, 1979, among freshwater teleosts of the poeciliid genera *Heterandria* and *Xiphophorus* (Figures 195 and 196).

Swedish entomologist Lars Brundin (1907–1993), a specialist in the systematics of southern hemisphere chironomid midges, was a pioneer in the use of cladistics in biogeographic analysis. He was among the first to use Hennig's phylogenetic trees to plot the distributional history of monophyletic groups of organisms. On a cladogram of the evolutionary relationships of midge species, published in 1966, he plotted the names of the various continents on which each species was found, resulting in what is called a "taxon-area cladogram" or an "area-cladogram" for short (Figure 197). As it turned out, the sequence of divergence of the taxa corresponded well with the pattern of the breakup of Gondwana to form the southern continents. Evidence from two seemingly disparate fields of study, systematics and biogeography, therefore reinforced each other. Among other things, Brundin's work showed that geological information could be used to evaluate which of several competing cladograms is likely to be correct.[10]

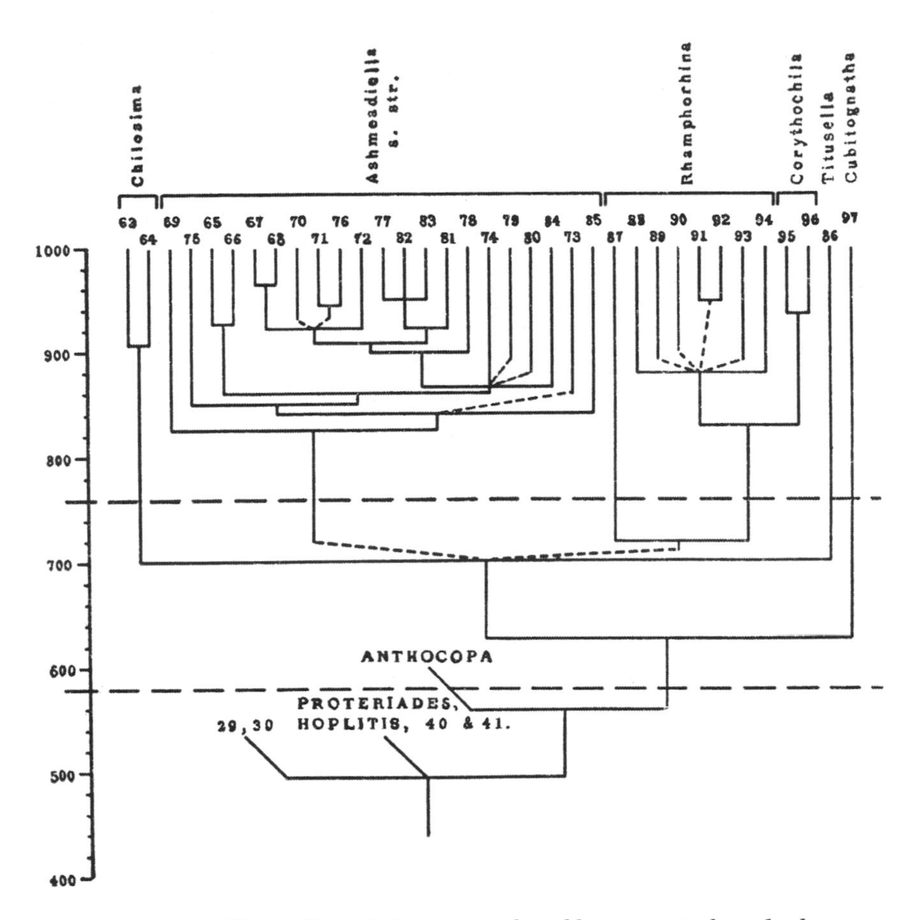

FIGURE 183. The earliest phylogeny produced by numerical methods, showing relationships among species of leaf-cutter bees, by Robert Reuben Sokal and Charles Duncan Michener.

Sokal and Michener, 1958, *University of Kansas Science Bulletin,* 38(22):1425, fig. 1; courtesy of Robert R. Sokal and Kirsten Jensen. Copyright © 1958, University of Kansas Museum of Natural History; used with permission.

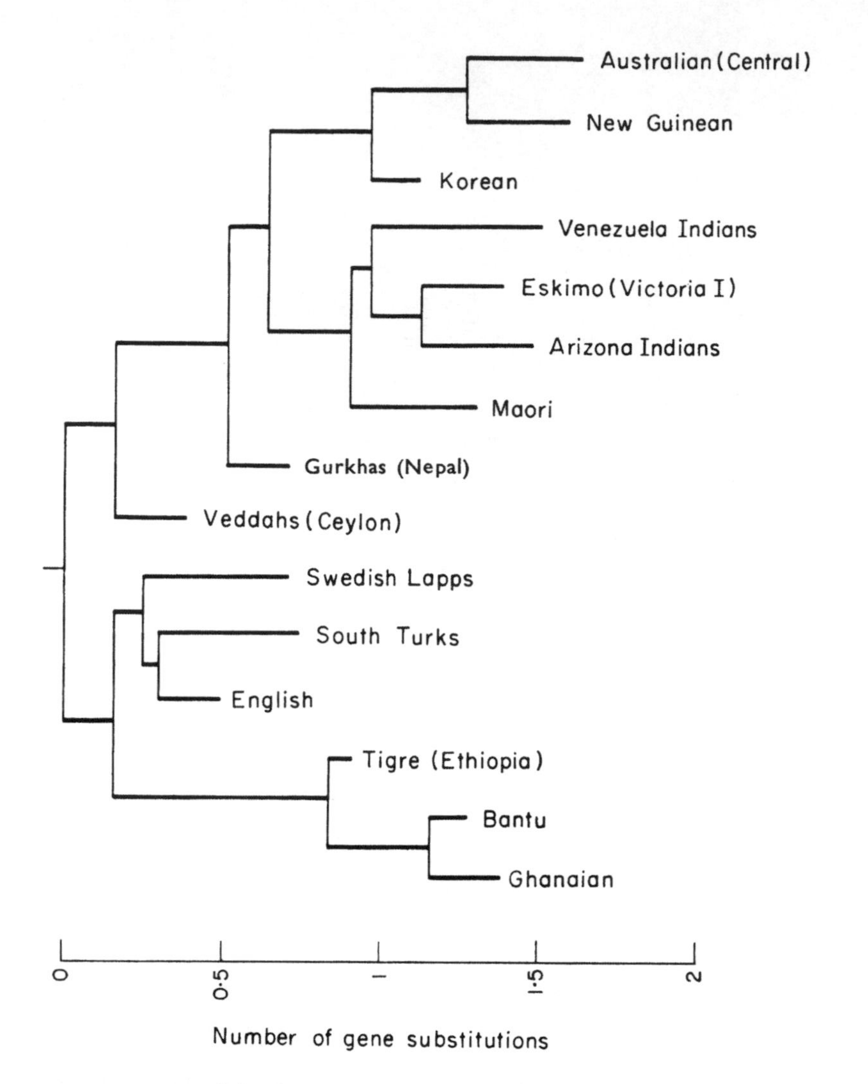

FIGURE 184. The first numerical phylogeny based on parsimony, showing relationships of human populations inferred from blood-group gene frequencies, presented at the 1963 International Congress of Genetics at The Hague, by Luigi Luca Cavalli-Sforza and Anthony William Fairbank Edwards.

Cavalli-Sforza and Edwards, 1965, "Analysis of human evolution," p. 929, fig. 5; courtesy of Laura Pritchard and Elsevier Limited. Used with permission.

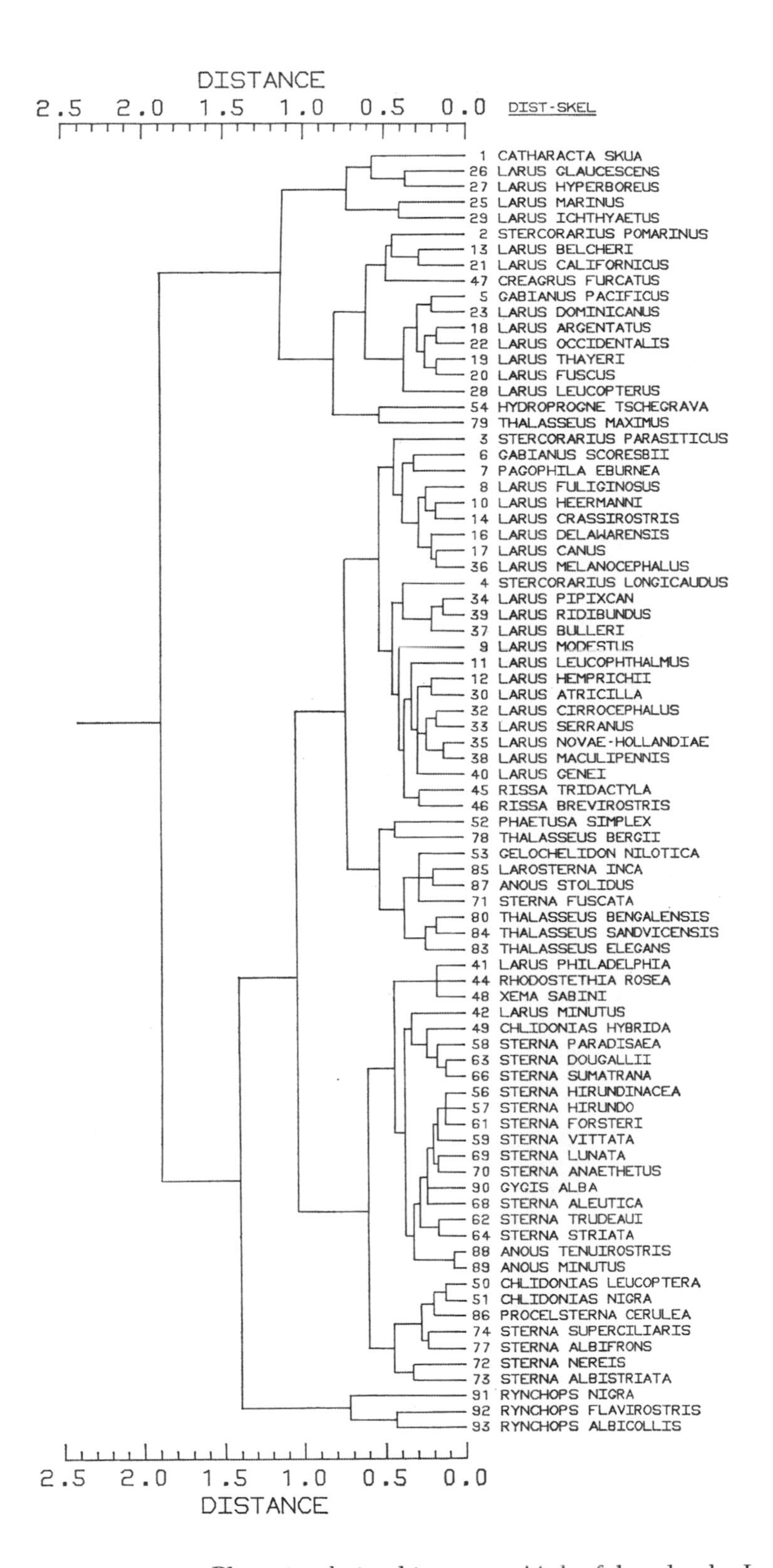

FIGURE 185. Phenetic relationships among birds of the suborder Lari (order Charadriiformes), the gulls and terns, by Gary Dean Schnell.

Schnell, 1970, *Systematic Zoology,* 19(3):266, fig. 12; courtesy of Gary D. Schnell and the Society of Systematic Biologists. Used with permission.

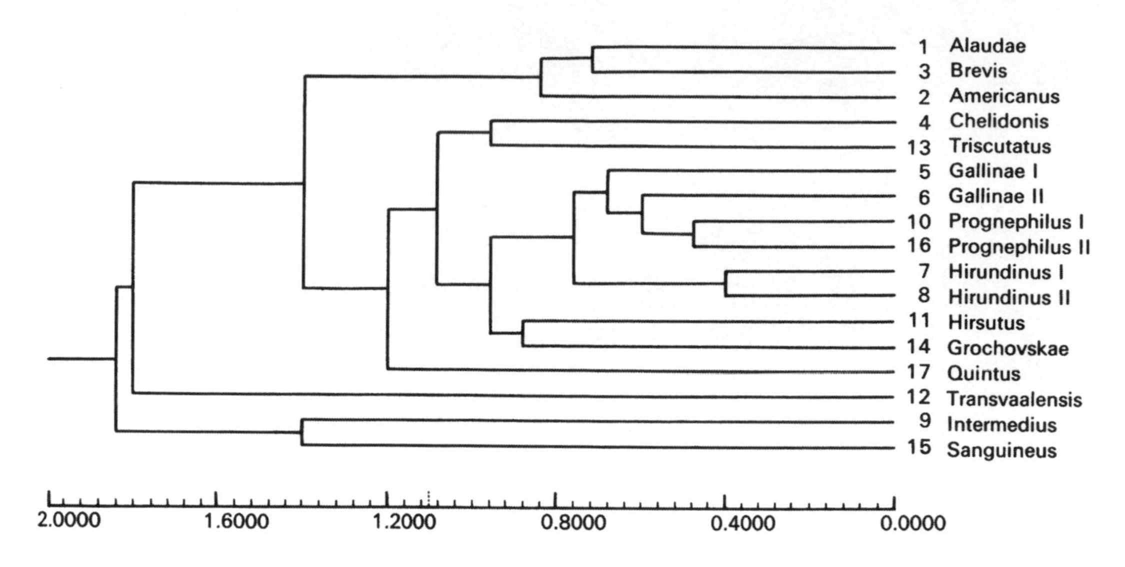

FIGURE 186. A phenogram by W. Wayne Moss showing the relationships among seventeen species of mites.

Moss, 1967, *Systematic Zoology*, 16(3):184, fig. 2; courtesy of Deborah Ciszek, Chris Payne, and the Society of Systematic Biologists. Used with permission.

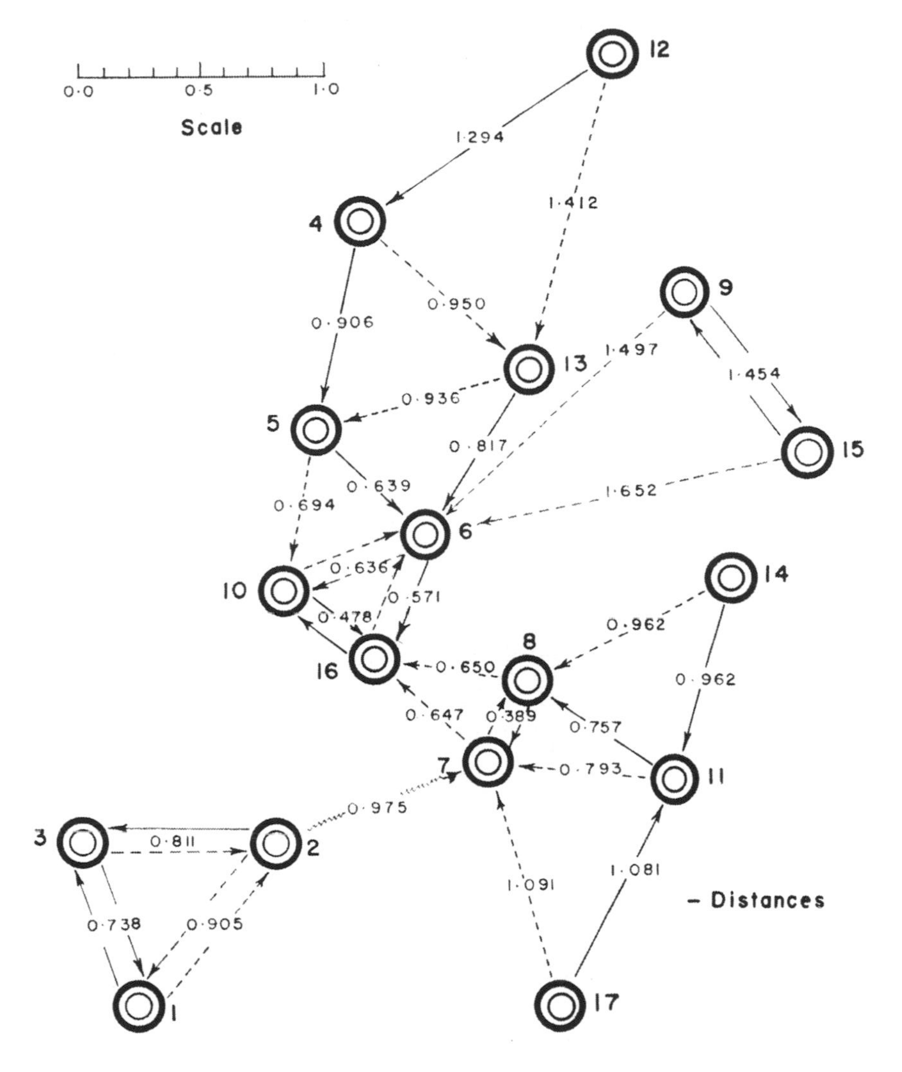

FIGURE 187. A mapping of relationships of mites, based on phenetic distances, by W. Wayne Moss. The placement of taxa, represented by circles, and numbered as they appear in the phenogram shown in figure 186, was obtained by triangulation of first-order and second-order vectors as indicated by the solid and dashed lines, respectively. The relatively isolated assemblage numbered 1–3 is connected to the other groups by a third-order vector as shown by a wavy line. Compare with the pairings of taxa given in figure 186.
Moss, 1967, *Systematic Zoology,* 16(3):190, fig.11; courtesy of Deborah Ciszek, Chris Payne, and the Society of Systematic Biologists. Used with permission.

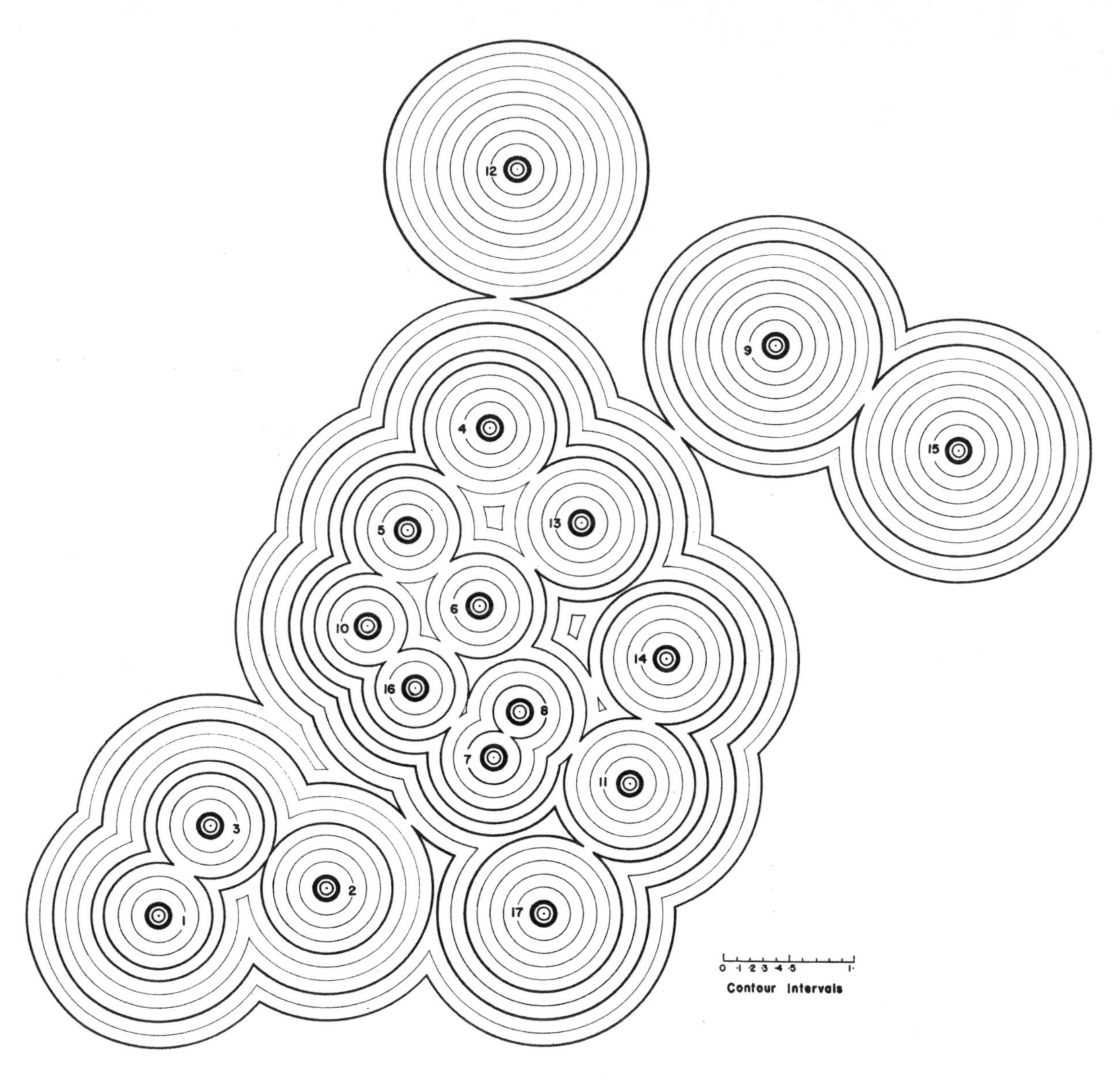

FIGURE 188. A "contour representation" of the distance phenogram and mapping of relationships of mites shown in figures 186 and 187, by W. Wayne Moss.

Moss, 1967, *Systematic Zoology,* 16(3):192, fig.13; courtesy of Deborah Ciszek, Chris Payne, and the Society of Systematic Biologists. Used with permission.

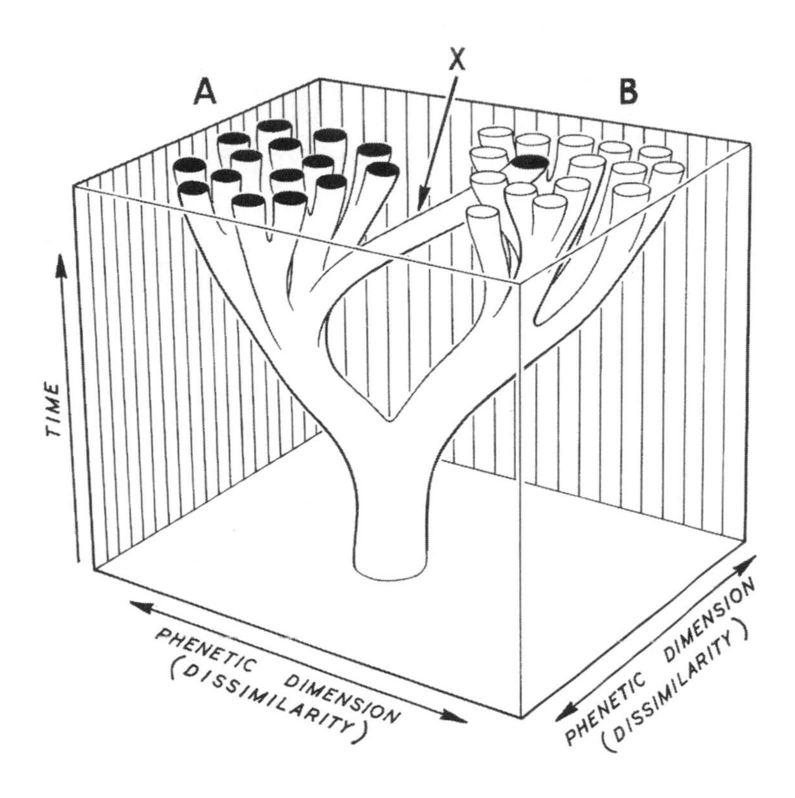

FIGURE 189. A tree illustrating the possible incongruence between phenetic and cladistic relationships, by Robert Reuben Sokal and Joseph Harvey Camin. Similarity is indicated in the two horizontal dimensions; the third dimension is time. Thinking cladistically, taxon X belongs to the monophyletic taxon A, but phenetically it would be considered part of taxon B as a result of evolutionary convergence.

Sokal and Camin, 1965, *Systematic Zoology,* 14(3):193, fig. 6; courtesy of Robert R. Sokal and the Society of Systematic Biologists. Used with permission.

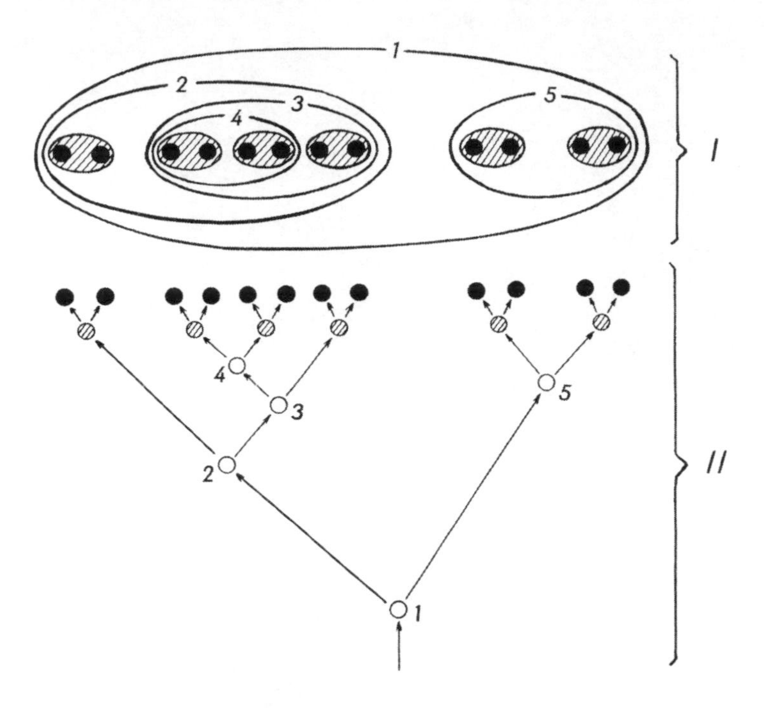

FIGURE 190. A phylogenetic tree of relationships between species of a monophyletic group, shown in lateral view and from above, by Willi Hennig.
Hennig, 1966, *Phylogenetic Systematics,* p. 71, fig. 18; copyright 1966, 1979, by the Board of Trustees of the University of Illinois. Used with permission of the University of Illinois Press.

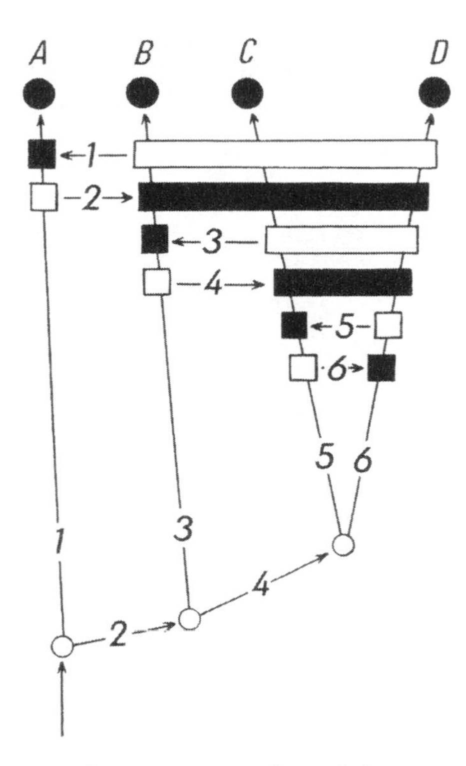

FIGURE 191. Primitive versus derived character states: Willi Hennig's classic demonstration that monophyletic groups are defined by shared derived stages of expression of at least one character. Derived character states are represented by solid rectangles; primitive character states, by open rectangles. Hennig, 1966, *Phylogenetic Systematics,* p. 91, fig. 22; copyright 1966, 1979, by the Board of Trustees of the University of Illinois. Used with permission of the University of Illinois Press.

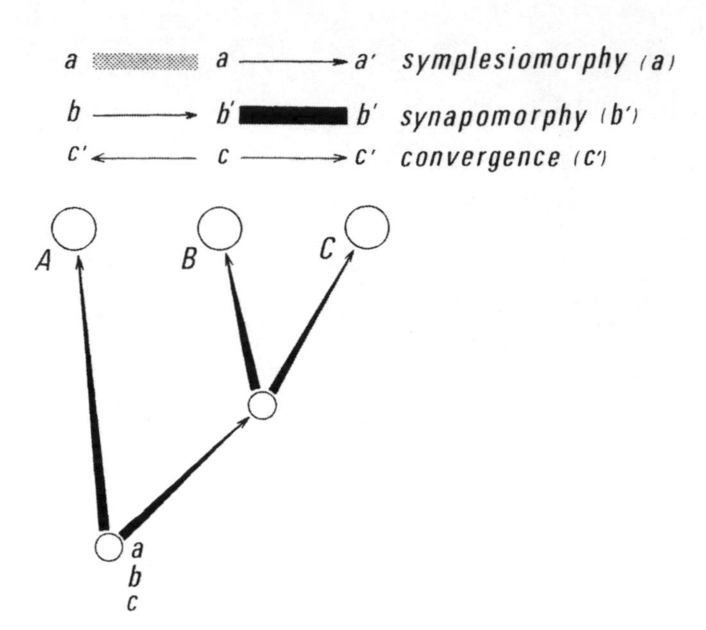

FIGURE 192. Willi Hennig's explanation of the "composite nature" of the concept of similarity, showing that similarity can be based on shared primitive character states (a), shared derived character states (b), or evolutionary convergence (c).

Hennig, 1966, *Phylogenetic Systematics,* p. 147, fig. 44; copyright 1966, 1979, by the Board of Trustees of the University of Illinois. Used with permission of the University of Illinois Press.

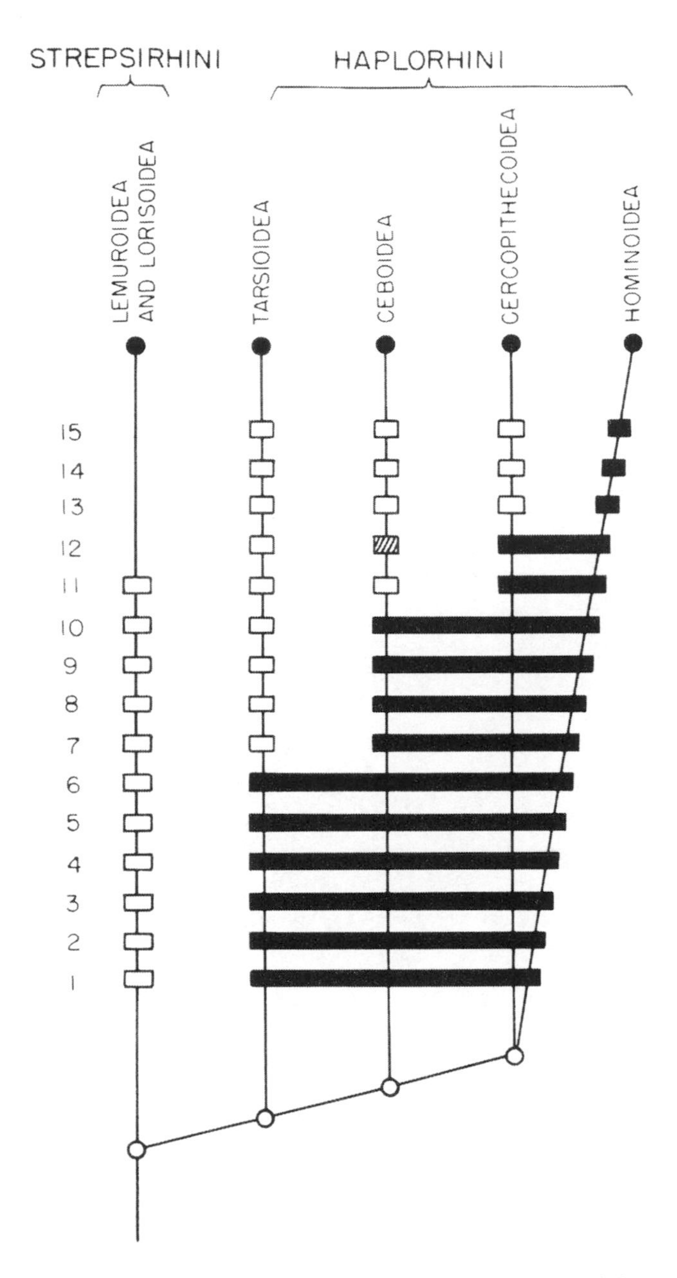

FIGURE 193. Relationships of the superfamilies of the primates, based on a cladistic analysis of fetal membrane and placental characters, by Winter Patrick Luckett. Shared derived features that connect the taxa are represented by black rectangles.

Luckett, 1975, "Ontogeny of the fetal membranes and placenta: Their bearing on primate phylogeny," p. 178, fig. 13; courtesy of Nel van der Werf and Springer Science and Business Media. Used with permission.

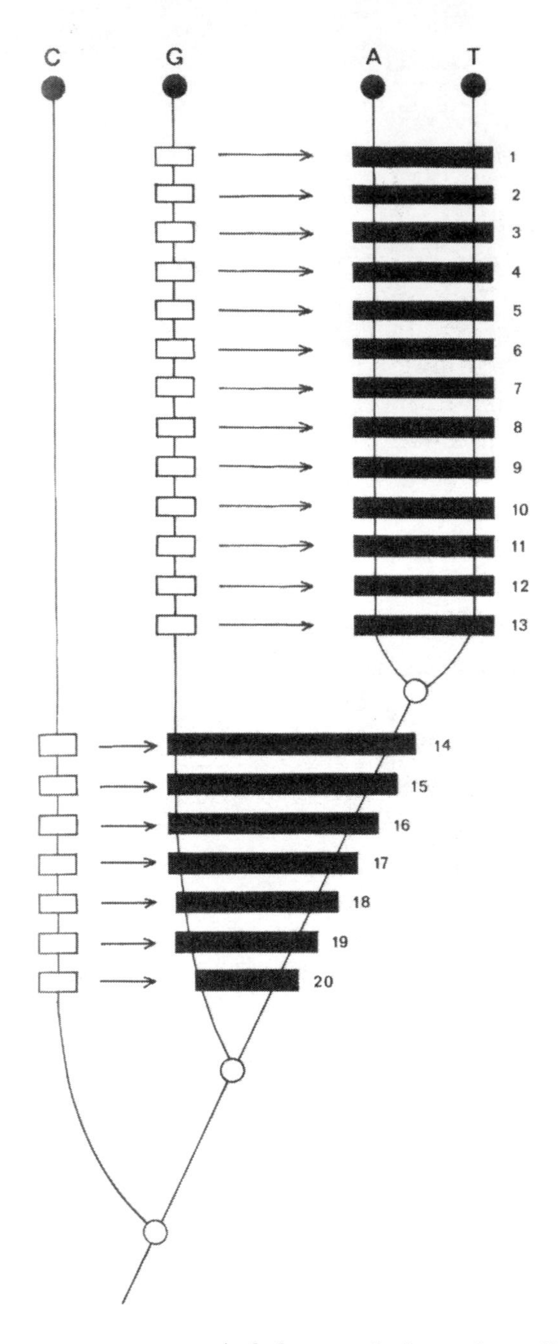

FIGURE 194. A cladogram of relationships of Recent actinopterygian fishes, the chondrosteans (C), gars (G), amiids (A), and teleosts (T), by Edward Orlando Wiley. Primitive character states, represented by open rectangles, point to derived states, represented by black rectangles.
Wiley, 1976, *University of Kansas, Museum of Natural History, Miscellaneous Publication,* 64:38, fig. 17a; courtesy of E. O. Wiley. Used with permission.

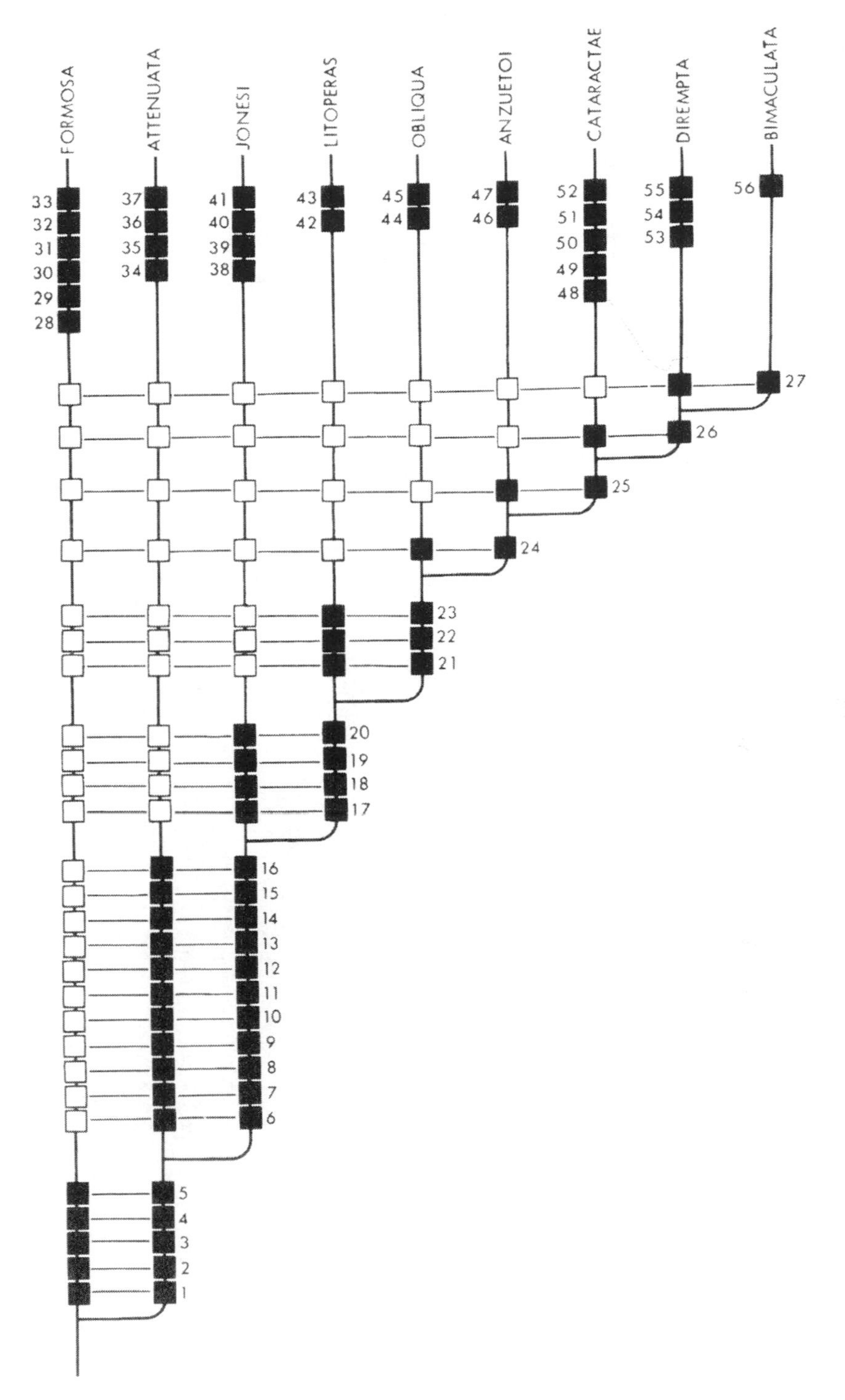

FIGURE 195. Relationships among species of fishes of the poeciliid genus *Heterandria,* by Donn Eric Rosen. Derived character states, represented by black squares, are joined to primitive states, represented by open squares. Unique diagnostic states of the taxa are represented by numbers 28–56.

Rosen, 1979, *Bulletin of the American Museum of Natural History,* 162(5):308, fig. 20; courtesy of Melanie L. J. Stiassny, Mai Qaraman, and the American Museum of Natural History. Used with permission.

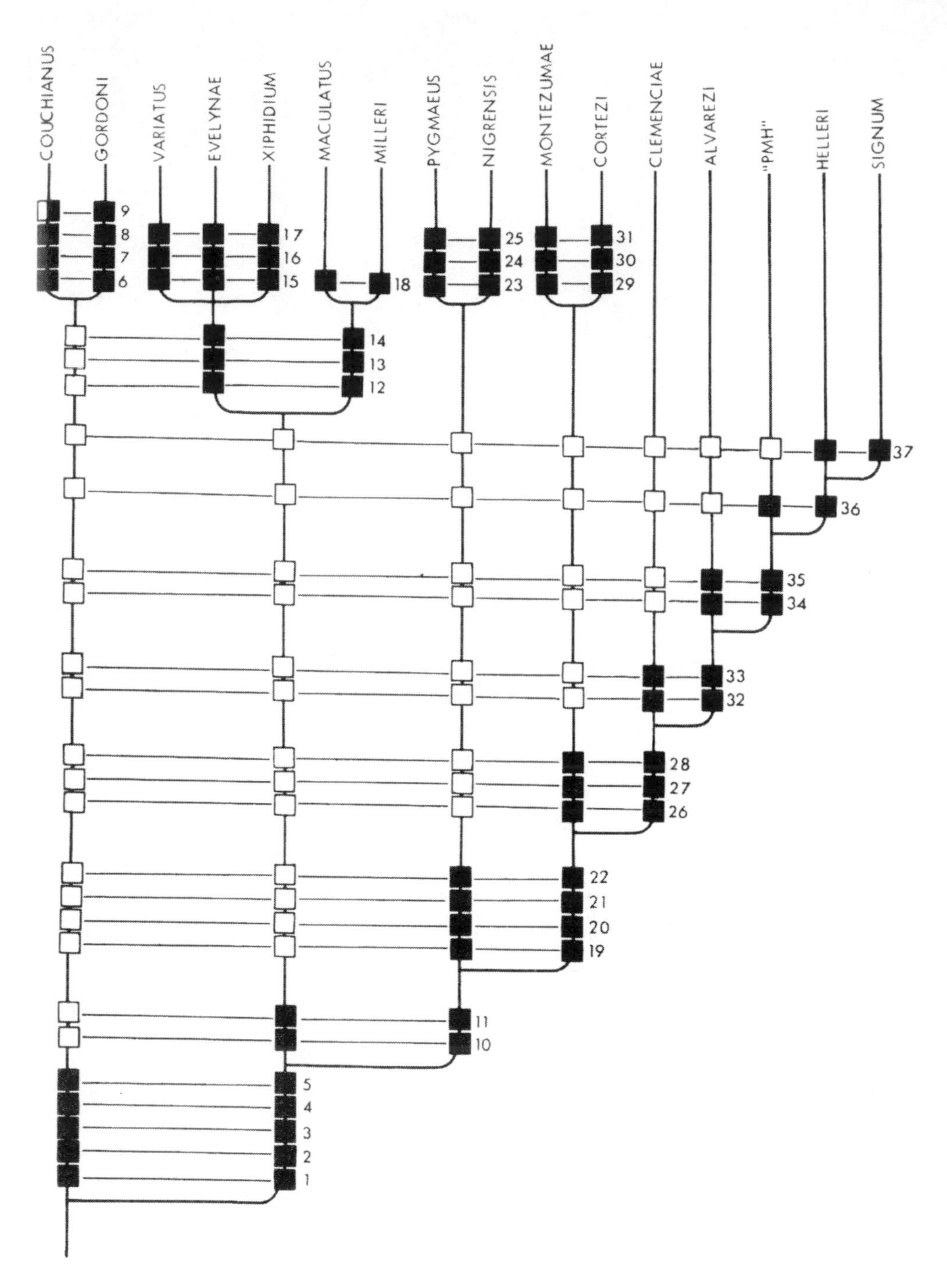

FIGURE 196. Relationships among species of fishes of the poeciliid genus *Xiphophorus,* by Donn Eric Rosen. Derived character states, represented by black squares, are joined to primitive states, represented by open squares. Unique diagnostic states of the taxa not shown.

Rosen, 1979, *Bulletin of the American Museum of Natural History,* 162(5): 348, fig. 37; courtesy of Melanie L. J. Stiassny, Mai Qaraman, and the American Museum of Natural History. Used with permission.

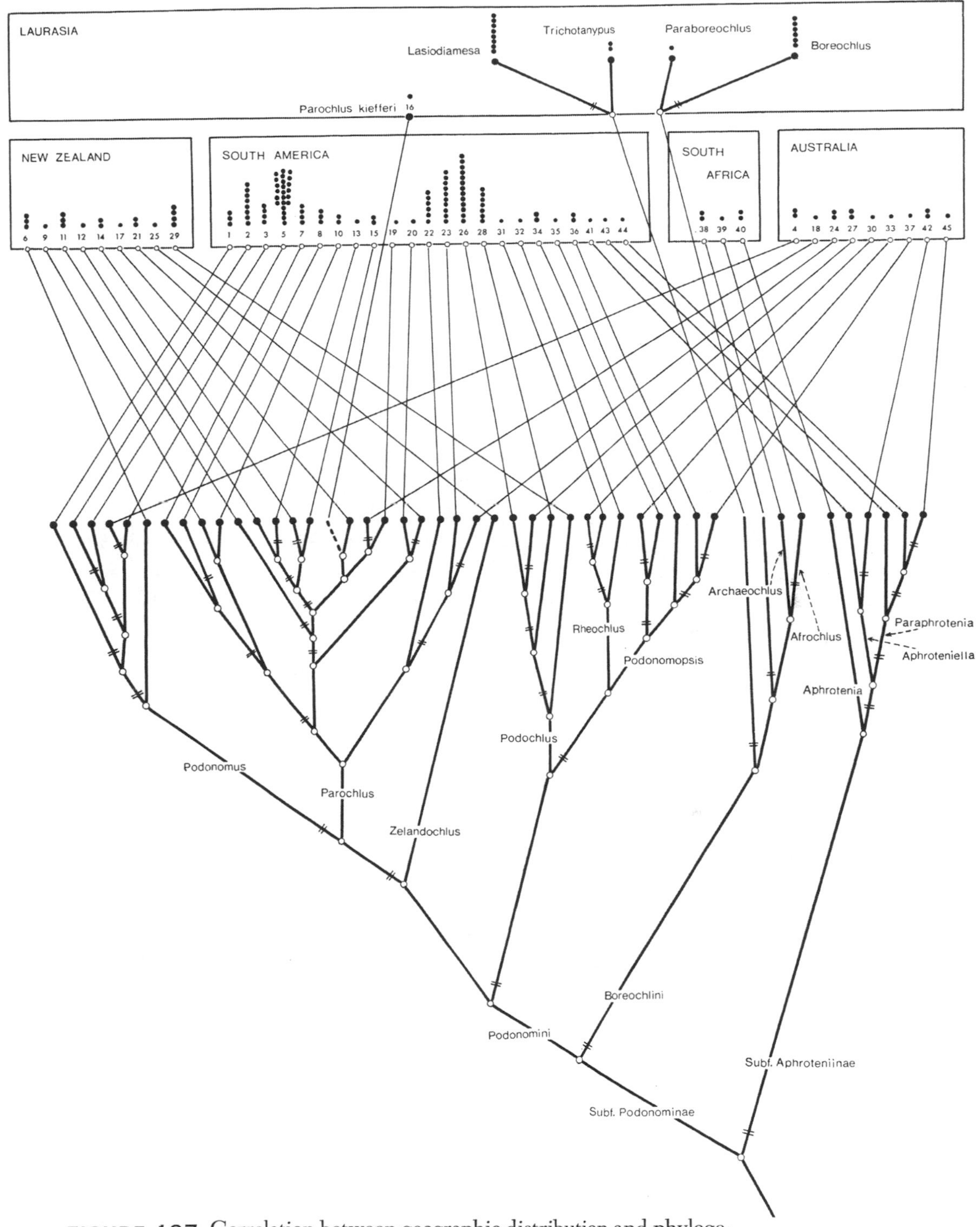

FIGURE 197. Correlation between geographic distribution and phylogenetic relationships of an assemblage of southern hemisphere chironomid midges, by Lars Brundin. In the rectangles above, the number of species in each group is indicated by small black dots.

Brundin, 1966, *Transantarctic Relationships and Their Significance, as Evidenced by Chironomid Midges,* p. 442, fig. 634; courtesy of Almqvist & Wiksell.

Early Molecular Trees

1962–1987

In the beginning of the 1960s, at about the same time that pheneticists and cladists were beginning to question the subjective approaches prevalent among systematists of the time, molecular biologists entered the field of evolutionary biology,[1] arguing that molecular evidence was relatively free of subjectivity and more direct than the morphological evidence on which classical systematists had previously relied.[2] By the mid-1960s, a host of different approaches to tree construction had been developed, all based on quantitative estimates of variance between macromolecules obtained from different species. Examples include the strength of immunological cross-reactions of blood plasma proteins, the number of differences in peptides from enzymatic digests of purified homologous proteins, the number of amino acid replacements between homologous proteins whose complete primary structures had been determined, mutation distances as estimated from sequences of Cytochrome c, and the degree of interspecific hybridization of DNA.[3] Notable among this early experimentation is a 1962 paper by Emile Zuckerkandl (1922–) and Linus Carl Pauling (1901–1994), two of the most prominent founders of the field of molecular evolution, titled "Molecular disease, evolution, and genic heterogeneity." It represents the first explicit statement of the then-unnamed "molecular clock hypothesis"—the idea that the rate of evolution in a given protein or DNA molecule is approximately constant over time and among evolutionary lineages and therefore that the number of amino acid substitutions can be used to make temporal divergence estimates.[4] In a follow-up to the 1962 paper, published in 1965, Zuckerkandl and Pauling coined the term "molecular evolutionary clock" and applied the hypothesis to show the probable relationships and divergence times of some mammalian hemoglobin chains, suggesting that humans and the horse are slightly more closely related to each other than humans are to oxen (Figure 198).

Also important is a 1967 paper by Walter Monroe Fitch (1929–2011) and Emanuel Margoliash (1920–2008) titled "Construction of Phylogenetic Trees," illustrated with a tree of animal relationships based on mutational distances as estimated from the protein Cytochrome c (Figure 199)—one of the earliest published distance matrix phylogenies.[5] Significant as well are the trees of Vincent Matthew Sarich (1934–2012), particularly the famous molecular-clock diagram coauthored with Allan Charles Wilson (1934–1991) in 1967 that showed that humans diverged from the chimps and gorillas only about five million years ago. This was a highly controversial claim to make at the time—when seemingly irrefutable paleontological evidence placed the split between hominids and hominoids closer to 15 million years (Figure 200)[6]—and it remains problematic today. Recent morphological evidence clearly shows that the divergence of humans was with the orangutan, not the African apes.[7] Another Sarich tree, this one published in 1976 with coauthor John Edward Cronin, shows the relationships of Old World monkeys based on blood protein sequences and immunological specificity (Figure 201).

Morris Goodman (1925–2010), one of the most productive and influential of the molecular phylogeneticists who emerged during the 1960s and 70s, published numerous phylogenetic trees based on protein sequence data. In one of these, appearing in 1975, he and his colleagues used sequence data to reconstruct the evolutionary history of hemoglobin (including possible ancestral sequences) and analyze which sites on the hemoglobin complex had evolved at which stages (Figure 202). Goodman called this the first "hard evidence of Darwinian evolution."[8] In 1982, he, in coauthorship with John Czelusniak and others, did the same for the DNA sequences of the hemoglobin genes (Figure 203).

In 1963, American ornithologist and molecular biologist Charles Gald Sibley (1917–1998) began to develop methods to hybridize DNA, and by the early 1970s, in collaboration with Jon Edward Ahlquist (1944–), he was applying the technique to resolve relationships among the living families of birds. Between early 1975 and mid-1986, the two had made more than 26,000 DNA-DNA comparisons among some 1,700 bird species, representing all the orders and 168 of the 171 families then recognized.[9] A landmark paper, coauthored with Jon Ahlquist and

Burt Leavelle Monroe Jr. (1930–1994), published in 1988, is illustrated with a number of trees, one of which is reproduced here (Figure 204).

In the mid-1980s, interest in DNA-DNA hybridization, which had been the dominant technique in molecular phylogenetics since 1974,[10] began to decline and the era of protein sequencing was waning as well, displaced by new methods of rapidly sequencing RNA and DNA itself.[11] Analyzing ribosomal RNA sequence data, Carl Richard Woese (1928–2012) and George Edward Fox (1945–) became famous almost overnight for their discovery in 1977 of the Archaea, a heretofore unknown kingdom or domain of life (Figure 205). Originally the Bacteria and Archaea (then called the Eubacteria and Archaebacteria) were thought to be one large diverse family of prokaryotes, but now, having defined Archaea as a new domain, Woese and Fox redrew the evolutionary tree of life, hypothesizing a three-domain system based on genetic relationships rather than morphological similarities. Their work is of primary importance in demonstrating the overwhelming diversity of microbial life—that single-celled organisms represent the vast majority of the Earth's genetic, metabolic, and ecosystem niche diversity.[12]

In a celebrated paper published in 1987, Rebecca Louise Cann (1951–) and colleagues generated a tree based on a world-wide survey of mitochondrial DNA from 147 people, from five geographic populations, Africa, Asia, Australia, New Guinea, and Europe (Figure 206). The results, which sparked a media frenzy when first reported, showed that all the mitochondrial DNA used in the study stemmed from one woman who was postulated to have lived about 200,000 years ago, probably in Africa.[13] Moreover, all the populations examined except the African population have multiple origins, implying that each area was colonized repeatedly.[14]

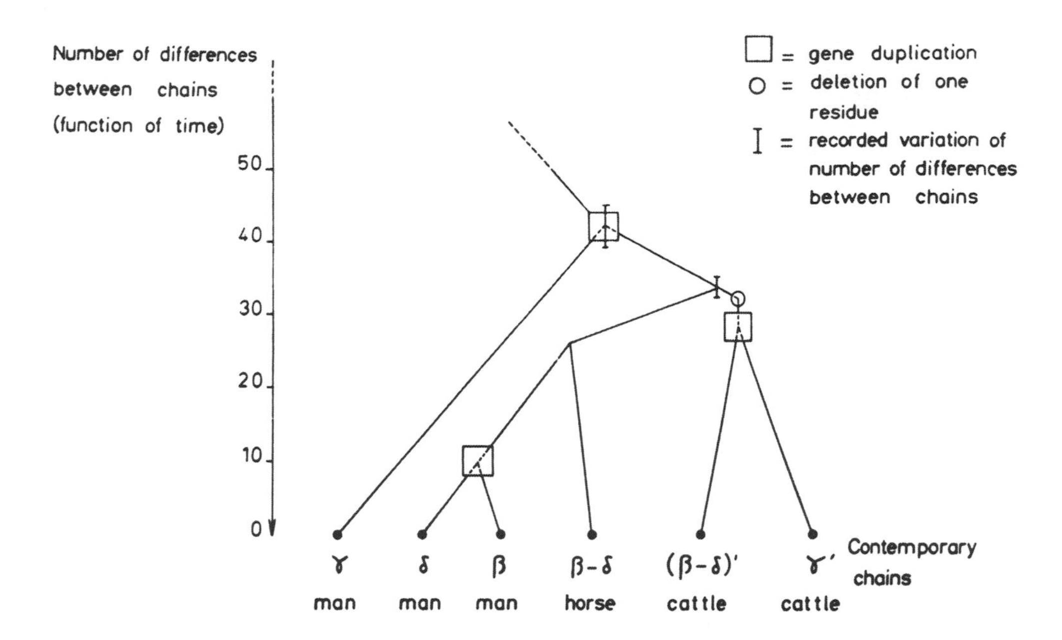

FIGURE 198. The probable relationships and divergence times of some mammalian hemoglobin chains, by Emile Zuckerkandl and Linus Carl Pauling.

Zuckerkandl and Pauling, 1965, "Evolutionary divergence and convergence in proteins," p. 156, fig. 4; courtesy of Laura Pritchard and Elsevier Limited. Used with permission.

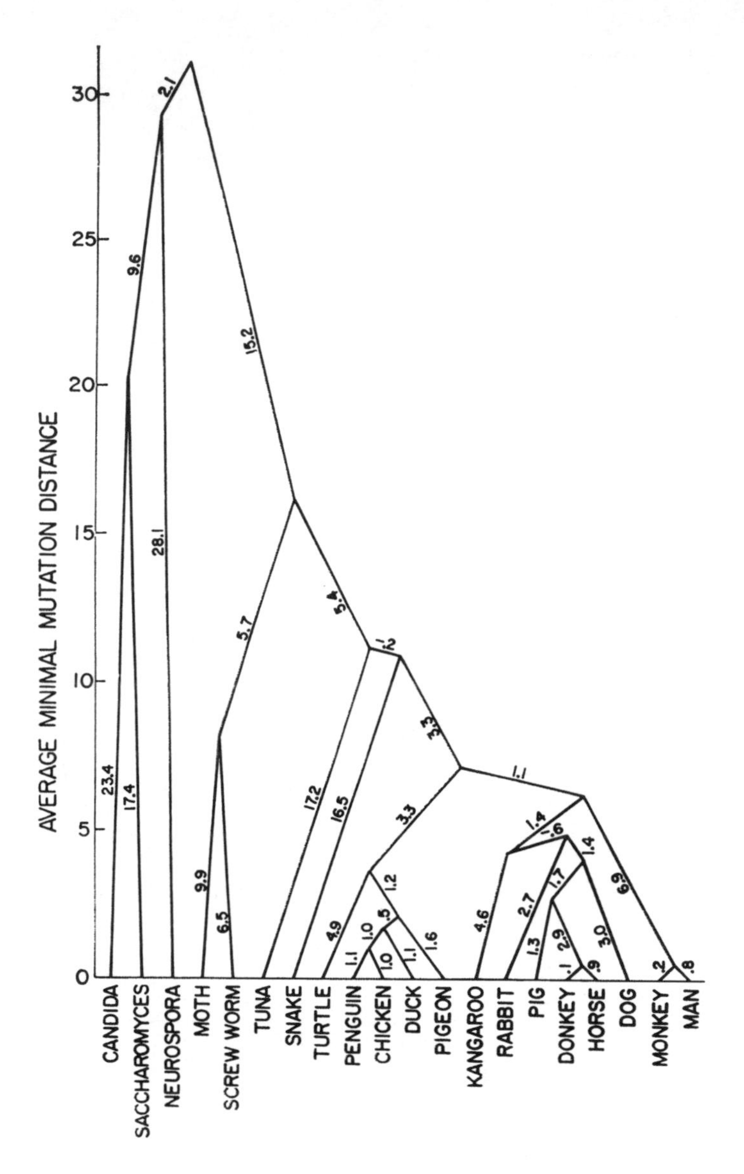

FIGURE 199. An early molecular tree of animal relationships based on mutational distances as estimated from the protein Cytochrome c, by Walter Monroe Fitch and Emanuel Margoliash.

Fitch and Margoliash, 1967, *Science,* 155(3760):282, fig. 2; courtesy of Walter M. Fitch and Thomas Dobrzeniecki. Used with permission.

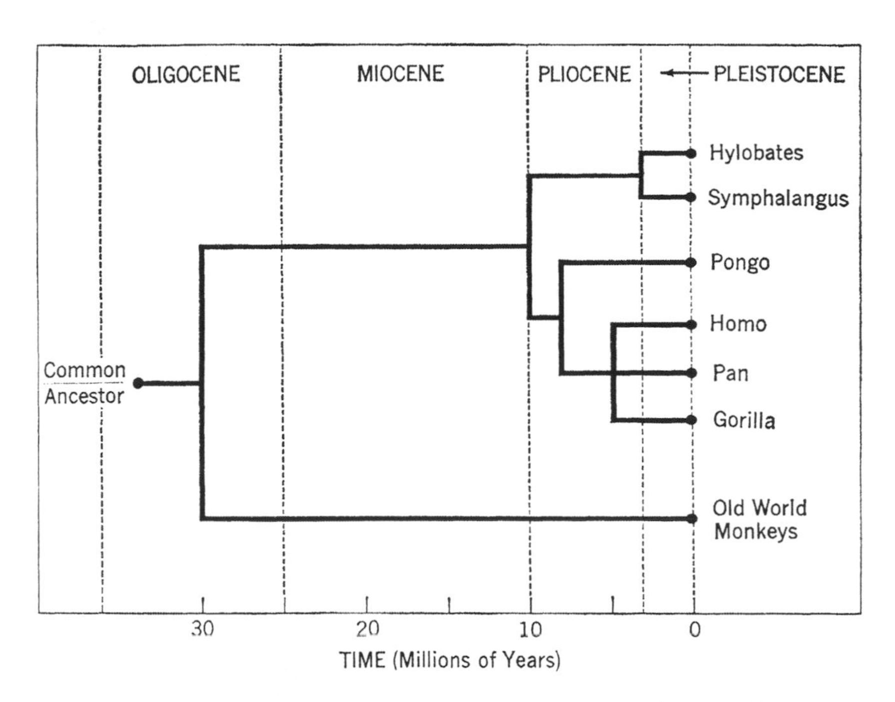

FIGURE 200. Times of divergence between the various hominoids estimated from immunological data, by Vincent Matthew Sarich and Allan Charles Wilson.

Sarich and Wilson, 1967, *Science,* 158:1201, fig. 1; courtesy of Vincent M. Sarich. Used with permission.

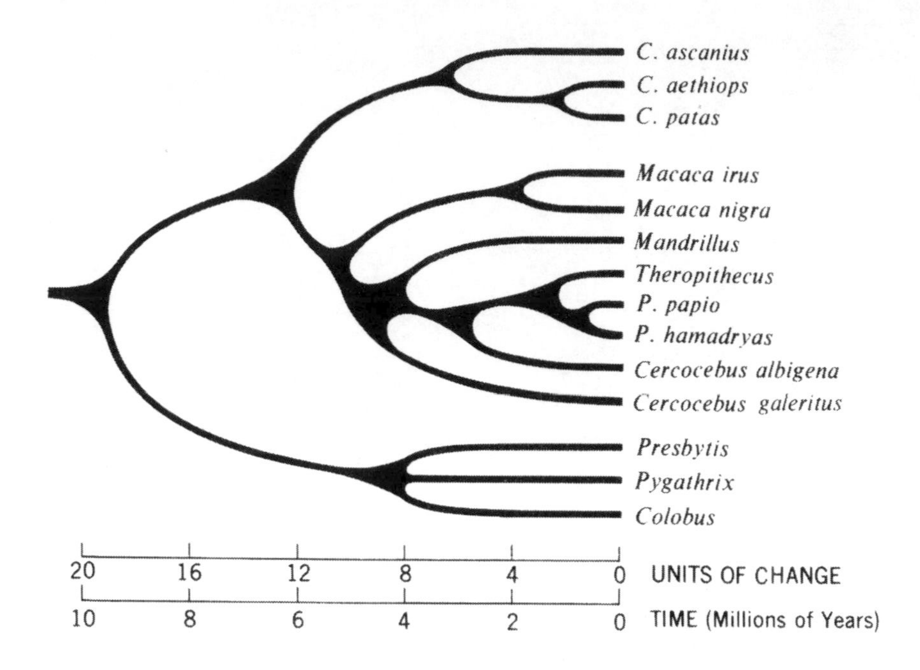

FIGURE 201. The relationships of Old World monkeys based on blood protein sequences, by Vincent Matthew Sarich and John Edward Cronin. Sarich and Cronin, 1976, "Molecular systematics of the primates," p. 151, fig. 6; courtesy of Vincent M. Sarich. Used with permission.

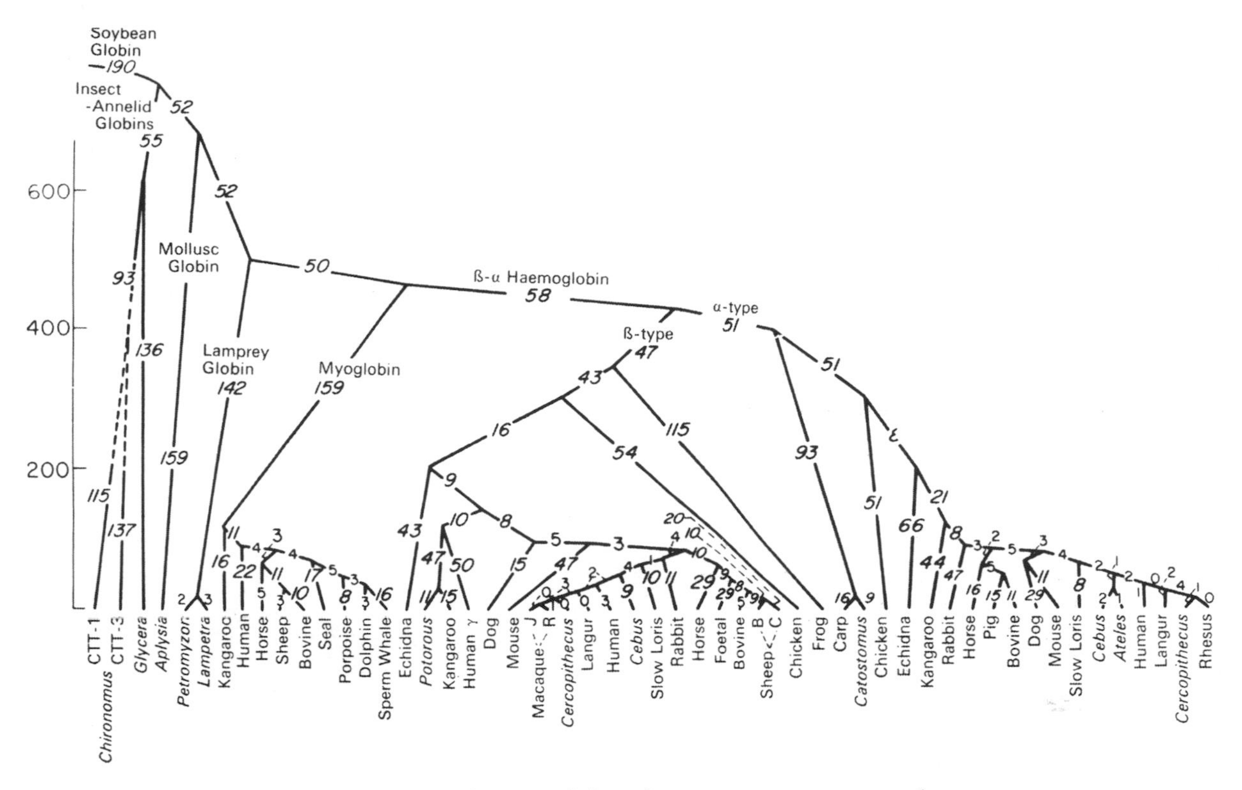

FIGURE 202. Relationships of mammals based on protein sequences and immunological specificity, by Morris Goodman and coauthors.

Goodman et al., 1975, *Nature,* 253:604, fig. 1; courtesy of the Nature Publishing Group. Used with permission.

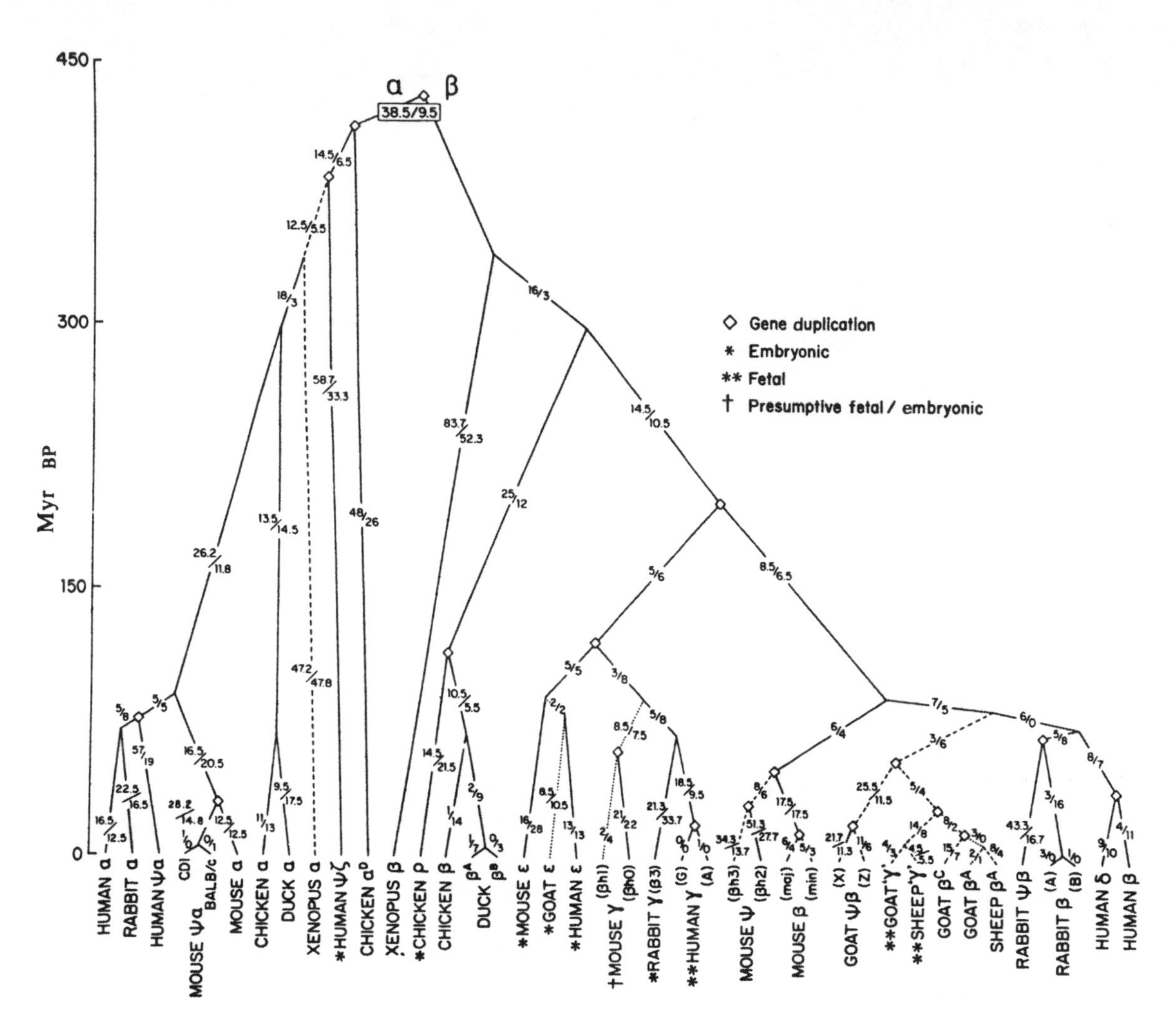

FIGURE 203. The evolutionary history of hemoglobin based on DNA sequences, by John Czelusniak and Morris Goodman.

Czelusniak et al., 1982, *Nature* 298(5871):299, fig. 2; courtesy of the Nature Publishing Group. Used with permission.

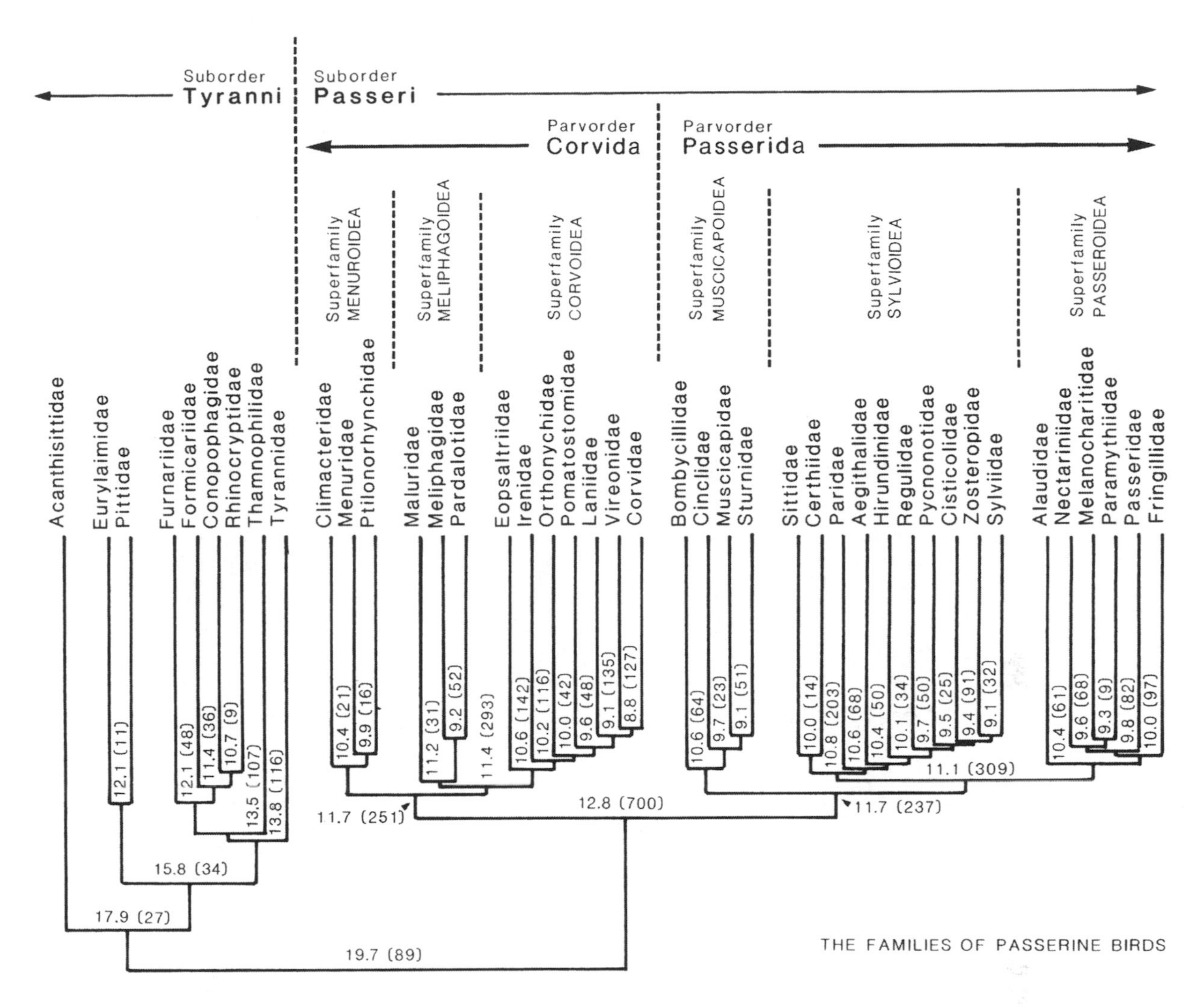

FIGURE 204. The families of passerine birds based on DNA-DNA hybridization, by Charles Gald Sibley and coauthors.

Sibley et al., 1988, *The Auk,* 105(3):412, fig. 5, courtesy of Scott Gillihan and the American Ornithologists' Union. Used with permission.

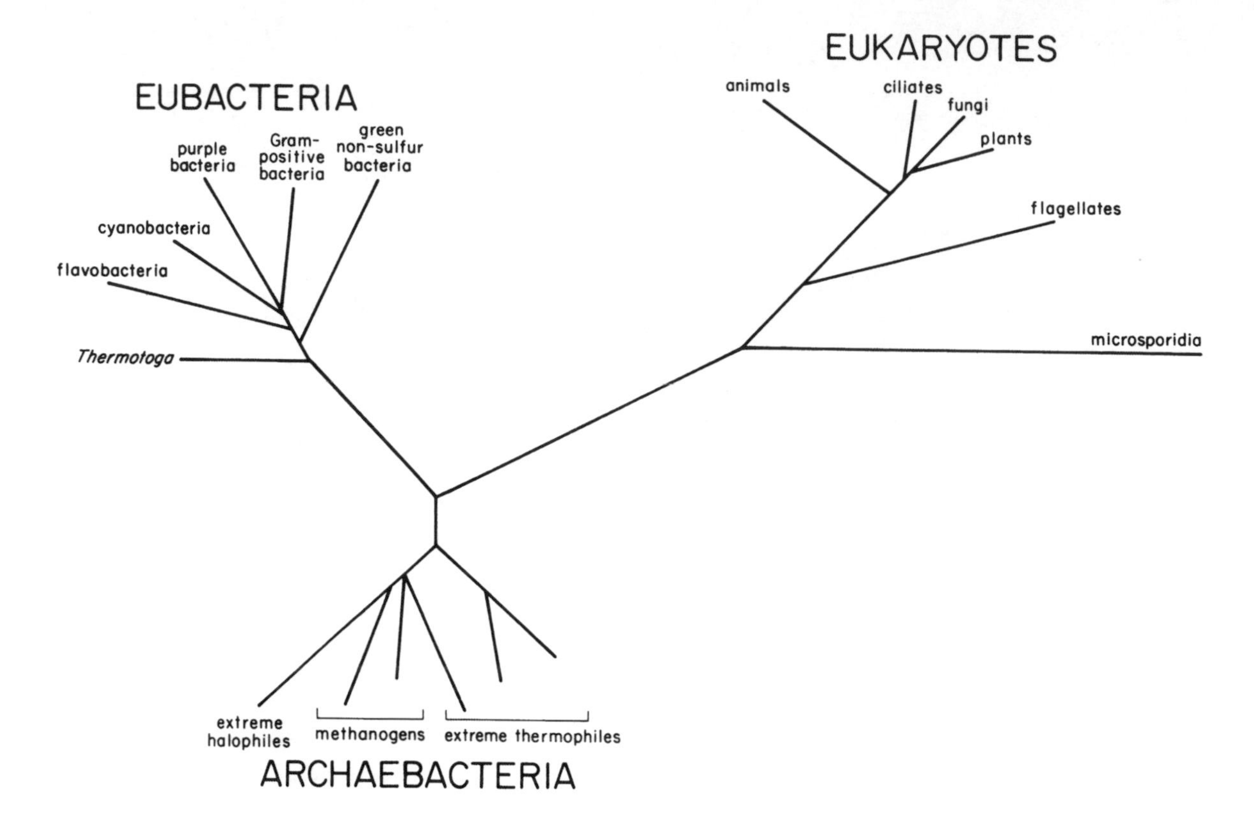

FIGURE 205. The discovery of Archaea: A tree of life, based on ribosomal RNA sequence data, showing that living systems represent three aboriginal lines of descent, published in 1987 by Carl Richard Woese, but based in large part on work initiated in 1977, with coauthor George Edward Fox.

Woese, 1987, *Microbiological Reviews,* 51(2):231, fig. 4, courtesy of Carl R. Woese. Used with permission.

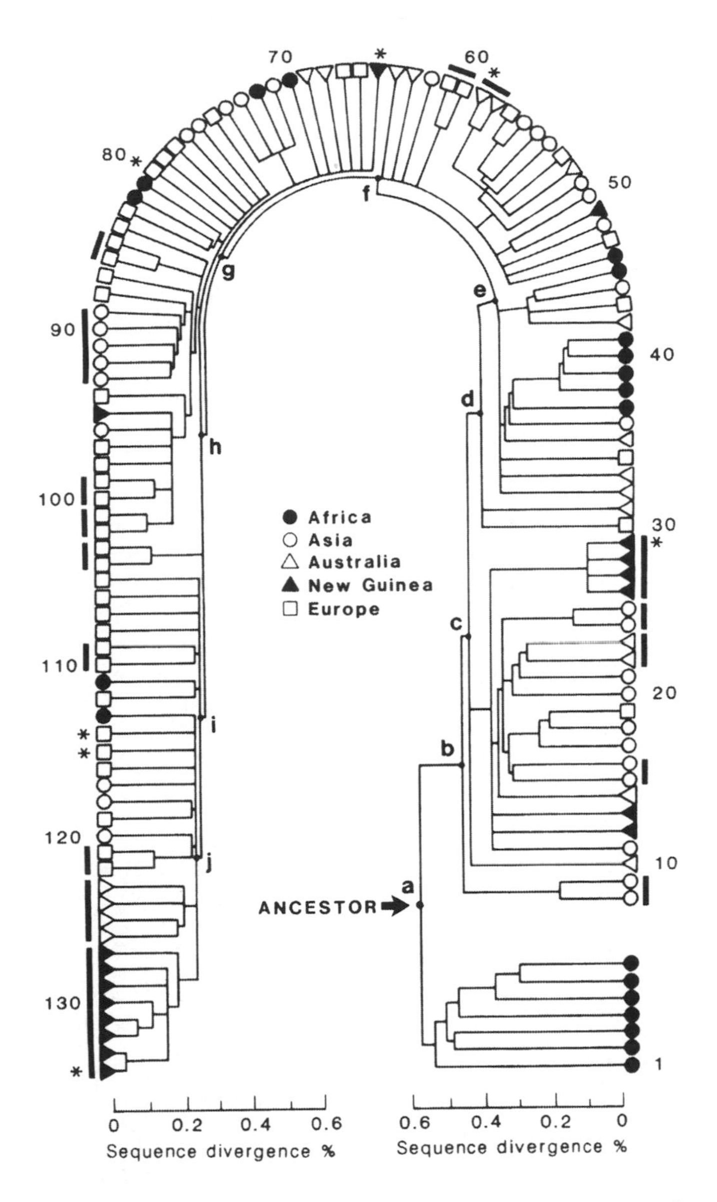

FIGURE 206. The mitochondrial "Eve" hypothesis, suggesting that all mitochondrial DNA stems from one woman who most likely lived in Africa about 200,000 years ago, by Rebecca Louise Cann and coauthors.

Cann et al., 1987, *Nature*, 325:34, fig. 3; courtesy of the Nature Publishing Group. Used with permission.

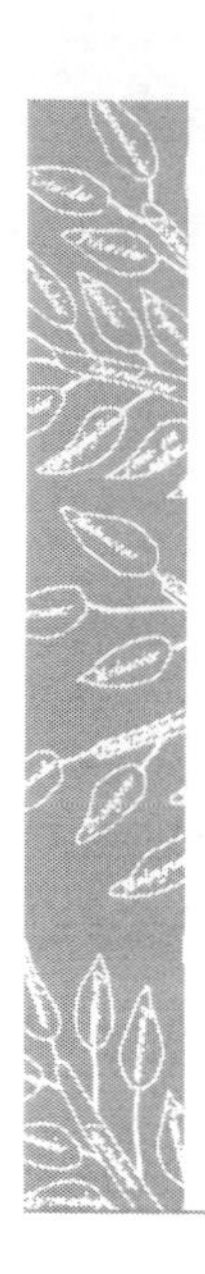

Notable Trees of the Past Four Decades

1970–2010

New World lizards of the large and complex genus *Cnemidophorus,* also known as the whiptail lizards, are notorious for their ability to hybridize and unusual also for their mode of reproduction. In about a third of the sixty or so recognized species of *Cnemidophorus* there are no males—the females reproduce through parthenogenesis, a form of asexual reproduction in which initial growth and development occurs without fertilization by a male. This process is rather well-known in some invertebrate groups, like aphids and bees and wasps, but very rare among vertebrates. The species that have no males originate through hybridization. Occasionally, a mating between a female of one species and a male of another produces a parthenogen, a female that is able to produce viable eggs that are genetically identical to her own cells. The lizards that hatch from these eggs are thus also parthenogens that can again produce identical eggs, resulting in an asexual, clonal population.

Parthenogenetic species resulting from a single hybridization are diploid (that is, they have two sets of chromosomes just as sexual species do), but sometimes these females mate with other males, producing offspring that are triploid (that is, they have three sets of chromosomes).[1] All of this complexity obviously results in confusion when it comes to sorting out the evolutionary relationships among the species. Charles Herbert Lowe (1920–2002) and coauthors provided a good example with a tree based on chromosome characters published in 1970 (Figure 207). The dashed lines (labeled 1 and 2) indicate two equally likely alternate routes for the derivation of the *Cnemidophorus tigris* group, either through common ancestry (1) or by way of hybridization with members of the *Cnemidophorus sexlineatus* group (2). The terminal *C. tesselatus* group is thought to have originated through hybridization between members of the *C. tigris* and *C. sexlineatus* groups. Note that the ancestral taxa are diploid, but some of the members of the more derived *C. sexlineatus* and *C. tesselatus* groups are triploid.

The explosive radiation of cichlid fishes in the rift lakes of East Africa was known to a few ichthyologists and evolutionary biologists in the mid-twentieth century, but it was the classic monograph of Geoffrey Fryer (1927–) and Thomas Derrick Iles (1927–), *The Cichlid Fishes of the Great Lakes of Africa,* published in 1972, that brought world attention to this spectacular biodiversity—some 1,750 species confined to and found only in three large lakes, Victoria, Tanganyika, and Malawi.[2] These "species flocks" encompass a wide array of morphologically and behaviorally specialized forms, all thought to have originated from just a few founder species in a relatively brief interval of geological time. One popular hypothesis to explain this diversification is a peculiar set of anatomical specializations of the feeding apparatus found only in cichlids that has allowed them to specialize on specific kinds of prey, thereby giving them a competitive advantage over other fishes that lack these anatomical modifications. A tree published by Fryer and Iles in 1972 demonstrates this enormous variety of feeding modes and prey types (Figure 208). From a "generalized ancestor," cichlids branch in all directions, the shape of their heads, the flexibility of their jaws, and the shape and pattern of their teeth, adapted for specific prey, from mollusk crushers to scale eaters, and from eye biters to rock and plant scrapers.

During the early 1850s, Charles Darwin spent a lot of time studying barnacles, both living and fossil.[3] His monographic series of papers on these animals are examples of the incredible attention to detail that he devoted to all his many publications.[4] Nevertheless, he was not always clear when it came to describing barnacle relationships, and as far as we know he never drew a tree to explain his thoughts about how they might have evolved. But through careful analysis of Darwin's work, philosopher and historian of science Michael Tenant Ghiselin (1939–), in coauthorship with Linda Jaffe, produced a reconstruction of Darwin's views on barnacle phylogeny in 1973 (Figure 209).[5] In Ghiselin and Jaffe's diagram, the limits between taxa are shown with lines, different kinds of lines indicating categorical rank. The straight lines indicate branching sequences, while dashed lines indicate uncertain affinity. Subfamilies and clusters of genera representing subfamilies are enclosed within ovals (a dotted oval for a basal assemblage of questionable species groups); families, within double-lined ovals; and orders, within thick-lined ovals. In attempting to elucidate Darwin's views on the role of phylogeny in classification, Ghiselin concluded that Darwin was careful not to include any polyphyletic groups, but his system does contain paraphyletic groups. Thus Darwin's system never took the form of a modern cladistic view.[6]

In the mid-1970s, evolutionary relationships among the perching birds and related orders presented almost insurmountable problems.[7] While some were attempting to resolve these difficulties by means of molecular methods—for example, the DNA-DNA hybridization protocols developed by Sibley and Ahlquist described in an earlier chapter (see Figure 204)—others were searching for new and untried morphological features. A tree published in 1977 by ornithologist John Alan Feduccia (1943–) is interesting because it hypothesizes a diphyletic Passeriformes based on anatomical differences in the structure of the stapes (Figure 210), one of the small bony ossicles of the middle ear. Thus, Feduccia shows the Tyranniformes on the far left, leading "toward advanced passerine morphology" and Passeriformes on the far right also bearing "advanced passerine morphology." In contrast, conventional phylogenies, based on other characters, indicate placement of the Tyranniformes between the Piciformes and Passeriformes and joined to the latter by a basal phyletic line, thus supporting passerine monophyly, that is, a single origin for all perching birds.

Paleontologist Thomas Stainforth Kemp (1943–), in his 1982 article, "The Reptiles That Became Mammals," used a phylogeny of the mammal-like reptiles (the "stem-mammals" or "proto-mammals" in cladistic nomenclature[8])—the group from which mammals evolved—to argue against the then-prevailing opinion that the fossil record is too incomplete to contribute in detail to evolutionary thought (Figure 211):

> Ideas about the ways in which evolutionary changes take place have come almost completely from studies of living organisms. . . . Fossils have played only a limited part in formulating evolutionary theory because the biologist can never be certain that all the critical stages are preserved in any particular part of the fossil record. So biologists have tended either to fit fossil evidence into theories derived from knowledge of living organisms or else—and these form the great majority of cases—assume that the fossil record is too incomplete to be helpful.[9]

Having a remarkably complete history over a span of some 120 million years and, at the same time, showing a wide and complex adaptive radiation, Kemp concluded that the fossil record of the mammal-like reptiles contributes to evolutionary theory in several ways that cannot be predicted from a study of living organisms alone.[10]

Lynn Alexander Margulis (1938–2011), well known for her now generally accepted symbiotic theory of how cell organelles (mitochondria and chloroplasts) evolved from bacteria, published (in coauthorship with Karlene Vila Schwartz, 1936–) a series of trees in the several editions of their *Illustrated Guide to the Phyla of Life on Earth,* based on the five kingdom system first proposed by Robert Whittaker in 1969 (see Figure 180). Three of them are reproduced here, all dating from the first 1982 edition. The first is a "phylogeny of life on Earth" (Figure 212); the second, a phylogeny of the Monera, autotrophic, fermenting, and respiring heterotrophic bacteria, all derived from anaerobic prokaryotic ancestors (Figure 213); and the third, a phylogeny of the Animalia derived from anaerobic protoctist ancestors, the latter comprising the eukaryotic microorganisms, such as algae, swimming water molds, slime molds, ciliates, amoebae, and many other groups (Figure 214).

In general support of the view presented by Lynn Margulis that mitochondria are derived from aerobic bacteria, and in keeping with the notion of a common origin of Bacteria, Archaea, and eukaryotic hosts, Raik-Hiio Mikelsaar (1939–) published a new concept of early cellular evolution in 1987 that he called the "archigenetic" hypothesis. He argued that mitochondria evolved, not from eubacteria, but from ancient predecessors of pro- and eukaryotes, which he called "protobionts" (Figure 215). He proposed further that animal, fungal, and plant mitochondria are endosymbionts derived from independent free-living cells called "mitobionts," which, having arisen at different developmental stages of protobionts, retained some of their ancient primitive features of the genetic code and the transcription-translation system.

A tree constructed by Dieter Korn (1958–), published in 1995 (Figure 216) demonstrates an important evolutionary phenomenon called "heterochrony"—a change in the timing of developmental events that leads to changes in size and shape. This can cause an organism to achieve sexual maturity and therefore adulthood at a juvenile stage of development, referred to as "neoteny," a kind of pedomorphosis. Alternatively, sexual maturity may be delayed while morphological development continues beyond the adult stage of the ancestor.[11] In Korn's example, the evolution of certain Late Devonian ammonites, the transition from *Kamtoclymenia endogona* to the genus *Parawocklumeria* involves neotenic retention of the juvenile triangular shell form in the adult of the descendant, while the evolution of the several genera derived from *Parawocklumeria* involves extension of the growth period of *Parawocklumeria* as indicated by the increasing complexity of the suture lines, a process called "hypermorphosis."[12]

No book of trees would be adequate without including some that depict dinosaur relationships. Among the most interesting of these are the diagrams produced by Paul Callistus Sereno (1957–), an American paleontologist from the University of Chicago who is widely known for his discoveries of numerous undescribed species on several continents. Two of the most spectacular Sereno trees, both published in 1999, are shown here. The first is a temporally calibrated phylogeny of the Dinosauria (Figure 217), showing that the ascendancy of dinosaurs on land occurred quite rapidly beginning about 215 million years ago, before the close of the Triassic, and quickly spread out across what was then the single consolidated landmass of Pangaea, ushering in the "dinosaur era"—a 150-million-year interval when virtually all animals a meter or more in length in dry land habitats of the Earth were dinosaurs.[13] The second Sereno diagram is a unique construction—a long, linear tree folded in upon itself to allow the full extent of dinosaur diversity to fit on the page (Figure 218). In showing the relationships among the two primary divisions of the dinosaurs—the ornithischian, or "bird-hipped," assemblage on the left and the saurischian, or "lizard-hipped," assemblage on the right—the thickened internal branches of the tree are scaled to reflect the number of supporting shared derived characters.

A pair of trees published in 2002 by Gregory Scott Paul (1954–) focuses on competing hypotheses of the placement of *Archaeopteryx* relative to modern birds and closely related non-flying dinosaur genera (Figure 219). In suggesting that some flightless dinosaurs were really "neoflightless dino-birds,"[14] Paul diagramed the conventional dinosaur-bird scheme shown on the left, which supports the notion that *Archaeopteryx,* a highly derived member of the order Saurischia,[15] was more closely related to modern birds (Avebrevicauda) than any other known dinosaur and that the flightless genera *Sinornithosaurus, Deinonychus,* and *Velociraptor* diverged earlier as a single clade. An alternative "neoflightless" hypothesis shown on the right, suggests that the latter three genera, originating from *Archaeopteryx,* underwent reduction or secondary loss of flight-related features.

A tree of chordate relationships published by vertebrate paleontologist Timothy Bryan Rowe (1953–) in 2004 attempts to present a "contemporary overview of chordate history by summarizing current views on relationships among the major chordate clades in light of a blossoming understanding of molecular, genetic, and developmental evolution, and a wave of exciting new discoveries from deep in the fossil record."[16] Rowe's beautifully designed diagram (Figure 220), shows the affinities of extant lineages as well as the oldest known fossils, superimposed on a geological time scale (the numbers at each node refer to subheadings in his text). Chordate history can now be traced across at least a half billion years of geological time. The taxon is unique among multicellular animals in diversifying in size across eight orders of magnitude and inhabiting virtually every terrestrial and aquatic environment on Earth.[17]

The Acanthomorpha, or spiny-rayed fishes, constitute the crown group of the teleosts, including some 16,000 species in more than three hundred families, roughly one-third of all living vertebrates.[18] While apparently monophyletic, united among other characters by the presence of true spines in the dorsal and anal fins, and derived features of anatomy that allow for extensive upper-jaw protrusion, the enormous anatomical diversity within the group has made it nearly impossible to reconstruct phylogenetic relationships. One of the better, more recent attempts is that published in 2005 by Agnès Dettai (1976–) and Guillaume Lecointre (1964–), based on a combined series of molecular datasets (Figure 221).

All flatfishes, which constitute the teleost order Pleuronectiformes, are characterized by having highly asymmetrical skulls as adults, with both eyes situated on one side of the head, an arrangement that arises through migration of one eye during late larval development.[19] Because there are no transitional forms to link flatfishes to their symmetrical relatives, the evolutionary origin of this extraordinary anatomical specialization has long been a mystery. But Matthew Scott Friedman (1980–), now at the University of Oxford, has recently shed new light on the problem. A phylogenetic analysis, based in part on the discovery of new fossil material from the Eocene of Europe, has provided the missing link (Figure 222). While retaining many primitive features unknown in extant forms, the orbital region of the skull of the fossils is strongly asymmetrical in large specimens, but eye migration is incomplete, with the eyes remaining on opposite sides of the head in smaller post-metamorphic individuals. Thus, the condition displayed by the fossils is clearly intermediate between living pleuronectiforms and the situation found in other fishes.

The phylum Cnidaria, which includes sea anemones, corals, sea pens, jellies, and hydrozoans, is unusual among animals in having two fundamental life cycles—an alternation of morphologically and ecologically divergent life stages, each produced by the previous stage through sexual or asexual reproduction.[20] In an attempt to cast light on the evolution of hydrozoan life history traits, Lucas Leclère (1982–) and colleagues inferred phylogenetic relationships of 142 species of the suborder Thecata (Leptomedusae), the most species-rich hydrozoan group, using ribosomal RNA (Figure 223). Comparing the results with trees based on characters of the life cycle (features of the medusa, medusoid, and fixed gonophore stages) and of colony shape (solitary polyp, stolonal, erect, and erect and branched forms), the authors found considerable corroboration, leading to the surprising conclusion that life-history traits in these animals were relatively stable over the long term, despite a high frequency of recent morphological character change.

Despite several decades of intense molecular phylogenetics, efforts to resolve the deep phylogenetic history of arthropods have been mixed at best.[21] Most analyses have been based on small samples of taxa and genes and have generated results that are inconsistent and weakly supported. Recently, however, Jerome Clifton Regier (1947–) and coauthors produced a tree of relationships of seventy-five arthropod and five outgroup species, based on an analysis of sixty-two nuclear protein-coding genes (Figure 224). The results provided a "statistically well-supported phylogenetic framework for the largest animal phylum and represent a step towards ending the often-heated, century-long debate on arthropod relationships."[22]

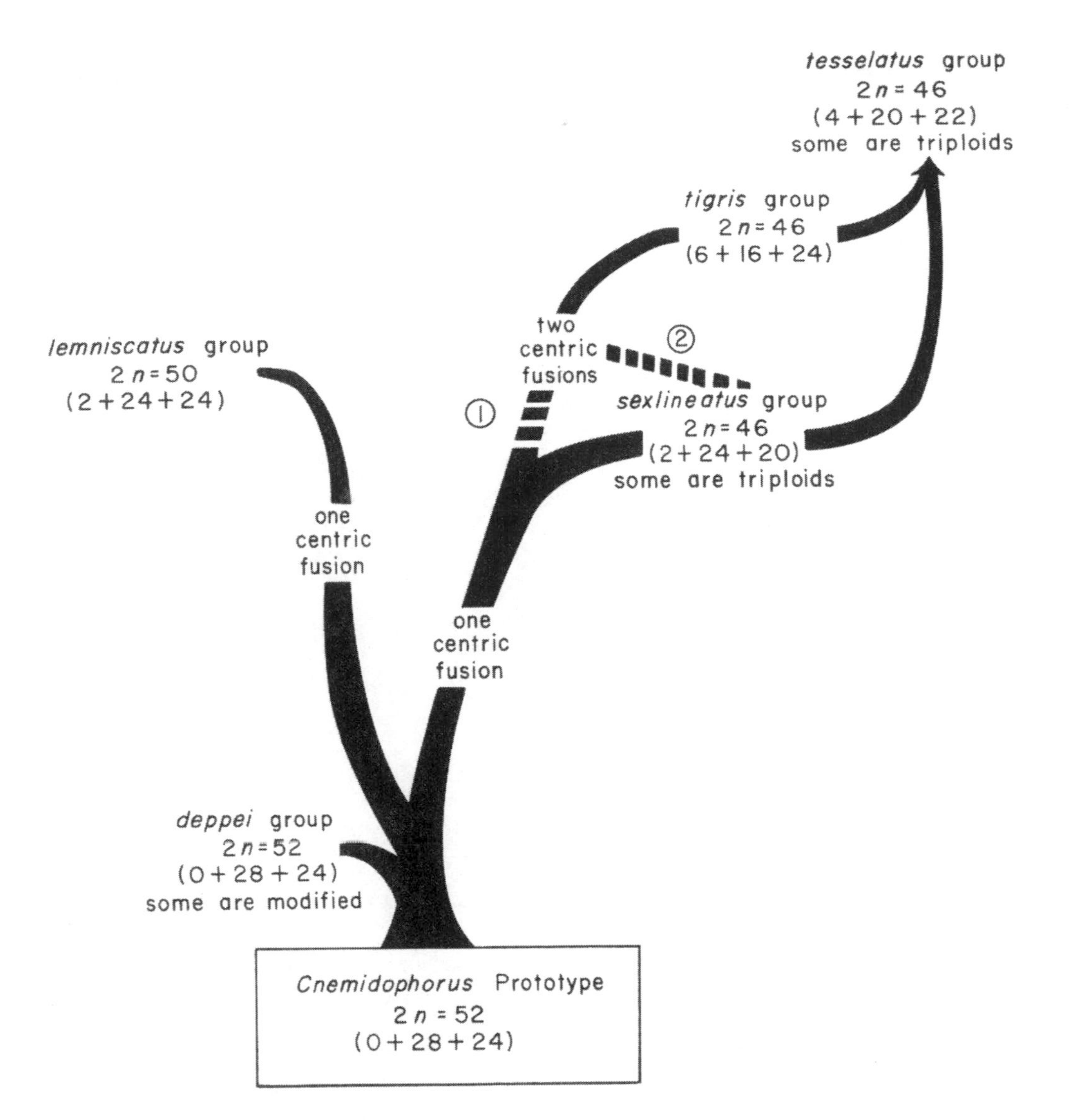

FIGURE 207. A phylogeny of the species groups of lizards of the genus *Cnemidophorus,* by Charles Herbert Lowe and colleagues. The dashed lines (labeled 1 and 2) indicate two equally likely alternate routes for the derivation of the *Cnemidophorus tigris* group, either through common ancestry (1) or by way of hybridization with members of the *C. sexlineatus* group (2).

Lowe et al., 1970, *Systematic Zoology,* 19(2):136, fig. 4; courtesy of Deborah Ciszek, Chris Payne, and the Society of Systematic Biologists. Used with permission.

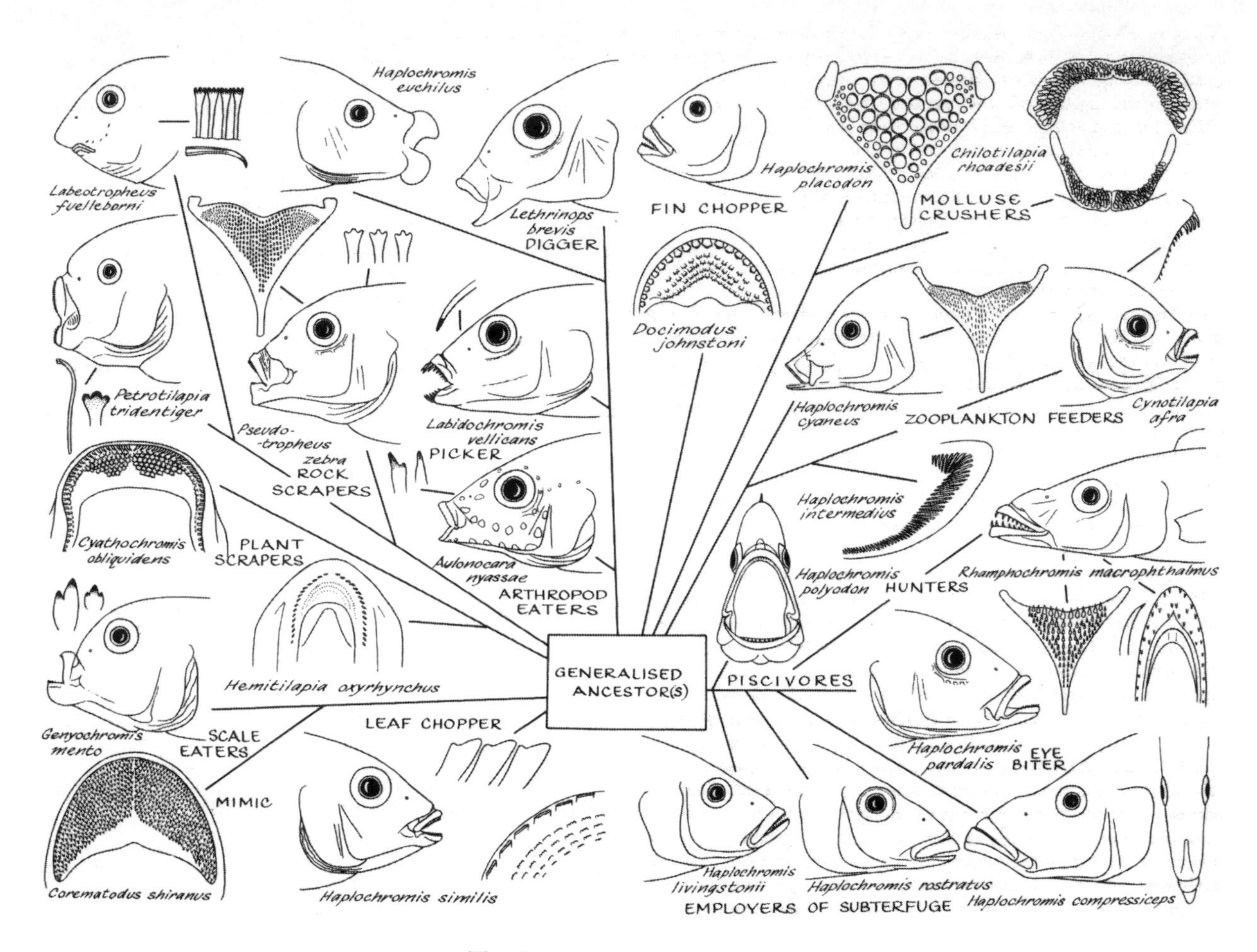

FIGURE 208. Trophic relationships within part of the adaptive radiation of the cichlids of Lake Malawi, by Geoffrey Fryer and Thomas Derrick Iles.

Fryer and Iles, 1972, *The Cichlid Fishes of the Great Lakes of Africa: Their Biology and Evolution,* pp. 488–489, fig. 333; courtesy of Geoffrey Fryer and T. Derrick Iles. Used with permission.

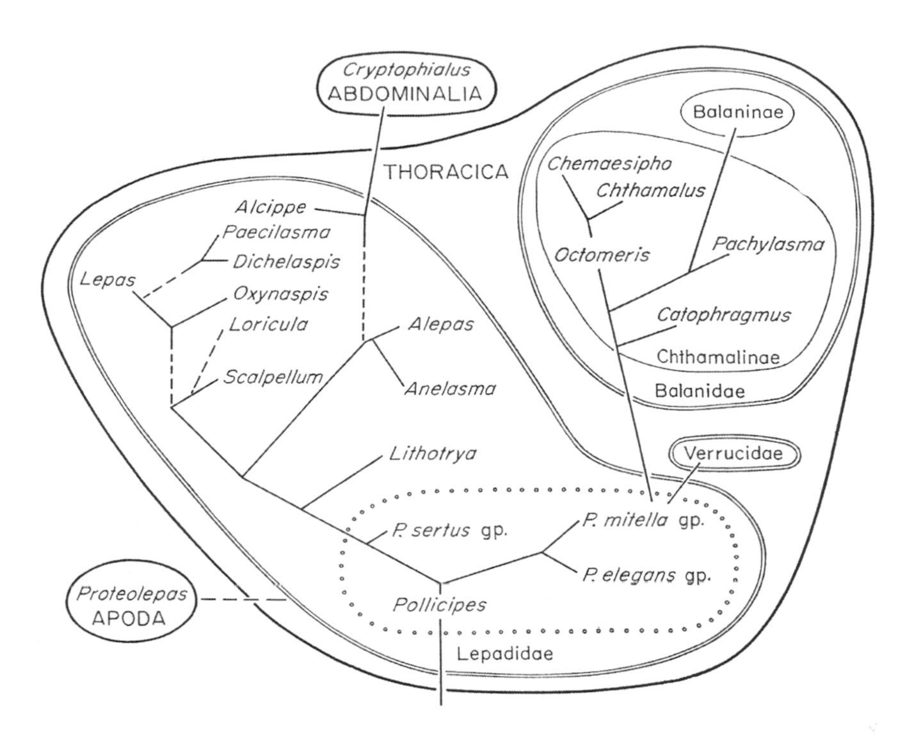

FIGURE 209. Charles Darwin's phylogeny of the barnacles, reconstructed by Michael Tenant Ghiselin and Linda Jaffe.

Ghiselin and Jaffe, 1973, *Systematic Zoology*, 22(2):137, fig. 1; courtesy of Michael T. Ghiselin and the Society of Systematic Biologists. Used with permission.

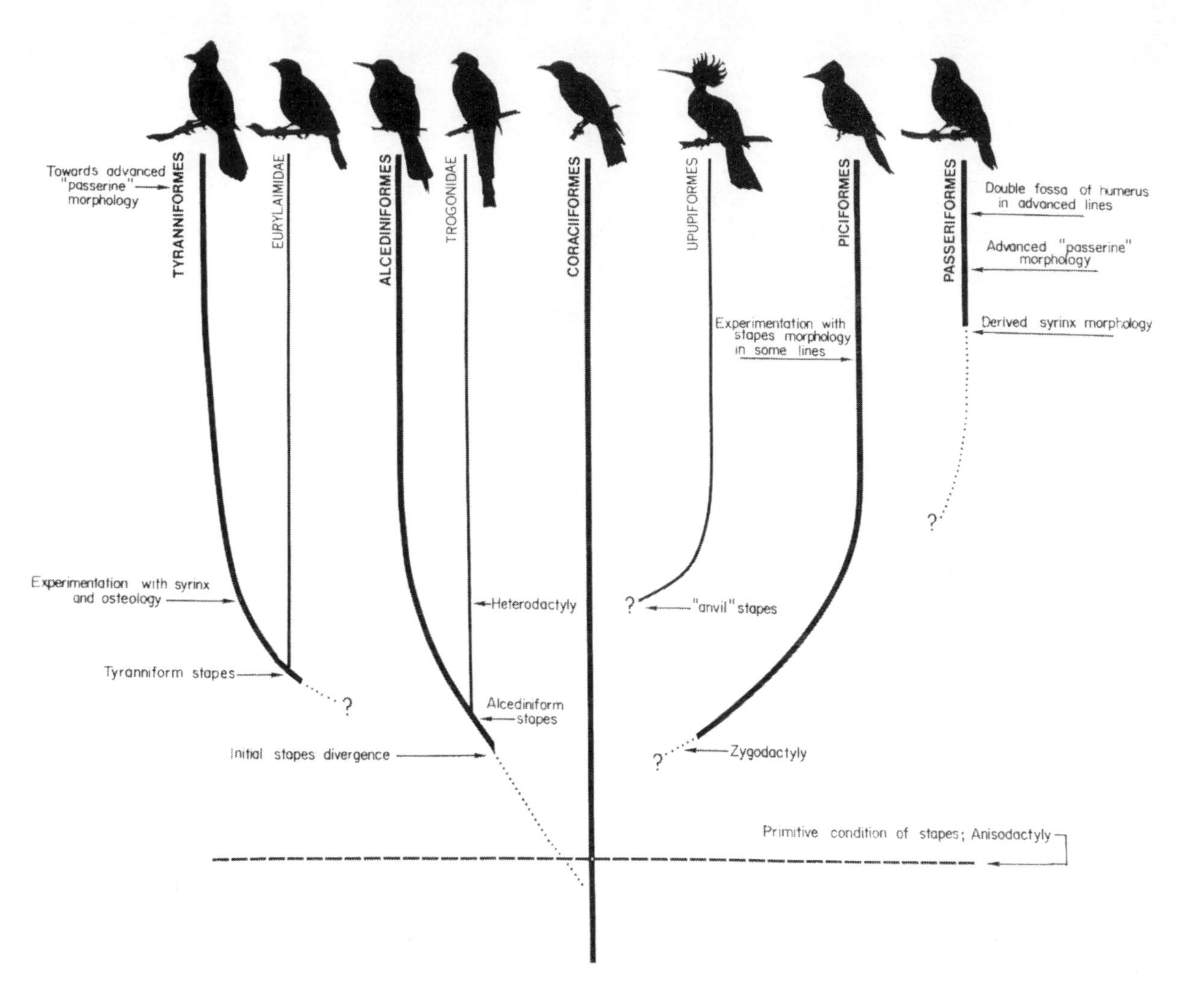

FIGURE 210. A diphyletic origin of perching birds based on anatomical differences in the stapes, one of the small bony ossicles of the middle ear, by John Alan Feduccia.

Feduccia, 1977, *Systematic Zoology,* 26(1):25, fig. 5; courtesy of Alan Feduccia and the Society of Systematic Biologists. Used with permission.

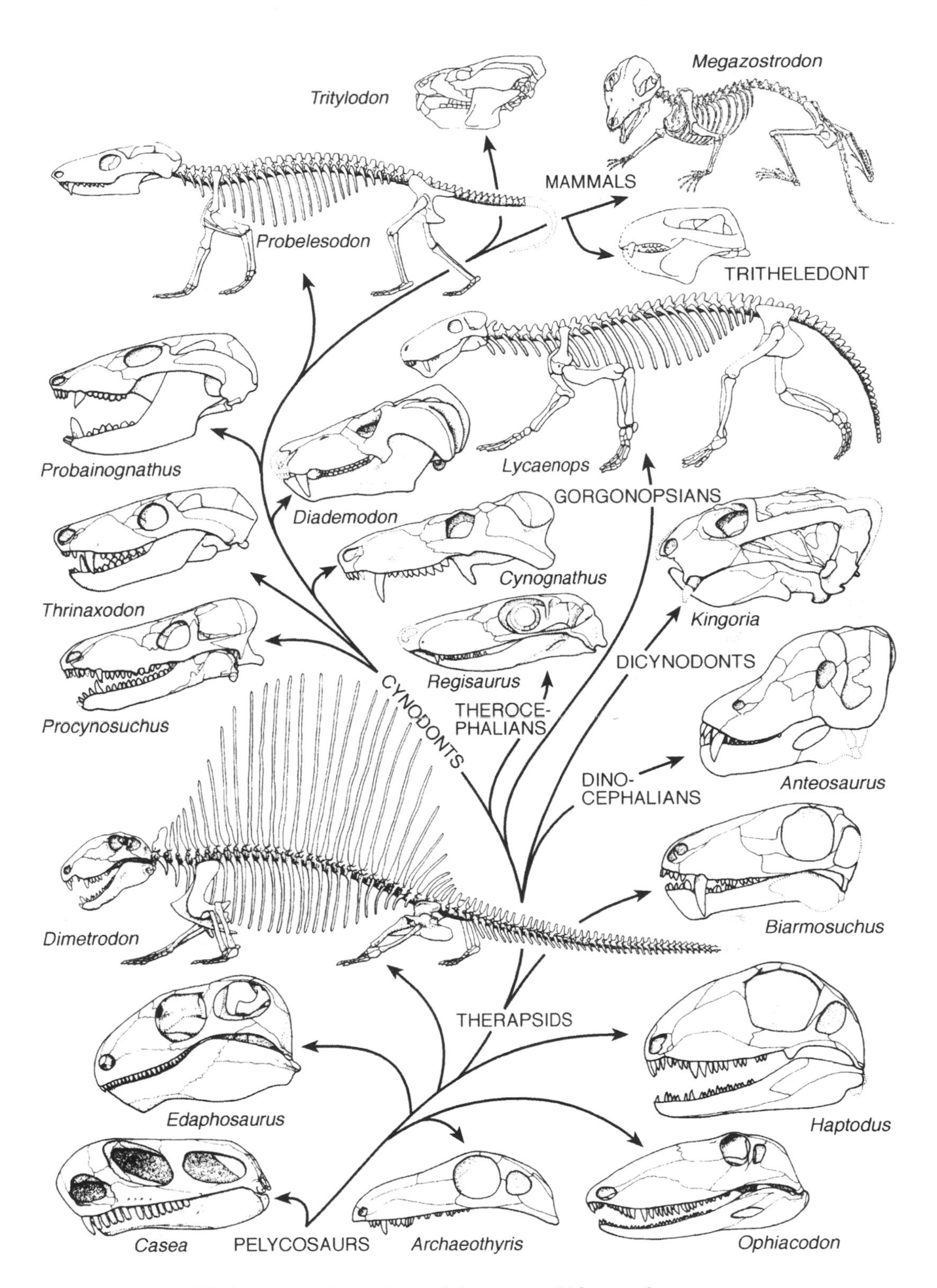

FIGURE 211. Phylogenetic relationships of the mammal-like reptiles, by Thomas Stainforth Kemp. "Although unconventionally attractive, this diagram is rigorously cladistic in content" (Kemp, 1999:229).

Kemp, 1982, *New Scientist,* 93:582, unnumbered figure; courtesy of Thomas S. Kemp and Małgosia B. Nowak-Kemp. Used with permission.

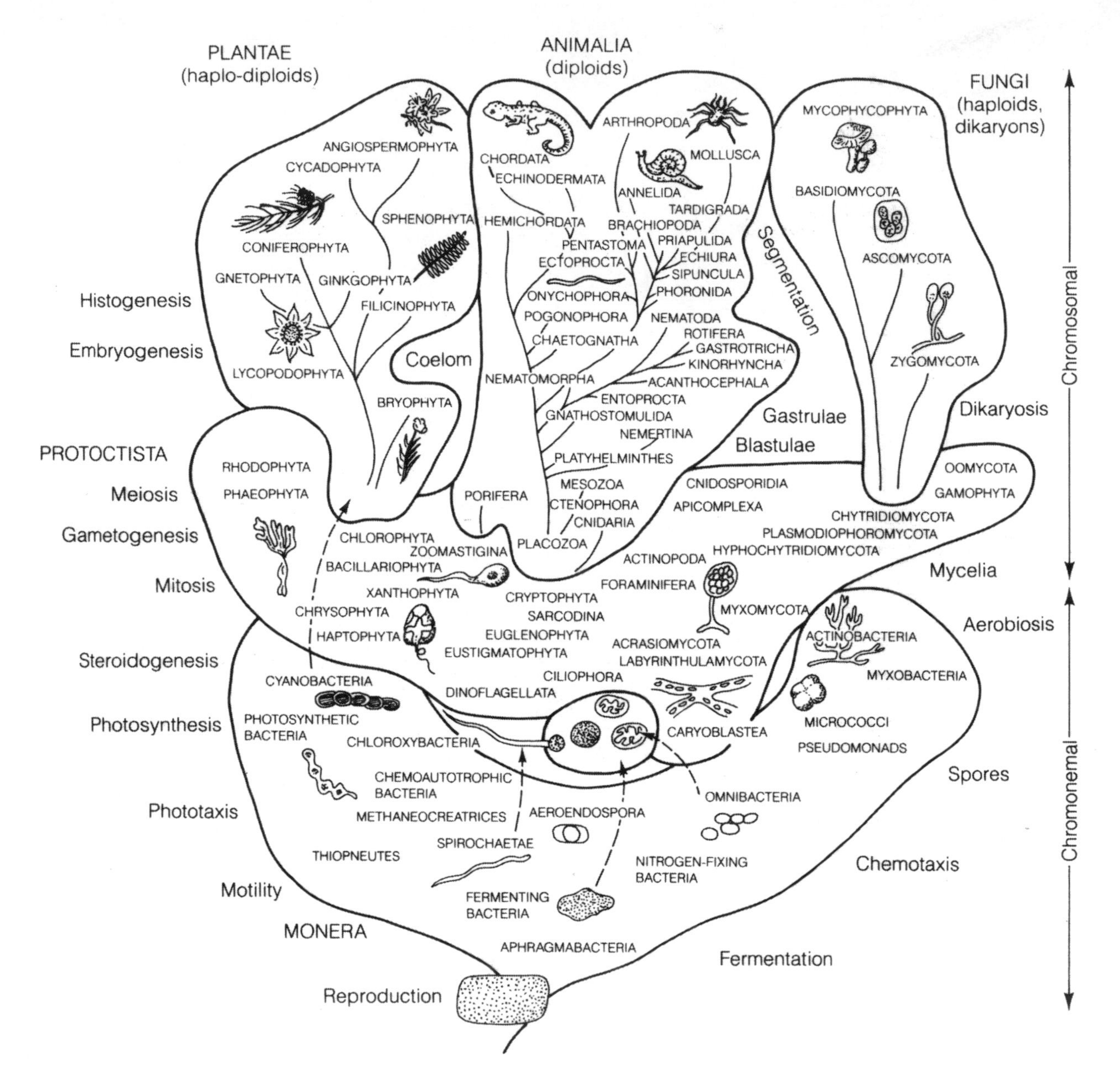

FIGURE 212. A phylogeny of life on earth by Lynn Alexander Margulis and Karlene Vila Schwartz, based on the Whittaker five-kingdom system and the symbiotic theory of the origin of eukaryotic cells.

Margulis and Schwartz, 1982, *Five Kingdoms: An Illustrated Guide to the Phyla of Life on Earth,* frontispiece; courtesy of Lynn A. Margulis and Karlene V. Schwartz. Used with permission.

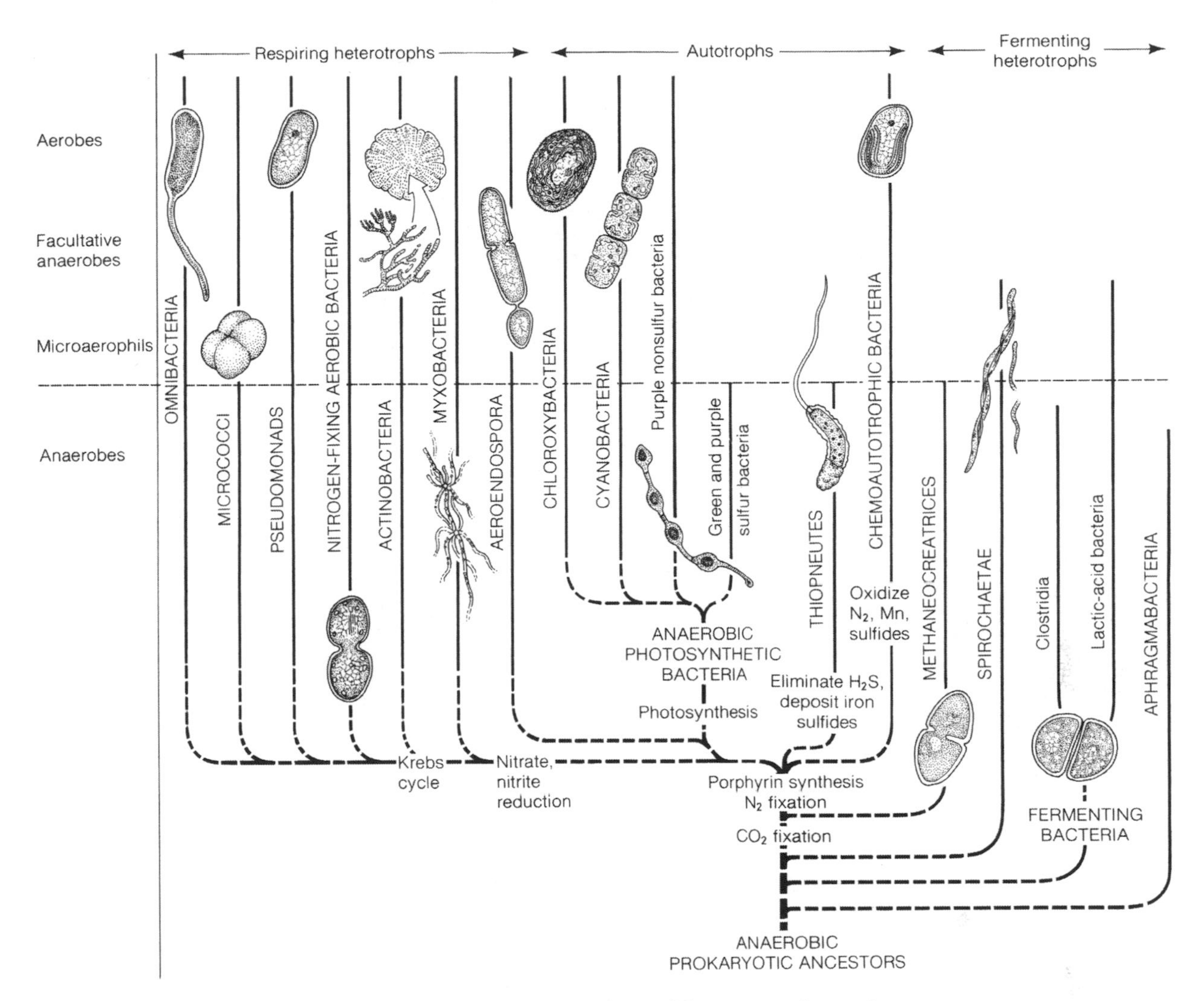

FIGURE 213. A phylogeny of the Monera, derived from anaerobic prokaryotic ancestors, by Lynn Alexander Margulis and Karlene Vila Schwartz.

Margulis and Schwartz, 1982, *Five Kingdoms: An Illustrated Guide to the Phyla of Life on Earth,* p. 24, unnumbered figure; courtesy of Lynn A. Margulis and Karlene V. Schwartz. Used with permission.

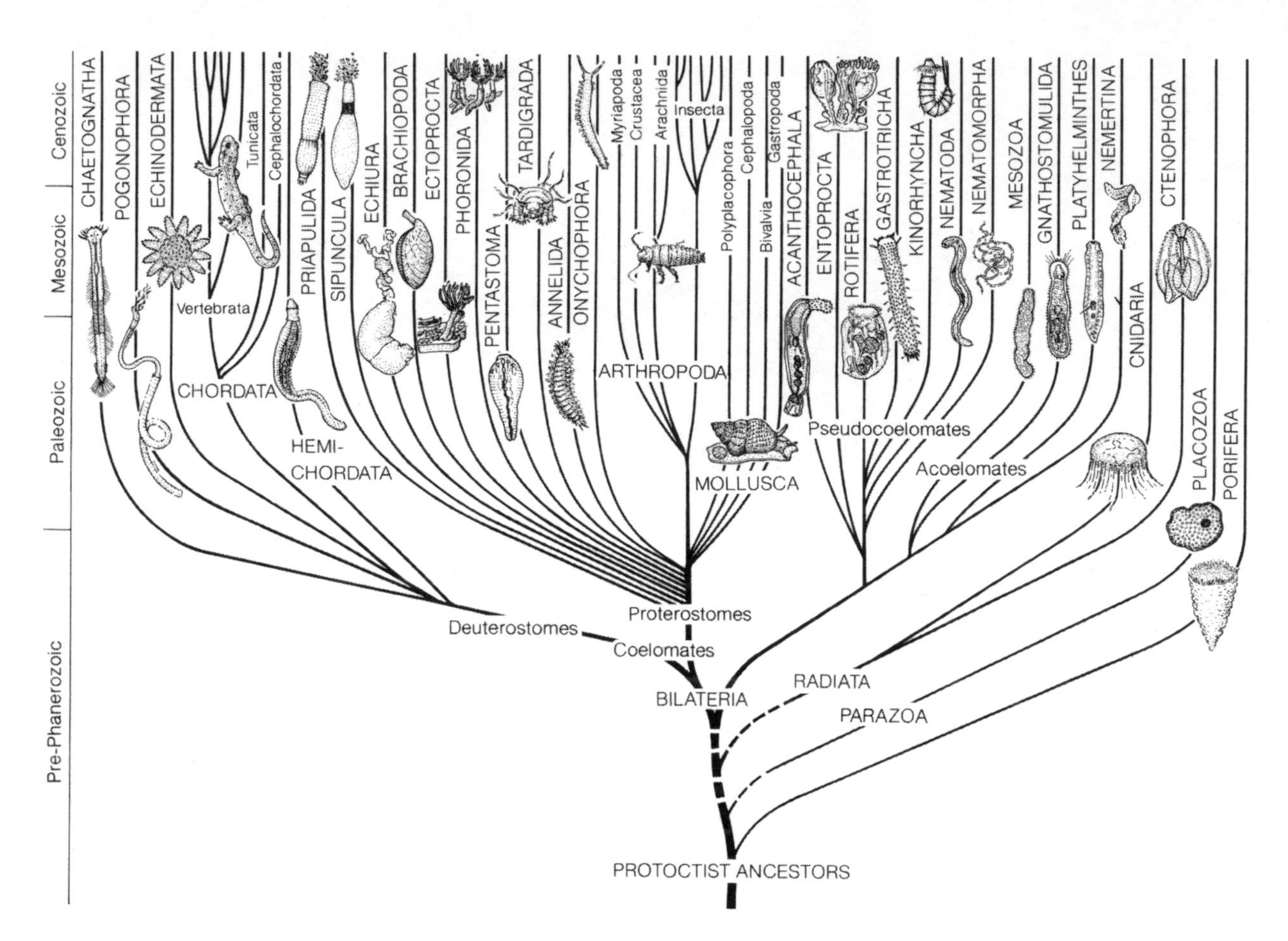

FIGURE 214. A phylogeny of the Animalia, derived from anaerobic protoctist ancestors, by Lynn Alexander Margulis and Karlene V. Schwartz.
Margulis and Schwartz, 1982, *Five Kingdoms: An Illustrated Guide to the Phyla of Life on Earth,* p. 160, unnumbered figure; courtesy of Lynn A. Margulis and Karlene V. Schwartz. Used with permission.

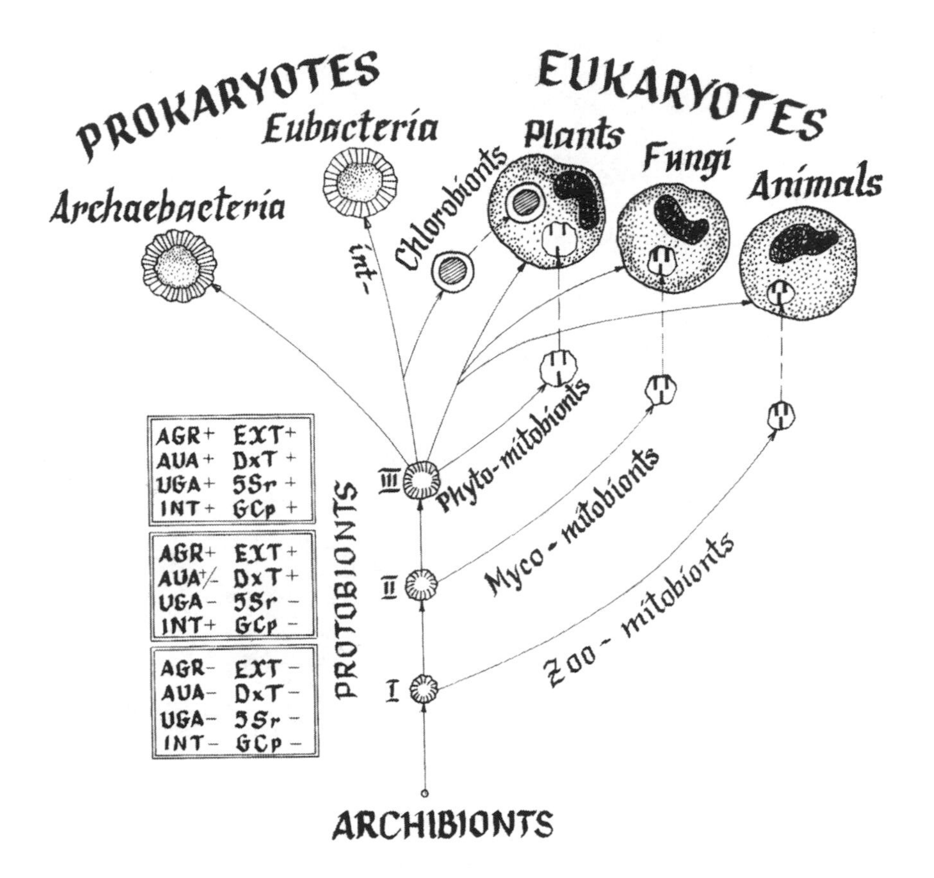

FIGURE 215. A tree demonstrating the principles of the archigenetic hypothesis of cellular evolution by Raik-Hiio Mikelsaar, which says that mitochondria have arisen neither from prokaryotes nor eukaryotes, but from free-living organisms called "mitobionts," which, in turn, were derived from primitive cells called "protobionts."

Mikelsaar, 1987, *Journal of Molecular Evolution,* 25(2):170, fig. 1; courtesy of Raik-Hiio Mikelsaar. Used with permission.

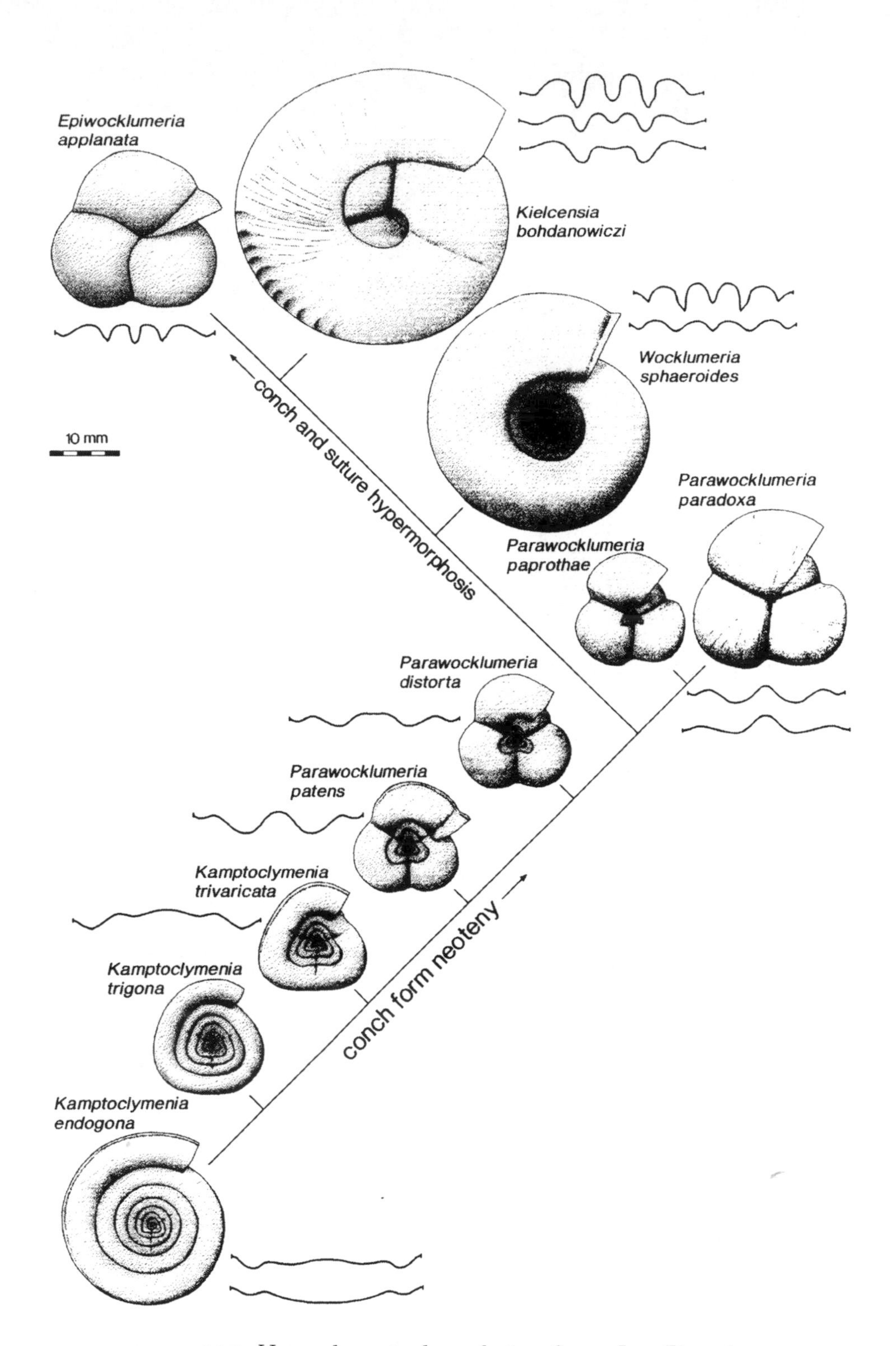

FIGURE 216. Heterochrony in the evolution of some Late Devonian ammonites, by Dieter Korn.

Korn, 1995, "Impact of environmental perturbations on heterochronic development in palaeozoic ammonoids," p. 252, fig. 12.4. In K. McNamamra, ed. *Evolutionary Change and Heterochrony,* Wiley & Sons, London; courtesy of Verity Butler and John Wiley & Sons Ltd. Used with permission.

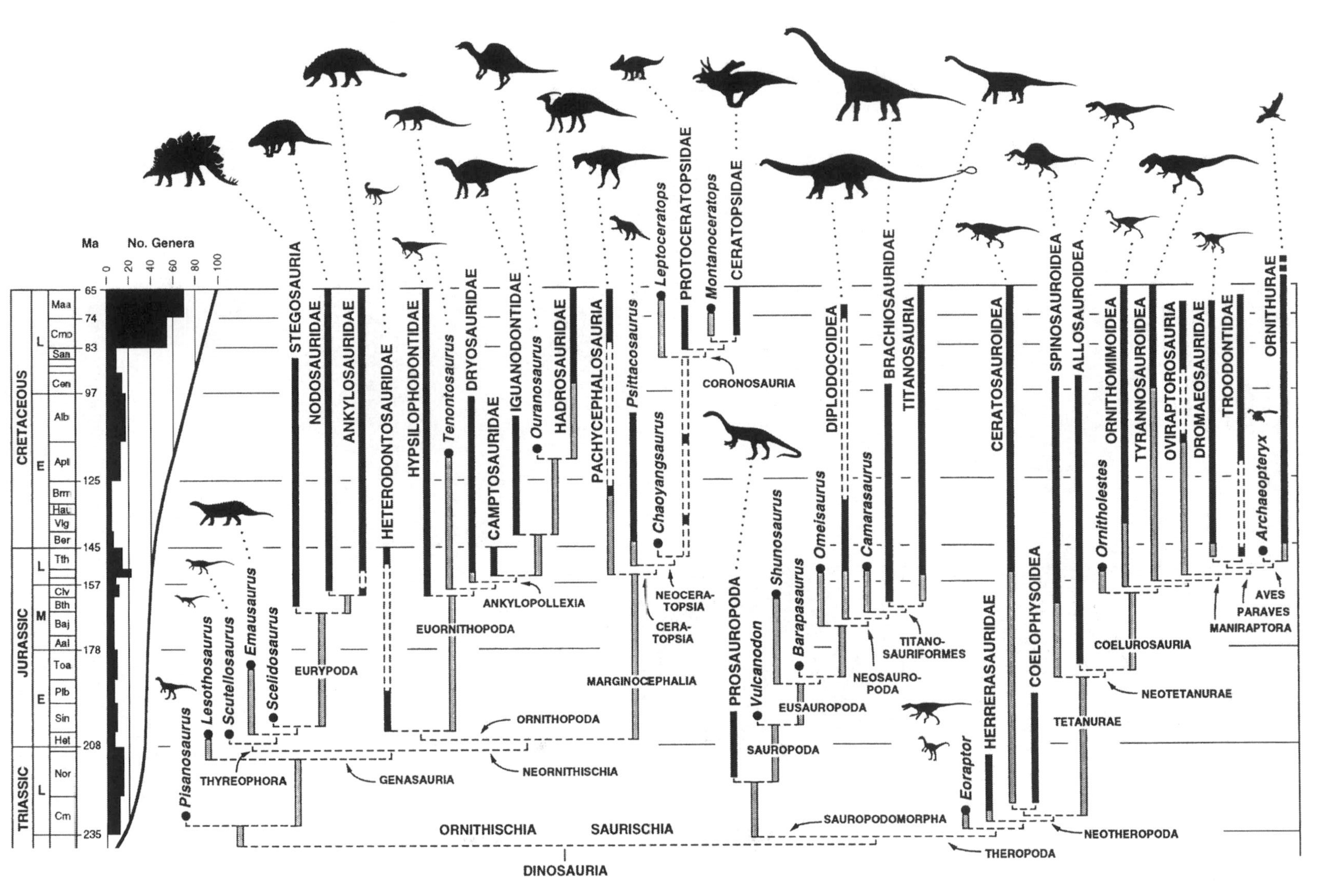

FIGURE 217. Relationships among dinosaurs, by Paul Callistus Sereno, demonstrating the rapid worldwide diversification of taxa during the Mesozoic.

Sereno, 1999, *Science,* 284:2138, fig. 1; courtesy of Paul C. Sereno. Used with permission.

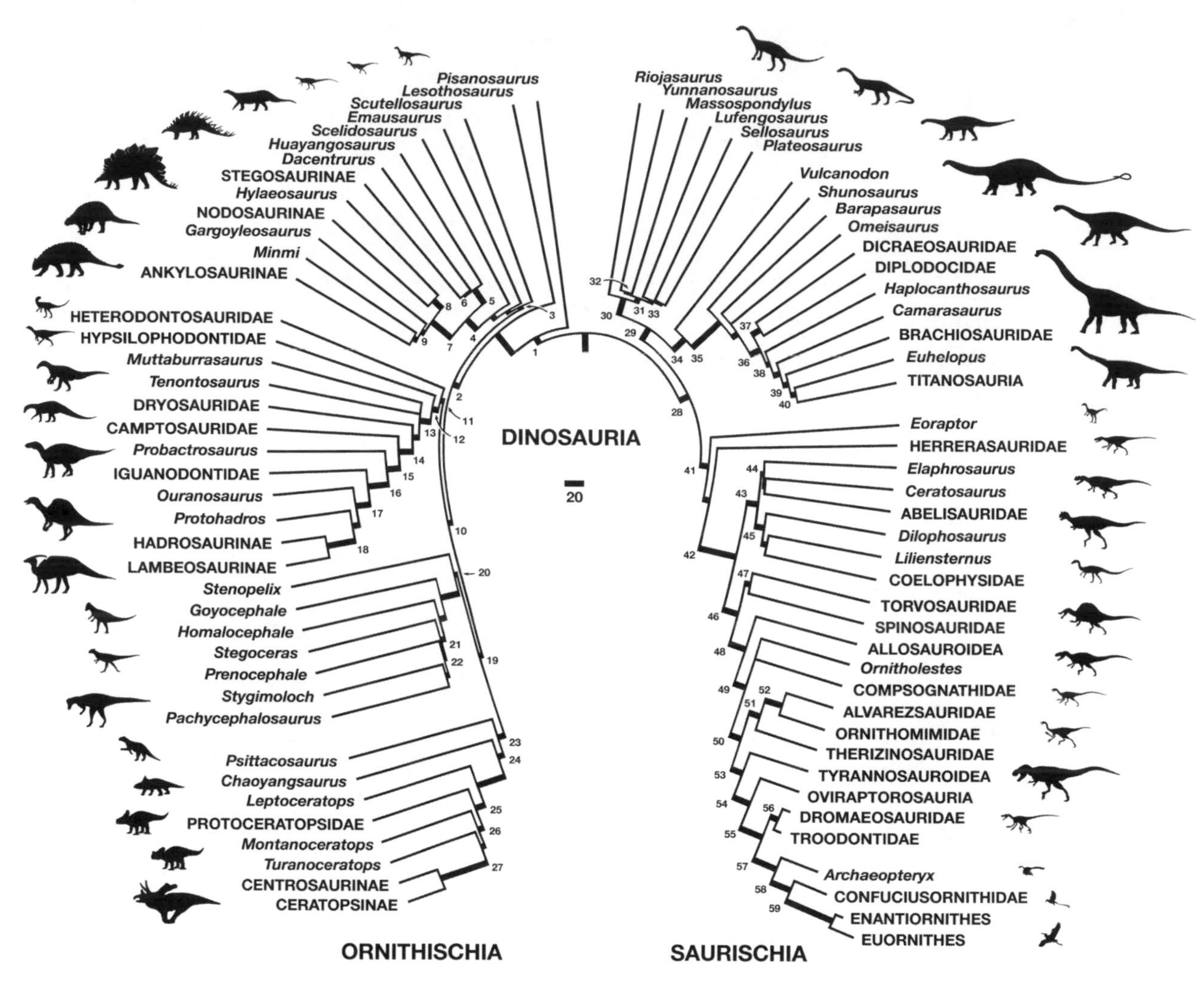

FIGURE 218. Dinosaur tree by Paul Callistus Sereno, emphasizing the sharp dichotomy between the two orders, the Ornithischia on the left and Saurischia on the right.

Sereno, 1999, *Science,* 284: 2139, fig. 2; courtesy of Paul C. Sereno. Used with permission.

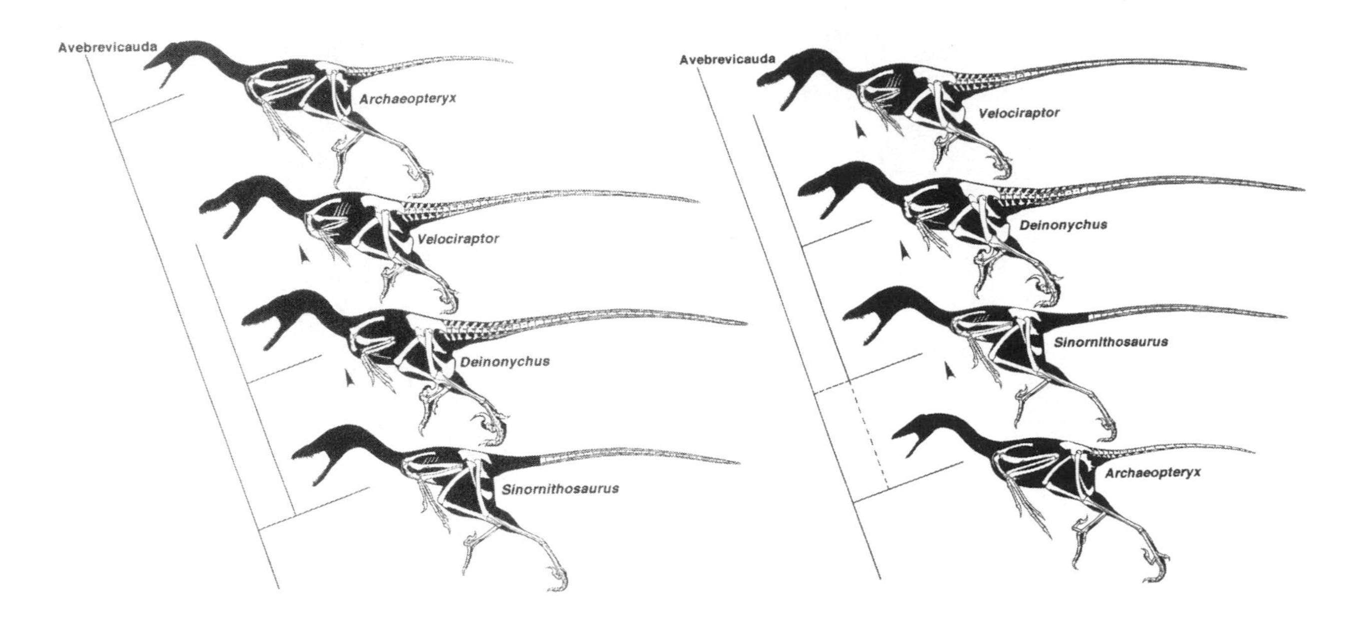

FIGURE 219. Competing hypotheses of the placement of *Archaeopteryx* relative to modern birds and closely related non-flying dinosaur genera, by Gregory Scott Paul.

Paul, 2002, *Dinosaurs of the Air: The Evolution and Loss of Flight in Dinosaurs and Birds,* pp. 240–241, fig. 11.1; courtesy of Gregory S. Paul and the Johns Hopkins University Press. Used with permission.

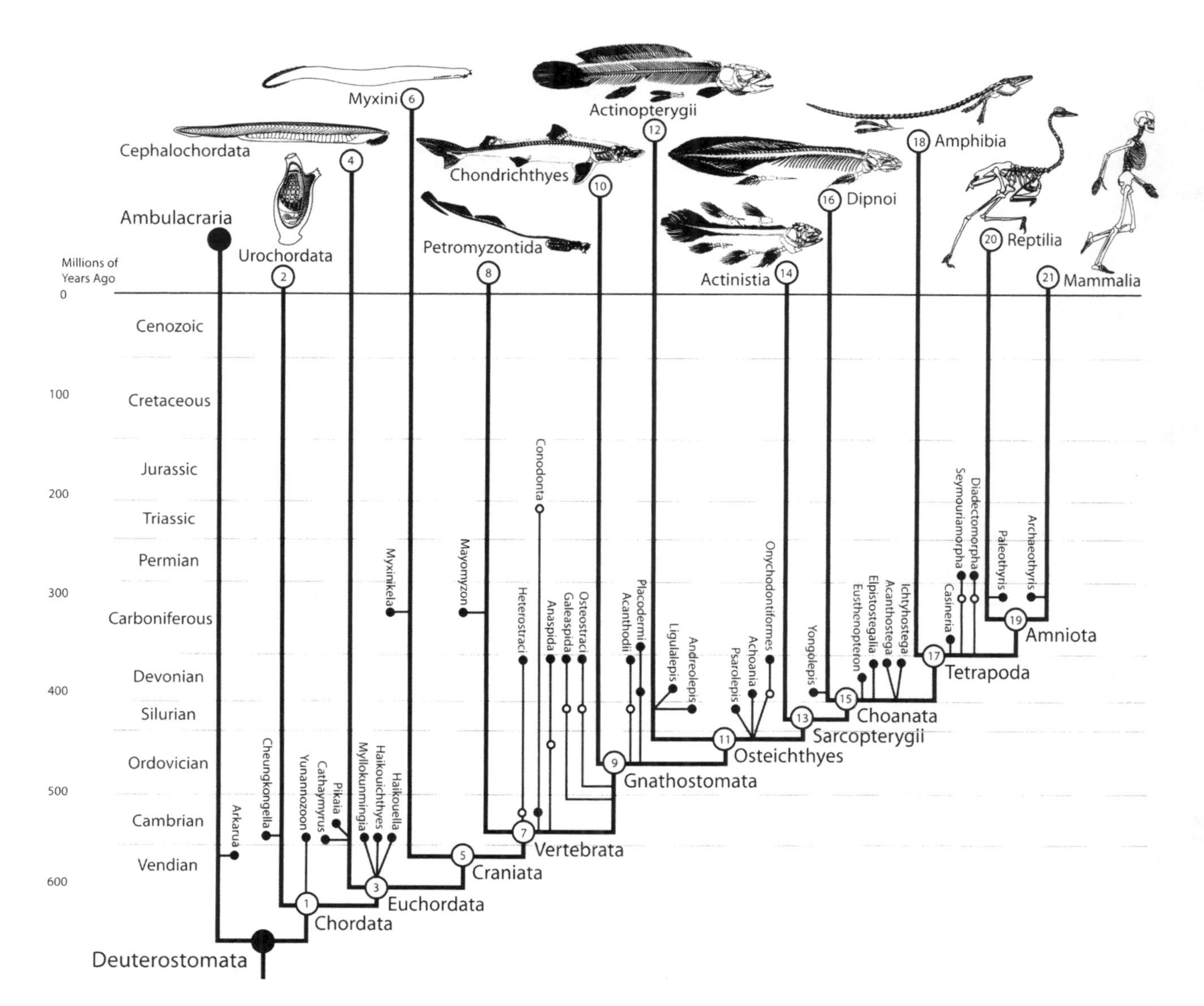

FIGURE 220. Timothy Bryan Rowe's phylogeny of chordates.

Rowe, 2004, "Chordate phylogeny and development," p. 385, fig. 23.1; courtesy of Timothy B. Rowe. Used with permission.

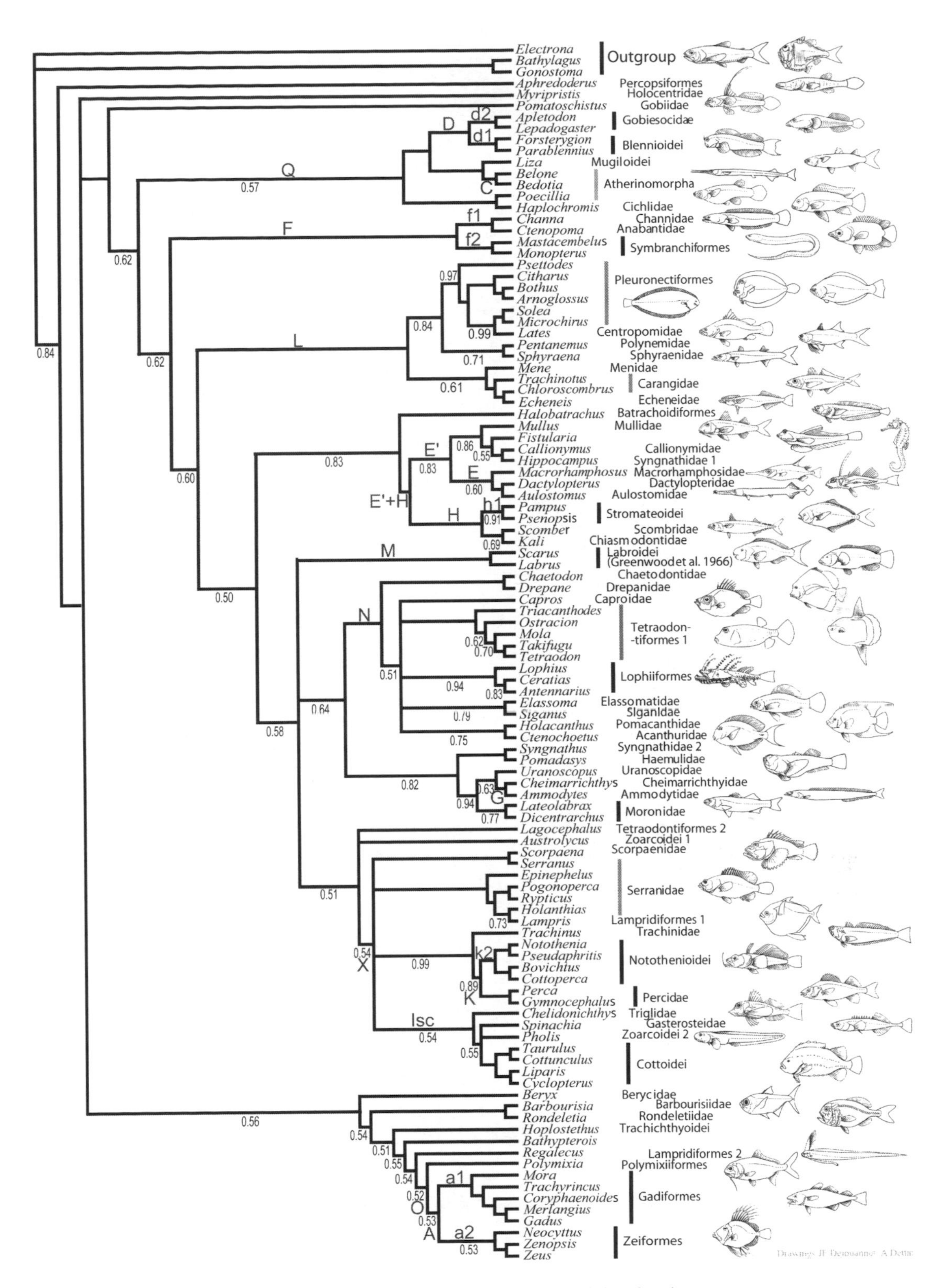

FIGURE 221. A molecular phylogeny of acanthomorph fishes by Agnès Dettai and Guillaume Lecointre.

Dettai and Lecointre, 2005, *C. R. Biologies,* 328:681, fig. 3; courtesy of Agnès Dettai and Guillaume Lecointre. Used with permission.

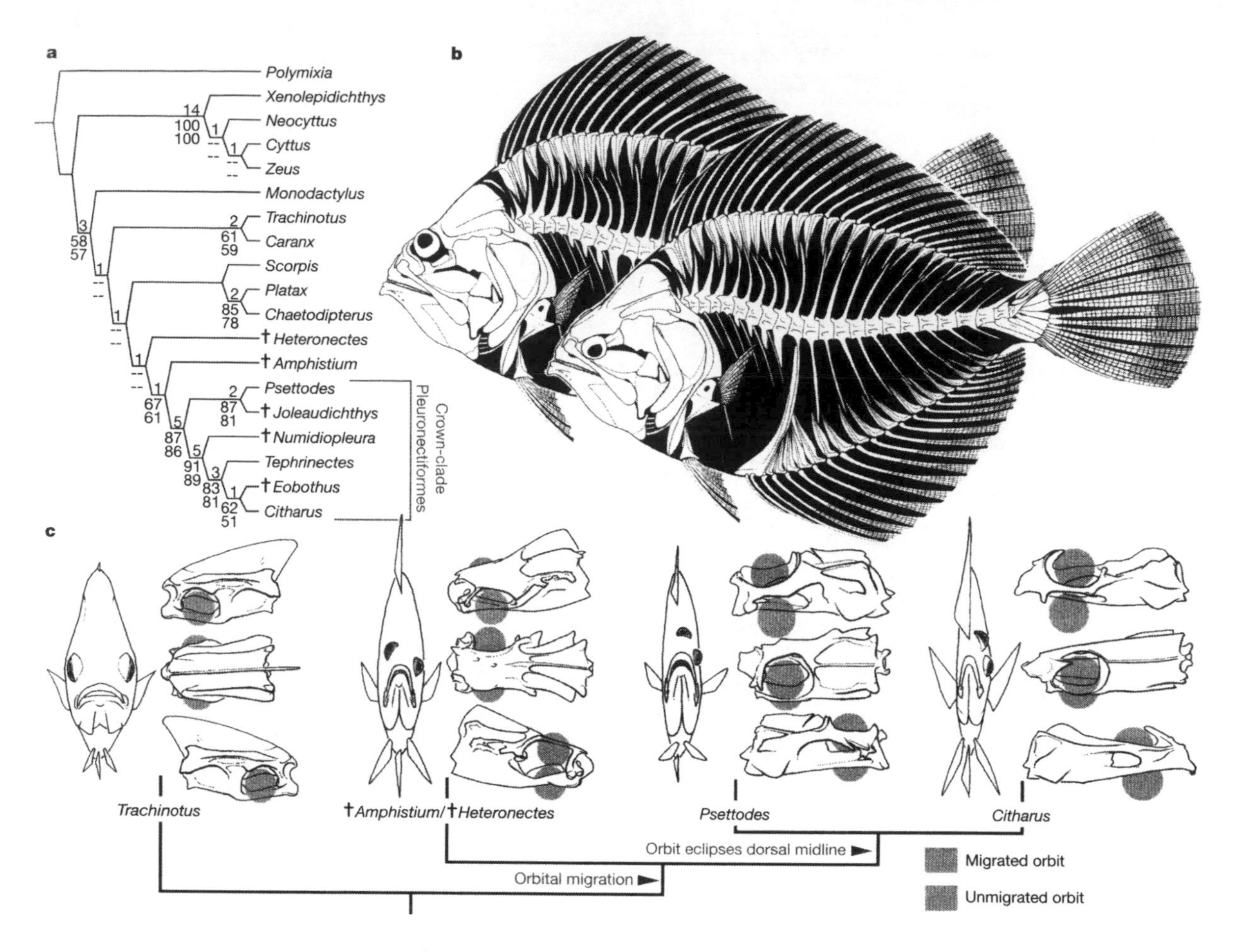

FIGURE 222. Phylogenetic placement of the Eocene genera *Heteronectes* and *Amphistium* and implications for the origin of cranial asymmetry in flatfishes, by Matthew Scott Friedman.

Friedman, 2008, *Nature,* 454:211, fig. 2; courtesy of Matthew S. Friedman and the Nature Publishing Group. Used with permission.

---

facing page

FIGURE 223. Two competing trees of relationships among hydrozoan cnidarians based on characters of the life cycle on the left and of colony shape on the right, by Lucas Leclère and coauthors.

Leclère et al., 2009, *Systematic Biology,* 58(5):517, fig. 4; courtesy of Lucas Leclère and the Society of Systematic Biologists. Used with permission.

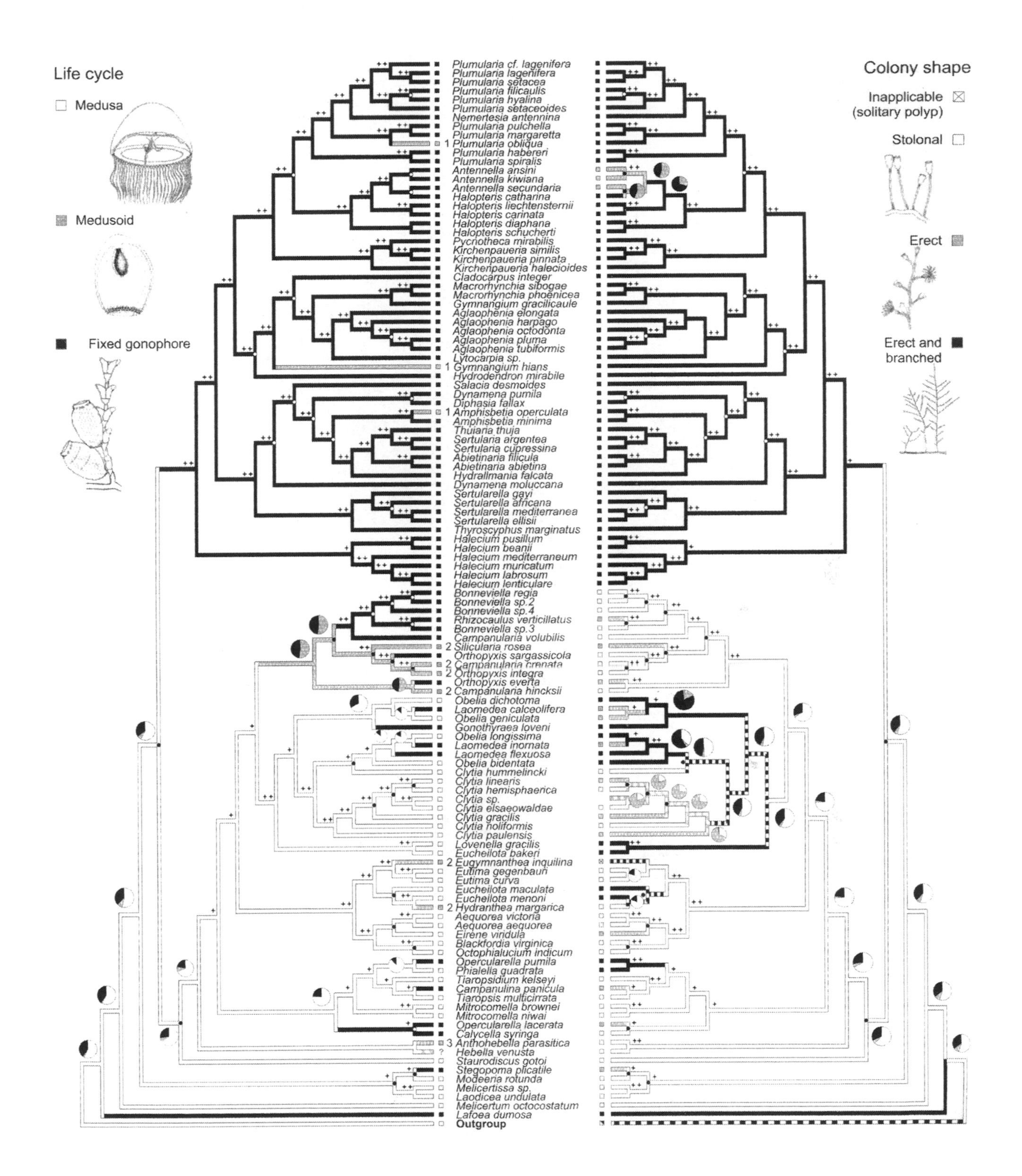

Life cycle
Medusa
Medusoid
Fixed gonophore
Colony shape
Inapplicable (solitary polyp)
Stolonal
Erect
Erect and branched
Plumularia cf. lagenifera
Plumularia lagenifera
Plumularia setacea
Plumularia filicaulis
Plumularia hyalina
Plumularia setaceoides
Nemertesia antennina
Plumularia pulchella
Plumularia margaretta
1 Plumularia obliqua
Plumularia habereri
Plumularia spiralis
Antennella ansini
Antennella kiwiana
Antennella secundaria
Halopteris catharina
Halopteris liechtensternii
Halopteris carinata
Halopteris diaphana
Halopteris schucherti
Pycnotheca mirabilis
Kirchenpaueria similis
Kirchenpaueria pinnata
Kirchenpaueria halecioides
Cladocarpus integer
Macrorhynchia sibogae
Macrorhynchia phoenicea
Gymnangium gracilicaule
Aglaophenia elongata
Aglaophenia harpago
Aglaophenia octodonta
Aglaophenia pluma
Aglaophenia tubiformis
Lytocarpia sp.
1 Gymnangium hians
Hydrodendron mirabile
Salacia desmoides
Dynamena pumila
Diphasia fallax
1 Amphisbetia operculata
Amphisbetia minima
Thuiaria thuja
Sertularia argentea
Sertularia cupressina
Abietinaria filicula
Abietinaria abietina
Hydrallmania falcata
Dynamena moluccana
Sertularella gayi
Sertularella africana
Sertularella mediterranea
Sertularella ellisii
Thyroscyphus marginatus
Halecium pusillum
Halecium beanii
Halecium mediterraneum
Halecium muricatum
Halecium labrosum
Halecium lenticulare
Bonneviella regia
Bonneviella sp.2
Bonneviella sp.4
Rhizocaulus verticillatus
Bonneviella sp.3
Campanularia volubilis
2 Silicularia rosea
Orthopyxis sargassicola
2 Campanularia crenata
2 Orthopyxis integra
Orthopyxis everta
2 Campanularia hincksii
Obelia dichotoma
Laomedea calceolifera
Obelia geniculata
Gonothyraea loveni
Obelia longissima
Laomedea inornata
Laomedea flexuosa
Obelia bidentata
Clytia hummelincki
Clytia linearis
Clytia hemisphaerica
Clytia sp.
Clytia elsaeowaldae
Clytia gracilis
Clytia noliformis
Clytia paulensis
Lovenella gracilis
Eucheilota bakeri
2 Eugymnanthea inquilina
Eutima gegenbauri
Eutima curva
Eucheilota maculata
Eucheilota menoni
2 Hydranthea margarica
Aequorea victoria
Aequorea aequorea
Eirene viridula
Blackfordia virginica
Octophialucium indicum
Opercularella pumila
Phialella quadrata
Tiaropsidium kelseyi
Campanulina panicula
Tiaropsis multicirrata
Mitrocomella brownei
Mitrocomella niwai
Opercularella lacerata
Calycella syringa
3 Anthohebella parasitica
? Hebella venusta
Staurodiscus gotoi
Stegopoma plicatile
Modeeria rotunda
Melicertissa sp.
Laodicea undulata
Melicertum octocostatum
Lafoea dumosa
Outgroup

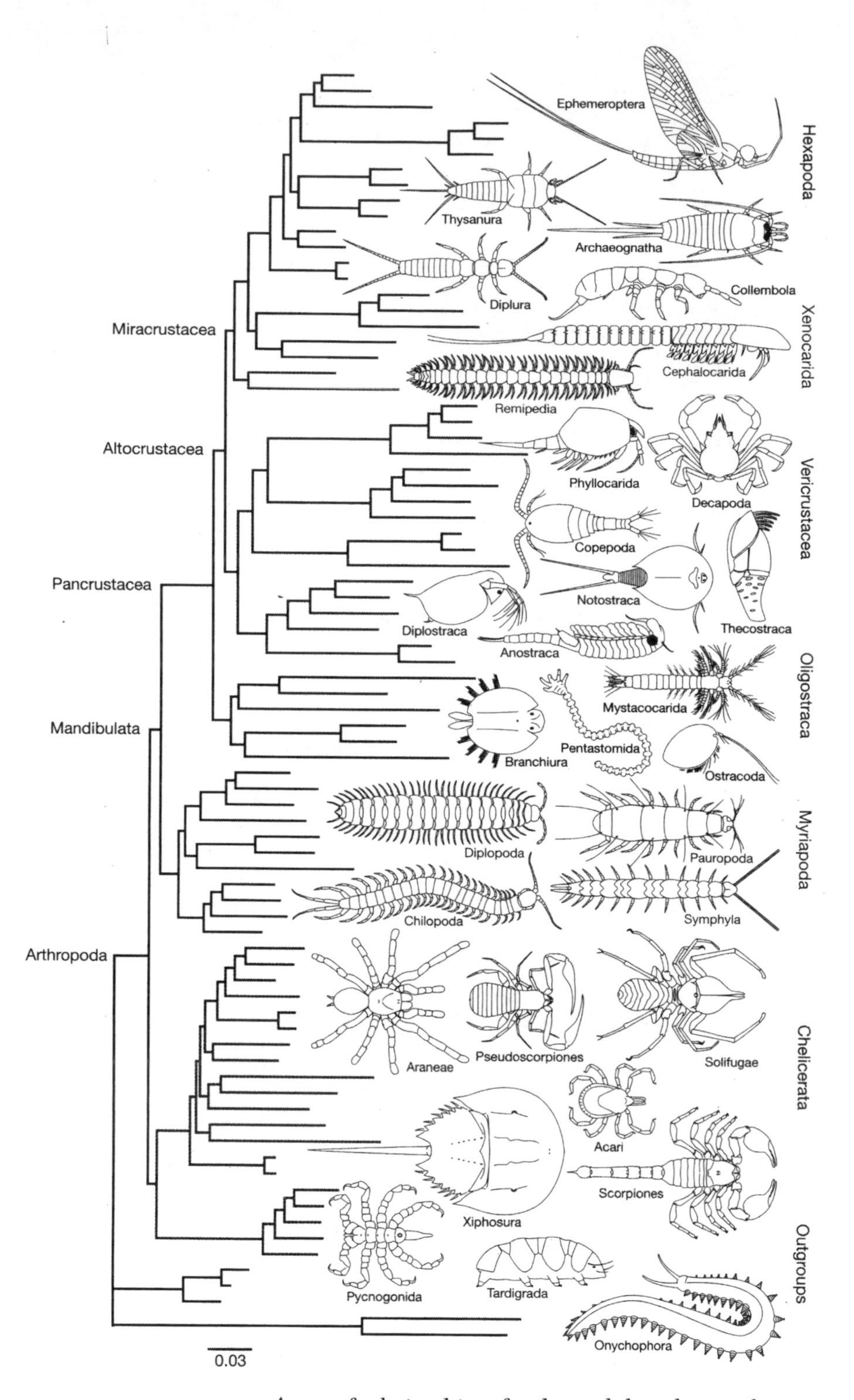

FIGURE 224. A tree of relationships of arthropods based on nuclear protein-coding genes, by Jerome Clifton Regier and colleagues. Line drawings of representatives of the major taxonomic groups emphasize the tremendous morphological diversity across the Arthropoda.

Regier et al., 2010, *Nature* 463:1081, fig. 2; courtesy of Jerome C. Regier and the Nature Publishing Group. Used with permission.

Primeval Branches and Universal Trees of Life

1997–2010

Following Carl Woese's 1977 discovery of the Archaea (see Figure 205) and the abandonment of Whittaker's five-kingdom system of life (see Figure 180), the stage was set for a new and closer look at the early branches of evolution. Important in that effort was molecular biologist Norman Richard Pace (1942–), famous for his 1997 "molecular view of microbial diversity"—the three phylogenetically distinct domains of life, a tree based on ribosomal RNA sequences (Figure 225). Pace demonstrated that the main diversity of life is microbial, distributed among three primary groups or domains: Archaea, single-celled microorganisms similar to Bacteria, but lacking internal organelles; Bacteria, also unicellular microorganisms, without organelles, but which differ, among other ways, in ribosomal RNA content and cell membrane structure; and Eucarya, organisms with cells that contain membrane-bound organelles such as a nucleus, mitochondria, and chloroplasts. Pace argued convincingly that microbial organisms are largely ignored, yet,

> the workings of the biosphere depend absolutely on the activities of the microbial world. Our texts articulate biodiversity in terms of large organisms: insects usually top the count of species. Yet, if we squeeze out any one of these insects and examine its contents under the microscope, we find hundreds or thousands of distinct microbial species. A handful of soil contains billions of microbial organisms, so many different types that accurate numbers remain unknown. At most only a few of these microbes would be known to us; only about 5000 non-eukaryotic organisms have been formally described (in contrast to the half-million described insect species). We know so little about microbial biology, despite it being a part of biology that looms so large in the sustenance of life on this planet.[1]

In 2004, Pace followed up his 1997 paper with more trees depicting the early branches of the tree of life, intended to summarize the general overall structure of the phylogenetic tree of life. Three of his branching diagrams, showing the primary subdivisions within each of the three phylogenetic domains of life, are reproduced here. The first is a diagrammatic view of the Bacteria, based on trees containing ribosomal RNA sequences of the indicated taxa, chosen to represent the broad diversity of Bacteria (Figure 226). The second and third diagrams represent similarly based phylogenies of the Archaea and Eucarya (Figures 227 and 228).

The three domains are nicely combined in a diagram published in 2004 by Sandra L. Baldauf and colleagues (Figure 229). Rooted near the base of the Bacteria, the tree is "our current best guess as to the composition and relationships among the major groups of living organisms, based on a large number of independent, partially overlapping studies."[2] A great tribute to the recent contributions of molecular systematics lies in the fact that the Bacteria, Archaea, and Eucarya are well defined and here to stay. Their integrity "is now confirmed by a tremendous body of data, including nearly 100 completely sequenced genomes. The identities of most of the major groups within these domains are also confirmed by many different data, both molecular and non-molecular."[3]

To finish off this celebration of trees of life through time, it seems most appropriate to devote the last page to David Mark Hillis's (1958–) universal tree of life (Figure 230). As described by Hillis and his colleagues, this radial branching diagram is based on an analysis of small subunit ribosomal RNA sequences sampled from about three thousand species from throughout the tree of life. The species were chosen based on their availability, but an attempt was made to include most of the major groups, sampled very roughly in proportion to the number of known species in each group—although, as Hillis points out, many groups remain over- or underrepresented. The number of species represented is approximately the square root of the number of species thought to exist on Earth (i.e., three thousand out of an estimated nine million species), or about 0.18% of the 1.9 million species that have been formally described and named.[4]

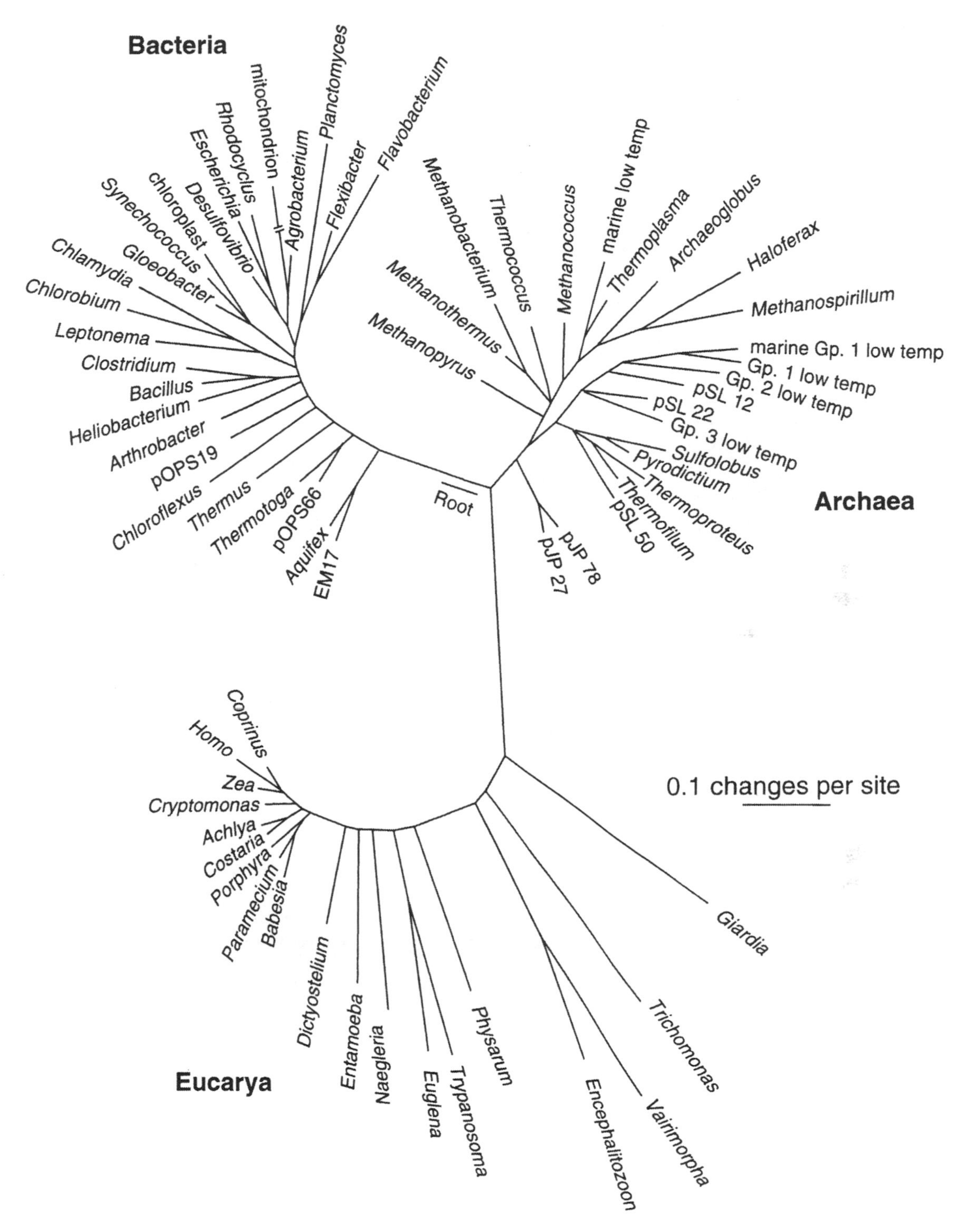

FIGURE 225. The three phylogenetically distinct domains of life, a tree based on ribosomal RNA sequences by Norman Richard Pace.

Pace, 1997, *Science*, 276:735, fig. 1; courtesy of Norman R. Pace. Used with permission.

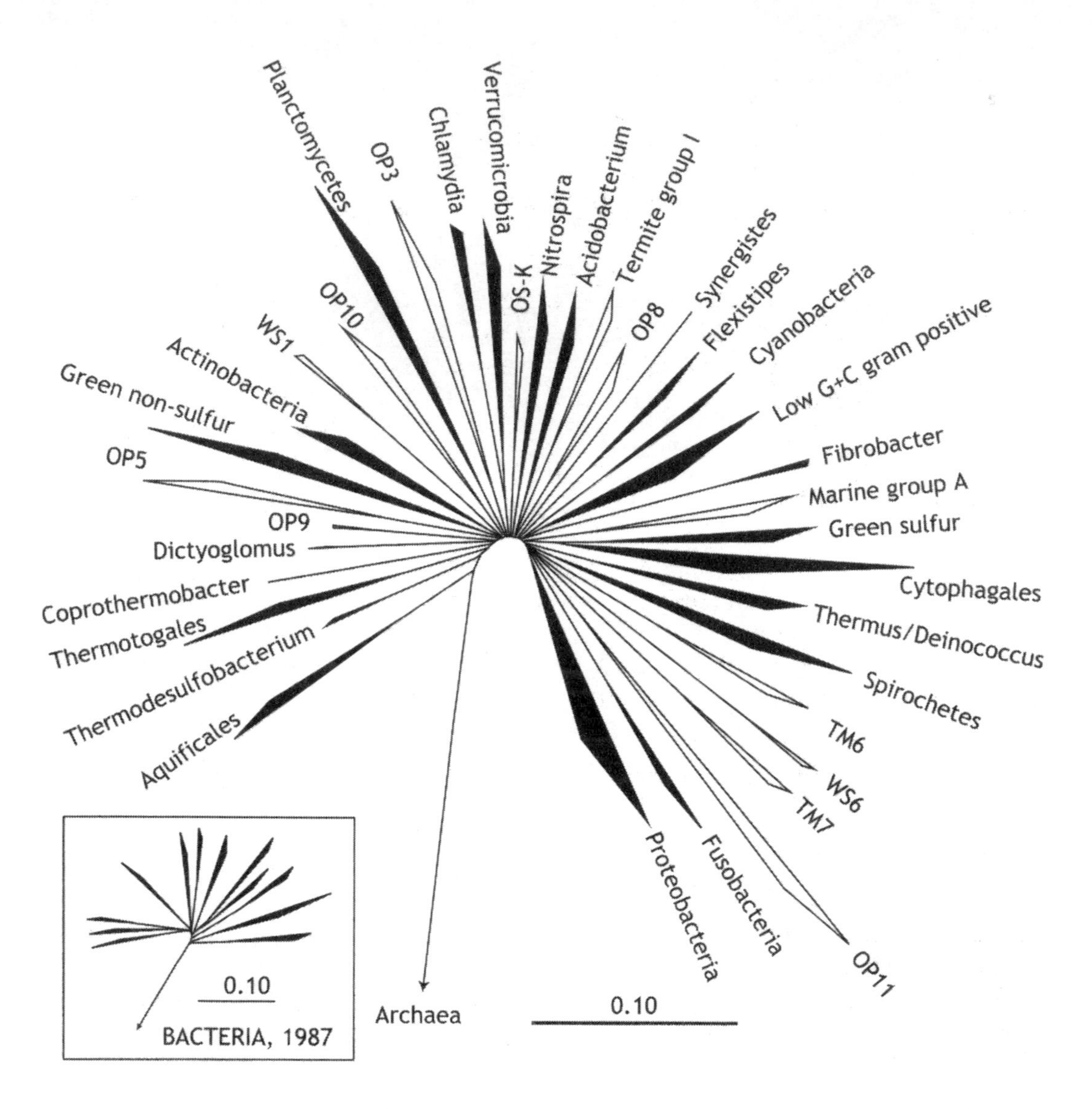

FIGURE 226. Phylogenetic divisions of the Bacteria, by Norman Richard Pace.

Pace, 2004, "The early branches in the tree of life," p. 80, fig. 5.2; courtesy of Norman R. Pace. Used with permission.

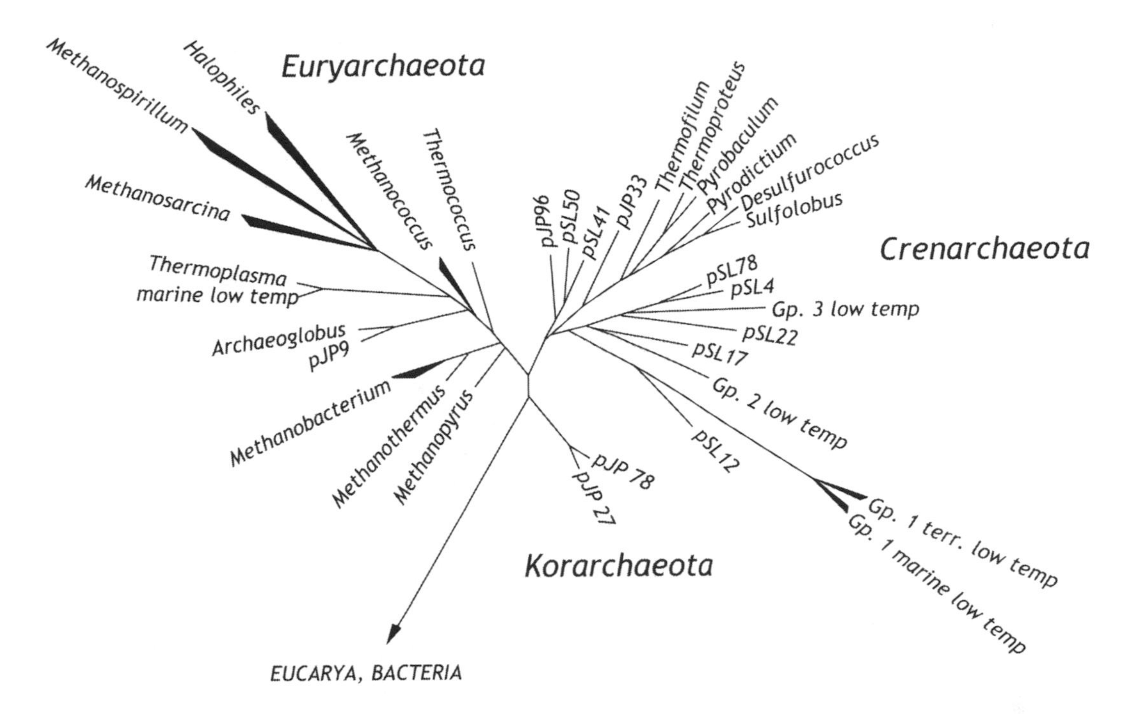

FIGURE 227. Phylogenetic divisions of the Archaea, by Norman Richard Pace.

Pace, 2004, "The early branches in the tree of life," p. 82, fig. 5.4; courtesy of Norman R. Pace. Used with permission.

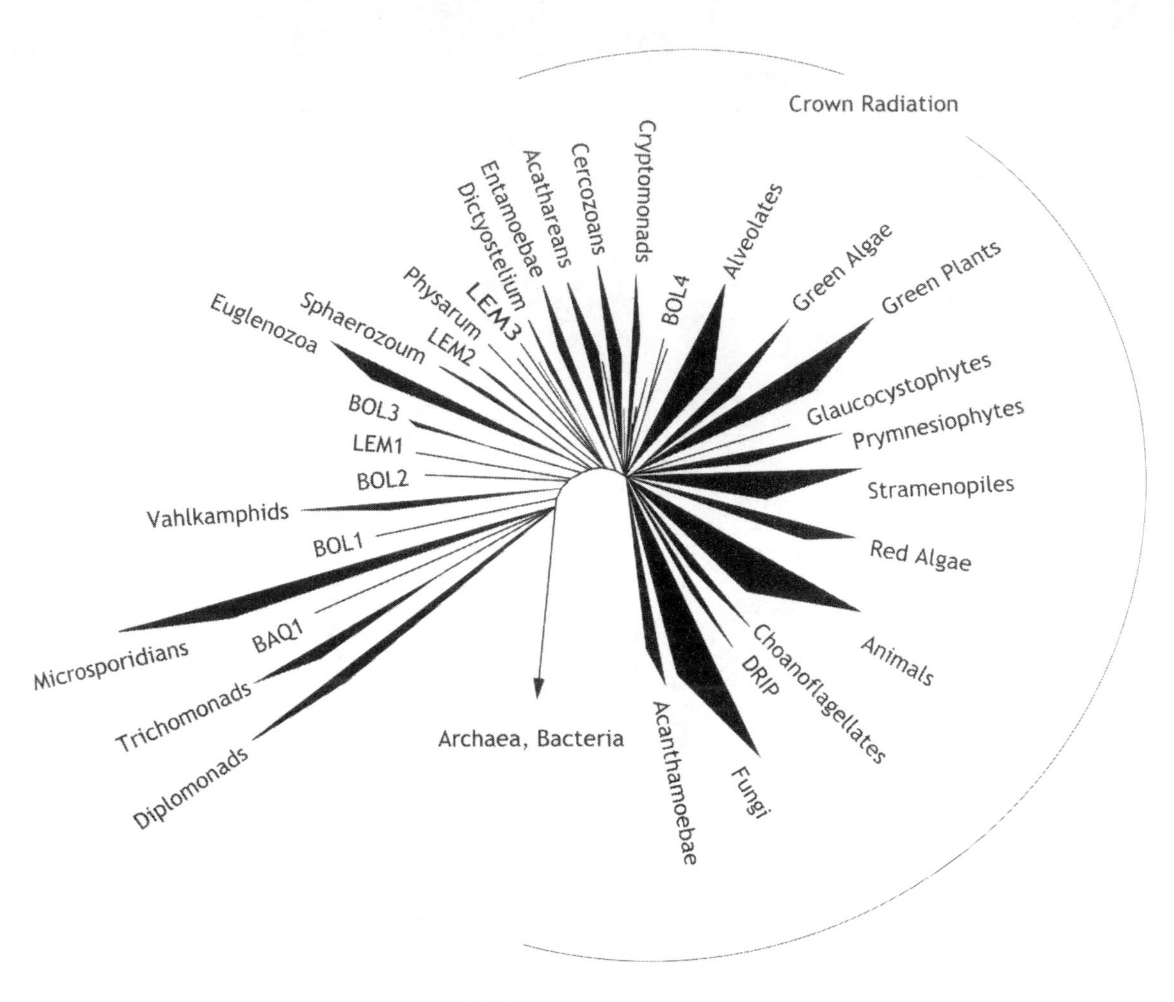

FIGURE 228. Phylogenetic divisions of the Eucarya, by Norman Richard Pace.

Pace, 2004, "The early branches in the tree of life," p. 83, fig. 5.5; courtesy of Norman R. Pace. Used with permission.

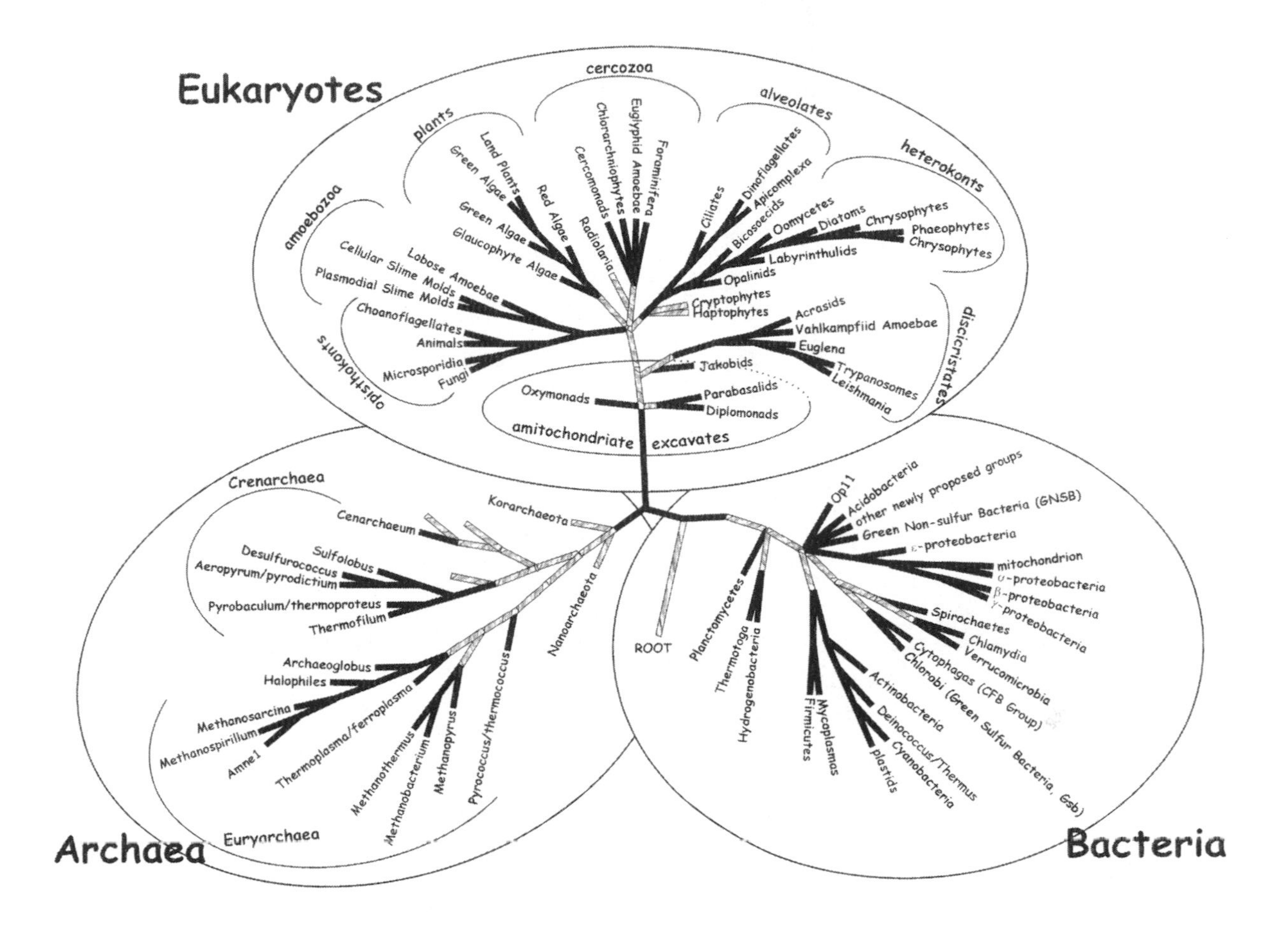

FIGURE 229. Major groups of life and their relationships to each other, by Sandra L. Baldauf and colleagues. Solid bars indicate groupings for which there is considerable molecular phylogenetic support. Shaded bars indicate tentative groupings with moderate, weak, or purely ultra-structural support.
Baldauf et al., 2004, "The tree of life: an overview," p. 45, fig. 4.1; courtesy of Sandra L. Baldauf. Used with permission.

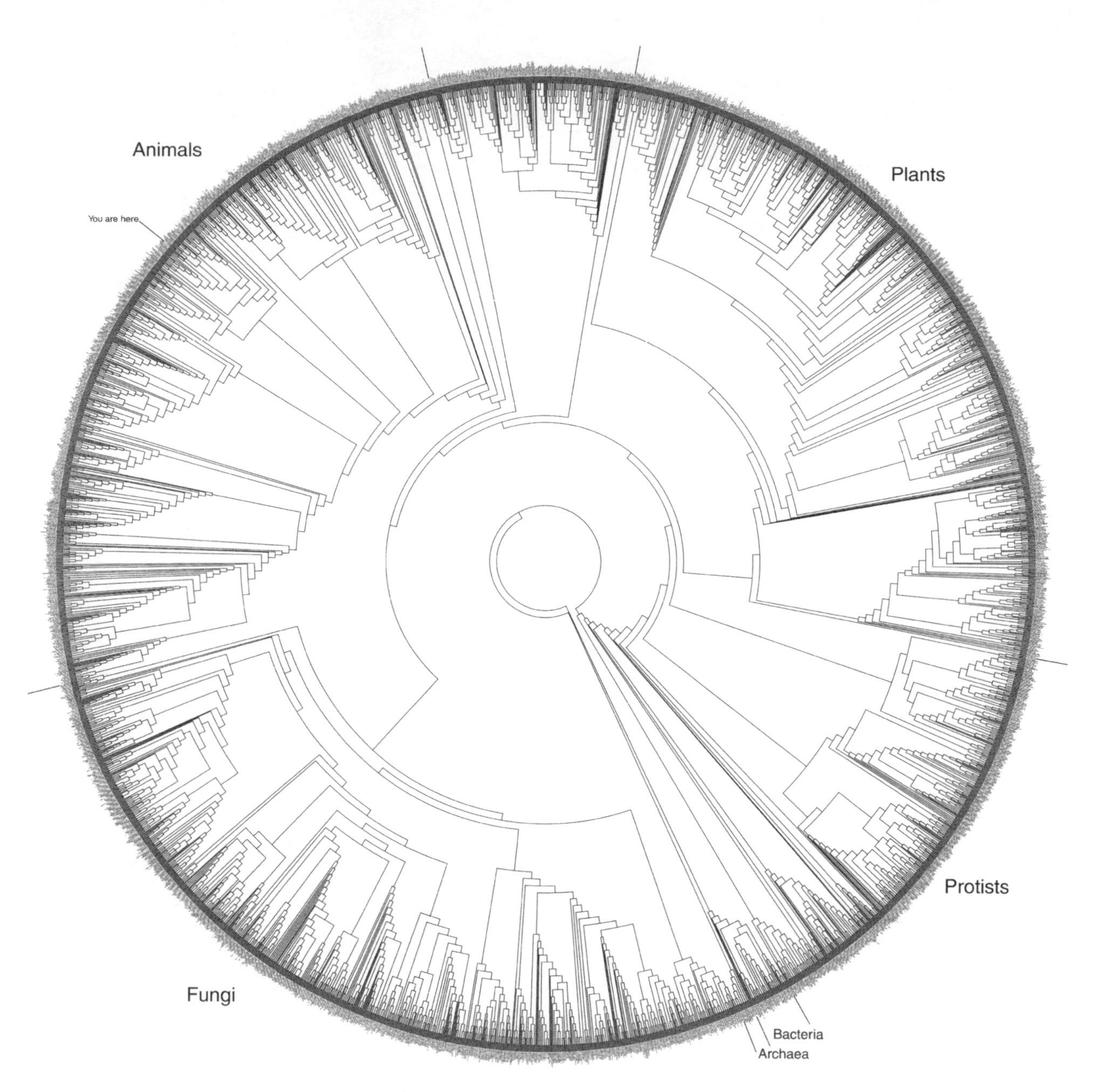

FIGURE 230. A universal tree of life based on ribosomal RNA sequences, sampled from about 3,000 species from throughout biodiversity, and constructed by David Mark Hillis and colleagues at the University of Texas at Austin.

Courtesy of David M. Hillis, Derrick J. Zwickl, Robin R. Gutell, University of Texas at Austin. Used with permission.

GLOSSARY

Terms are defined as they are used in biology and may have different meanings in other fields.

ADAPTIVE RADIATION: The evolutionary diversification of a monophyletic lineage (organisms derived from a single common ancestor), leading to a variety of forms each adapted to particular environmental conditions.

AFFINITY: A relationship or resemblance in structure between taxa that suggests a common origin.

ANALOGY, ANALOGOUS: Characters, or traits, in different taxa having the same or similar function but different evolutionary origins. Antonym of *homologous*.

APOMORPHY: In cladistic analysis, a derived character state, or trait. Antonym of *plesiomorphy*.

AREA-CLADOGRAM: A branching diagram that uses phylogenetic relationships to trace geographical changes, either dispersal events or barriers to dispersal.

ARTICULATA: Animals with different portions of their bodies composed of moveable parts that are connected (articulated) to each other. No longer considered scientifically valid, the group originally contained annelid worms, crustaceans, arachnids, and insects.

ASSEMBLAGE: A group of taxa.

BASAL: A term commonly used to describe a lineage of organisms that branches off near the root or base of a tree, but criticized by many scientists as nonsensical. Antonym of *crown*.

BIOGEOGRAPHY: The study of the geographic distribution of organisms.

CHARACTER: Any heritable attribute or feature of an organism (including morphological, behavioral, developmental, and molecular data) that can be used for recognizing, differentiating, or classifying a taxon. See also *character state.*

CHARACTER STATE: The state or value of an attribute or feature of an organism. For example, some character states for eye color are brown, green, and blue.

CHARACTER-STATE TREE: A branching diagram that shows the way character states have evolved but not necessarily the way the organisms that possess those character states have evolved.

CLADE: A group of organisms (or branch of a tree) containing all the descendents of a common ancestor. See also *monophyletic.*

CLADISTICS: A method of phylogenetic analysis by which groups containing all the descendents of a common ancestor are formed by shared-derived characters, or traits, first described in detail by Willi Hennig in the mid-twentieth century.

CLADOGRAM: A branching diagram showing the sequence of evolutionary divergence of organisms through time. See also *cladistics.*

CLASSIFICATION: A concise list of organisms grouped or ranked according to the pattern of branching seen in a branching diagram. Also the act or process of classifying.

CONVERGENT EVOLUTION: The acquisition of the same or similar biological trait in unrelated lineages.

CROWN: A term commonly used to describe a lineage of organisms that branches most distantly from the root or base of a tree, but criticized by many scientists as nonsensical. Antonym of *basal.*

DERIVED: Modified relative to a primitive condition, or trait. Antonym of *primitive.*

DICHOTOMOUS TREE: A branching diagram in which all branch points lead to only two immediate descendents, in contrast to a polytomous tree, which has three or more descendents at one or more branch point.

DICOTYLEDONS, DICOTYLEDONOUS: Flowering plants with seeds typically bearing two embryonic leaves, or cotyledons, in contrast to monocotyledons, which typically have only one embryonic leaf.

DIPHYLETIC: A group of organisms composed of two distinct phylogenetic lineages.

DNA-DNA HYBRIDIZATION: A molecular technique that measures the degree of genetic similarity between pools of DNA sequences, usually used to determine the genetic distance between two species. When several species are compared in this way, the similarity values allow the species to be arranged in a phylogenetic tree.

EVE HYPOTHESIS: The notion of a matrilineal most recent common ancestor, that is, the hypothetical woman ("Mitochondrial Eve") from whom all living humans today descend, on their mother's side, and through the mothers of those mothers and so on, back until all lines converge on one person.

HETEROCHRONY: A change in the timing of developmental events, leading to changes in size and shape of structural parts of an organism.

HOMOLOGY, HOMOLOGOUS: Characters, or traits, in different taxa that are structurally similar due to a common evolutionary origin. Antonym of *analogous*.

HOMOPLASY: Characters, or traits, that are similar due to convergent, parallel, or reversed evolution rather than shared ancestry.

INARTICULATA: A term no longer considered scientifically valid for animals with bodies not composed of interconnected moveable parts. Antonym of *Articulata*.

KEY: A device for identifying taxa using logical choices. At each point in the decision process, two or more alternatives are offered, each leading to a result or a further choice.

MITOCHONDRIAL DNA: The circular, double-stranded genome of eukaryotic mitochondria.

MOLECULAR CLOCK: Also known as gene clock or evolutionary clock. The use of molecular data (usually nucleotide or amino acid sequences) to calculate rates of molecular change and thus estimate the time in geological history when two taxa diverged.

MONOCOTYLEDONS, MONOCOTYLEDONOUS: Flowering plants with seeds typically bearing only a single embryonic leaf, or cotyledon, in contrast to dicotyledons, which typically have two embryonic leaves.

MONOPHYLY, MONOPHYLETIC: A natural group of organisms that includes all the descendents of a common ancestor. Antonym of *polyphyletic*. See also *clade*.

NODE: A branching point in a phylogenetic tree, representing the common ancestor of the lineages descending from that branching point.

NONHOMOLOGOUS: Characters, or traits, in different taxa that appear structurally similar due to convergent evolution rather than shared ancestry.

NUMERICAL TAXONOMY: The use of numerical methods in taxonomic or systematic classification. See also *phenetics*.

ONTOGENETIC: Pertaining to developmental change through the life cycle of an organism.

ONTOGENY: The development of an organism from its origin to its mature form and eventual death.

OUT-GROUP: A closely related taxon outside the study group used in cladistic analysis to provide information about the direction of character-state change. See also *character state; cladistics.*

PARALLELISM: When one or more characters, or traits, arise independently in two or more unrelated taxa.

PARAPHYLY, PARAPHYLETIC: A group of organisms that includes some, but not all, of the descendants of a common ancestor.

PARSIMONY: In cladistic analysis, the principle that determines the choice of hypotheses requiring the fewest *ad hoc* assumptions about character convergence, parallelism, and reversal.

PHENETICS: An approach to classifying organisms based on overall similarity, usually in morphology or other outwardly observable traits, regardless of their phylogenic or evolutionary relationships. This approach has been largely superseded by cladistics.

PHENOGRAM: A branching diagram used to visually represent the relationships of organisms by estimates of overall similarity among terminal taxa without reference to evolution.

PHYLOGENESIS: See *phylogeny.*

PHYLOGENETIC SYSTEMATICS: The field within biology that reconstructs evolutionary history and studies the patterns of relationships among organisms. See also *cladistics.*

PHYLOGENY: The evolutionary development and history of a taxon, or of a heritable attribute of an organism (including morphological, behavioral, developmental, and molecular data).

PHYLOGEOGRAPHY: The study of historical processes responsible for the present-day geographic distributions of organisms.

PLESIOMORPHY: In cladistic analysis, a primitive character state, or trait. Antonym of *apomorphy.*

POLYPHYLY, POLYPHYLETIC: A taxonomic group that does not contain the common ancestor of the members of the group and whose members therefore have two or more separate origins. Antonym of *monophyletic.*

PRIMITIVE: Unmodified relative to a derived condition. Antonym of *derived.*

RADIATA: Radially symmetrical animals, or animals that appear symmetrical on all sides, as opposed to bilaterally symmetrical animals in which the two sides are mirror images of each other. This term was much more broadly applied in the past but now encompasses the phyla Cnidaria (e.g., corals, anemones, jellies) and Ctenophora (comb jellies).

RETICULATE EVOLUTION: Evolution characterized by occasional hybridization and combination of two species.

REVERSAL: When a character, or trait, present in the ancestor of a taxon, disappears in its descendents, and then evolves again.

SYMBIOGENESIS: The merging of two separate organisms to form a single new organism.

SYMPLESIOMORPHY: In cladistic analysis, a shared primitive character state. Antonym of *synapomorphy*.

SYNAPOMORPHY: In cladistic analysis, a shared derived character state. Antonym of *symplesiomorphy*.

SYSTEMATICS: The study of biological diversity or, more specifically, the ordering of the diversity of nature through construction of a classification that can serve as a general reference system.

TAXON (PL. TAXA): Any of the formal categories used in classifying organisms.

TAXONOMY: That branch of biological sciences that deals with the discovery, recognition, definition, and naming of groups of organisms.

TRANSMUTATION OF SPECIES: A term first used by Jean-Baptiste Lamarck in 1809 to describe his theory of transformation of one species into another, and commonly used later for evolutionary ideas in the 19th century before Darwin published his *Origin of Species* in 1859.

NOTES

Preface

1 See Voss, 1952.
2 The map metaphor, as first articulated by Linnaeus: "All plants show affinities on all sides, like the territories on a geographical map" (Linnaeus, 1751:27; translation by Nelson and Platnick, 1981:95).
3 Nelson and Platnick, 1981:121.
4 German naturalist August Batsch writing in 1787 (pp. 296–298) seemed resigned to the problem: "It is very difficult to construct a natural system which is at once true and self-consistent; until now no one has succeeded in doing it. . . . Nature has her true and correct system; that we do not yet know it, or perhaps never will know it completely, does not prove, however, that we cannot approach it and therefore gain in our knowledge of the truth."

Introduction

1 Cook, 1974:6; see also Kuntz and Kuntz, 1987.
2 DeVarco and Clegg, 2010.
3 For example, see Craw, 1992:68.
4 Voss, 1952:17.
5 Bonnet, 1764:59; translated by Voss, 1952:16.
6 Pallas, 1766:23–24; translated by E. N. Genovese, in Archibald, 2009:563.
7 Buffon, 1766:335; translated by Voss, 1952:17.
8 Augier, 1801:2; translated by Stevens, 1983:206.

9 Lamarck, 1809:59; translated by Elliot, in Lamarck, 1914:37.
10 Wallace, 1855:187.
11 Darwin, 1859:129–130.
12 Lovejoy, 1936:88–92, 202
13 Yates, 1954:143.
14 Ragan, 2009:2.

Bracket and Tables, Circles and Maps, 1554–1872

1 For example, see Voss, 1952:3; Nelson and Platnick, 1981:73.
2 Greene, 1983:786, 787.
3 Adler, 1989:7; Attenborough et al., 2007:17.
4 Voss, 1952:8.
5 Wilkins, 1668:22.
6 Voss, 1952:6.
7 Jordan, 1905:405.
8 Linnaeus, 1751:27; see Nelson and Platnick, 1981:95.
9 Linnaeus, 1751.
10 Buffon, 1755:255; translated by Nelson and Platnick, 1981:95–96.
11 Ibid.
12 Roger, 1997:322–323, 414.
13 Ibid., 299–304; see also Ragan, 2009:10.

Early Botanical Networks and Trees, 1766–1815

1 Ragan, 2009:10.
2 Stevens, 1994:185.
3 See Lenoir, 1978:64, 67.
4 Staudt, 2003:18, 364; Burkhardt, 1995:79.
5 Ibid.
6 Duchesne, 1766:13–14; see also Duchesne, 1792:343; Staudt, 2003:18, 364.
7 Duchesne, 1766:220–221; Burkhardt, 1995:79.
8 Stevens, 1983:203.
9 Augier, 1801:2, translated by Stevens, 1983:206.
10 Augier, 1801:5.
11 Stevens, 1983:203, 210; Archibald, 2009:564.
12 Stevens, 1984:178, 179.

The First Evolutionary Tree, 1786–1820

1 Augier did not accept an evolutionary mechanism for his tree; in fact, several times he made reference to the Creator (1801:i, 6); see Stevens, 1983:210, 1984:178; Ruse, 1996:542 n.2.2; Archibald, 2009:564.
2 Ragan, 2009:8.
3 Mayr, 1972:61.
4 Lamarck's botanical and zoological series as presented in the article "Classes" in *Encyclopédie méthodique, Botanique,* 2:33, 1786; see also Burkhardt, 1995:56.
5 Lamarck, 1809:462, translation by Hugh Elliot, in Lamarck, 1914:178; see also Mayr, 1972:77; Gould, 1999b.
6 Burkhardt, 1995:162.
7 Burkhardt, 1995; Gould, 1999a, b; Ruse and Travis, 2009:932–934.

Diverse and Unusual Trees of the Early Nineteenth Century, 1817–1834

1 Nelson and Platnick, 1981:100.
2 Ragan, 2009:5.
3 Ibid.
4 See Buffon's 1755 dog genealogy.
5 Candolle, 1828b:11; translation by Nelson and Platnick, 1981:10.
6 Stevens, 1994:166.

The Rule of Five, 1819–1854

1 Hull, 1988:93–96; O'Hara, 1988:2747.
2 Ibid, 2749.
3 Gould, 1984:21; O'Hara, 1991:256.
4 Macleay, 1821:395.
5 Gould, 1984:14.
6 Vigors, 1824:509.
7 Swainson, 1835:129.
8 See Farber, 1985:56.
9 Coggon, 2002:26, 29.

Pre-Darwinian Branching Diagrams, 1828–1858

1 Ruse, 1996:123–125.
2 Ruse, 1996:111; Ospovat, 1981:11–12, 124–125; Voss, 2010:93.
3 Barry, 1837:121.

4 Voss, 2010:93.
5 Carpenter, 1841:196–197.
6 Chambers's "Vestiges" brought together various ideas of evolution and progressive transmutation of species governed by God-given laws in an accessible narrative that tied together numerous speculative scientific theories of the age. It was initially well received by polite Victorian society and became a bestseller, but its unorthodox themes contradicted the natural theology of the time and were reviled by orthodox clergymen and scientists, who readily found fault in its deficiencies (see Secord, 2000; Bowler, 2003).
7 Voss, 2010:97–98.
8 Ospovat, 1981:159.
9 Nelson and Platnick, 1981:118.
10 See Rosen et al., 1981:166–167.
11 Quinarianism was rejected by most naturalists by the early 1840s, but Irish clergyman William Hincks (1773/4–1871) continued to promote it and was teaching circular systems as late as 1870 (see Coggon, 2002:5, 7).
12 Strickland, 1841:189–190.
13 Wallace, 1856:195, 206, 212.
14 Ibid., 206; Coggon, 2002:31.
15 Ragan, 2009:14.
16 Wallace, 1856:206.
17 Wallace, 1855:187.
18 Archibald, 2009:561.
19 Ibid., 572–576.
20 Windsor, 1991:44, 196.
21 Agassiz, 1844:170.
22 Patterson, 1981:214.
23 Eldredge and Cracraft, 1980:149; Archibald, 2009:587. "Agassiz's example shows clearly that belief in evolution is not necessary for the production of such diagrams. . . . The information contained in these diagrams is therefore not necessarily concerned with evolution or phylogeny" (Patterson, 1977:580; see also Patterson, 1981).
24 Agassiz's 1844 "spindle" diagram foreshadows the trees made popular by Alfred Romer in the 1940s, 50s, and 60s; see Figures 132–140).
25 Also published by Agassiz and Gould, 1851.
26 Agassiz and Gould, 1848: frontispiece.
27 Archibald, 2009:567.
28 Ragan, 2009:17, fig. 20.
29 Gliboff, 2008:123–154.
30 Junker, 1995:271.

Evolution and the Trees of Charles Darwin, 1837–1868

1 Voss, 2010:61–63.
2 Darwin, 2008:180.
3 Voss, 2010:63.
4 Ibid., 108.
5 Ibid., 114.
6 Ibid., 114–115, fig. 32, Darwin Archive 10.2.26r–s.
7 Stauffer, 1975:236–237.
8 Darwin, in Stauffer, 1975:238–239.
9 Darwin, 1859:116–117.
10 Ibid., 116.
11 Darwin, 1887:341–344.
12 Ibid., 342.
13 Voss, 2010:181–182.
14 William C. L. Martin to Darwin, undated; Darwin Correspondence, 13:402. Available at www.darwinproject.ac.uk/home.
15 "As the larger ground-feeding birds seldom take flight except to escape danger, I believe that the nearly wingless condition of several birds, which now inhabit or have lately inhabited several oceanic islands, tenanted by no beast of prey, has been caused by disuse" (Darwin, 1859:134).
16 Darwin Correspondence, 13:402.

The Trees of Ernst Haeckel, 1866–1905

1 Dayrat, 2003:515, 517.
2 Gliboff, 2008:4, 156.
3 Ibid., 155–156.
4 Haeckel, vol. 2, 1876:52, 278.
5 Ragan, 2009:21.
6 "Man has developed gradually, and step by step, out of the lower Vertebrata, and more immediately out of Ape-like Mammals. That this doctrine is an inseparable part of the Theory of Descent . . . is recognized by all thoughtful adherents of the theory, as well as by all its opponents who reason logically" (Haeckel, vol. 2, 1876:263–264).
7 Darwin, 1871:199.
8 Haeckel, vol. 2, 1876:325–326.
9 Ibid., 326; see also, in this same volume (p. 308), Haeckel's "Systematic Survey of the 12 Species of Men and Their 36 Races."
10 Gliboff, 2008:157.
11 Richards, 2005:100; 2008.

12 No fossils of human ancestors were known at the time, other than some unexplained Neanderthal remains (see Berra, 2009:74).
13 Haeckel, 1868:492, 493; 1874:491.
14 Dubois, 1894.
15 Haeckel, 1910:543–544.

Post-Darwinian Nonconformists, 1868–1896

1 Warner, 1979:7; see also O'Hara, 1991:265.
2 Lewis, in Warner, 1979:666.
3 Stevens, 1994:236, 240.
4 Saville-Kent, vol. 1, 1880:35.
5 Ibid., 38; see also Corliss, 1959:180–182.
6 Stevens, 1994:240.
7 Ibid., 236.

More Late Nineteenth-Century Trees, 1874–1897

1 Stevens, 1994:240.
2 Lam, 1936:159.
3 It appears that Engler even had the shape of the cashew tree—typically a low, broad structure, with a short, often irregular shaped trunk, and long, bare, nearly horizontal primary limbs—in mind when he designed his diagram.
4 Sharpe, 1891:30–31, 37–43.
5 O'Hara, 1991:264, 265.
6 Walters, 2003:142.
7 O'Hara, 1988:2753.
8 Ibid., 2754.
9 Ibid.
10 Dobell, 1951:20–22.
11 Willmann, 2003:455.
12 Ibid.
13 Lankester, 1881:504; see Ruse, 1996:235–237.
14 As originally hypothesized by Dollo in 1893, his law states that "An organism is unable to return, even partially, to a previous stage already realized in the ranks of its ancestors" (Dollo, as quoted in Anonymous, 1970:1102).
15 Rosen et al., 1981:170.
16 Carroll, 1988:148–153; Pough et al., 2009:123–128, 198–200.
17 Lam, 1936:166–167.
18 Richard G. Olmstead, personal communication, 5 January 2011.

Trees of the Early Twentieth Century, 1901–1930

1 O'Hara, 1988:2754.
2 Mitchell, 1901:270.
3 Gaffney, 1984:289, 291.
4 Lam, 1936:157.
5 Sapp, 2009:116–118.
6 Margulis, 1970.
7 Goldthwait, 1936:568; Mitman, 1990:457.
8 Ibid.
9 Patten, 1924:635.
10 Lam, 1936:157–158.
11 Stevens, 1984:190, 191; Richard G. Olmstead, personal communication, 5 January 2011.
12 Windsor, 1991:228–231.
13 Eigenmann, 1917:47–48.
14 See also Small, 1922.
15 Small, 1919:201; 1922:128.
16 Small, 1919:221.
17 Olson, 1971:408–411.
18 Patterson, 1977:580.
19 Osborn, 1921:233.
20 Osborn, 1926, in Osborn, 1934a:212; see also Osborn, 1934b:178.
21 Osborn, 1933, in Osborn, 1934a:214; see also Osborn, 1934b:180.
22 Shoshani, 1998:480.
23 Hennig, 1950; see Moody, 1985:216.
24 Ibid., 221.
25 Patten, 1923:49.
26 Gould, 1994:433.
27 Ibid., 434.
28 Penland, 1924:63; Richard G. Olmstead, personal communication, 9 July 2010.
29 See Garrod's 1874 tree of parrot relationships, Figures 100 and 101. Although Camp (1923), in his classification of lizards, recognized four criteria by which characters states could be polarized, he did not explicitly suggest out-group comparison (Moody, 1985:216).
30 Penland, 1924:64.
31 Weiner, 2003.
32 Spencer, 1990.
33 Reproduced by Storer, 1943:194, 195.

The Trees of Alfred Sherwood Romer, 1933–1966

1 Romer, 1933a:16.
2 These paleontological trees of Romer appeared in his 1933 first edition of *Vertebrate Paleonotology* (1933b) but in less-finished form.

Additional Trees of the Mid-Twentieth Century, 1931–1943

1 Craw, 1992:74–75; Donoghue and Cracraft, 2004:1.
2 Garstang, 1931:241.
3 Mayr et al., 1953:175.
4 Knight, 1980:164.
5 Schaffner, 1934:134.
6 Tilden, 1935:40.
7 Lam, 1936:171.
8 Ibid., 173.
9 Ibid., 173–174.
10 Copeland, 1938:416.
11 Mayr et al., 1953:170.
12 Milne and Milne, 1939:541.
13 Raymond, 1939:309.
14 Ibid., 306.
15 Stirton, 1940:165; Mayr et al., 1953:172.
16 "Knowing animal behavior as I did, and being instructed in the methods of phylogenetic comparison as I was, I could not fail to discover that the very same methods of comparison, the same concepts of analogy and homology, are as applicable to characters of behavior as they are in those of morphology" (Lorenz, 1974:231).
17 Willmann, 2003:476; see also Williams and Ebach, 2008:63–64.
18 Lorenz, 1941:287–288.
19 Storer, 1943:2.
20 Ibid.

The Trees of William King Gregory, 1938–1951

1 Gregory, 1942, in Gregory, 1951:86.
2 Gregory, 1945, in Gregory, 1951:378, 468, 547.
3 Gregory, 1946, in Gregory, 1951:757.
4 Ibid., 758.
5 Gregory, 1948, in Gregory, 1951:998-999.
6 Gregory, 1946, in Gregory, 1951:718.
7 Originally published in less-finished form by Gregory, 1933:405.

Hints of New Approaches, 1954–1969

1 For a brief historical discussion of reticulate evolution as a result of hybridization, see Stevens, 1984:194–195.
2 Richard G. Olmstead, personal communication, 9 December 2010.
3 Good, 1956:87; see also Stevens, 1984:194–195.
4 Sneath and Sokal, 1973:260.
5 Whittaker, 1959:210; see also Whittaker, 1957:536–538.
6 See Sapp, 2009:107–111, 265.
7 Leppik, 1965:12.
8 Ibid., 10.
9 Greenwood et al., 1966:345.
10 Ibid., 346.

Phenograms and Cladograms, 1958–1966

1 Suárez-Díaz and Anaya-Muñoz, 2008:452; see also Hull, 1988:117.
2 Suárez-Díaz and Anaya-Muñoz, 2008:452.
3 Sneath, 1961:136–137.
4 Sokal and Michener, 1958:1409.
5 Joseph Felsenstein, personal communication, 12 January 2011.
6 Ibid.
7 Sneath and Sokal, 1973:260–262.
8 Moss, 1967:188–191.
9 Donoghue and Cracraft, 2004:1.
10 Nelson and Platnick, 1981:473–480; McCravy, 2008:485–486.

Early Molecular Trees, 1962–1987

1 It should not be forgotten, however, that molecular methods were being applied on a small scale at a much earlier date; for example, see Nuttall, 1904.
2 Suárez-Díaz and Anaya-Muñoz, 2008:452; see also Moritz and Hillis, 1996:3, 5–6.
3 Fitch and Margoliash, 1967:279.
4 Zuckerkandl and Pauling, 1962:200–201; see also Morgan, 1998:155, 164.
5 Felsenstein, 2004:132; see also Atchley, 2011:804.
6 See Cherfas and Gribbin, 1981:520; Gibbons, 2006:74–76.
7 John R. Grehan, personal communication, 13 August 2012.
8 In an interview of Morris Goodman conducted by Joel Hagen, 28

July 2004, in Detroit, Michigan (see: http://authors.library.caltech.edu/5456/1/hrst.mit.edu/hrs/evolution/public/goodman.html).

9 Sibley, 1994:87; Corbin and Brush, 1999:813; Schodde, 2000:75–76.
10 Ahlquist, 1999:858.
11 Patterson, 1987:2.
12 Woese and Fox, 1977; Woese et al., 1978; Woese, 1987.
13 The "mitochondrial Eve hypothesis"; see Templeton, 1993:51–72.
14 Cann, 1987:31.

Notable Trees of the Past Four Decades, 1970–2010

1 Reeder et al., 2002.
2 Barlow, 2000:2.
3 Berra, 2009:52–53, 56–57.
4 Darwin, 1851a, b; 1854a, b.
5 Ghiselin and Jaffe, 1973:137.
6 Ibid., 132.
7 Feduccia, 1977:19.
8 Donoghue, 2005:555.
9 Kemp, 1982:581.
10 Ibid., 584; see also Kemp, 1999:227, 229.
11 Kemp, 1999:219.
12 Ibid., 220.
13 Sereno, 1999:2137.
14 Paul, 2002:224
15 It is a rather confusing fact that modern birds evolved from "lizard-hipped" saurischians rather than "bird-hipped" ornithischians; see node 57 in Paul Sereno's tree shown in Figure 215.
16 Rowe, 2004:384.
17 Ibid.
18 Stiassny et al., 2004:419.
19 Friedman, 2008:209.
20 Leclère et al., 2009:509.
21 Regier et al., 2010:1079.
22 Ibid.

Primeval Branches and Universal Trees of Life, 1997–2010

1 Pace, 1997:734.
2 Baldauf et al., 2004:44.
3 Ibid.
4 Pennisi, 2003:1692–1697.

REFERENCES

Adler, K. 1989. Contributions to the history of herpetology. *Contributions to Herpetology,* no. 5. Society for the Study of Amphibians and Reptiles, Oxford, Ohio.

Agassiz, L. 1844. *Recherches sur les poissons fossiles.* Published by the author, printed in Neuchâtel, Switzerland, vol. 1.

Agassiz, L., and A. A. Gould. 1848. *Principles of Zoology: Touching the Structure, Development, Distribution and Natural Arrangement of the Races of Animals, Living and Extinct; with Numerous Illustrations. For the Use Schools and Colleges.* Part 1, *Comparative Physiology.* Gould, Kendall, and Lincoln, Boston.

Agassiz, L., and A. A. Gould. 1851. *Outlines of Comparative Physiology: Touching the Structure and Development of the Races of Animals, Living and Extinct. For the Use Schools and Colleges.* H. G. Bonn, London.

Ahlquist, J. E. 1999. Charles G. Sibley: A commentary on 30 years of collaboration. *The Auk,* 116(3):856–860.

Anonymous. 1970. Evolution: Ammonites indicate reversal. *Nature,* 225:1101–1102.

Archibald, J. D. 2009. Edward Hitchcock's pre-Darwinian (1840) "tree of life." *Journal of the History of Biology,* 42:561–592.

Atchley, W. R. 2011. Retrospective: Walter M. Fitch (1929–2011). *Science,* 332:804.

Attenborough, D., S. Owens, M. Clayton, and R. Alexandratos. 2007. *Amazing Rare Things: The Art of Natural History in the Age of Discovery.* Yale University Press, New Haven.

Augier, A. 1801. *Essai d'une nouvelle classification des végétaux.* Bruyset Ainé, Lyon, France.

Baer, K. E. von. 1828. *Über Entwicklungsgeschichte der Thiere: Beobachtung und Reflexion.* 2 vols. Gebrüder Bornträger, Königsberg, Germany.

Baldauf, S. L., D. Bhattacharya, J. Cockrill, P. Hugenholtz, J. Pawlowski, and A. G. B. Simpson. 2004. The tree of life: An overview. Pp. 43–75, In: J. Cracraft and M. J. Donoghue (editors), *Assembling the Tree of Life.* Oxford University Press, Oxford.

Barlow, G. W. 2000. *The Cichlid Fishes: Nature's Grand Experiment in Evolution.* Perseus Publishing, Cambridge, Massachusetts.

Barry, M. 1837. Further observations on the unity of structure in the animal kingdom, and on congenital anomalies, including "hermaphrodites"; with some remarks on embryology, as facilitating animal nomenclature, classification, and the study of comparative anatomy. *Edinburgh New Philosophical Journal,* 22:345–364.

Batsch, A. J. G. C. 1787. *Versuch einer Anleitung zur Kenntniss und Geschichte der Pflanzen, für academische Vorlesungen entworfen und mit den nöthigsten Abbildungen versehen,* vol. 1. Gebauer, Halle, Germany.

Batsch, A. J. G. C. 1802. *Tabula affinitatum regni vegetabilis.* Landes-Industrie-Comptoir, Weimar, Germany.

Bennett, A. W. 1887. On the affinities and classification of algae. *Journal of the Linnean Society of London, Botany,* 24(158):49–61.

Bentham, G. 1873. Notes on the classification, history, and geographical distribution of Compositae. *Journal of the Linnean Society of London,* 13:335–577.

Berra, T. M. 2009. *Charles Darwin: The Concise Story of an Extraordinary Man.* Johns Hopkins University Press, Baltimore.

Bessey, C. E. 1897. The phylogeny and taxonomy of the angiosperms. *Botanical Gazette,* 24:145–178.

Bessey, C. E. 1915. The phylogenetic taxonomy of flowering plants. *Annals of the Missouri Botanical Garden,* 2:109–164.

Blainville, H. M. D. de. 1822. *De l'organisation des animaux, ou principes d'anatomie comparée.* F. G. Levrault, Paris.

Bonnet, C. 1764. *Contemplation de la Nature.* Marc-Michel Rey, Amsterdam, 2 vols.

Bovelles, C. de. 1512. *Physicorum elementorum libri decem denis capitibus distincti, quae capita denis sunt propositionibus exornata, unde libri sunt decem, capita centum, propositiones mille.* In aedibus Ioannis Parui & Iodoci Badii Ascensii, Paris.

Bowler, P. J. 2003. *Evolution: The History of an Idea,* 3rd edition. University of California Press, Berkeley and Los Angeles.

Bronn, H. G. 1858. *Untersuchungen über die Entwickelungs-Gesetze der organischen Welt während der Bildungs-Zeit unserer Erd-Oberfläche: eine von der Französischen Akademie im Jahre 1857 gekrönte Preisschrift*. E. Schweizerbart'sche Verlagshandlung und Druckerei, Stuttgart, Germany.

Brundin, L. 1966. *Transantarctic Relationships and Their Significance, as Evidenced by Chironomid Midges, with a Monograph of the Subfamilies Podonominae and Aphroteniinae and the Austral Heptagyiae*. Almqvist & Wiksell, Stockholm.

Buffon, G. L. L., comte de. 1755. *Histoire naturelle, générale et particulière, avec la description du cabinet du Roy*, vols. 5 and 14. Imprimerie Royale, Paris.

Burkhardt, R. W., Jr. 1995. *The Spirit of System: Lamarck and Evolutionary Biology*. Harvard University Press, Cambridge.

Bütschli, J. A. O. 1876. Untersuchungen über freilebende Nematoden und die Gattung *Chaetonotus*. *Zeitschrift für Wissenschaftliche Zoologie*, 26:363–413.

Caesius, F. 1651. Phytosophicae tabulae. Pp. 901–952, In: F. Hernández (editor), *Rerum medicarum Novae Hispaniae thesaurus seu Nova plantarum, animalium et mineralium mexicanorum historia*. Mascardi, Rome.

Camp, C. L. 1923. Classification of the lizards. *Bulletin of the American Museum of Natural History*, 48(11):289–481.

Candolle, A.-P. de. 1827. *Mémoire sur la famille des Légumineuses*. A. Belin, Paris.

Candolle, A.-P. de. 1828a. *Mémoire sur la famille des Mélastomacées*. Treuttel and Würtz, Paris.

Candolle, A.-P. de. 1828b. *Mémoire sur la famille des Crassulacées*. Treuttel and Würtz, Paris.

Cann, R. L., M. Stoneking, and A. C. Wilson. 1987. Mitochondrial DNA and human evolution. *Nature*, 325:31–36.

Carpenter, W. B. 1841. *Principles of General and Comparative Physiology*, 2nd edition. John Churchill, London.

Carroll, R. L. 1988. *Vertebrate Paleontology and Evolution*. W. H. Freeman, New York.

Cavalli-Sforza, L. L., and A. W. F. Edwards. 1965. Analysis of human evolution. Pp. 923–933, In: S. J. Geerts (editor), *Genetics Today, Proceedings of the XI International Congress of Genetics, The Hague, The Netherlands, September 1963*, vol. 3. Pergamon Press, Oxford.

[Chambers, R.] 1844. *Vestiges of the Natural History of Creation*. Churchill, London.

Cherfas, J., and J. Gribbin. 1981. The molecular making of man: The DNA of living species tells a story about the origin and evolution of humans that is very different from the conventional interpretation of the fossil record. *New Scientist,* 91(1268):518–521.

China, W. E. 1933. A new family of Hemiptera-Heteroptera, with notes on the phylogeny of the suborder. *Annals and Magazine of Natural History,* series 1, 12:180–196.

Coggon, J. 2002. Quinarianism after Darwin's *Origin*: The circular system of William Hincks. *Journal of the History of Biology,* 35:5–42.

Cook, R. 1974. *The Tree of Life: Image for the Cosmos.* Avon, New York.

Copeland, H. F. 1938. The kingdoms of organisms. *Quarterly Review of Biology,* 13(4):383–420.

Corbin, K. W., and A. H. Brush. 1999. In memoriam: Charles Gald Sibley, 1917–1998. *The Auk,* 116(3):806–814.

Corliss, J. O. 1959. Comments on the systematics and phylogeny of the Protozoa. *Systematic Zoology,* 8(4):169–190.

Craw, R. 1992. Margins of cladistics: Identity, difference, and place in the emergence of phylogenetic systematics, 1864–1975. Pp. 65–107, In: P. Griffiths (editor), *Trees of Life: Essays in Philosophy of Biology.* Kluwer Academic Publishers, Dordrecht, the Netherlands.

Czelusniak, J., M. Goodman, D. Hewett-Emmett, M. L. Weiss, P. J. Venta, and R. E. Tashian. 1982. Phylogenetic origins and adaptive evolution of avian and mammalian haemoglobin genes. *Nature* 298(5871): 297–300.

Darwin, C. 1851a. *A Monograph of the Fossil Lepadidae; or, Pedunculated Cirripedes of Great Britain.* Palaeontographical Society, London.

Darwin, C. 1851b. *A Monograph of the Sub-class Cirripedia, with Figures of all the Species. The Lepadidae; or, Pedunculated Cirripedes.* Ray Society, London.

Darwin, C. 1854a.*The Balanidae (or Sessile Cirripedes); the Verrucidae, &c.* Ray Society, London.

Darwin, C. 1854b. *A Monograph of the Fossil Balanidae and Verrucidae of Great Britain.* Palaeontographical Society, London.

Darwin, C. 1859. *On the Origin of Species by Means of Natural Selection, or the Preservation of Favoured Races in the Struggle for Life.* John Murray, London. [Facsimile, 1964, Harvard University Press, Cambridge, Massachusetts.]

Darwin, C. 1860. *Über die Entstehung der Arten im Thier- und Pflanzen-Reich durch natürliche Züchtung, oder Erhaltung der vervollkommneten Rassen im Kampfe um's Daseyn.* Translated by Heinrich G. Bronn. E. Schweizerbart, Stuttgart, Germany.

Darwin, C. 1871. *The Descent of Man and Selection in Relation to Sex.* 2 vols., D. Appleton and Co., New York.

Darwin, C. 1887. *The Life and Letters of Charles Darwin, Including an*

Autobiographical Chapter. Edited by his son Francis Darwin, in three volumes. D. Appleton and Co., New York.

Darwin, C. 2008. *Charles Darwin's Notebooks, 1836–1844: Geology, Transmutation of Species, Metaphysical Enquiries.* Transcribed and edited by P. H. Barrett, P. J. Gautrey, S. Herbert, D. Kohn, and S. Smith. Natural History Museum, London; Cambridge University Press, Cambridge.

Dayrat, B. 2003. The roots of phylogeny: How did Haeckel build his trees? *Systematic Biology,* 52(4):515–527.

Dettai, A., and G. Lecointre. 2005. Further support for the clades obtained by multiple molecular phylogenies in the acanthomorph bush. *C. R. Biologies,* 328:674–689.

DeVarco, B., and E. Clegg, 2010. *ReVisioning trees. Shape of Thought: An Introduction to the Emergent and Ancient Art of Visual Communication,* 27 July 2010, http://shapeofthought.typepad.com/shape_of_thought/revisioning-trees

Dobell, C. 1951. In memoriam: Otto Bütschli (1848–1920) "architect of protozoology." *Isis,* 42(1):20–22.

Dollo, L. 1896. Sur la phylogénie des dipneustes. *Bulletin de la Societe Belge de Geologie, Paleontology et d'Hydrologie,* 9(2):79–128.

Donoghue, M. J., and J. Cracraft. 2004. Introduction: Charting the tree of life. Pp. 1–4, In: J. Cracraft and M. J. Donoghue (editors), *Assembling the Tree of Life.* Oxford University Press, Oxford.

Donoghue, P. C. J. 2005. Matters of the record: Saving the stem group—a contradiction in terms? *Paleobiology,* 31(4):553–558.

Dubois, E. 1894. Pithecanthropus erectus, *eine menschenähnliche Übergangsform aus Java.* Landesdruckerei, Batavia, the Netherlands.

Duchesne, A.-N. 1766. *Histoire naturelle des fraisiers.* Didot le Jeune et Panckoucke, Paris.

Duchesne, A.-N. 1792. Sur le fraisier de Versailles. *Journal d'histoire naturelle,* 2:343–347.

Dunal, M.-F. 1817. *Monographie de la famille des Anonacées.* Treuttel and Würtz, Paris.

Eichwald, C. E. von. 1821. *De regni animalis limitibus atque evolutionis gradibus. Specimen quod consentiente amplissimo philosophorum ordine Univers. Caes. Dorpat. Ut veniam legendi rite sibi acquibat mens. Octobr. Publicae disceptationi submittit.* Joannis Christian Schünmann, Dorpat [Tartu, Estonia].

Eichwald, C. E. von. 1829. *Zoologia specialis quam expositis animalibus tum vivis, tum fossilibus potissimum Rossiae in universum,* et *Poloniae in species, in usum lectionum publicarum in Universitate Caesarea Vilnensi habendarum. Pars prior. Propaedeuticam zoologiae atque specialem He-*

terozoorum expositionem continens. Josephus Zawadzki, Vilnae [Vilnius, Lithuania].

Eigenmann, C. H. 1917. The American Characidae. *Memoirs of the Museum of Comparative Zoology,* Harvard University, 43(1):3–102.

Eldredge, N., and J. Cracraft. 1980. *Phylogenetic Patterns and the Evolutionary Process: Method and Theory in Comparative Biology.* Columbia University Press, New York.

Engler, H. G. A. 1874. Über Begrenzung und systematische Stellung der natürliche Familie der Ochnaceae. *Nova Acta Physico-Medica Academiae Caesareae Leopoldino-Carolinae Germanicae Naturae Curiosorum,* 37(2):1–28.

Engler, H. G. A. 1881. Über die morphologische Verhältnisse und die geographische Verbreitung der Gattung *Rhus* wie der mit ihr verwandten, lebenden und ausgestorbenen Anacardiaceen. *Botanische Jahrbücher für Systematik, Pflanzengeschichte und Pflanzengeographie,* 1:364–426.

Farber, P. L. 1985. Aspiring naturalists and their frustrations: The case of William Swainson (1789–1855). Pp. 51–59, In: A. Wheeler and J. H. Price (editors), *From Linnaeus to Darwin: Commentaries on the History of Biology and Geology,* Papers from the Fifth Easter Meeting of the Society for the History of Natural History, 28–31 March 1983, Natural History in the Early Nineteenth Century. *Society for the History of Natural History, Special Publication,* no. 3.

Feduccia, A. 1977. A model for the evolution of perching birds. *Systematic Zoology,* 26(1):19–31.

Felsenstein, J. 2004. *Inferring Phylogenies.* Sinauer Associates, Sunderland, Massachusetts.

Fitch, W. M., and M. Margoliash. 1967. Construction of phylogenetic trees. *Science,* 155(3760):279–284.

Friedman, M. 2008. The evolutionary origin of flatfish asymmetry. *Nature,* 454:209–212.

Fryer, G., and T. D. Iles. 1972. *The Cichlid Fishes of the Great Lakes of Africa: Their Biology and Evolution.* Oliver and Boyd, Edinburgh.

Fürbringer, M. 1888. *Untersuchungen zur Morphologie und Systematik der Vögel, zugleich ein Beitrag zur Anatomie der Stütz- und Bewegungsorgane.* 2 vols. T. J. Van Holkema, Amsterdam.

Gaffney, E. S. 1984. Historical analysis of theories of chelonian relationship. *Systematic Zoology,* 33(3):283–301.

Garrod, A. H. 1874. On some points in the anatomy of the parrots which bear on the classification of the suborder. *Proceedings of the Zoological Society of London,* 42(1):586–598.

Garstang, W. 1931. The phyletic classification of Teleostei. *Proceedings of the Leeds Philosophical Society* (*Scientific Section*), 2(5):240–260.

Garstang, W. 1951. *Larval Forms, and Other Zoological Verses.* Blackwell, Oxford.

Gemma, C. 1576. [Classification of orchids]. Pp. 94–95, In: M. de Lobel, *Plantarum seu stirpium historia. . . . Cui annexum est adversariorum volumen.* Plantini, Antwerpiae [Antwerp].

Gessner, C. 1551–1587. *Historiae animalium.* Five books in three folio volumes: Vol. 1, Liber I. *De quadrupedibus viviparis,* 1551; Liber II. *De quadrupedibus oviparis,* 1554. Vol. 2, Liber III. *De avium natura,* 1555. Vol. 3, Liber IV. *De piscium et aquatilium animantium natura,* 1558; Liber V. *De serpentium natura,* 1558. Froschoverum, Zurich.

Ghiselin, M. T., and L. Jaffe. 1973. Phylogenetic classification in Darwin's monograph on the sub-class Cirripedia. *Systematic Zoology,* 22(2):132–140.

Gibbons, A. 2006. *The First Human: The Race to Discover Our Earliest Ancestors.* Doubleday, New York.

Gill, T. N. 1872. Arrangement of the families of fishes, or classes Pisces, Marsipobranchii, and Leptocardii. *Smithsonian Miscellaneous Collections,* 247.

Giseke, P. D. 1792. *Praelectiones in ordines naturals plantarum.* Benj. Gottl. Hoffmann, Hamburg, Germany.

Gliboff, S. 2008. *H. G. Bronn, Ernst Haeckel, and the Origins of German Darwinism.* The MIT Press, Cambridge.

Goldfuss, G. A. 1817. *Über die Entwicklungsstufen des Thieres.* Leonhard Schrag, Nürnberg.

Goldthwait, J. W. 1936. William Patten (1861–1932). *Proceedings of the American Academy of Arts and Sciences,* 70(10):566–568.

Good, R. D. 1956. *Features of Evolution in the Flowering Plants.* Longmans, Green, and Company, London.

Goodman, M., G. W. Moore, and G. Matsuda. 1975. Darwinian evolution in the genealogy of Haemoglobin. *Nature,* 253:603–608.

Goodspeed, T. H. 1954. *The Genus* Nicotiana; *Origins, Relationships and Evolution of its Species in the Light of their Distribution, Morphology and Cytogenetics.* Chronica Botanica Company, Waltham, Massachusetts.

Gould, S. J. 1984. The rule of five. *Natural History,* 108(1):18–23.

Gould, S. J. 1994. *Eight Little Piggies: Relections in Natural History.* Norton, New York.

Gould, S. J. 1999a. A division of worms: Jean Baptiste Lamarck's contributions to evolutionary theory, part 1. *Natural History,* 108(1):18–23.

Gould, S. J. 1999b. Branching through a wormhole: Reclassifying the types of "worms." Jean Baptiste Lamarck's contributions to evolutionary theory, part 2. *Natural History,* 108(2):24–27.

Greene, E. L. 1983. *Landmarks of Botanical History,* part 2. Stanford University Press, Stanford, California.

Greenwood, P. H., D. E. Rosen, S. H. Weitzman, and G. S. Myers. 1966. Phyletic studies of teleostean fishes, with a provisional classification of living forms. *Bulletin of the American Museum of Natural History,* 131(4):339–456.

Gregory, W. K. 1933. Fish skulls: A study of the evolution of natural mechanisms. *Transactions of the American Philosophical* Society, 23(2):75–481.

Gregory, W. K. 1935. Winged sharks. *Bulletin of the New York Zoological Society,* 38:129–133.

Gregory, W. K. 1951. *Evolution Emerging: A Survey of Changing Patterns from Primeval Life to Man.* Vol. 1, text; vol. 2, atlas. The Macmillan Company, New York.

Gregory, W. K., and G. M. Conrad. 1938. The phylogeny of the characin fishes. *Zoologica,* 23(17):319–360.

Gruenberg, B. C. 1919. *Elementary Biology: An Introduction to the Science of Life.* Ginn and Co., Boston.

Hackel, E. 1889. *Monographiae Phanerogamarum prodromi nunc continuatio, nunc revisio editoribus et pro parte auctoribus Alphonso et Casimir de Candolle, volume sextum Andropogoneae auctore.* G. Masson, Paris.

Haeckel, E. 1866. *Generelle morphologie der organismen. Allgemeine grundzüge der organischen formen-wissenschaft, mechanisch begründet durch die von Charles Darwin reformirte descendenztheorie.* 2 vols. G. Reimer, Berlin.

Haeckel, E. 1868. *Natürliche Schöpfungsgeschichte: gemeinverständliche wissenschaftliche Vorträge über die Entwickelungslehre im Allgemeinen und diejenige von Darwin, Goethe und Lamarck, im Besonderen über die Anwendung derselben auf den Ursprung des Menschen und andere damit zusammenhhangende Grundfragen der Naturwissenschaft.* G. Reimer, Berlin.

Haeckel, E. 1870. *Natürliche Schöpfungsgeschichte: gemeinverständliche wissenschaftliche Vorträge über die Entwickelungslehre im Allgemeinen und diejenige von Darwin, Goethe und Lamarck, im Besonderen über die Anwendung derselben auf den Ursprung des Menschen und andere damit zusammenhhangende Grundfragen der Naturwissenschaft.* Second edition. G. Reimer, Berlin.

Haeckel, E. 1874. *Anthropogenie oder Entwickelungsgeschichte des menschen: gemeinverständliche wissenschaftliche Vorträge über die Grundzüge der menschlichen Keimes- und Stammes-Geschichte.* Wilhelm Engelmann, Leipzig, Germany.

Haeckel, E. 1876. *The History of Creation: Or the Development of the Earth and its Inhabitants by the Action of Natural Causes, A Popular Exposition of the Doctrine of Evolution in General, and of that of Darwin, Goethe, and Lamarck in Particular*. Translated by E. Ray Lankester, in 2 vols. Henry S King & Co., London.

Haeckel, E. 1905. *Der Kampf um den Entwickelungs-Gedanken*. G. Reimer, Berlin.

Haeckel, E. 1910. *The Evolution of Man, A Popular Scientific Study*. Vol. 2, *The Evolution of the Species or Phylogeny*. Translated from the 5th edition by Joseph McCabe. G. P. Putnam's Sons, New York.

Hay, O. P. 1908. The fossil turtles of North America. *Carnegie Institute of Washington Publication*, 75:1–568.

Hennig, W. 1950. *Grundzüge einer Theorie der phylogenetischen Systematik*. Deutscher Zentralverlag, Berlin.

Hennig, W. 1966. *Phylogenetic Systematics*. University of Illinois Press, Urbana and Chicago.

Hernández, F. 1651. *Rerum medicarum Novae Hispaniae thesaurus seu nova plantarum, animalium et mineralium mexicanorum historia*. J. Mascardi, Rome.

Hitchcock, E. 1840. *Elementary Geology*. J. S. & C. Adams, Amherst.

Horaninow, P. F. 1834. *Primae lineae systematis naturae, nexui naturali omnium evolutionique progressivae per nixus reascendentes superstructui*. Krajanis, Petropoli [St. Petersburg, Russia].

Hull, D. L. 1988. *Science as a Process: An Evolutionary Account of the Social and Conceptual Development of Science*. University of Chicago Press, Chicago.

Jardine, W. 1858. *Memoirs of Hugh Edwin Strickland, M.A.* John van Voorst, London.

Jordan, D. S. 1905. The history of ichthyology. Pp. 387–428, In: *A Guide to the Study of Fishes*. Henry Holt and Co., New York.

Junker, T. 1995. Darwinism, materialism and the revolution of 1848 in Germany: On the interaction of politics and science. *History and Philosophy of the Life Sciences*, 17(2):271–302.

Jussieu, A.-H. de. 1825. Mémoire sur le groupe des Rutacées, seconde partie. *Mémoires du muséum d'histoire naturelle* (Paris) 12: 449–542.

Kaup, J. J. 1854. Einige Worte über die systematische Stellung der Familie der Raben, Corvidae. *Journal für Ornithologie*, 2, *Jahresversammlung*, xlvii–lvii.

Keith, A. 1897. *An Introduction to the Study of Anthropoid Apes*. Page & Pratt, London.

Keith, A. 1925. *The Antiquity of Man*. 2 vols. Williams and Norgate, London.

Kemp, T. S. 1982. The reptiles that became mammals. *New Scientist,* 93:581–584.

Kemp, T. S. 1999. *Fossils and Evolution.* Oxford University Press, Oxford.

Knight, W. J. 1980. Obituary and bibliography. *Entomologist's Monthly Magazine,* 115:164–175.

Korn, D. 1995. Impact of environmental perturbations on heterochronic development in palaeozoic ammonoids. Pp. 245–260, In: K. J. McNamara (editor), *Evolutionary Change and Heterochrony.* Wiley and Sons, London.

Kuntz, M. L., and P. G. Kuntz (editors). 1987. *Jacob's Ladder and the Tree of Life: Concepts of Hierarchy and the Great Chain of Being.* Peter Lang, New York.

Kusnezov, N. I. 1896. Subgenus *Eugentiana* Kusnez. generis *Gentiana* Tournef. *Acta Horti Petropolitani,* 15:1–507.

Lam, H. J. 1936. Phylogenetic symbols, past and present. *Acta Biotheoretica,* 2(3):153–194.

Lamarck, J.-B.-P.-A. de M. de. 1778. *Flore françoise, ou, Description succincte de toutes les plantes qui croissent naturellement en France: disposée selon une nouvelle méthode d'analyse, & à laquelle on a joint la citation de leurs vertus les moins équivoques en médecine, & de leur utilité dans les arts par M. le Chevalier de Lamarck,* vol. 2. Imprimerie Royale, Paris.

Lamarck, J.-B.-P.-A. de M. de. 1786. *Encyclopédie méthodique. Botanique,* vol. 2. Panckoucke, Paris.

Lamarck, J.-B.-P.-A. de M. de. 1809. *Philosophie zoologique, ou exposition des considérations relatives à l'histoire naturelle des animaux; la diversité de leur organisation et des facultés qu'ils en obtiennent; aux causes physiques qui maintiennent en eux la vie et donnent lieu aux mouvemens qu'ils exécutent j enfin; à celles qui produisent, les unes le sentiment, et les autres l'intelligence de ceux qui en sont doués.* Dentu et chez l'auteur, Paris.

Lamarck, J.-B.-P.-A. de M. de. 1815. *Histoire naturelle des animaux sans vertèbres . . . précédée d'une introduction offrant la détermination des caractères essentiels de l'animal, sa distinction du végétal et des autres corps naturels, enfin, l'exposition des principes fondamentaux de la Par M. de Lamarck.* Verdière, Paris.

Lamarck, J.-B.-P.-A. de M. de. 1820. *Système analytique des connaissances positives de l'homme, restreintes à celles qui proviennent directement ou indirectement de l'observation.* Imprimerie de A. Belin, Paris.

Lamarck, J.-B.-P.-A. de M. de. 1914. *Zoological Philosophy, An Exposition with Regard to the Natural History of Animals: The Diversity of Their Organisation and the Faculties which They Derive from It; The Physical*

Causes which Maintain Life within Them and Give Rise to Their Various Movements; Lastly, Those which Produce Feeling and Intelligence in Some Among Them. Translated, with an introduction, by Hugh Elliot. Macmillan and Co., London.

Lankester, E. R. 1881. *Limulus* an arachnid. *Quarterly Journal of Microscopical Science,* 21:504–548, 609–649.

Leclère, L., P. Schuchert, C. Cruaud, A. Couloux, and M. Manuel. 2009. Molecular phylogenetics of Thecata (Hydrozoa, Cnidaria) reveals long-term maintenance of life history traits despite high frequency of recent character changes. *Systematic Biology,* 58(5):509–526.

Lenoir, T. 1978. Generational factors in the origin of *Romantische Naturphilosophie. Journal of the History of Biology,* 11(1):57–100.

Leppik, E. E. 1965. Some viewpoints on the phylogeny of rust fungi. Part 5, Evolution of biological specialization. *Mycologia,* 57:6–22.

Lewis, G. 1868. *Natural History of Birds: Lectures on Ornithology in Ten Parts,* part 1. J. A. Bancroft & Co., Philadelphia.

Lindley, J. 1838. Exogens. *Penny Cyclopaedia,* 10:130.

Linnaeus, C. 1735. *Systema naturae, sive regna tria naturae systematice proposita per classes, ordines, genera, & species*. Theodorum Haak, Leiden.

Linnaeus, C. 1751. *Philosophia botanica, in qua explicantur fundamenta botanica, cum definitionibus partium, exemplis terminorum, observationibus rariorum, adjectis figuris aeneis*. Godofr. Kiesewetter, Stockholm.

Lobel, M. de. 1576. [*Observationes.*] *Plantarum seu stirpium historia. . . . Cui annexum est adversariorum volumen*. Plantini, Antwerpiae [Antwerp].

Loesener, L. E. T. 1908. Monographia aquifoliacearum II. *Nova Acta Physico-Medica Academiae Caesareae Leopoldino-Carolinae Germanicae Naturae Curiosorum,* 89:1–313.

Lorenz, K. 1941. Vergleichende Bewegungsstudien an Anatinen. *Journal für Ornithologie,* 3:194–293.

Lorenz, K. 1953. Comparative studies on the behaviour of Anatinae. *Avicultural Magazine,* 59:80–91.

Lorenz, K. 1974. Analogy as a source of knowledge. *Science,* 185(4147): 229–234.

Lovejoy, A. O. 1936. *The Great Chain of Being: A Study of the History of an Idea*. The William James lectures delivered at Harvard University, 1933, by Arthur O. Lovejoy. Harvard University Press, Cambridge.

Lowe, C. H., J. W. Wright, C. J. Cole, and R. L. Bezy. 1970. Chromosomes and evolution of the species groups of *Cnemidophorus* (Reptilia: Teiidae). *Systematic Zoology,* 19(2):128–141.

Luckett, W. P. 1975. Ontogeny of the fetal membranes and placenta: Their bearing on primate phylogeny. Pp. 157–182, In: W. P. Luckett and F. S.

Szalay (editors), *Phylogeny of the Primates, a Multidisciplinary Approach.* Plenum Press, New York.

Lull, R. 1512. *De nova logica, de correllativis, necnon et de ascensu et descensu intellectus.* Jorge Costilla, Valencia.

Lysenko, O., and P. H. A. Sneath. 1959. The use of models in bacterial classification. *Journal of General Microbiology,* 20:284–290.

Macleay, W. S. 1819. *Horae entomologicae: or Essays on the Annulose Animals.* vol. 1, part 1. S. Bagster, London.

Macleay, W. S. 1821. *Horae entomologicae: or Essays on the Annulose Animals.* vol. 1, part 2. S. Bagster, London.

Margulis, L. 1970. *Origin of Eukaryotic Cells.* Yale University Press, New Haven.

Margulis, L., and K. V. Schwartz. 1982. *Five Kingdoms: An Illustrated Guide to the Phyla of Life on Earth.* Freeman and Co., San Francisco.

Matthew, W. D. 1930. The phylogeny of dogs. *Journal of Mammalogy,* 11(2):117–138.

Mayr, E. 1972. Lamarck revisited. *Journal of the History of Biology,* 5:55–94.

Mayr, E., E. G. Linsley, and R. L. Usinger. 1953. *Methods and Principles of Systematic Zoology.* McGraw-Hill, New York.

McCravy, K. W. 2008. Biogeography. Pp. 481–487, In: J. L. Capinera (editor), *Encyclopedia of Entomology,* 2nd edition, 4 vols., Springer Science+Business Media B.V., Dordrecht, the Netherlands.

Merezhkowsky, C. 1910. Theorie der zwei Plasmaarten als Grundlage der Symbiogenese, einer neuen Lehre von der Entstehung der Organismen. *Biologisches Centralblatt,* 30:277–303, 321–347, 353–367.

Mikelsaar, R. 1987. A view of early cellular evolution. *Journal of Molecular Evolution,* 25(2):168–183.

Milne, M. J., and L. J. Milne. 1939. Evolutionary trends in caddis worm case construction. *Annals of the Entomological Society of America,* 32(3):533–542.

Milne-Edwards, H. 1844. Considérations sur quelques principes relatifs à la classification naturelle des animaux, et plus particulièrement sur la distribution méthodique des mammifères. *Annales des sciences naturelles,* series 3, 1:65–99.

Mitchell, P. C. 1901. On the intestinal tract of birds; With remarks on the valuation and nomenclature of zoological characters. *Transactions of the Linnean Society of London, Zoology,* series 2, 8:173–275.

Mitman, G. 1990. Evolution as gospel: William Patten, the language of democracy, the Great War. *Isis,* 81(3):446–463.

Moody, S. M. 1985. Charles L. Camp and his 1923 classification of lizards: An early cladist? *Systematic Zoology,* 34(2):216–222.

Morgan, G. J. 1998. Emile Zuckerkandl, Linus Pauling, and the molecular clock, 1959–1965. *Journal of the History of Biology,* 31:155–178.

Morison, R. 1672. *Plantarum umbelliferarum distributio nova, per tabulas cognationis et affinitatis ex libro naturæ observata & detecta.* Sheldonian Theater, Oxford.

Moritz, C., and D. M. Hillis. 1996. Molecular systematics: context and controversies. Pp. 1–13, In: D. M. Hillis, C. Moritz, and B. K. Mable (editors), *Molecular Systematics,* 2nd edition, Sinauer Associates, Sunderland, Massachusetts.

Moss, W. W. 1967. Some new analytic and graphic approaches to numerical taxonomy, with an example from the Dermanyssidae (Acari). *Systematic Zoology,* 16(3):177–207.

Nelson, G., and N. Platnick. 1981. *Systematics and Biogeography: Cladistics and Vicariance.* Columbia University Press, New York.

Newman, E. 1837. Further observations on the Septenary System. *Entomological Magazine,* 4:234–251.

Nuttall, G. H. 1904. *Blood Immunity and Blood Relationship: A Demonstration of Certain Blood-Relationships Amongst Animals by Means of the Precipitin Test for Blood.* Cambridge University Press, Cambridge, England.

O'Hara, R. J. 1988. Diagrammatic classifications of birds, 1819–1901: Views of the natural system in 19th-century British ornithology. Pp. 2746–2759, In: H. Ouellet (editor), *Acta XIX Congressus Internationalis Ornithologici,* National Museum of Natural Sciences, Ottawa.

O'Hara, R. J. 1991. Representations of the natural system in the nineteenth century. *Biology and Philosophy,* 6:255–274.

Olson, E. C. 1971. *Vertebrate Paleontology.* Wiley-Interscience, New York.

Osborn, H. F. 1917. *The Origin and Evolution of Life: On the Theory of Action Reaction and Interaction of Energy.* Charles Scribner's Sons, New York.

Osborn, H. F. 1921. Adaptive radiation and classification of the Proboscidea. *Proceedings of the National Academy of Sciences,* 7(8):231–234.

Osborn, H. F. 1927a. Recent discoveries relating to the origin and antiquity of man. *Science,* 65:481–488.

Osborn, H. F. 1927b. *Man Rises to Parnassus; Critical Epochs in the Prehistory of Man.* Princeton University Press, Princeton; Oxford University Press, London.

Osborn, H. F. 1934a. Aristogenesis, the Creative Principle in the Origin of Species. *American Naturalist,* 68(716):193–235.

Osborn, H. F. 1934b. Evolution and geographic distribution of the Proboscidea: Moeritheres, Deinotheres and Mastodonts. *Journal of Mammalogy,* 15(3):177–184.

Osborn, H. F. 1836–1842. *The Proboscidea: A Monograph of the Discovery, Evolution, Migration and Extinction of the Mastodonts and Elephants of the World.* Trustees of the American Museum of Natural History, American Museum Press, New York.

Osporat, D. 1981. *The Development of Darwin's Theory: Natural History, Natural Theology, and Natural Selection, 1838–1959.* Cambridge University Press, Cambridge, England.

Pace, N. R. 1997. A molecular view of microbial diversity and the biosphere. *Science,* 276:734–740.

Pace, N. R. 2004. The early branches in the tree of life. Pp. 76–85, In: J. Cracraft and M. J. Donoghue (editors), *Assembling the Tree of Life.* Oxford University Press, Oxford.

Pallas, P. S. 1766. *Elenchus zoophytorum sistens generum adumbrationes generaliores et specierum cognitarum succintas descriptiones, cum selectis auctorum synonymis.* Apud Petrum van Cleef, Hagae-Comitum [The Hague, the Netherlands].

Patten, W. 1912. *The Evolution of the Vertebrates and Their Kin.* Blakiston's Son & Co., Philadelphia.

Patten, W. 1923. *Evolution, Part Two: The Evolution of Plant and Animal Life.* The Dartmouth Press, Hanover, New Hampshire.

Patten, W. 1924. Why I teach evolution. *Scientific Monthly,* 19(6):635–647.

Patterson, C. 1977. The contribution of paleontology to teleostean phylogeny. Pp. 579–643, In: M. K. Hecht, P. C. Goody, and B. M. Hecht, *Major Patterns in Vertebrate Evolution.* Plenum Press, New York.

Patterson, C. 1981. Agassiz, Darwin, Huxley, and the fossil record of teleost fishes. *Bulletin of the British Museum of Natural History,* 35(3):213–224.

Patterson, C. 1987. Introduction. Pp. 1–22, In: C. Patterson (editor), *Molecules and Morphology in Evolution: Conflict or Compromise.* Cambridge University Press, Cambridge, England.

Paul, G. S. 2002. *Dinosaurs of the Air: The Evolution and Loss of Flight in Dinosaurs and Birds.* Johns Hopkins University Press, Baltimore.

Pena, P., and M. de Lobel. 1570 [1571]. *Stirpium adversaria nova, perfacilis vestigatio, luculentaque accessio ad priscorum, praesertim Dioscorides, & recentiorum, materiam medicam. Quibus propediem accedet altera pars. Qua conjectaneorum de plantis appendix, de succis medicatis et metallicis sectio, antiquae & novatae medicine lectiorum remediorum thesaurus opulentissimus, de succedaneis libellus continentur.* Purfcetii, Londini [London].

Penland, C. W. 1924. Notes on North American scutellarias. *Rhodora, Journal of the New England Botanical Club,* 26(304):61–79.

Pennisi, E. 2003. Modernizing the tree of life. *Science,* 300(5626): 1692–1697.

Pough, F. H., C. M. Janis, and J. B. Heiser. 2009. *Vertebrate Life,* 8th edition. Benjamin Cummings, San Francisco.

Ragan, M. A. 2009. Trees and networks before and after Darwin. *Biology Direct,* 4:43, doi:10.1186/1745-6150-4-43.

Raymond, P. E. 1939. *Prehistoric Life.* Harvard University Press, Cambridge.

Reeder, T., H. C. Dessauer, and C. J. Cole. 2002. Phylogenetic relationships of whiptail lizards of the genus *Cnemidophorus* (Squamata, Teiidae): A test of monophyly, reevaluation of karyotypic evolution, and review of hybrid origins. *American Museum Novitates,* 3365:1–61.

Regier, J. C., J. W. Shultz, A. Zwick, A. Hussey, B. Ball, R. Wetzer, J. W. Martin, and C. W. Cunningham. 2010. Arthropod relationships revealed by phylogenomic analysis of nuclear protein-coding sequences. *Nature,* 463:1079–1083.

Reichenow, A. 1882. *Die Vögel der zoologischen Gärten: Leitfaden zum Studium der Ornithologie mit besonderer Berücksichtigung der in Gefangenschaft gehaltenen Vögel, ein Handbuch für Vogelwirthe,* vol. 1. Verlag von L. U. Kittler, Leipzig, Germany.

Reichenow, A. 1913. *Die Vogel, Handbuch der systematischen Ornithologie.* 2 vols. Ferdinand Enke, Stuttgart, Germany.

Richards, R. J. 2005. Ernst Haeckel and the struggles over evolution and religion. *Annals of the History and Philosophy of Biology,* 10: 89–115.

Richards, R. J. 2008. *The Tragic Sense of Life: Ernst Haeckel and the Struggle over Evolutionary Thought.* University of Chicago Press, Chicago.

Roger, J. 1997. *Buffon: A Life in Natural History.* Translated by S. L. Bonnefoi. Cornell University Press, Ithaca, New York.

Romer, A. S. 1933a. *Man and the Vertebrates.* University of Chicago Press, Chicago [second edition, 1937; third edition, 1941; fourth edition, retitled *The Vertebrate Story,* 1949b].

Romer, A. S. 1933b. *Vertebrate Paleontology.* University of Chicago Press, Chicago [second edition, 1945; third edition, 1966].

Romer, A. S. 1949a. *The Vertebrate Body.* W. B. Saunders, Philadelphia [second edition, 1955; third edition, 1962; fourth edition, 1970].

Romer, A. S. 1949b. *The Vertebrate Story.* University of Chicago Press, Chicago [the fourth edition of *Man and the Vertebrates*].

Romer, A. S. 1961. Synapsid evolution and dentition. Pp. 9–56, In: G. Vandebroek (editor), *International Colloquium on the Evolution of*

Lower and Non-specialized Mammals, vol. 1. Koninklijke Vlaamse Academie Voor Wetenschappen, Letteren en Schone Kunsten van Belgie, Brussels.

Rosen, D. E. 1979. Fishes from the uplands and intermontane basins of Guatemala: Revisionary studies and comparative biogeography. *Bulletin of the American Museum of Natural History,* 162(5):267–376.

Rosen, D. E., P. L. Forey, B. G. Gardiner, and C. Patterson. 1981. Lungfishes, tetrapods, paleontology, and plesiomorphy. *Bulletin of the American Museum of Natural History,* 167(4):159–276.

Rowe, T. B. 2004. Chordate phylogeny and development. Pp. 384–409, In: J. Cracraft and M. J. Donoghue (editors), *Assembling the Tree of Life,* Oxford University Press, Oxford.

Rüling, J. P. 1766. *Commentatio botanica, de ordinibus naturalibus plantarum.* Litteris Frider. Andr. Rosenbusch, Göttingae [Göttingen, Germany]. [Reprinted in P. Usteri, ed., 1793, *Delectus opusculorum botanicorum,* vol. 2, 431–462.]

Ruse, M. 1996. *Monad to Man: The Concept of Progress in Evolutionary Biology.* Harvard University Press, Cambridge.

Ruse, M., and J. Travis. 2009. *Evolution: The First Four Billion Years.* Belknap Press of Harvard University Press, Cambridge.

Sapp, J. 2009. *The New Foundations of Evolution: On the Tree of Life.* Oxford University Press, Oxford.

Sarich, V. M., and J. E. Cronin. 1976. Molecular systematics of the primates. Pp. 141–170, In: M. Goodman, R. E. Tashian, and J. H. Tashian (editors), *Molecular Anthropology: Genes and Proteins in the Evolutionary Ascent of the Primates.* Plenum Press, New York.

Sarich, V. M., and A. C. Wilson. 1967. Immunological time scale for hominid evolution. *Science,* 158:1200–1203.

Saville-Kent, W. 1880. *A Manual of the Infusoria: Including a Description of All Known Flagellate, Ciliate, and Tentaculiferous Protozoa, British and Foreign, and an Account of the Organization and Affinities of the Sponges.* 3 vols. David Bogue, London.

Schaffner, J. H. 1934. Phylogenetic taxonomy of plants. *Quarterly Review of Biology,* 9(2):129–160.

Schimper, W.-P. 1869–1874. *Traité de paléontologie végétale, ou, La flore du monde primitif dans ses rapports avec les formations géologiques et la flore du mond actuel.* 3 vols. J. B. Baillière et Fils, Paris.

Schnell, G. D. 1970. A phenetic study of the suborder Lari (Aves) II. Phenograms, discussion, and conclusions. *Systematic Zoology,* 19(3):264–302.

Schodde, R. 2000. Obituary: Charles G. Sibley 1911–1998. *Emu,* 100:75–76.

Secord, J. A. 2000. *Victorian Sensation: The Extraordinary Publication,*

Reception, and Secret Authorship of Vestiges of the Natural History of Creation. University of Chicago Press, Chicago.

Sereno, P. C. 1999. The evolution of dinosaurs. *Science,* 284:2137–2147.

Seringe, N. C. 1815. *Essai d'une monographie des saules de la Suisse.* Maurhofer and Dellenbach, Berne.

Sharpe, R. B. 1891. *A Review of Recent Attempts to Classify Birds: An Address Delivered before the Second International Ornithological Congress on the 18th of May, 1891.* Office of the Congress, Budapest.

Shoshani, J. 1998. Understanding proboscidean evolution: A formidable task. *Trends in Ecology & Evolution,* 13(12):480–487.

Sibley, C. G. 1994. On the phylogeny and classification of living birds. *Journal of Avian Biology,* 25(2):87–92.

Sibley, C. G., J. E. Ahlquist, and B. L. Monroe Jr. 1988. A classification of the living birds of the world based on DNA-DNA hybridization studies. *The Auk,* 105(3):409–423.

Small, J. 1919. The origin and development of the Compositae, chapter 8, general conclusions. *New Phytologist,* 18(7):201–234.

Small, J. 1922. Age and area, and size and space, in the Compositae. Pp. 119–136, In: J. C. Willis, *Age and Area: A Study in Geographical Distribution and Origin of Species,* chapter 13. Cambridge University Press, Cambridge, England.

Sneath, P. H. A. 1961. Recent developments in theoretical and quantitative taxonomy. *Systematic Zoology,* 10(3):118–139.

Sneath, P. H. A., and R. R. Sokal. 1973. *Numerical Taxonomy: The Principles and Practice of Numerical Classification.* Freeman and Company, San Francisco.

Sokal, R. R., and J. H. Camin. 1965. The two taxonomies: areas of agreement and conflict. *Systematic Zoology,* 14(3):176–195.

Sokal, R. R., and C. D. Michener. 1958. A statistical method for evaluating systematic relationships. *The University of Kansas Science Bulletin,* 38(22):1409–1438.

Spencer, F. 1990. *Piltdown: A Scientific Forgery.* Natural History Museum Publications, Oxford University Press, Oxford.

Staudt, G. 2003. *Les dessins d'Antoine Nicolas Duchesne pour son Histoire naturelle des fraisiers.* Publications Scientifiques du Muséum, Paris.

Stauffer, R. C. 1975. *Charles Darwin's Natural Selection, Being the Second Part of His Big Species Book Written from 1856 to 1858.* Cambridge University Press, Cambridge, England.

Stevens, P. F. 1983. Augustin Augier's "Arbre Botanique" (1801), a remarkable early botanical representation of the natural system. *Taxon,* 32(2):203–211.

Stevens, P. F. 1984. Metaphors and typology in the development of botanical systematics 1690–1960, or the art of putting new wine in old bottles. *Taxon,* 33(2):169–211.

Stevens, P. F. 1994. *The Development of Biological Systematics: Antoine-Laurent de Jussieu, Nature, and the Natural System.* Columbia University Press, New York.

Stiassny, M. L. J., E. O. Wiley, G. D. Johnson, and M. R. de Caravalho. 2004. Gnathostome fishes. Pp. 410–429, In: J. Cracraft and M. J. Donoghue (editors), *Assembling the Tree of Life.* Oxford University Press, Oxford.

Stirton, R. A. 1940. Phylogeny of North American Equidae. *Bulletin of the Department of Geological Sciences, University of California Publications,* 25(4):165–198.

Stirton, R. A. 1959. *Time, Life, and Man: The Fossil Record.* Wiley and Sons, New York.

Storer, T. I. 1943. *General Zoology.* McGraw-Hill, New York.

Strickland, H. E. 1841. On the true method of discovering the natural system in zoology and botany. *Annals and Magazine of Natural History,* 6:184–194.

Suárez-Díaz, E., and V. H. Anaya-Muñoz. 2008. History, objectivity, and the construction of molecular phylogenies. *Studies in History and Philosophy of Biological and Biomedical Sciences,* 39(4):451–468.

Swainson, W. J. 1835. *A Treatise on the Geography and Classification of Animals.* Longman, London.

Swainson, W. J. 1836. *On the Natural History and Classification of Birds,* vol. 1. Longman, Rees, Orme, Brown, Green, and Longman, London.

Swainson, W. J. 1837. *On the Natural History and Classification of Birds.* vol. 2. Longman, Rees, Orme, Brown, Green, and Longman, London.

Swingle, D. B. 1928. *A Textbook of Systematic Botany.* McGraw-Hill, New York.

Templeton, A. R. 1993. The "Eve" hypotheses: A genetic critique and reanalysis. *American Anthropologist,* new series, 95(1):51–72.

Tilden , J. E. 1935. *The Algae and Their Life Relations: Fundamentals of Phycology.* University of Minnesota Press, Minneapolis.

Vigors, N. A. 1824. Observations on the natural affinities that connect the orders and families of birds. *Transactions of the Linnean Society of London,* 14(3):395–517.

Voss, E. G. 1952. The history of keys and phylogenetic trees in systematic biology. *Journal of the Scientific Laboratories, Denison University,* 43(1–2):1–25.

Voss, J. 2010. *Darwin's Pictures: Views of Evolutionary Theory, 1837–1874.* Yale University Press, New Haven.

Wallace, A. R. 1855. On the law which has regulated the introduction of new species. *Annals and Magazine of Natural History*, second series, 16:184–196.

Wallace, A. R. 1856. Attempts at a natural arrangement of birds. *Annals and Magazine of Natural History*, second series, 18:193–216.

Walters, M. 2003. *A Concise History of Ornithology: The Lives and Works of its Founding Figures.* Yale University Press, New Haven.

Warner, D. J. 1979. *Graceanna Lewis, Scientist and Humanitarian.* Smithsonian Institution Press, Washington, D.C.

Weiner, J. S. 2003. *The Piltdown Forgery: The Classic Account of the Most Famous and Successful Hoax in Science.* Oxford University Press, Oxford.

Wernham, H. F. 1914. *A Monograph of the Genus* Sabicea. Trustees of the British Museum of Natural History, London.

Whittaker, R. H. 1957. The kingdoms of the living world. *Ecology*, 38(3):536–538.

Whittaker, R. H. 1959. On the broad classification of organisms. *Quarterly Review of Biology*, 34(3):210–226.

Whittaker, R. H. 1969. New concepts of kingdoms of organisms. *Science*, 163(3863):150–160.

Wiley, E. O. 1976. The phylogeny and biography of fossil and recent gars (Actinopterygii: Lepisosteidae). *University of Kansas, Museum of Natural History, Miscellaneous Publication*, 64, 111.

Wilkins, J. 1668. *An Essay towards a Real Character, and a Philosophical Language.* Gellibrand & Martyn, London.

Williams, D. M., and M. C. Ebach. 2008. *Foundations of Systematics and Biogeography.* Springer Science+Business Media, New York.

Willmann, R. 2003. From Haeckel to Hennig: The early development of phylogenetics in German-speaking Europe. *Cladistics*, 19:449–479.

Willughby, F. 1686. *De historia piscium libri quatuor, jussu & sumptibus Societatis Regiae Londinensis editi. . . . Totum opus recognovit, coaptavit, supplevit, librum etiam primum & secundum integros adjecit Johannes Raius e Societate Regia.* Theatro Sheldoniano, Oxonii [Oxford].

Windsor, M. P. 1991. *Reading the Shape of Nature: Comparative Zoology at the Agassiz Museum.* University of Chicago Press, Chicago.

Woese, C. R. 1987. Bacterial Evolution. *Microbiological Reviews*, 51(2): 221–271.

Woese, C. R., and G. E. Fox. 1977. Phylogenetic structure of the prokaryotic domain: The primary kingdoms. *Proceedings of the National Academy of Sciences*, 74(11):5088–5090.

Woese, C. R., L. J. Magrum, and G. E. Fox. 1978. Archaebacteria. *Journal of Molecular Evolution*, 11(3):245–252.

Yates, F. A. 1954. The art of Ramon Lull: An approach to it through Lull's theory of the elements. *Journal of the Warburg and Courtauld Institutes,* 17(1–2):115–173.

Zimmermann, W. 1931. Arbeitsweise der botanischen Phylogenetik und anderer Gruppierubgswissenschaften. Pp. 941–1053, In: E. Abderhalden (editor), *Handbuch der biologischen Arbeitsmethoden,* Abt. 3, 2, Teil 9, Urban & Schwarzenberg, Berlin.

Zuckerkandl, E., and L. Pauling. 1962. Molecular disease, evolution, and genetic heterogeneity. Pp. 189–225, In: M. Kasha and B. Pullman (editors), *Horizons in Biochemistry: Albert Szent-Györgyi Dedicatory Volume.* Academic Press, New York.

Zuckerkandl, E., and L. Pauling. 1965. Evolutionary divergence and convergence in proteins. Pp. 97–166, In: V. Bryson and H. Vogel (editors), *Evolving Genes and Proteins.* Academic Press, New York.

INDEX